in our literature, we can mark off areas

Ion Implantation

Ion Implantation

Heiner Ryssel
Chair of Electron Devices
University Erlangen

and

Ingolf Ruge

Fraunhofer Institute of Solid State Technology
Munich

Translated by Jerome-Allan Buckman and Heiner Ryssel
Fraunhofer Institute of Solid State Technology
Munich

A Wiley–Interscience Publication

JOHN WILEY & SONS

Chichester · New York · Brisbane · Toronto · Singapore

This edition is published by permission of B. G. Teubner, Stuttgart, and is the sole authorized English translation of the original German edition.

Library of Congress Cataloging in Publication Data:

Ryssel, Heiner, 1941–
Ion implantation.

Translation of: Ionenimplantation.
Includes index.
1. Ion implantation. 2. Semiconductors. I. Ruge, Ingolf. II. Title.
TK7871.85.R9713 1986 621.3815′2 84-3723

ISBN 0 471 10311 X

British Library Cataloguing in Publication Data:

Ryssel, Heiner
Ion implantation.
1. Ion implantation
I. Title II. Ruge, Ingolf
660.2′9723 TK7871.85

ISBN 0 471 10311 X

Printed and bound in Great Britain

Contents

List of Frequently Used Symbols

(Symbols which appear only once are explained where they occur in the text)

A	Area, atomic weight
a	Length, distance, screening parameter
a_0	Bohr radius ($a_0 \approx 0.053$ nm)
B	Magnetic induction
C	Concentration (index V for vacancies); capacitance
C_{ox}	Oxide capacitance per unit area
c_0	Velocity of light ($c_0 = 2.998 \times 10^8$ ms^{-1})
D	Diffusion coefficient (index i for intrinsic, v for vacancies, n for electrons, p for holes); diameter
d	distance
d_f	Field enhancement factor
d_v	Vacancy enhancement factor
E	Energy, acceleration energy; field strength
E_a	Activation energy
E_G	Band gap
E_t	Trap level
E_i	Intrinsic Fermi level
E_d	Displacement energy
E_R	Rydberg energy ($E_R = 13.6$ eV)
f	Frequency
g	Generation rate, degeneracy factor
h	Planck constant
h_{FE}	Current gain
I	Current
j	Current density (index Diff for diffusion current density, S for saturation current density, R-G for recombination-generation current density, Tot for total current density)
K	Constant

k	Boltzmann constant ($k = 1.38 \times 10^{-23}$ J/K $= 8.617 \times 10^{-5}$ eV/K)
M	Mass (index 1 for ion, index 2 for matrix atom)
m	Mass of light ions; segregation coefficient
m_a	Atomic mass unit ($m_a = 1.66 \times 10^{-27}$ kg)
m_e	Electron mass ($m_e = 9.11 \times 10^{-31}$ kg)
m^*	Effective mass
N	Concentration, atomic density
$N(x)$	Concentration
N_A	Acceptor concentration
N_B	Background doping
N_D	Donor concentration
N_S	Sheet carrier concentration (index eff for effective sheet carrier concentration)
N_c	Density of states in the conduction band
N_d	Number of displaced atoms
N_t	Trap concentration
$N_\square$	Dose (index a for amorphizing, c for critical dose)
n	Electron concentration
n_{eff}	Effective electron concentration
n_i	Intrinsic carrier concentration
P	Power
p	Impact collision parameter
Q	Charge
Q_B	Space charge per unit area; integral base doping
Q_{ss}	Interface charge per unit area
q	Elementary charge ($q = 1.602 \times 10^{-19}$ As)
R	Range; resistance; radius
ΔR	Range straggling
R_H	Hall coefficient (index S for sheet Hall coefficient)
R_p	Mean projected range
ΔR_p	Mean projected range straggling (= standard deviation; index L for lateral straggling)
r	Distance of colliding particles
S	Sputtering yield; stopping cross section (index e for electronic, index n for nuclear collisions)
s	Order of power potential; recombination velocity; distance
T	Temperature (index A for annealing, index I for implantation temperature)
T	Transferred energy (index e for electronic, index m for maximum, index n for nuclear interaction)
t	Time; normalized scattering parameter, thickness, distance
t_{ox}	Oxide thickness
V	Voltage, potential (index D for diffusion voltage, index G for gate voltage, index H for Hall voltage, index T for threshold voltage); interaction potential

v	Velocity (index th for thermal velocity)
w	Space charge width
X_D	Mean range of radiation damage
ΔX_D	Standard deviation of radiation damage
x	x-direction
x_j	Depth of p–n junction
Z	Atomic number
Δ	Difference
ε	Normalized energy; emission factor
ε_0	Free space permittivity ($\varepsilon_0 = 8.854 \times 10^{-12}$ F/m)
ε_r	Relative dielectric constant, relative permittivity
λ_0	Light wave length
μ	Mobility (index C for conductivity, index H for Hall mobility)
μ_s	Sheet mobility
ψ	Angle; potential
$\varphi(a/r)$	Screening function
Φ	Potential; barrier height; deflection angle in centre of mass system
ρ	Normalized range; density; resistivity
ρ_s	Sheet resistivity
$\Delta\rho$	Normalized range straggling
σ	Cross-section; standard deviation; black-body radiation constant ($\sigma = 5.67 \times 10^{-12}$ W cm^{-2} K^{-4}) trapping cross-section (index n for electrons, index p for holes)
σ_s	Sheet conductivity
τ	Lifetime (index V for vacancies, indices n, p for minority carrier lifetime of electrons, holes)
τ_R	Relaxation time constant
θ	Scattering angle
ω	Angular frequency

Preface

Several books on the topic of ion implantation have been published in the last few years; almost all of these have been addressed to scientists active in this field, and so have dealt to a very great extent with theory. In contrast, this book is directed less to ion-implantation experts, and more to research-and-development persons in industry, research laboratories and universities who are interested in ion implantation as a method of changing material properties, and who want to know whether implantation techniques can be used to solve their special problems.

We believe that this mainly requires, apart from a short description of the theoretical background, a consideration of the various practical matters that arise when using ion implantation, such as the electrical activation of implanted ions and diffusion effects, as well as a discussion of standard measurement methods, requirements concerning accelerators, and in addition—of course—many examples of the use of ion implantation.

The main emphasis of this book is placed on the doping of semiconductors through ion implantation, since this is at present, and presumably will be for a long time to come, its main application. From case to case, however, further applications outside the semiconductor field will be considered.

In order to make the book easier to use, SI units or units derived from them have been utilized. An exception is the second chapter which deals with the fundamental equations of ion implantation, in which the original equation forms have been retained for historical reasons. This seems justifiable since the book is not intended as an aid to theoretical work on ion stopping in matter, but rather, as a guide to the practical uses of ion implantation.

We would especially like to thank Dr H. Betz, Dr P. Eichinger, Mr H. Kranz, and Dr. K. Wiedeburg for their help during manuscript preparation in the form of discussions, calculations, and a critical examination of the book. We are additionally grateful to Dr Eichinger for his help in writing several parts of section 5.8. Furthermore, we would like to thank Miss B. Schmiedt, Miss M. Schenk, Miss E. Traumüller, and Miss H. Zeller, as well as Mrs M. Podstowka and Mr M. Bleier, for their technical assistance during measurements and manuscript preparation.

Munich, Autumn 1977 H. Ryssel, I. Ruge

Preface to English Edition

The first edition of this book in German was well received, and so it was decided to publish an English translation. This was the beginning of a long and arduous process, since it can be more work to translate a technical book than to write one. Finally, though, it was done. Thanks to many readers, printing errors, incorrect references, and miswritten equations were able to be corrected. The book has been completely updated and several sections have been rewritten; the sections on laser annealing are new and additional references up to 1983 have been included. I would especially like to thank Mr J. Buckman, who accomplished the translation with my assistance. In addition, I would like to thank Miss B. Schmiedt, Mrs M. Podstowka, and Miss H. Zeller for their technical assistance during the preparation of this edition.

Munich, Spring 1984 H. Ryssel

1
Introduction

1.1 ION IMPLANTATION AND ITS APPLICATIONS

During the process of ion implantation, atoms or molecules are ionized, accelerated in an electrostatic field, and implanted into a solid. In this way, a great many combinations of ions and substrates are possible (e.g., boron in silicon, silicon in silicon, tellurium in gallium arsenide, platinum in iron). The acceleration energy can be between a few thousand electron volts (keV) and several million electron volts (MeV). The penetration depth of the ions depends not only upon the energy, but also upon the mass of the ions and the atomic mass of the solid. Thus, for example, the average range of 10-keV phosphorus ions in silicon is about 14 nm; that of 1-MeV boron ions is about 1756 nm. As a result of such an ion bombardment, it is possible to alter nearly all properties of the surface layer of the solid.

In recent years, ion implantation has been developed as a new process for the doping of semiconductor crystals. The most common processes used in the past have been doping during epitaxy, diffusion, and alloying. In the case of epitaxy, layers having the desired doping are grown on a wafer; during the diffusion process, doping atoms are diffused from the surface into the semiconductor. In the case of alloying, the semiconductor layer nearest to the surface is dissolved and doped during recrystallization.

According to its nature, the process of implantation is basically independent of chemical solubility limits, the temperature during implantation, and the concentration of the dopant at the surface of the semiconductor. The doping atoms introduced have a concentration profile which is generally described by a Gaussian distribution, with an average projected range R_p and a standard deviation ΔR_p.

Implantation offers a number of technological advantages which are important in the fabrication of electronic devices:

1. Speed, homogeneity, and reproducibility of the doping process.
2. Exact controllability of the number of doping atoms introduced, by simple current integration, which is especially important for low concentrations (e.g., threshold-voltage adjustment of MOS transistors).

3. Lower requirements for the purity of the dopants, since they are separated according to their mass.
4. Avoidance of high processing temperatures during implantation.
5. Simple masking methods, for example, with the use of thick layers of oxide, nitride, metal or photoresist.
6. Possibility of doping through thin passivating layers (e.g., SiO_2, Si_3N_4).
7. Low penetration depth of the ions (in general, less than a few micrometres); it is possible to dope shallow layers with very high doping gradients (e.g., IMPATT diodes, microwave transistors).
8. Multiple implantation by changing the accelerator voltage during implantation makes possible a relatively free choice of the doping profile, whereby one is not limited to the Gaussian shape.
9. Because of the minimal lateral scattering, it is possible to fabricate devices with very small dimensions and to keep the parasitic capacitance low.

However, ion implantation also has several disadvantages which limit its application possibilities:

1. As a result of bombardment with heavy particles, radiation damage is produced, which is generally undesirable because of changes in the semiconductor's electrical properties; furthermore, most of the doping atoms do not come to rest after implantation on regular lattice sites, and are therefore not electrically active. By a suitable high-temperature treatment (annealing), it is necessary to restore the crystal lattice and to bring the introduced atoms on to electrically-active lattice sites.
2. Doping by implantation is limited to layers near the surface. Although greater penetration depths can be achieved with higher acceleration voltages, the electrical properties of the layers thus produced are not yet satisfactory.
3. Due to additional effects during or after implantation (e.g., channelling, diffusion), it is not possible in general to obtain the theoretically predicted profile. In most cases, such effects are manifested by a deeper penetration of the doping atoms.

The most important questions which therefore have to be answered concern the annealing of radiation damage, the electrical activation of the implanted atoms, the shape of the doping profile, and the influence of implantation upon the electrically important material parameters, i.e., the mobility and the lifetime.

The use of ion implantation is not restricted to the field of semiconductors; it is also coming into application in other areas of physics and chemistry of solids, though efforts here are still in the research stage. With the development of high-current implantation machines, it has become possible not only to dope semiconductors, but also to alter chemical, mechanical and optical properties of solids. Using implantations with high doses ($\geqslant 10^{17}$ cm^{-2}), a transformation of the layer near the surface into another chemical compound is possible, e.g., silicon by implantation of carbon into silicon carbide. Electrochemical surface passivation by ion bombardment is possible for several metals. A few other fields of application are superconduction (increase in the transition temperature), hardening of metals, fabrication of light guides by implantation into quartz and GaAs, and investigations in nuclear physics. Thus, a wide new field for the application of ion implantation has been opened; within the scope of this book, however, it can be only treated in outline form.

1.2 HISTORICAL REVIEW

The beginnings of ion implantation are now more than twenty years in the past. Long before then, solids were irradiated by ions; there were also theories regarding the range of the penetrating particles [92], [93] as well as investigations on radiation damage in solids, and even in semiconductors [179]. However, a purposeful alteration of material properties by the implantation of ions was not taken into consideration until, in 1952, Ohl irradiated point-contact diodes with helium and achieved a reduction of the reverse current [546]. Later on, Cussins [166] tried to dope germanium by implantation of various ions, but the radiation damage which was produced simultaneously prevented success.

In 1957, Shockley [673] obtained the first patent for the technique of ion implantation. In this patent, it was pointed out for the first time that annealing after implantation is necessary for recrystallization of the crystal lattice. This patent covers practically all aspects of implantation.

From then on, more and more articles dealing with implantation were published. In 1962, the first nuclear-radiation detectors were produced by implantation of phosphorus into silicon [37] and in 1963, Ion Physics in Massachusetts started with the implantation of solar cells [411].

At the same time, the theoretical bases regarding the range distribution of low-energy ions in solids were established in Denmark by Lindhard, Scharff and Schiøtt [443], [442], proceeding from the work of Bohr [92]. The different penetration behaviour of ions into amorphous solids and into single crystals (channelling) was also investigated from about 1963 on [594], [536].

The final breakthrough began in the middle of the 1960s, when numerous centres for nuclear research (Chalk River Nuclear Laboratories, Oak Ridge National Laboratories, AERE Harwell) began to take an interest in this new technology.

For the production of MOS transistors, the use of implantation has become generally accepted; hardly any integrated MOS circuit is produced nowadays without the use of one or more implantation steps. In the field of bipolar transistors, the use of implantation is steadily increasing, while it is already being employed as standard technology for several special devices. Integrated circuits in GaAs are routinely implanted.

Since 1967, several conferences dealing with ion implantation have been held. Following the 'Conference on Applications of Ion Beams to Semiconductor Technology' (Grenoble, France, 1967) [19], the 'First International Conference on Ion Implantation in Semiconductors' was held in 1970 at Thousand Oaks, USA; this was the first conference in a series which was continued at

Garmisch, Germany in 1971, Yorktown Heights, USA in 1972, Osaka, Japan in 1974, Boulder, USA in 1976, Budapest, Hungary in 1978, Albany, USA in 1980 and Grenoble, France in 1982 [21], [22], [23], [24], [25], [25a], [25b]. In between, there were two European conferences on ion implantation at Reading, England in 1971 [20] and at Warwick, England in 1975 [30], as well as an American–Japanese seminar at Kyoto, Japan in 1971 [26]. Also of significance were two implantation conferences at Lublin, Poland in 1974 [27] and Budapest, Hungary in 1975 [28]. Several conferences have been devoted to the equipment aspects of ion implantation. The first was held in 1977 at Salford, England, followed by conferences at Trento, Italy in 1978, Kingston, Canada in 1980 and Berchtesgaden, Germany in 1982 [28a], [28b], [28c], [28d], [28e]. During several other conferences, there have been special meetings dealing with ion implantation or ion-implanted devices.

In addition to the conference proceedings, three books dealing exclusively with ion implantation have already been published. The first book, concentrating particularly on basic research, has been written by Mayer, Erikson and Davies [9], while a more recent work also emphasizing the technical-equipment aspects has been prepared by Dearnaley, Freeman, Nelson, and Stephen [2]. The book by Wilson and Brewer [13] concentrates almost exclusively on the technical equipment aspects. Carter and Grant [1a], as well as Townsend *et al.* [12a], similarly to Mayer *et al.*, focus on theoretical aspects. Several special bibliographies on the topic of ion implantation have also been published [14], [14a], [15], [16], [17], [18].

2
Theoretical Background of Ion Implantation

The theoretical background of ion implantation has mainly been developed during the last two decades, basing upon the work of Bohr [92], [93] and Rutherford [604]. These foundations include theories on the range distribution of implanted ions, on the energy loss of ions during implantation, and on the distribution of the radiation damage produced by this irradiation, as well as theories on the different ion-penetration behaviour in amorphous versus crystalline solids (channelling effect).

In this chapter, the basic aspects of the most important theories will be discussed, together with a number of measurements for illustration purposes. A more detailed treatment can be found in the specialized literature, for example, References [2], [6], [9], [442] and [443], which also contain extensive literature quotations.

2.1 RANGE OF IONS IN SOLIDS

The first investigations regarding the range and scattering of charged particles in solids were made around 1900 by Lenard as well as by Rutherford, who thus obtained first indications that atoms consist of a nucleus and a shell.

Later, Bohr [92], [93] calculated the energy loss per path length for heavy charged particles, on the basis of collisions between particles and bound electrons. He obtained, for the energy loss:

$$-\frac{dE}{dx}=\frac{4\pi z_1^2 q^4 N Z_2}{m_e v^2}\left[\ln\frac{1.123\, M_1 m_e v^3}{\bar{\omega} Z_1 q^2 (M_1+m_e)}-\ln(1-\beta^2)-\beta^2\right] \tag{2.1}$$

where Z_1 and v are the charge number and the ion velocity, Z_2 and N are the atomic number and the atomic density of the target atoms, m_e and q are the electron mass and the electron charge, $\ln\bar{\omega}=\Sigma f_i \ln\omega_i$, where f_i and ω_i are the strength of the oscillator and the frequency of the ith electron, and $\beta=v/c_0$, where c_0 is the velocity of light.

The last two terms contain a relativistic correction. The classical view is only permissible for:

$$z_1/137\beta \gg 1 \tag{2.2}$$

where 137β is the ion velocity in units of the electron velocity in a hydrogen atom. Further limits for equation (2.1) can be found in Reference [755].

Through quantum-mechanical considerations, Bethe and Bloch [74], [75], [84] later arrived at expressions which describe the binding conditions of electrons more realistically.

Within the scope of this book, it is not possible to consider these theories in more detail, with their various improvements and modifications [76], [84], [91] as well as their areas of validity [755], [766]. However, all of these works have in common that they only consider stopping by collisions with electrons.

Basically, there are several different effects which occur during the bombardment of solids with heavy charged particles, and which stop or scatter them. Classified in a somewhat general manner, these are as follows [701]:

1. Inelastic collisions with bound electrons of the stopping medium. The energy loss in such collisions takes place by excitation or ionization of atoms or molecules.
2. Inelastic collisions with nuclei. These lead to bremsstrahlung, nuclear excitation or nuclear reactions.
3. Elastic collisions with bound electrons.
4. Elastic collisions with nuclei or whole atoms, whereby a part of the kinetic energy is transferred to the particles absorbing the impact.
5. Cerenkov radiation. This is produced by particles which pass through the medium faster than the phase velocity of light.

Inelastic nuclear collisions and elastic collisions with electrons play an insignificant role in the stopping of particles, in comparison with inelastic collisions with electrons (electronic stopping) and elastic nuclear collisions.

Therefore, only the latter two mechanisms will be considered in the following. Which of the two effects predominates depends upon the energy and the mass of the accelerated particles, as well as upon the mass and the atomic number of the medium. In the energy range important for ion implantation, both effects are to be considered. For the calculation of the stopping, one defines cross-sections for electronic and nuclear stopping, $S_{e,n}$, as follows:

$$S_{e,n} = -\frac{1}{N}\left(\frac{dE}{dx}\right)_{e,n} \tag{2.3}$$

In Figure 2.1, the general behaviour of both stopping cross sections, dependent on the energy, is given. Values of energies E_1, E_2, E_3 of Figure 2.1 for various masses are shown in Table 2.1 [9]. The energies of ion implantation are lower than one MeV, therefore, they are always in the region below E_3. E_3 is determined by the condition of equation (2.2). Beyond E_3 is the region where the

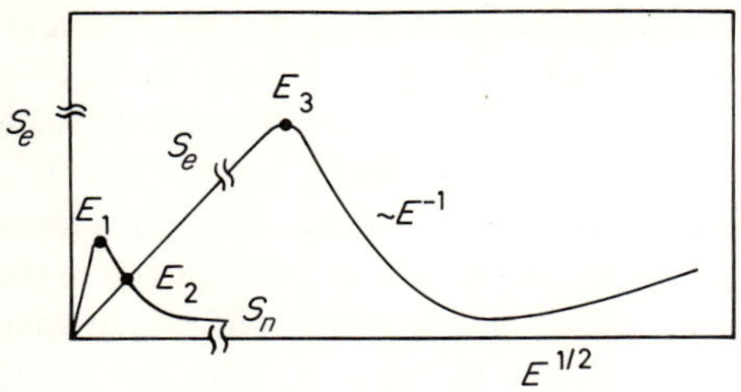

Figure 2.1 Stopping power S_e and S_n as a function of energy (schematic)

theory of Bethe and Bloch is valid; the ascending curve at very high energies is obtained by a relativistic correction.

Only in recent years has there been any interest in the energy range of ion implantation; this was first theoretically investigated by Lindhard and Scharff [443], and by Lindhard, Scharff and Schiøtt (LSS) [442]. In general, one refers to their results as the 'LSS theory'. According to this theory, the ions have a Gaussian range distribution around an average projected range R_p, with a standard deviation ΔR_p. In a further article [622], Schiøtt discussed in detail the calculation of the projected range. A more suitable formulation for the mathematical treatment was provided by Sanders [615].

Differing calculations of the range parameters have been made by Brice [104] as well as by Furukawa [266]. As their starting point, they used the LSS arguments; however, instead of the integral expressions for range distribution and standard deviation, they calculated the energy distribution during the stopping process of the ions. Therefore, these theories are more useful for the calculation of range distributions in multi-layer structures, and do not take for granted a Gaussian range profile.

All these theories have in common their limitation to an amorphous medium, which certainly does not apply generally to semiconductors, as well as their neglect of such secondary effects as diffusion. Furthermore, they only deal with the stopping of particles, and not with the effects of this stopping process upon the target material. The changes which are described by the rather general concept of

Table 2.1 Energies E_1, E_2 and E_3 of Figure 2.1 [9] for silicon, germanium and tin

	E_1 (keV)			E_2 (keV)			E_3(keV)
Ion	Si	Ge	Sn	Si	Ge	Sn	
B	3	7	12	17	13	10	3×10^3
P	17	29	45	140	140	130	3×10^4
As	73	103	140	800	800	800	2×10^5
Sb	180	230	290	2000	2000	2000	6×10^5
Bi	530	600	700	6000	6000	6000	2×10^6

'radiation damage' are manifested in a more or less significant rearrangement of the structure, or in the case of crystals, in a destruction of the lattice. They are especially apparent due to their effects upon the electrical parameters, although they can also alter many other properties of matter (density, elasticity, etc.). As far as the electrical parameters are concerned, they especially affect the mobility, the lifetime, and under some circumstances the carrier concentration.

Radiation damage has a distribution in the solid which generally deviates from the distribution of the ions, since the position at which the maximum energy is deposited to the target atoms is not the position at which the ions also come to rest. It can therefore be expected that the maximum of the radiation-damage distribution lies nearer to the surface than that of the ions. Theoretical discussions of radiation-damage distributions have been published by Brice [101], Sigmund and Sanders [685], Pavlov [558] and Furukawa [268a], who calculated either the energy deposited into nuclear collisions per unit distance, or the concentration of vacancies.

The following paragraphs will now deal with the basic aspects of the most important theories on the stopping and distribution of ions, as well as on the distribution of radiation damage.

As has already been mentioned, two mechanisms for the loss of energy are to be considered at low energies; on the one hand, collisions of ions with free and bound electrons, and on the other, collisions with atomic nuclei. On the assumption that both processes are independent of each other, one obtains for the energy loss per distance unit:

$$-\frac{\mathrm{d}E}{\mathrm{d}x} = N[S_\mathrm{n}(E) + S_\mathrm{e}(E)] \tag{2.4}$$

If $S_\mathrm{n}(E)$ and $S_\mathrm{e}(E)$ are known, it is possible to integrate equation (2.4), and one obtains:

$$R = \frac{1}{N}\int_0^E \frac{\mathrm{d}E}{S_\mathrm{n}(E) + S_\mathrm{e}(E)} \tag{2.5}$$

R is the average range of a particle of the energy E in an amorphous medium.

2.1.1 Nuclear Stopping

The energy loss $\mathrm{d}E$ of an ion by elastic nuclear interaction, in a layer $\mathrm{d}x$, is proportional to the atomic density N as well as to the total energy transferred in all individual collisions T_n, i.e.,

$$S_\mathrm{n}(E) = -\frac{1}{N_\mathrm{n}}\left(\frac{\mathrm{d}E}{\mathrm{d}x}\right)_\mathrm{n} = \int_0^\infty T_\mathrm{n}(E, p) 2\pi p \, \mathrm{d}p = \int_0^{T_m} \mathrm{d}\sigma(E, T_\mathrm{n}) \tag{2.6}$$

where $\mathrm{d}\sigma$ is the differential cross section ($\mathrm{d}\sigma = 2\pi p \, \mathrm{d}p$) and p is the impact parameter. T_m is the maximum transferable energy for a central collision. Figure

2.2 shows a typical scattering process between an ion and an atom. The incident ion is deflected by the angle θ_1, and transfers the energy T_n to the target atom, which leaves its original position at the angle θ_2. For values of p between ∞ and 0, one obtains values for $T_n(E, p)$ between 0 and T_m. T_m amounts to:

$$T_m = 4 \frac{M_1 M_2}{(M_1 + M_2)^2} E \tag{2.7}$$

M_1 amd M_2 are the masses of ion and target.

The calculation of $T_n(E, p)$ is a well known problem of classical mechanics |116|:

$$T_n(E, p) = E \frac{2M_1 M_2}{(M_1 + M_2)^2} (1 - \cos \Phi) \tag{2.8}$$

where Φ is the deflection angle in the centre-of-mass system. When calculated according to Reference |296|, the following is obtained:

$$\Phi = \pi - 2p \int_0^{u_{max}} \frac{du}{(1 - V(u)/E_r - p^2 u^2)^{1/2}} \tag{2.9}$$

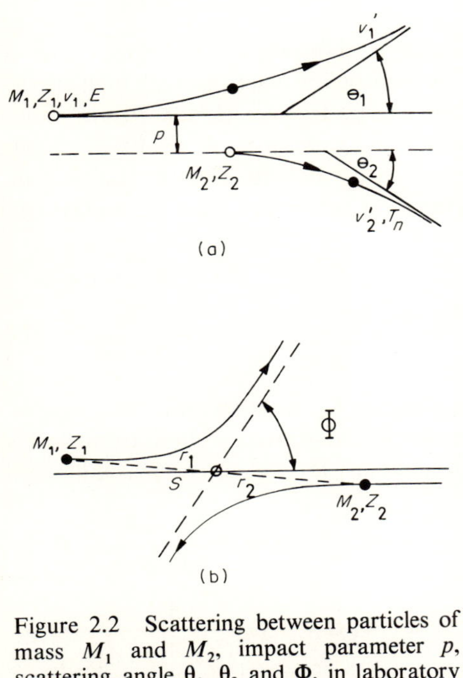

Figure 2.2 Scattering between particles of mass M_1 and M_2, impact parameter p, scattering angle θ_1, θ_2 and Φ, in laboratory system (a) and centre-of-mass system (b)

where $u = 1/r$, $r = r_1 + r_2$ (see Figure 2.2) is the distance between particles in the centre-of-mass system, $V(u)$ is the interaction potential, $E_r = EM_2/(M_1 + M_2)$ is the energy of the ion in the centre-of-mass system, and u_{max} is the reciprocal value of the minimum distance between the particles.

In order to calculate equation (2.9), $V(u)$ has to be known. If the screening effect of the electrons were to be neglected, the classical case of Rutherford for the scattering of α-particles would be obtained, and $V(u)$ could be directly written as the Coulomb potential.

For heavy particles of low velocity, and in the case of relatively distant collisions, the screening effect of the electrons cannot be neglected and the following is written:

$$V(r) = \frac{Z_1 Z_2 q^2}{r} \varphi\left(\frac{r}{a}\right) \tag{2.10}$$

where $\varphi(r/a)$ is a suitable screening function and a is a screening parameter (of the order of magnitude of the Bohr radius). Because of the use of the cgs system in the case of all the cited classical derivations, the factor $4\pi\varepsilon_0$ before r is missing in equation (2.10).

Analytical solutions can be given for power potentials (Lindhard and Scharff |441|):

$$V(r) = \frac{Z_1 Z_2 q^2}{r} \frac{1}{s} \left(\frac{a}{r}\right)^{s-1} \tag{2.11}$$

However, the simple analytical form is lost for other values of s than 1 and 2. The approximation $s = 1$ is realistic for collisions with high energy transfer, as is $s = 2$ for collisions with low energy transfer (Lindhard and Scharff |441|).

The best agreement with experimental results is obtained with use of the statistical atomic model according to Thomas and Fermi |240|, |725|, from which the screening function can be derived as a solution of the differential equation:

$$\varphi''(x) = \varphi^{3/2}(x) x^{-1/2} \tag{2.12}$$

In this case, a is obtained by:

$$a = \tfrac{1}{2}(\tfrac{3}{4}\pi)^{2/3} a_0 (Z_1^{2/3} + Z_2^{2/3})^{-1/2} \tag{2.13}$$

where a_0 is the Bohr radius ($a_0 \approx 0.053$ nm). With use of this potential, LSS calculated the differential effective cross-section for nuclear stopping. Using suitable normalization (reduction) for range, energy, and scattering parameter:

$$\begin{aligned} \rho &= C_R R = 4\pi a^2 N \frac{A_1 A_2}{(A_1 + A_2)^2} R \\ \varepsilon &= C_E E = \frac{a}{q^2} \frac{A_2}{Z_1 Z_2 (A_1 + A_2)} E \\ t &= \varepsilon^2 \sin^2 (\Phi/2) \end{aligned} \tag{2.14}$$

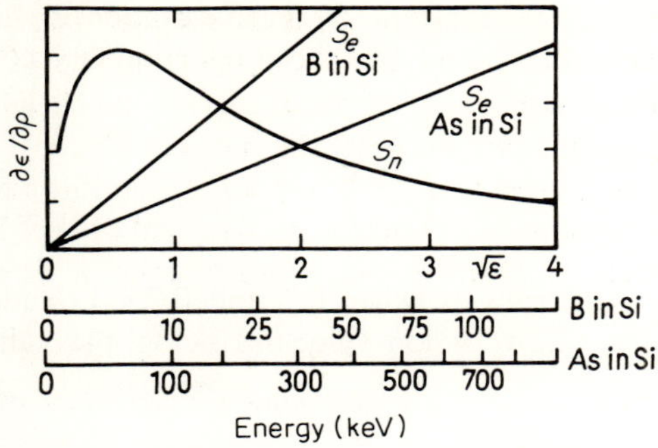

Figure 2.3 S_n and S_e in normalized form for boron and arsenic in silicon

($A_{1,2}$ = atomic weights of ion and target), they succeeded in obtaining a generally valid dependence of the stopping power S_n upon the energy, for all ion-target combinations, which is dimensionless in the units used.

The differential cross-section for nuclear scattering is given by the approximating equation:

$$\frac{d\sigma}{dt} = 2\pi a^2 t^{3/2} f(t^{1/2}) \tag{2.15}$$

Values for the universal scattering function $f(t^{1/2})$ have been given by LSS in table form.

A numerical calculation of the reduced stopping power $(d\varepsilon/d\rho)_n$ is given in Figure 2.3, together with values for electronic stopping.

2.1.2 Electronic Stopping

Electronic stopping decreases with increasing particle velocity at high energies in the validity region of the Bethe theory, as shown in Figure 2.1. The low-energy region was first investigated by Lindhard and Winter [438]. They assumed that the electrons form a free electron gas, and were able to demonstrate that the electronic-stopping cross-section is proportional to the velocity of the ions, and therefore proportional to the root of the energy:

$$S_e(E) = -\frac{1}{N}\left(\frac{dE}{dx}\right)_e = k'E^{1/2}$$

$$k' = \frac{kC_R}{C_E^{1/2}N} \tag{2.16}$$

$$k = \xi_e \frac{0.0793 \cdot Z_1^{1/2} Z_2^{1/2} (A_1 + A_2)^{3/2}}{(Z_1^{2/3} + Z_2^{2/3})^{3/4} A_1^{3/2} A_2^{1/2}}$$

where ξ_e is a dimensionless constant of the magnitude $Z_1^{1/6}$; C_R and C_E are obtained from equation (2.14), and k is the proportionality constant in the normalized representation, i.e.,

$$-\left(\frac{d\varepsilon}{d\rho}\right)_e = k\varepsilon^{1/2} \tag{2.17}$$

The values for k are generally between 0.1 and 0.25. Equation (2.17) is valid (according to equation (2.2)) for ion velocities below the value:

$$v < Z_1^{2/3} q^2 / h \tag{2.18}$$

For higher energies, S_e runs through a wide maximum and then decreases like $v^{-1,\ldots,2}$ in the Bethe–Bloch region. In Figure 2.3, the nuclear-stopping cross-section and two curves of the electronic-stopping cross-section for boron and arsenic are shown in normalized form.

A somewhat different model has been developed by Firsov for the electronic stopping [242]. In contrast to LSS, he assumes that the ion and the target atom, during the collision, form a quasi-molecule. This model has the advantage that it can be generalized to explain oscillations which occur in measured values of S_e [228]. In the case of amorphous media, the calculation is carried out similarly to that of the nuclear-stopping cross-section.

One first calculates the energy loss, T_e, resulting from the electronic interaction between particles having the nuclear charges Z_1 and Z_2, and then integrates over all impact parameters

$$S_e = \int_0^{\infty} T_e(E, p) 2\pi p \, dp \tag{2.19}$$

The energy loss of the incident particle 1 during formation of a quasi-molecule takes place through the transfer of momentum and energy to electrons of the solid particle 2. The entire energy loss of particle 1, when it passes from $-\infty$ to $+\infty$ via particle 2, is approximately:

$$T_e = \frac{4.3 \times 10^{-8}(Z_1 + Z_2)^{5/2}}{|1 + 3.1 \times 10^7 (Z_1 + Z_2)^{1/2} p|^5} v \qquad \text{(eV)} \tag{2.20}$$

where v is in cm/s and p is in cm.

For the stopping cross-section, one obtains by integration:

$$S_e(v) = 2.34 \times 10^{-23}(Z_1 + Z_2)v \qquad (\text{eV cm}^2) \tag{2.21}$$

Ranges are obtained according to this model together with a suitable S_n, similarly to the LSS model.

2.1.3 Range Distribution

By means of the values calculated in Sections 2.1.1 and 2.1.2 for $S_e(E)$ and

$S_n(E)$ as well as the corresponding reduced values, one can calculate the range according to equation (2.4). In Figure 2.4, curves of the reduced ranges are shown, depending on the reduced energy, with k as a parameter [442]; these are valid for all ion-target combinations.

The result of equation (2.4) is the average total range of an ion. The range straggling (ΔR) relating to this range is given, according to LSS, for an approximation of the Thomas–Fermi potential with a power potential [442] (see also equation (2.11)), as:

$$\left(\frac{\Delta\rho}{\rho}\right)^2 = \frac{(\Delta R)^2}{R^2} = \frac{s-1}{s(2s-1)} \frac{4M_1 M_2}{(M_1 + M_2)^2} \tag{2.22}$$

The projection of R on the incidence direction of the ion beam, the so-called projected average range R_p, and likewise a projected standard deviation ΔR_p (which

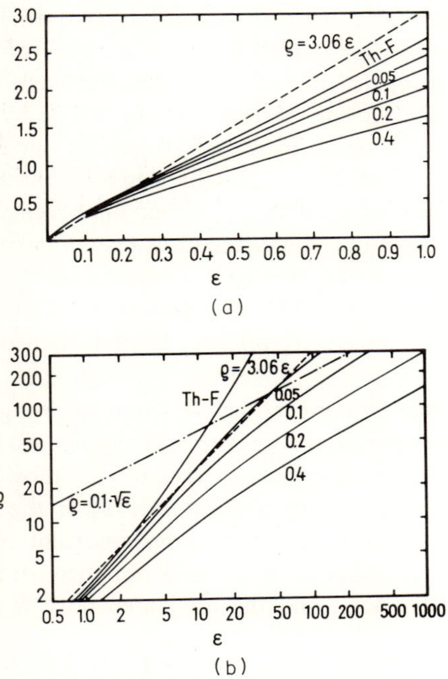

Figure 2.4 Normalized range–energy curves as a function of parameter k for electronic stopping [442]: (a) for $\varepsilon \leqslant 1$; Th–F is calculated neglecting electronic stopping. $\rho = 3.06\varepsilon$ is calculated for a power potential with $s = 2$ (constant nuclear stopping); (b) for $\varepsilon \geqslant 1$; the dot–dash line applies for $k = 0.1$ neglecting nuclear stopping

in the following are abbreviated as 'range' and 'standard deviation'), are the factors which are of practical interest, and also the only ones which can be measured. The solution of integro-differential equations, derived from LSS [442], is necessary for their calculation. LSS themselves calculated approximate values for a simplified interaction potential; Sigmund and Sanders [685] directly solved the integral equation under the assumption of an r^{-s} potential. Johnson and Gibbons [387] solved the differential equation by series expansion, which led in some cases to significant errors, as shown by Brice [104] and Furukawa [266]. More recently, Furukawa [268] and Brice [102] have calculated the differential equation of the second order, and obtained values which agree with experiments better than Reference [387]. The most extensive calculations and tables for practically all ion-substrate combinations occurring in semiconductor technology have been published by Gibbons *et al.* [5]. This work estimates the influence of higher moments, and also gives values for electronic stopping and stopping through nuclear collisions. A more general collection of tables, which is mainly important for implantation into metals and also deals with the distribution of radiation damage as well as that of ionization, has been compiled by Brice [1]. Simple rules of interpolation, which have an accuracy of about 10 per cent and are especially important for ion-substrate combinations for which no accurate calculations are available, can be found in an article by Schiøtt [623].

Calculations of the stopping processes of ions at low energies ($\leqslant$1 keV) have been carried out by Winterbon [779]. Starting with the LSS arguments, Monte Carlo calculations can be used for a further improved calculation of stopping processes. One thus obtains more accurate results, particularly for multilayer structures and for the calculation of back-scattered ions, which are not considered in the classical LSS theory [373]. However, these calculations are very time-consuming, and results have therefore only been published for a few special cases [373], [347].

In Figure 2.5, values for the range (R_p) and the range straggling (ΔR_p) are given for boron, phosphorus, and arsenic in silicon, depending on the acceleration energy of the ions. In the Appendix, values for the most common ion-substrate combinations are given according to Biersack [79]. In the next chapter in Figures 3.25 to 3.27, further values are given for some important masking layers.

The range distribution of implanted ions is obtained from the average range R_p, the standard deviation ΔR_p, and the implanted dose $N_\square$, as follows:

$$N(x) = \frac{N_\square}{\sqrt{(2\pi)}\Delta R_p} \exp\left[-\frac{(x-R_p)^2}{2\Delta R_p^2}\right] \tag{2.23}$$

The maximum doping concentration is:

$$N_{max} = \frac{N_\square}{\sqrt{(2\pi)}\Delta R_p} \tag{2.24}$$

Equation (2.23) is derived under the assumption that the integral from $-\infty$ to $+\infty$

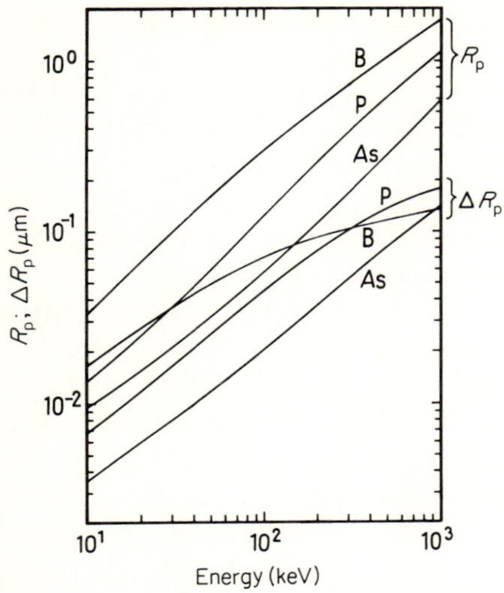

Figure 2.5 Range R_p and range straggling ΔR_p of arsenic, boron, and phosphorus in silicon, according to Reference [5]

versus $N(x)$ equals the implanted dose $N_\square$, and neglects back-scattered ions. The error caused by the assumption of infinite semiconductor dimensions is minimal; an exact expression can easily be arrived at:

$$N(x) = \frac{2N_\square}{\sqrt{(2\pi)}\Delta R_p(1 + \operatorname{erf} R_p/\sqrt{(2)}\Delta R_p)} \exp\left[-\frac{(x-R_p)^2}{2\Delta R_p^2}\right] \qquad (2.25)$$

However, the number of back-scattered ions can only be determined by Monte Carlo calculations, and is therefore always neglected during practical investigations, although it can be considerable in low-energy implantations.

Profiles calculated according to equation (2.23) for boron in silicon at various implantation energies, using values for R_p and ΔR_p from Figure 2.5, are shown in Figure 2.6. With increasing energy and penetration depth, the maximum doping decreases, because of the increasing range straggling. In semiconductor physics, one does not normally use a linear plot of doping versus depth as shown in this figure, but rather a semilogarithmic scale, since one is interested in the doping profile over several orders of magnitude. The theoretical profiles according to equation (2.23) are parabolas in this representation. If, in the calculation of the range, moments of higher order are taken into account, then more complicated distributions are obtained. With the use of the third-order moment, one obtains a distribution resulting from two Gaussian curves, according to Gibbons *et al.* [288]. The use of functions with four moments, for example, according to Edgeworth [5] or Pearson [338], leads to a better profile description. However,

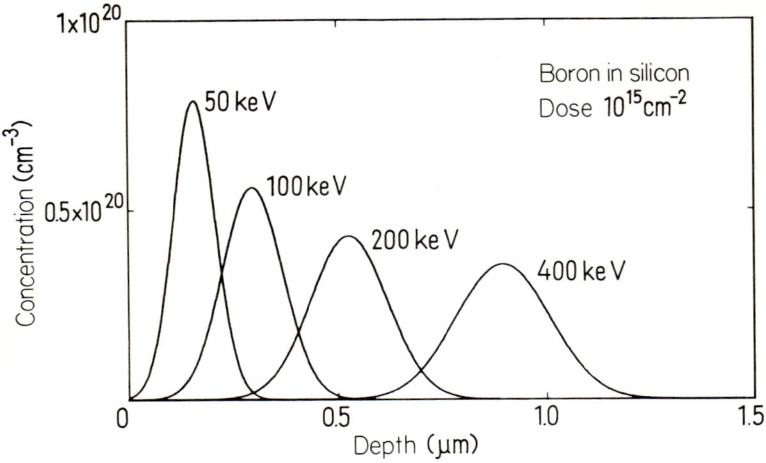

Figure 2.6 Theoretical range distribution of boron in silicon (linear scale)

because of the many physical effects which prevent ideal range distributions (e.g., channelling effect, diffusion), this does not lead to any significant improvement, so that the Gaussian approximation appears suitable for nearly all kinds of problems.

The correlation between the LSS theory and measurements of range distributions is relatively good in general, especially when experimental values for the electronic stopping are used for the calculations. This is notably the case with boron, for which it has been possible to eliminate a deviation of about 30 per cent between experimental results and range calculations according to LSS with use of stopping-power data measured by Eisen [221]. Unfortunately, corresponding S_e values are available only for a few ion-substrate combinations. Measurements of range and range straggling will be compared with theoretical values in Chapter 6.1.

2.2 RADIATION DAMAGE IN SOLIDS

In the preceding discussion, only the stopping of ions has been considered, without taking into account any effects upon the crystal lattice. Atoms can be displaced from their lattice sites, depending on the energy and the mass of the implanted ions as well as the mass of the target. The displaced atoms themselves can also displace other atoms, so that a collision cascade results. This leads to an accumulation of vacancies and interstitial atoms (Frenkel defects), as well as to complex lattice defects along the ion path (clusters). Heavy ions can transfer more energy to lattice atoms than lighter ones. These atoms, according to their energy, can displace additional lattice atoms. A diagram showing formation of radiation damage can be seen in Figure 2.7, for light as well as heavy ions.

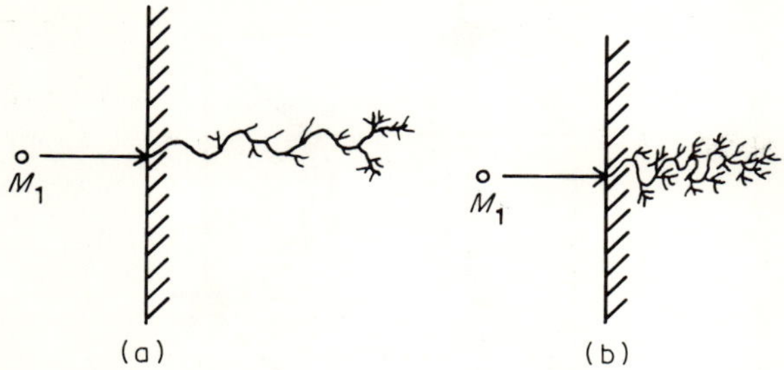

Figure 2.7 Schematic representation of radiation-damage formation: (a) $M_1 < M_2$ (light ion); (b) $M_1 > M_2$ (heavy ion)

The damaged areas begin to overlap with increasing dose, and finally form an amorphous layer extending to a certain depth, i.e., a layer in which there is no longer any long-range order in the lattice. The amount of radiation damage and its distribution depend on the ion species, the temperature, the energy, the dose, and the channelling which may possibly occur (see Section 2.3). As a measure for the concentration of radiation damage, one uses the part of the energy which is transferred to the solid in the form of nuclear collisions [101], or the number of vacancies produced [558], [268a]. It has been demonstrated, for 500-keV oxygen ions in silicon [703], that more complex defects too are approximately proportional to this. A detailed survey on radiation damage and its annealing behaviour has been provided by Gibbons [7]. Fundamental work on defects, mainly in silicon, has been done (among others) by Vook [749], as well as by Kimerling and Poate [409]. The next chapter will deal with the practical aspects of radiation-damage effects.

2.2.1 Structure of Defects

The simplest defect, a Frenkel defect, is produced by the displacement of a lattice atom on to an interstitial site; a vacancy and an interstitial atom are the result. The crystallographic structure of a vacancy is shown in Figure 2.8, according to Corbett [153]. Vacancies can have various charge states (e.g., neutral, positive, negative, double-negative); they can form a complex with impurity atoms, and can influence their diffusion. Double vacancies can be formed when an incident ion displaces two neighbouring lattice atoms. Their formation from two simple vacancies is also possible. Double vacancies are stable up to about 550 K.

Dislocations can be formed by accumulation of simple defects, or can grow during annealing, starting from unannealed radiation damage, into the undamaged crystal. Dislocation lines only anneal at high temperatures (>1000 °C), and sometimes do not anneal at all. Further defects can be formed by an accumulation of vacancies or interstitial atoms, as well as by a combination of impurity atoms with vacancies or interstitial atoms.

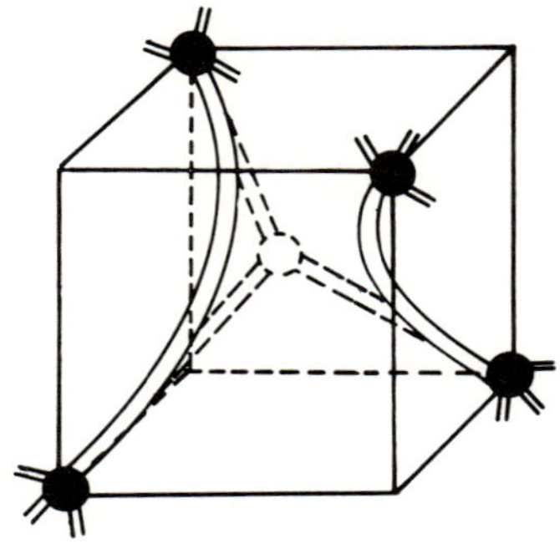

Figure 2.8 Structure of a silicon vacancy; according to Corbett [153]

If many lattice atoms are displaced by an incident ion, and if this occurs within a sufficiently small volume, then a local amorphous zone is produced, which is usually referred to as a cluster; there are various assumptions about the precise structure of these clusters. Due to the high mass and energy of the incident particles, it must always be expected that this will occur during ion implantation. The processes which may take place during implantation, as well as during the subsequent annealing treatment, are accordingly complex and difficult to explain theoretically. In the next chapter, these problems will be dealt with in more detail.

2.2.2 Number of Displaced Atoms

The number of atoms displaced in a primary collision ($N_{d,p}$) per incident ion can be calculated, based on the Kinchin–Pease theory [410], by:

$$N_{d,p} = \frac{E}{2E_d} \tag{2.26}$$

where E is the energy given off by the incident particle and E_d is the displacement energy of a lattice atom. In the case of silicon, E_d is approximately 14 eV [52], [543], [694]; for other semiconductors, E_d is between 8 eV and 30 eV. Equation (2.26) is only valid when the total energy of the primary particle is less than a critical value E_A, which is given by:

$$E_A = 2E_R Z_1 Z_2 (Z_1^{2/3} + Z_2^{2/3})^{1/2} \frac{M_1 + M_2}{M_2} \tag{2.27}$$

where $E_R = 13.6$ eV (Rydberg energy). In this energy range, the model for collisions between hard spheres applies. Values for E_A are given in Table 2.2 for arsenic, antimony, boron and silicon in silicon. Since equation (2.26) is linear in energy, the total number of atoms displaced by a particle is calculated by:

Table 2.2 Critical energies E_A, E_B, E_C in keV calculated according to equations (2.26) and (2.29), for boron, silicon, arsenic and antimony in silicon

	B	Si	As	Sb
E_A (keV)	7.8	36.3	185.3	459.4
E_B (keV)	871.7	2.36×10^4	4.85×10^5	2.3×10^6
E_C (keV)	17	120	800	2000

$$N_d = \frac{E_n}{2E_d} \tag{2.28}$$

E_n is the total energy loss of a particle in primary and secondary nuclear collisions. A modified calculation by Sigmund [682] results in a similar expression. Equation (2.28) does not have the limitation of equation (2.26), and is also valid in regions where there are screened Coulomb collisions, and where the Thomas–Fermi potential can be used.

At higher energies, only some of the nuclear collisions cause displacement of atoms; in addition, the mechanism of electronic stopping becomes dominant for the stopping of the ions. Two critical energies, E_B and E_C, are significant in this region |7|, |410|. For E_B, the following applies:

$$E_B = \frac{M_1 M_2}{(M_1 + M_2)^2} \frac{E_A^2}{E_d} \tag{2.29}$$

E_C is approximately the energy for which $S_n = S_e$, i.e., E_2 as mentioned in Section 2.1. Values for E_B and E_C are given in Table 2.2. For energies above E_B, one can assume the number of displaced atoms to be about half of the number calculated using equation (2.26) |7|. The total number of displaced atoms per primary particle, in the energy region greater than E_B and E_C, is given by:

$$N_d = \frac{P(E - E_c) + bE_c}{E_d} \tag{2.30}$$

P is approximately 10^{-3}; b is 1/2 for $E_B \geqslant E_C$ and 1/4 for $E_B \leqslant E_C$. If $E_B \gg E_C$,

Table 2.3 Number of displaced atoms in silicon ($E_d = 14$ eV), according to Kinchin–Pease [410]

Ion	10 keV	50 keV	200 keV
Antimony	357	1 785	7 143
Arsenic	357	1 785	28 500*
Boron	606*	609	620
Silicon	357	4 280	4 290

* $E < E_C$.

which is generally the case for silicon, equation (2.29) is always valid for $E > E_C$. Several examples for the number of displaced atoms in silicon, for various ions and energies according to equations (2.26) and (2.30), appear in Table 2.3. Better results are obtained with use of equation (2.28) and calculated values of E_n; *cf.* the following section.

2.2.3 Range Distribution of Radiation Damage

As a measure of the radiation damage, one either uses that part of the energy transferred to the solid by nuclear interaction [101], or the number of vacancies produced [268a], [558], assuming a certain energy expenditure for vacancy production. Since ions require a certain energy for the production of radiation damage, the maximum of the radiation-damage distribution is always closer to the surface than the maximum of the ion distribution.

Sigmund and Sanders [685] have calculated the spatial distribution of radiation damage by means of an approximation for the Thomas–Fermi potential, without consideration of electronic stopping and without use of a displacement energy. In other words, they calculated the energy transferred in nuclear collisions for the implanted ion as well as for the atom with which it collides. This calculation proceeds similarly to that in the LSS theory [442], and is carried out with the use of power potentials (*cf.* equation (2.10)). Although higher moments have been calculated, the results are generally used only for determination of Gaussian distributions of radiation damage. Figure 2.9 shows the relationship between the average projected range (R_p) and the range of the radiation damage (X_D), as well as the corresponding relationship between the standard deviations ($\Delta R_p/\Delta X_D$), for various power potentials in the case of silicon. For boron in silicon, as an example, one obtains $X_D = 0.8\,R_p$; $\Delta X_D = 0.75\,\Delta R_p$.

In a more recent study, Sigmund *et al.* also take electronic stopping into

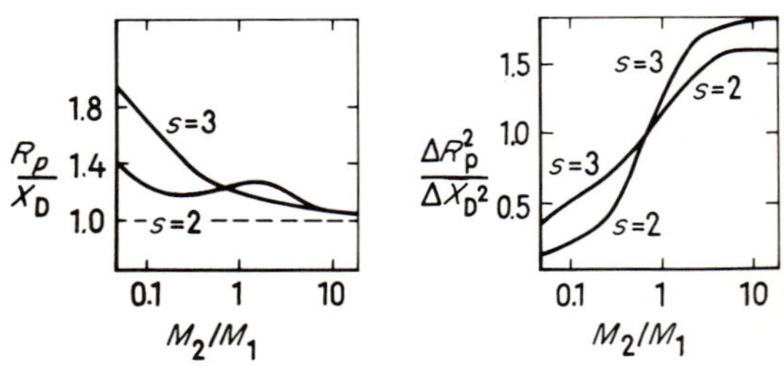

Figure 2.9 Ratio between the average projected ion range and the range of radiation damage R_p/X_D, and ratio between the projected standard deviations of the ion distribution and the damage distribution $\Delta R_p^2/\Delta X_D^2$, as functions of the mass ratio M_2/M_1 [685]. $s = 3$ applies for $\varepsilon \leqslant 0.2$, $s = 2$ applies for $0.8 \leqslant \varepsilon \leqslant 2.0$

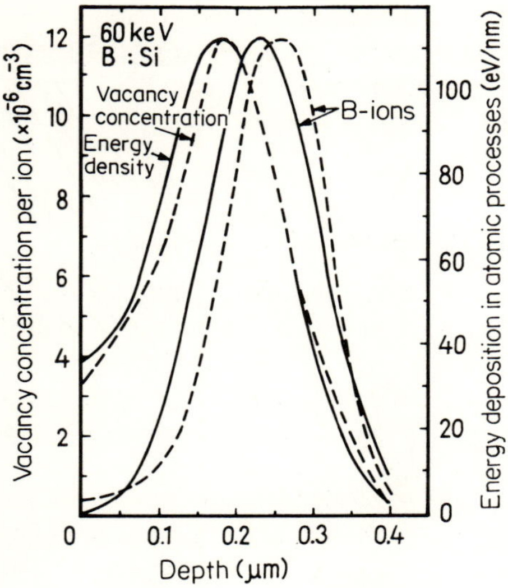

Figure 2.10 Comparison between energy distribution deposited into atomic processes according to Brice [101] and Monte Carlo calculation of vacancy concentration according to Pavlov [558], in both cases for 60 keV. The ion distributions based on both theories are also given. Continuous lines according to Brice [101], broken lines according to Pavlov [558]

account [681], but only report results for identical mass of ion and target atom. Pavlov *et al.* [558] have obtained, by means of a Monte Carlo calculation, distributions of vacancies as well as ions for boron, aluminium, phosphorus and arsenic at 20, 40, and 60 keV. Their calculations include electronic stopping, but use the unrealistic Bohr potential. As a measure of the radiation-damage distribution, Pavlov uses the distribution of the vacancies produced, assuming a displacement energy of 30 eV.

Brice [101], [104] uses a different method to likewise take the electronic interaction into account. This calculation is carried out in two steps. First, the spatial distribution of the ions during the stopping process is calculated, i.e., the ion distribution at an average energy (E') between their initial energy (E) and 0 is determined. If the position of the ions, the cross-section, and the experimentally measured ratio of energy loss in electronic and atomic processes are all known, then one can calculate the spatial distribution of the energy deposited in atomic processes. This work is based on the use of somewhat modified LSS equations [442]. For his calculations, Brice uses experimental data on energy loss in collisions between silicon atoms as compiled by Sattler [620]. In Figure 2.10, the

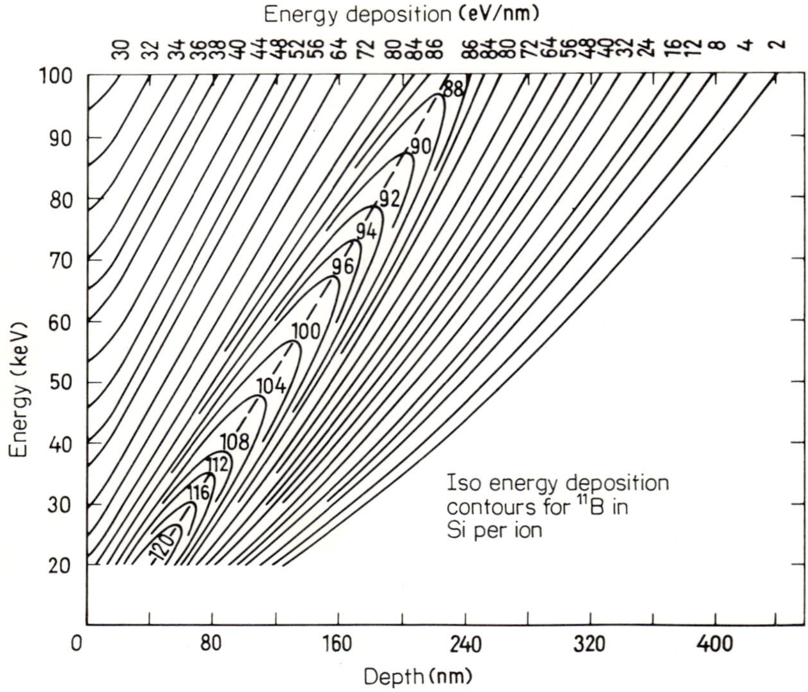

Figure 2.11 Curves of constant energy deposition into atomic processes per ion for boron in silicon, according to Brice [104]. Reproduced by permission of Gordon and Breach

results of Brice and Pavlov for 60 keV boron ions in silicon are compared. In Reference [104], Brice has calculated curves of constant energy deposition in nuclear collisions for many ion species, within the energy region between 20 and 400 keV. An example of these results for boron is shown in Figure 2.11. By drawing a horizontal line through the initial energy, one can determine the energy deposited in nuclear collisions. Figure 2.12 shows a few such profiles at various implantation energies, for boron in silicon. The theories of both Pavlov and Brice yield an asymmetrical distribution, with less decrease of the radiation-damage concentration towards the surface. The qualitative agreement between the theories is good.

Tsurushima and Tanoue [739] have developed a further method, which can also be used for substances consisting of different atoms. In order to calculate the energy transferred in nuclear collisions, an energy-distribution function of the penetrating ions at a certain depth is derived, and is integrated together with a weighting function. This is dependent upon the total cross-section for the displacement of atoms, as well as upon the average energy deposition in those collisions which can displace atoms of the substrate material. Using this theory, one can calculate distributions of the energy deposition in nuclear collisions $(\mathrm{d}E/\mathrm{d}x)_\mathrm{n}$ as well as in atomic displacements $(\mathrm{d}E/\mathrm{d}x)_{\mathrm{n,d}}$.

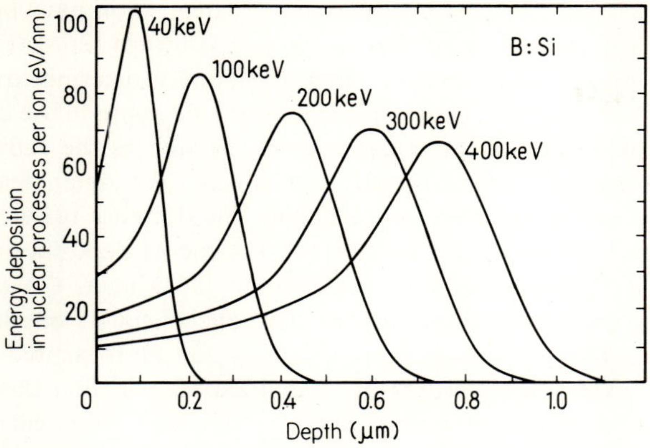

Figure 2.12 Depth distribution of energy deposited into atomic processes for boron in silicon, according to Brice [104]. Reproduced by permission of Gordon and Breach

Examples of the energy depositions $(dE/dx)_n$ and $(dE/dx)_{n,d}$, for arsenic in silicon, are shown in Figure 2.13. The results of this theory agree well with the calculations of Brice, but are more generally applicable. Unfortunately, in contrast to Brice, no comparably extensive calculations are provided.

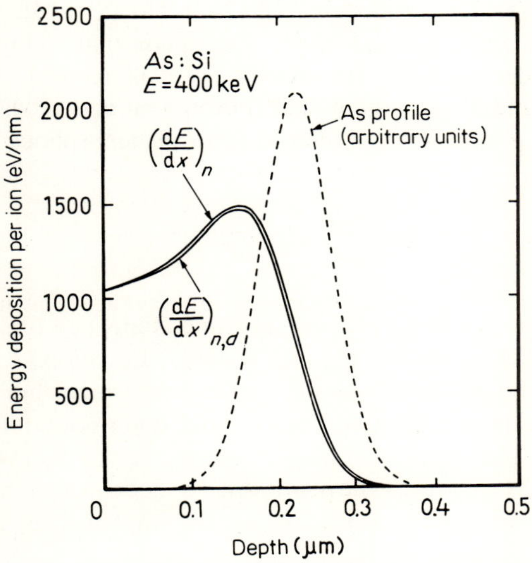

Figure 2.13 Total energy deposition and energy deposition into nuclear collisions for a 400 keV arsenic implantation into silicon [739]. Reproduced by permission of the Physical Society of Japan

Further models and theories regarding energy deposition have been published by Gibbons [287], as well as by Furukawa and Ishiwara [266]. The qualitative results agree with the other theories cited. It has been determined recently [79] that the works cited do not accurately indicate the decrease in the energy deposition towards the surface in atomic processes, as well as the radiation-damage distribution, since the back-scattered ions which once again leave the semiconductor have not been taken into account. The decrease towards the surface should therefore be considerably greater, and the energy deposition at the surface should be zero. An experimental confirmation is difficult. On the one hand, various defects can occur according to the concentration; on the other, only relatively high radiation-damage concentrations can be measured, and this can never be done over several orders of magnitude, as with ion distributions.

The number of displaced atoms can be calculated from energy-deposition distributions according to the modified Kinchin–Pease formula, equation (2.28), if it is taken into account that E_n is the integral over the distribution function of energy deposition in nuclear collisions.

2.2.4 Formation of Amorphous Layers

If the implanted ion dose, and thus the concentration of radiation damage, is sufficiently high, the damage clusters overlap and an amorphous layer is formed. A number of models have been postulated to estimate the dependence of this effect upon mass, temperature, and dose [7], [267], [513], [514], [737].

The simplest method is based on the assumption that all target atoms must be displaced in order that an amorphous layer is formed. By means of the Kinchin–Pease formula (equation (2.26)) for the number of displaced atoms, one obtains, for the critical dose needed to form an amorphous layer:

$$N_{\square,a} = \frac{2E_d N}{(dE/dx)_n} \tag{2.31}$$

where N is the atomic density of the substrate atoms and $(dE/dx)_n$ is the energy deposited in nuclear collisions per unit distance. Equation (2.31) underestimates the critical dose for practically all ion species, since annealing effects upon radiation damage during implantation, for example, due to out-diffusion of vacancies and collisions with already-displaced atoms, are not taken into account. Morehead and Crowder [514] have proposed a model which does take account of out-diffusion of vacancies, depending on the implantation temperature. They obtained:

$$N_{\square,a} = N^{\circ}_{\square,a}(1 - \delta R N^{\circ}_{\square,a}{}^{2})^{-2} \tag{2.32}$$

$N^{\circ}_{\square,a}$ is an expression as in equation (2.31), but without the factor 2; δR is the reduction of the damage cluster by out-diffusion, and is given by:

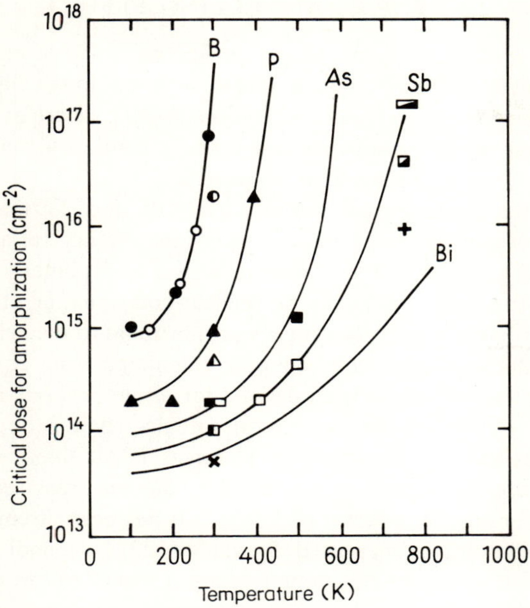

Figure 2.14 Temperature dependence of the critical dose for silicon amorphization by boron, arsenic, bismuth, antimony and phosphorus |514|. Experimental values by Crowder |162| (●, ▲, ■); Westmoreland |527|, |766| (○, □); Mayer |9| (◐, ▲, ◧, ■□, ×); Eriksson |229| (◪, ◪, +). Reproduced by permission of Gordon and Breach

$$\delta R = 2(D_v t)^{1/2} \tag{2.33}$$

where D_v is the diffusion constant of vacancies and t is the time in which the vacancies diffuse.

In Figure 2.14, the temperature dependence of the critical dose is shown for antimony, boron, phosphorus, arsenic, and bismuth. The experimental values were obtained by electro-paramagnetic-resonance and back-scattering measurements |9|, |162|, |229|, |572|. Gibbons |7| has suggested, in the case of light ions, that one must assume an overlapping of the clusters in order to obtain realistic conclusions. In comparing measurements, one must keep in mind that the term 'amorphous' has different meanings in various measurement methods, and that the dependence on current density during implantation (dose-rate effect), as well as annealing effects during and after implantation prior to measurement, can be significant. With an increase in the dose rate, the critical dose decreases in most cases.

2.3 THE CHANNELLING EFFECT

Up to the present, all theories on range in solids have had in common that they assume an amorphous target, i.e., a target with no long-range lattice order. However, almost all semiconductors are monocrystals, or are at least crystalline. Since crystals have highly anisotropic properties, greater or lesser deviations from the theoretical profile shape must be expected.

Because of the symmetrical arrangement of the lattice atoms in the monocrystal, ions can penetrate more deeply into the crystal along major axes (e.g., ⟨110⟩, ⟨111⟩, ⟨100⟩ in the diamond lattice) and planes. Within these so-called channels, there are practically no nuclear collisions; i.e., the stopping only takes place electronically, and the range is proportional to the velocity of the ions. However, the values for electronic stopping calculated from equation (2.17) are not applicable, since they have been derived for an amorphous medium. Using a modification by Firsov of the expression for electronic stopping in equation (2.21), an expression for the range in channels is obtained which accurately describes the stopping dependence upon the channel cross-section, but which overestimates the range by a factor of 1.5 to 2. It has been attempted to take into account the oscillations in measured S_e values [222] depending on Z_1 (which have a great influence upon the range), using a modification of the theory by Firsov [136], [223a], but this does not yet appear completely successful [9]. A comparison between experimental S_e values in channels, depending on mass, and theoretical values [136], [222] is shown in Figure 2.15.

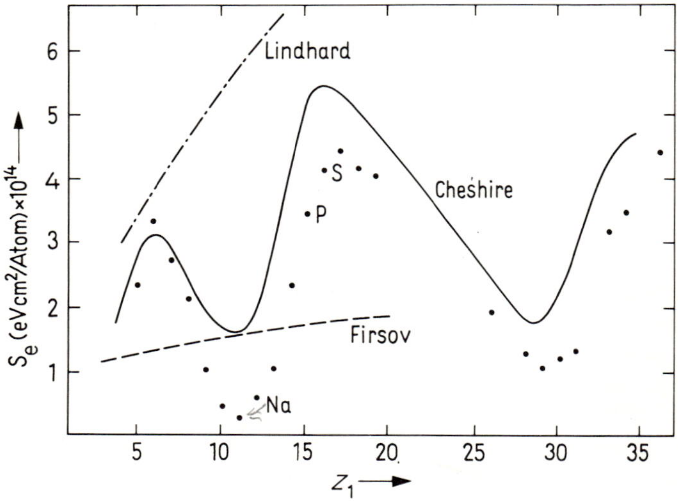

Figure 2.15 Comparison between experimental S_e data according to Eisen [222] and theoretical calculations according to Cheshire *et al.* [136]. Values for S_e according to LSS [422] and Firsov [242] are given for comparison, after Dearnaley [2]. Reproduced by permission of North Holland Physics Publishing

2.3.1 Critical Angle

The critical angle, below which an ion can enter into a channel without leaving it again, results (Lindhard [440]) from the following considerations: one divides the energy of the ion into components which are vertical, $E_\perp$, and parallel, $E_\parallel$, to the axis of the channel. As long as the vertical component is smaller than the repelling potential, $V(r)$, of the atomic chain (Figure 2.16), the ion remains in the channel. The following applies [440]:

$$V(r) = \frac{Z_1 Z_2 q^2}{d} \ln \left[\left(\frac{Ca}{r} \right)^2 + 1 \right] \tag{2.34}$$

where d is the distance between the atoms, ($C \approx \sqrt{3}$), and the logarithmic expression is an approximation of the Thomas–Fermi potential. For $E_\perp = V(r)$, the minimum distance between the ion and the atomic row ($r_{\min}$) can be calculated:

$$E_\perp = E \sin^2 \psi = V(r_{\min}) \tag{2.35}$$

The condition that a particle does not leave the channel between two atoms is:

$$r_{\min} = d \frac{v_\perp}{v_\parallel} = d \tan \psi \approx d\psi \tag{2.36}$$

By substituting equation (2.36) into equation (2.35), two expressions are obtained for the critical angle, depending on the energy:

$$\psi_{c1} = \left(\frac{2Z_1 Z_2 q^2}{Ed} \right)^{1/2} \qquad \text{for } E \geqslant \frac{2Z_1 Z_2 q^2}{a^2} \tag{2.37}$$

$$\psi_{c2} = \left(\psi_{c1} \frac{Ca}{\sqrt{(2)}d} \right)^{1/2} \qquad \text{for } E \leqslant \frac{2Z_1 Z_2 q^2}{a^2} \tag{2.38}$$

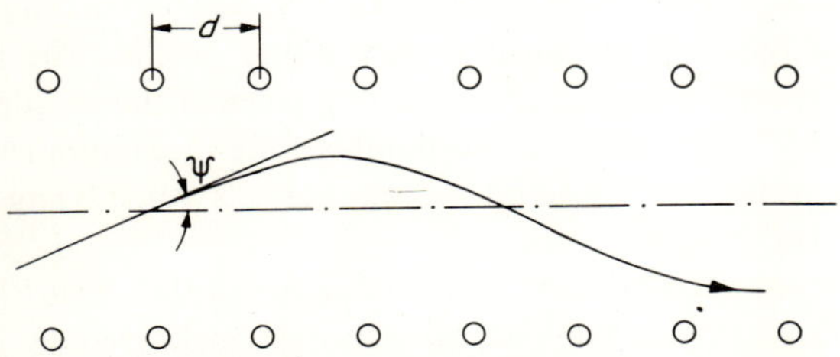

Figure 2.16 Schematic description of an ion path in a channel for $\psi \leqslant \psi_c$

the greater the energy, the smaller the critical angle
ie; the less channelling

Table 2.4 Critical angle for channelling of boron, phosphorus and antimony in silicon [9]

Ion	Energy (keV)	Critical angle ⟨100⟩	⟨110⟩	⟨111⟩
Boron	10	4.76	6.97	5.30
	100	2.67	3.47	2.98
	300	2.03	2.98	2.26
Phosphorus	10	5.79	7.51	6.45
	100	3.26	4.22	3.63
	300	2.47	3.21	2.76
Antimony	10	6.95	9.01	7.74
	100	3.91	5.07	4.35
	300	2.97	3.84	3.31

Equation (2.38) applies for the range of energies normally used in ion implantation. Values of the critical angle for boron, phosphorus, and antimony in silicon, at various energies, are given in Table 2.4.

Even if one precisely implants in the direction of a channel, a certain fraction of the ions will be scattered by the atoms at the surface, and the profile will have an 'amorphous' component. If the incidence direction of the ion beam deviates by more than the critical angle from a channel direction, then only a correspondingly reduced portion of the ions can enter and remain in the channel. A minimum channelling effect is obtained at angles between 7° and 10°. At higher angles, planar channelling is possible.

2.3.2 Channelling Profiles

As already stated, the channelling effect is extremely direction-dependent. In addition, it is also dependent upon the ion dose, the temperature, and the presence of an amorphous scattering layer. In the following, only experimental examples for these dependences will be discussed, since theoretical interpretations exist only in rudimentary form, if at all. A very distinct profile is shown in Figure 2.17 [228]. It was obtained with implantation of 500 keV ^{42}K ions into tungsten, and serves well to illustrate the three parts of a channelling profile. The regions 1 and 3 correspond to the amorphous and channelling parts of the profile; region 3 is terminated by a steep slope, which corresponds to the maximum range. Region 2 is formed by ions which are scattered out of channels while being stopped, due to irregularities in the crystal lattice.

Most previous investigations of semiconductors in this regard have been made using silicon layers implanted with phosphorus, measured by the radio-tracer method [183] in order to retain the original ion distribution, and to avoid secondary effects through annealing.

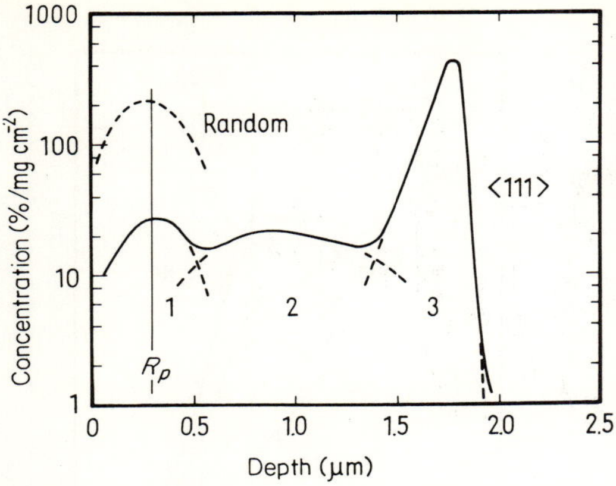

Figure 2.17 Range distribution of 500 keV ^{42}K ions in tungsten implanted at 25 °C in the ⟨111⟩ direction; according to [228]. Reproduced by permission of the American Physical Society

Orientation

The dependence of the channelling effect upon the orientation is shown in Figure 2.18, for silicon implanted with phosphorus. The maximum range remains constant, and even at 8° misorientation there is a considerable channelling component. In order to avoid the channelling effect, the sample is generally tilted by 7° to 10°. As can be seen in Figure 2.18 as well as Table 2.4, this is not always sufficient. Still more misorientation can cause scattering into higher-indexed chan-

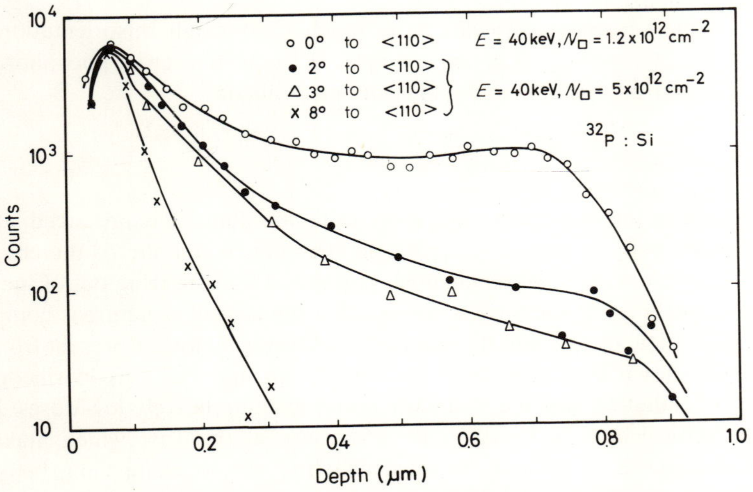

Figure 2.18 Dependence of the channelling effect on tilting for 40 keV ^{32}P ions in silicon [183]. Reproduced by permission of the National Research Council of Canada

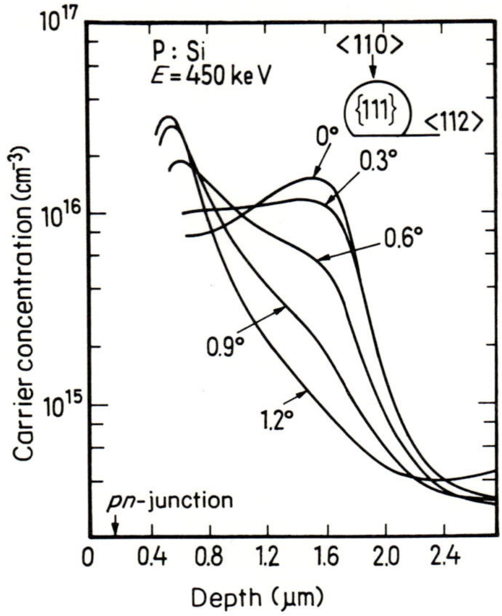

Figure 2.19 Dependence of the channelling effect on tilting for 450 keV P ions in ⟨111⟩-oriented silicon [585]; measured by the capacitance-voltage method. Reproduced by permission of Solid State Technology, Technical Publishing, a Company of Dun & Bradstreet

nels. A further example, showing how much even small misorientations can influence the channelling, is given in Figure 2.19. In this case, phosphorus has been implanted at 450 keV into ⟨111⟩-oriented silicon.

Ion Dose

With increasing ion dose, more and more radiation damage is produced, and the channels are finally destroyed. In principle, the range is not altered thereby, since only the number of ions within channels is reduced by scattering out of the channels (dechannelling). If the dose is so high that the crystal structure is completely destroyed (critical dose), then the channelling effect is no longer present from then on. This is shown in Figure 2.20, for 40-keV phosphorus ions in silicon. The influence of radiation damage is already noticeable at relatively low doses. In this case, the critical dose is $5 \cdot 10^{14}$ cm^{-2} [9]. Even at the dose which makes the crystal amorphous, a tail can be seen, which is due to channelling before the amorphous layer was formed. If the crystal is made amorphous before implantation of the dopant atoms by implantation of an inert gas or of silicon (in the case of a silicon substrate), etc., then no channelling at all occurs.

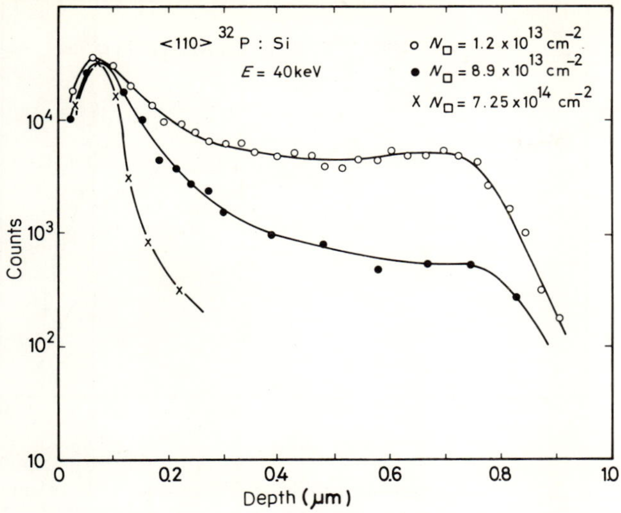

Figure 2.20 Concentration profile of ^{32}P ions implanted into silicon with 40 keV in the ⟨110⟩ direction, as a function of dose |183|. Reproduced by permission of the National Research Council of Canada

Temperature

If the implantation temperature is increased, two effects occur. First, the radiation damage anneals completely or partly during implantation, through which a smaller fraction of the ions is scattered out of the channels. Second, the amplitude of lattice vibrations increases, and thus also the number of ions scattered at the surface ('minimum yield'). This scattered fraction χ of the ions can be calculated, according to Lindhard |440|, from the atomic density N, the atomic distance d, the Thomas–Fermi radius a, and the average vibration amplitude of the lattice $\rho_\perp$, as follows:

$$\chi = \pi N d(\rho_\perp^2 + a^2) \tag{2.39}$$

Similarly to the specific heat, these effects depend on the Debye temperature. The relationship between the mean square of the vibration amplitude vertical to an axis, and the temperature, is |41a|:

$$\rho_\perp^2 = \frac{2}{3}\rho^2 = \frac{2}{3}\left[\frac{9\hbar^2}{kM_2\Theta_D}\left(\frac{\varphi(x)}{x} + \frac{1}{4}\right)\right] \tag{2.40}$$

with $x = \Theta_D/T$, and with the Debye function:

$$\varphi(x) = \frac{1}{x}\int_0^x \frac{t\,dt}{e^t - 1} \tag{2.41}$$

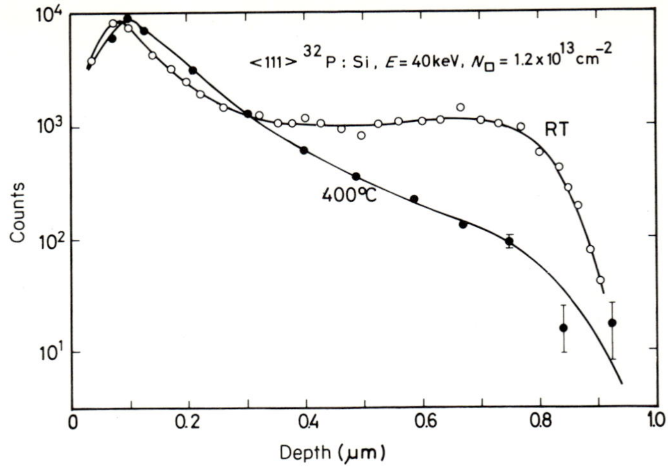

Figure 2.21 Concentration profile of ^{32}P ions in ⟨111⟩-oriented silicon implanted at room temperature as well as 400 °C [183]. Reproduced by permission of the National Research Council of Canada

In Figure 2.21, for illustration, profiles are shown which were obtained by phosphorus implantation at room temperature as well as at 400 °C into ⟨111⟩-oriented silicon. The maximum range remains constant, only the number of ions scattered out of the channels increases. The profile has a logarithmic slope over a wide range; this is very frequently measured, especially in the case of samples which have been misoriented, in order to avoid channelling. A discussion of these effects, based on back-scattering measurements, has been compiled by Fujimoto *et al.* [264].

Scattering Layer

If a crystal is covered by an amorphous layer (e.g., on oxide), then the ions (depending on their velocity) are scattered, and the number of ions which can penetrate into channels is thus decreased. Expressions for the angular distribution of ions after multiple scattering in thin layers can be found in works by Moline [509] and Meyer [492]; these permit qualitative conclusions on the reduction of the channelling effect by an oxide layer. At small scattering angles Θ, the following expression is valid for the angular distribution:

$$P(\Theta, t) = \frac{\alpha}{2\pi} [\Theta^2 + \alpha^2]^{-3/2} \tag{2.42}$$

where $\alpha = 2\pi CNt$, $C = 0.33\, a^2/\varepsilon$, a is the Fermi radius, ε the reduced energy according to the LSS theory and t the thickness of the oxide.

The full width half maximum (FWHM) of the angular distribution is inversely

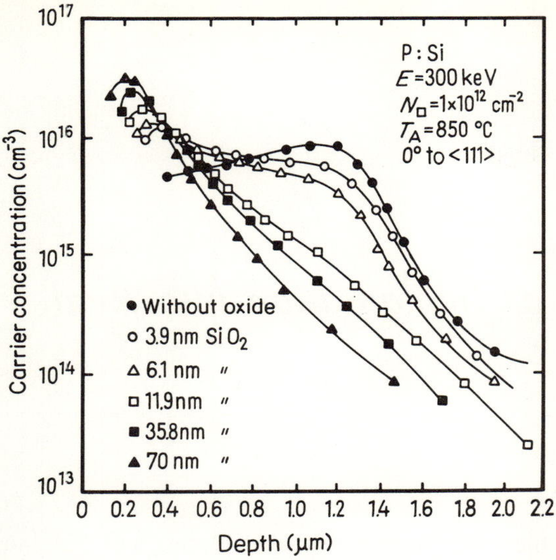

Figure 2.22 Dependence of the channelling effect on oxide thickness for 300 keV ^{31}P ions in silicon [509], measured by the capacitance-voltage method. Reproduced by permission of Springer-Verlag, Heidelberg

proportional to the energy, and approximately proportional to the thickness of the scattering layer. According to Reference [509], a misorientation of 2 per cent for phosphorus ions at 300 keV corresponds to an SiO_2 layer of 12 nm, but at 100 keV, of only 4 nm. The experimental relationship between oxide thickness and channelling, for 300 keV phosphorus ions, is shown in Figure 2.22.

However, the channelling effect cannot be avoided completely by the use of scattering layers, by implantation at increased temperatures or by misorientation between the crystal axis and the ion beam; it is only possible to avoid it through a previous amorphizing implantation. On the other hand, this effect could be purposefully taken advantage of to produce deeper profiles, since the channelling range can be higher than the amorphous range by up to a factor of 10. According to studies by Dearnaley [183], though, the orientation must be accurate to within 0.1° for a parallel beam to obtain reproducible results. In addition, there are still unexplained influences due to varying crystal-growth methods; so that purposeful use of the channelling effect, as in the fabrication of semiconductor devices, cannot be considered at present.

300 keV shows less channelling than 100 keV

3
Problems of Implantation into Real Solids

The processes which take place in real solids during and after implantation, where the emphasis is once again placed on semiconductors, will be considered in this chapter; therefore, only implantation into crystals will be considered. Implantation into amorphous and polycrystalline solids will only be mentioned where significant differences occur.

During implantation, ions come to rest at random positions in the crystal lattice; many different types of radiation damage and dislocations are produced by nuclear collisions and collision cascades, sometimes going as far as the formation of amorphous areas. Therefore it is necessary, by a suitable heat treatment (annealing), to restore the crystal lattice and to place the implanted ions on electrically-active lattice sites.

A number of deviations from the theoretical profile (in general, this term refers to a Gaussian distribution according to the LSS theory [442]) occur in the case of implanted distributions. Regarding the cause of the deviations from the theoretical ion distribution, one must distinguish between primary effects caused by inadequacies in the development of theories, and effects which occur during or after implantation and have nothing to do with the stopping of the ions. Special examples of the first case are the neglect of terms of higher order in theoretical development based on LSS—this effect has already been discussed—and the assumption of an amorphous medium in all theoretical treatment, although the usual semiconductors (germanium, silicon, II–VI and III–V compounds) are single crystalline. Some examples of the second case are effects such as thermal diffusion during the necessary annealing or during implantation, radiation-enhanced diffusion, diffusion enhanced by radiation damage, the suppression of diffusion by radiation damage, and electrical activation dependent on radiation damage; there may also be other effects which are not yet known. Not to be neglected, also, are the influences of profile measurement, which can additionally simulate a deviation from the theoretical profile. These influences are discussed in detail in Chapter 5.

The use of masking layers is unavoidable in the manufacture of semiconductor devices, in which case implantation through thin layers can be advantageous, and in some circumstances even necessary, because of passivation. Effects related to this will also be discussed below.

3.1 EFFECTS AND ANNEALING OF RADIATION DAMAGE

The recrystallization of layers damaged during implantation is a critical matter in the use of this process for semiconductor doping. If recrystallization is not possible, then implantation is not feasible except in a few special cases (see, for example, Section 6.3.3), since mobility and carrier lifetime would be negatively influenced by defects. Therefore, soon after the first experiments in semiconductor doping through implantation, this undesirable defect production was already being investigated. For this purpose, the Rutherford back-scattering technique [174], [484] has become most preferred as the method of measurement, although several other methods are suitable to quantitatively determine the effects of radiation damage. Among others, there are the methods of electron-spin resonance, paramagnetic resonance, electron-microscopic investigation, infrared absorption, optical reflection, Raman scattering, delineation of dislocations, volume changes due to radiation damage and, as an indirect method, the electrical activation of implanted ions. Mayer *et al.* [9] have compiled a more detailed survey of all physical methods for determination of the radiation-damage concentration and the formation of amorphous layers.

Radiation damage also occurs in amorphous and polycrystalline solids after implantation [215], [262]. However, its effect is even more difficult to determine than in the case of monocrystals, and one is mainly dependent on phenomenological descriptions. In this case, too, one uses annealing for restoration of the original structure. There are a number of applications which depend on the effects of radiation damage, for example, the implantation of light guides in glass and quartz [700], the compensation of GaAs [210], the *n*-doping of InSb [250], and the alteration of metallurgical properties. In the semiconductor field, apart from a few exceptions, radiation damage is an undesirable side-effect of ion implantation.

Many publications deal with radiation damage in semiconductors and other solids (e.g. [154], [189]), and this problem has also been the subject of several conferences during the last few years [365], [34a], [742a]. However, the effects of radiation damage are often treated separately from the electrical properties, which are the most important in the case of semiconductors; and ion implantation is sometimes only the means of purposely producing radiation damage, in order to investigate this damage itself rather than its effects on the electrical parameters. During the last few years, articles by Vook [749], Nelson [533], [535], as well as Kimerling and Poate [409], among others, have dealt with radiation damage in semiconductors.

Various theories which were discussed in Chapter 2 deal with the deposition of energy in atomic processes, from which the number of displaced atoms can be calculated. Although practically only the elastic impacts discussed above displace atoms of the lattice, the influence of electronic stopping can be important, as shown by Picraux and Vook [571], due to the dependence of defect annealing upon the charge. Recrystallization of amorphous layers is possible by irradiation with light ions [535]; whether this effect is only due to ionization is not yet clear. Following the more theoretical discussion in Chapter 2, the following will deal with some practical aspects of the production and annealing of radiation damage. In this context, it cannot be avoided that experiments are often described without providing theoretical explanations. This applies especially for the transition from layers which are only slightly damaged (where many defects can be clearly identified, for example by electron-spin resonance experiments) to layers which are so heavily damaged that an amorphous phase has been formed. The above applies also for annealing behaviour, its dependence on temperature, crystal orientation, and masking layers, as well as for the influence of ionization and other effects.

3.1.1 Defect Generation

Many basic experiments in defect generation have been carried out at low temperatures with low doses, in order to avoid premature annealing and permit examination of the defects. However, for practical uses, ion implantation is carried out at room temperature or at higher temperatures; furthermore, the doses are often so high that individual defects can no longer be identified, and only integral measurement methods are practical.

Simple Defects

At low doses, and with relatively light ions compared to the mass of the target atoms, simple defects are mainly obtained. For example, in the case of silicon one finds, among others (Brower and Beezhold [107]), Si–G7 (negatively-charged divacancy, V^{2-}), Si–Sl (neutral vacancy–oxygen complex, V–O), Pi–P3 (neutral 4-fold vacancy) and Si–SL2 (negative 4-fold vacancy). Vacancies can have different charge states in silicon; usually, they are negatively charged in *n*-type material, and neutrally charged in *p*-type material. Many different kinds of such simple defects could be found in numerous semiconductors, by means of electron-spin resonance measurements (ESR), paramagnetic resonance measurements (EPR), and other methods. At higher doses, there appear more complex defects and amorphous zones, which gradually overlap and finally form an amorphous layer. Measurements of the change in the refractive index and the absorption by Baranowa [55] are given in Figure 3.1, for various ions in silicon; this figure shows the saturation of divacancy formation at higher doses, as well as their decrease during the formation of amorphous layers. The abrupt change in the slope of the refractive index curves indicates the beginning of overlapping of amorphous areas. In the case of heavy ions, complicated defects (clusters) already form at low doses; these can no longer be described by simple models. Very detailed examinations were carried out for SiO_2 by EerNisse and Norris [214],

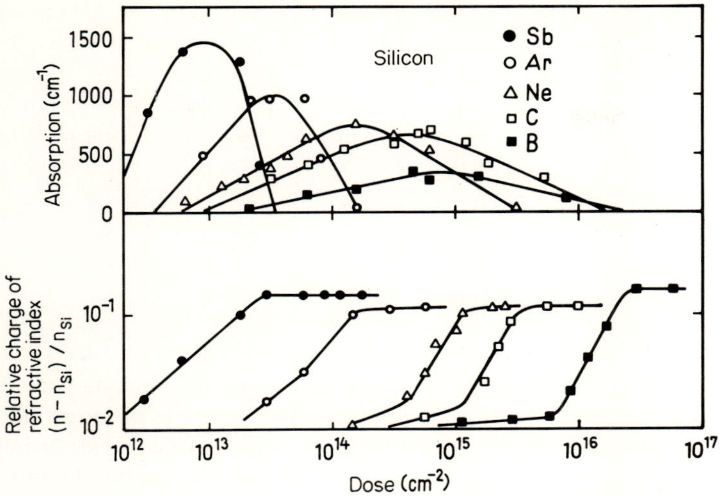

Figure 3.1 Dose dependence of the absorption coefficient of divacancies at 1.8 microns, and of the refractive index at 1.2 microns, for silicon implanted at 80 keV with various ion species [55]

[215]. The structural damage due to ion bombardment takes the form of broken Si–O bonds, which are annealed at 650 °C. In the case of SiO_2, structural defects are not only produced by nuclear collisions, but also by electronic stopping. Additionally, charged defects are caused by ionization [215].

Complex Defects

At high implantation temperatures, significantly higher doses are necessary to produce the same quantity of defects. This is due to an annealing *in situ*, because of the generally higher mobility of the defects. An interaction between defects and impurity atoms can also take place. In the case of silicon, formation of a defect type consisting of a vacancy and a donor (E-centre), after implantation of group V elements, is of importance [106]. Hirata *et al.* [335] determined the significance of a complex consisting of a vacancy and two neighbouring substitutional group V atoms. Electrical interaction between defects and doping atoms, for example, between negatively-charged vacancies and positively-charged group V atoms, can also occur. These effects not only influence the defect concentration and its annealing behaviour, but can also influence diffusion effects.

Dislocation lines and dislocation loops can be formed from a combination of simple defects, or as a result of stress caused by unannealed radiation damage. An important type is the Frank–Read source [7], [253], which consists of a line of simple defects (e.g. vacancies) with impurity atoms or radiation damage clusters at the ends (Figure 3.2(a)). The dislocation line can be bent by stress, and a dislocation loop is thus formed (Figure 3.2(b)–(d)). This loop can increase

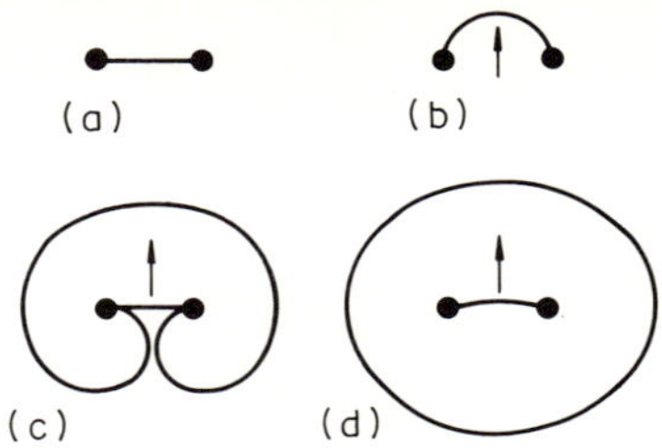

Figure 3.2 Formation of dislocation loops via the Frank–Read mechanism [253]. The arrows indicate the growth direction

further in size under the influence of additional stress. Stacking faults can be formed along crystal planes by an accumulation of vacancies or interstitial atoms; these stacking faults can grow by the addition of further vacancies or interstitial atoms, or can be saturated by the addition of impurity atoms. In this way, a gettering effect of the radiation damage can occur.

Dose Rate

Where implantation is carried out at a high dose rate (current density), two competing effects can occur. If the sample is heated by the ion irradiation, some of the radiation damage can thereby be annealed immediately (however, measurements on silicon show that defects which are difficult to anneal can remain [158]). If the sample is not heated, higher defect concentrations are usually obtained with increasing dose rate [220], [730], [750]. In Figure 3.3, an example of this

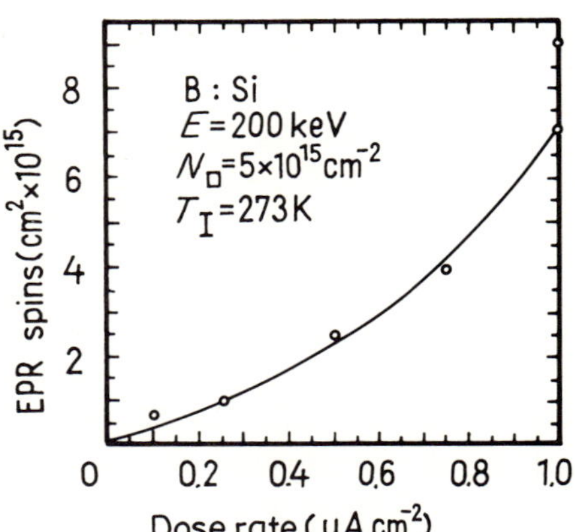

Figure 3.3 Dependence of radiation damage on dose rate for boron implantation in silicon [161]. Implantation energy 200 keV, dose 5×10^{15} cm^{-2}

phenomenon is shown, according to Crowder [161]. EPR spins, depending on the dose rate for a dose of $5 \cdot 10^{15}$ cm^{-2} boron ions at 200 keV, are indicated. The difference in the damage amounts to almost a factor of ten. Further measurements in this regard were carried out by Picraux and Vook [570], [571]. Davies *et al.* [176], basing upon experiments by Gyulai (cited in [176]) as well as by Müller and Ryssel [523], found a pronounced dependence on the energy density in the collision cascade, with the use of polyatomic molecular beams. Tinsley *et al.* [730] carried out a very precise study for indium-implanted GaAs. They explained their results with a model by Vook and Stein [750] and concluded from their measurements that GaAs has an annealing stage at room temperature.

Amorphous Layers

Amorphous layers result from high-dose implantations, depending on the mass of the ions as well as the implantation temperature. In Section 2.2, the model by Morehead and Crowder was discussed [514]. The criteria for the radiation-damage concentration in monocrystals depend on the measurement method. According to the method used, therefore, a different value for the 'amorphizing dose' will be obtained. Müller [517] has presented a discussion of these problems. The fact that the height of the channelling spectrum agrees with the yield of the

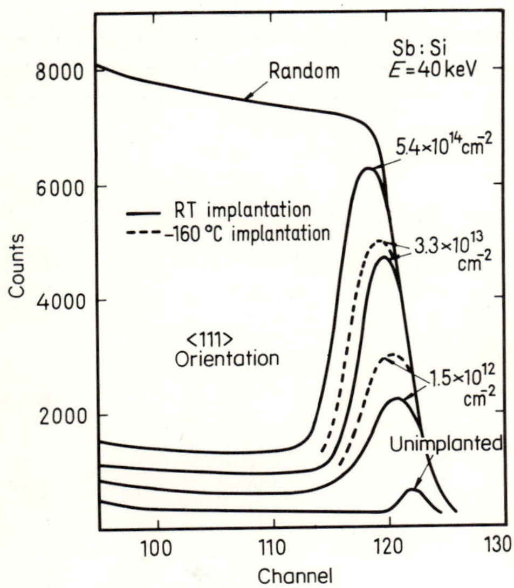

Figure 3.4 Back-scattering spectra in random and channelling direction measured after Sb implantation in silicon at 40 keV with varying doses at room temperature as well as −160 °C [572]. Reproduced by permission of the American Institute of Physics

random spectrum in back-scattering measurements (see Section 5.8), or in the case of silicon, the fact that a very steep annealing step occurs in the electrical activation around 600 °C is generally used as the criterion for the formation of an amorphous layer in monocrystals. As an example, back-scattering spectra are shown in Figure 3.4, from which varying damage, depending on the implanted dose and the implantation temperature, can be seen. Here, no amorphous layer has been formed. Examples showing the formation of amorphous layers can be found in Figures 3.6 and 3.8 to 3.10.

The structure of amorphous layers still remains mostly unclear, but at least at low temperatures (about $\leqslant$125 K), the formation of amorphous layers is independent of the mass of the implanted ions, if the same energy is given off in atomic processes [749]. Values of the amorphous dose, depending on implantation temperature for various ions in silicon, have been given in Figure 2.15. Table 3.1 shows the dose necessary to make silicon, GaAs, and GaP amorphous, using different ions at room temperature. At high implantation temperatures ($\geqslant$600 °C for silicon, $\geqslant$450 °C for GaAs), amorphous layers are no longer formed; under some conditions, however, defects can form which negatively influence the electrical properties [158]. It can be seen from Table 3.1 that, with increasing mass, a lower ion dose is required to obtain amorphization. During implantation with high current densities, the samples can be heated to a considerable degree, thus reducing this dose. For the occurrence of an amorphous layer, characteristic

Table 3.1 Values of the critical dose for the formation of amorphous layers for several elements in silicon, GaAs and GaP at room temperature

Semiconductor	Element	Mass of the main isotope	Dose (cm^{-2})	References
Si	B	11	8×10^{16}	[513]
	N	14	2×10^{15}	[187]
	Ne	20	10^{14}	[487a]
	Al	27	$\geqslant 5 \times 10^{14}$	[9]
	P	31	6×10^{14}	[513]
	Ar	40	4×10^{14}	[187]
	Ga	70	2×10^{14}	[513]
	As	75	2×10^{14}	[513]
	Kr	84	2×10^{14}	[187]
	Sb	122	10^{14}	[9]
	In	204	10^{14}	[9]
	Tl	204	5×10^{13}	[9]
	Bi	209	5×10^{13}	[9]
GaAs	C	12	10^{15}	[316]
	Si	28	2×10^{14}	[316]
	Zn	64	3×10^{13}	[767]
	Cd	114	3×10^{13}	[767]
GaP	Te	130	10^{14}	[325]

is a change in the colour of the semiconductor. While silicon and GaAs become lighter in colour (milky appearance), GaP changes from orange to black. Based on the condensation of water vapour, one can easily recognize the amorphous area by breathing on the sample, or by bringing a cold sample into humid air. This method is also generally useful to distinguish implanted areas from unimplanted ones, and even allows judgement of the implantation homogeneity. For reasons of sample purity, however, breathing on the sample is only recommendable in special cases.

3.1.2 Alteration of Material Properties

Through the production of more or less complex defects, radiation damage has very distinct effects upon various material parameters. These can be desirable, can have hardly any influence, or can also be very disadvantageous. In the case of semiconductors they are almost always undesirable, since they generally very greatly reduce the mobility and lifetime of the minority carriers. Only in a few special cases are damage effects desirable, as with the compensation of GaAs and the *n*-type doping of InSb, as well as some lead chalcogenides (see Chapters 6 and 7). In connection with non-semiconductors, the effects of radiation damage can sometimes be used advantageously, for example, for surface hardening of metals and the fabrication of light guides in glass and quartz through alteration of the refractive index. Within the scope of this section, however, only the most important effects and their results will be discussed.

Electrical Properties

Radiation damage has a very great influence on the electrical properties of semiconductors. Even very small implantation doses reduce the lifetime of the minority carriers considerably, for example, an implantation of boron into silicon at 100 keV with a dose of 10^{12} cm^{-2} reduces the lifetime below 10^{-9} s [409]. The mobility of the carriers is also greatly influenced, though not to the same degree. After implanting with high doses ($\geqslant 10^{13}-10^{14}$ cm^{-2}, depending on the mass), the mobility for silicon and GaAs is lower than 1 cm^2/Vs. The carrier concentration after implantation is usually low, because the ions are partly trapped at defects and are therefore electrically inactive. Exceptions are InSb and a few chalcogenides, which can be *n*-doped by radiation damage and also exhibit a sufficiently high mobility [250], [249]. More details on this will be given in Chapters 6 and 7. In the case of some other III–V semiconductors, such as GaAs, GaP and $Ga_{1-x}Al_xAs$, radiation damage compensates the carrier concentration; i.e., the conductivity and mobility are greatly reduced [210], [238], and specific resistivities of 10^6 to 10^9 Ωcm have been achieved. With the exceptions of the two effects just mentioned, the *n*-doping and the compensation of some semiconductors, radiation damage has only negative effects on the electrical parameters of semiconductors, so that annealing is absolutely necessary to reduce such damage following implantation.

Optical Properties

Due to radiation damage, the refractive index is changed, and thereby the reflectivity of implanted semiconductors and insulators as well. The refractive index is increased with the concentration of radiation damage (if a chemical reaction is not involved), and light can be conducted in the implanted area. Light guides and other opto-electronic components have already been thus produced in GaAs [279] and quartz [700]. This effect can also be used for measuring purposes in materials where it would normally be undesirable. By measuring the reflectivity together with a layer-removal technique in the case of silicon, for example, it is possible to measure radiation-damage profiles [790], a field which is actually the domain of back-scattering technique. In Figure 3.5, examples of such measurements for silicon, following implantation with 400 keV arsenic ions, are given for varying doses [790].

Volume Change

During high-dose implantations which lead to the formation of amorphous layers, there is almost always a volume change, especially with monocrystalline semiconductors. Generally, there occurs an increase in volume (e.g. silicon), but a decrease is also possible, i.e. an increase in density (e.g. GaAs [261]). The volume change can be used for a measurement method to indicate the formation of amorphous layers [212], [720]. The stress produced in this way can be of such magnitude that the substrate is bent [489]. This is especially seen in the case of thin samples for electron-microscopy examination. If masking layers, which are necessary with sensitive semiconductors (see Section 3.3), are used, this effect can adversely influence the adhesion. In the case of extreme high-dose implantations ($\geq 10^{17}$ cm^{-2}), blistering as well as destruction of the surface layer may occur.

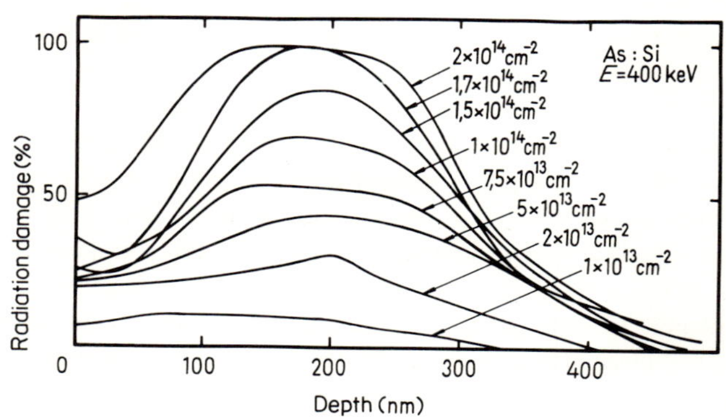

Figure 3.5 Depth distribution of radiation damage in silicon obtained by reflectivity measurements in combination with layer-removal technique, after arsenic implantations at 400 keV with doses between 10^{13} cm^{-2} and 2×10^{14} cm^{-2} [790]

Other Material Properties

In addition to the already mentioned alteration of material parameters due to radiation damage, a great many additional effects can occur, for example, passivation against oxidation (metals), increasing or lowering of the critical temperature in the case of superconductors, changing of elastic properties, and surface hardening of solids. In connection with semiconductors, two interesting points are the capture of mobile sodium ions, which are extremely harmful to electrical properties, on defect centres in SiO_2 [260] (especially in the case of MOS transistors), as well as the increase in the etch rate of SiO_2 [511] (see Figure 3.19).

3.1.3 Recrystallization of Radiation Damage

With a suitable heat treatment, it is almost always possible to reduce the amount of radiation damage in monocrystalline material, or to largely eliminate the negative effects of such damage. In some cases, however, a stable phase is formed, for example in GaP [670], which can no longer be eliminated by annealing. Also in the case of originally amorphous solids, annealing serves to eliminate undesirable side effects. Because of the complexity of the latter topic, as well as its relative unimportance for semiconductor devices, only monocrystalline material will be discussed below. A complete annealing of radiation damage is not easy to determine. Even with a relatively high number of dislocations and other defects, a semiconductor shows electrical parameters which can hardly be distinguished from those of perfect crystalline material. This applies especially for carrier concentration and mobility; a significantly more sensitive indicator of defects is the minority carrier lifetime. In this section, once again, experimental results will only be presented for the most part; it would generally not be possible to offer consistent explanations.

Temperature Dependence

Simple defects anneal at low temperatures; in silicon, for example, vacancies at 70 K and 150 K (double-negatively charged and neutral, respectively), vacancy-group V defects (complexes between vacancies and group V doping atoms) at 400 to 500 K, and vacancy-group III defects at about 500 K. Double vacancies are probably stable up to 550 K, and therefore account for a large portion of the simple defects after ion implantation. Vacancy clusters probably transform at 200 to 300 K into dislocations. The recrystallization of a highly-damaged or amorphous layer starts in general from the undamaged crystal. Back-scattering measurements for two samples implanted with different doses in silicon are shown in Figure 3.6. Both samples were implanted at 77 K. After implantation with the lower dose, no amorphous layer is formed and all defects anneal at 600 °C. At the higher dose, an amorphous layer results. With increasing annealing temperature, the amorphous area at first becomes narrower (a so-called epitaxial

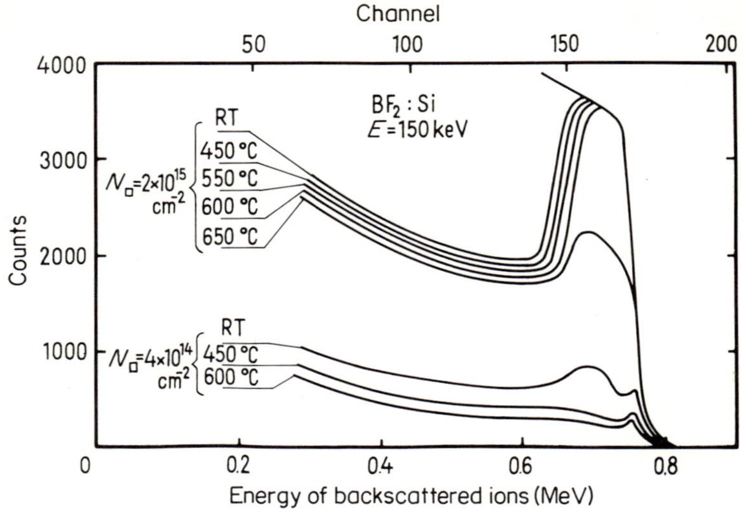

Figure 3.6 Recrystallization of amorphous layers starting from the undamaged substrate after BF_2 implantations into silicon at 150 keV [10]

recrystallization, starting from the monocrystalline substrate, takes place) until, finally, the height of the amorphous-distribution curve decreases. At higher temperatures, one obtains the same back-scattering spectrum as in the case of an undamaged crystal. For this purpose, in the case of silicon, temperatures between 600 °C and 1000 °C are generally necessary. The transition area between amorphous and crystalline material is not abrupt, and possesses a high density of defects. During epitaxial recrystallization, defects can therefore migrate from this area of high damage density into the recrystallized area and the undisturbed crystal.

If the implantation has taken place at high energies (MeV range), then the damage is isolated below the surface of the semiconductor, while the surface region is not damaged. The reason for this is the low extent of ion stopping due to nuclear collisions at high energies. Only when the energy has been reduced by electronic stopping (cf. Figure 2.1) can a great number of nuclear collisions occur within the semiconductor, producing radiation damage. During annealing, an epitaxial recrystallization takes place, starting from the undamaged surface and from the undamaged substrate, which is not correlated and therefore does not result in a perfectly recrystallized crystal. Better preconditions exist when the amorphous layer extends up to the surface. Recrystallization can then begin at the crystalline substrate (see Figure 3.6), and can continue up to the surface.

If the implantation was performed with a dose which does not produce an amorphous layer, but only a great many defects and radiation-damage clusters, then recrystallization, under some circumstances, can require significantly more energy than in the case of completely disordered, amorphous material. Thus, a complete restoration of the lattice will only be possible at higher temperatures. On the other hand, at very low implantation doses, the effect of radiation damage as opposed to the doping effect (carrier concentration, mobility) is negligible, or can easily be eliminated by annealing at low temperatures (in the case of silicon, 400–500 °C).

Dislocation lines and loops which have formed out of simple defects, or as a result of the annealing of complicated defects, anneal completely only at extremely high temperatures (in the case of silicon, ⩾1000 °C), and are therefore of particular importance.

The simplified description given in this section is not generally applicable. It is quite valid for elemental semiconductors such as silicon and germanium, but is partly or completely non-valid for compound semiconductors, for example, those produced from compounds of the groups III and V of the periodic system. The reason for this probably lies mainly in the fact that, in the case of compound semiconductors, one must not only bring one component on to a suitable position in the lattice, but rather two or (in the case of more complex compounds) even more.

However, even in the case of silicon, these simple considerations are only valid as long as one does not use sensitive methods of measurement, for example, measurement of leakage currents at *p–n* junctions, lifetime measurements or the etching of dislocations. Therefore, in the following sections, a number of examples will be presented, without making any attempt to find a common pattern in the varying annealing behaviour of the layers.

Influence of the Implanted Ions

The implanted ions can also have a significant influence on the recrystallization behaviour. This is not surprising when one considers the relatively high concentration which is necessary for the production of heavily damaged or amorphous layers. Their influence can mainly be explained by:

(a) Segregation at the recrystallization boundary which can lead to irregular growth,
(b) Precipitations and formation of other phases (chemical compounds),
(c) Atomic radius.

These points, and especially the last one, are not only important for the recrystallization of the damaged lattice, but also for the incorporation of the implanted ions on to lattice sites, which is a precondition for their use in semiconductor technology. An example in which the dependence of the annealing on the ion species is especially clear is the so-called reverse annealing of implanted boron in silicon, which was discussed in Section 3.2, dealing with the electrical activation of implanted ions. The annealing of radiation damage can, under certain circumstances, be completely prevented at very high implantation doses by exceeding the solubility limit, or by the formation of chemical compounds. For example, by implantation of high doses of nitrogen into silicon, a silicon–nitride

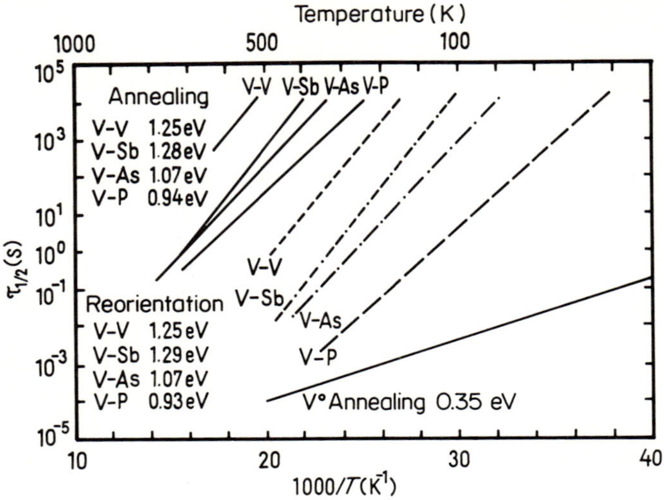

Figure 3.7 Characteristic recovery times to 50 per cent of various defects in silicon depending on the reciprocal temperature. Annealing of E-centres according to Hirata [335], reorientation according to Elkin and Watkins [224]. The data for divacancies are by Cheng *et al.* [134]; those for vacancies are by Watkins [760] (according to Vook [749])

layer (Si_3N_4) is formed which remains amorphous; while at lower nitrogen doses, the nitrogen is incorporated as a donor on to regular lattice sites, with complete lattice recrystallization [797].

The nature of numerous complex defects has been clarified by electron-spin resonance measurements, and their annealing behaviour was measured. Figure 3.7 shows various measurement results according to Vook [749]. The occurrence of these defects, and their annealing at relatively low temperatures, can only be observed after implantations with low doses. Layers which have been made amorphous by implantation show an annealing behaviour characterized by recrystallization of the layer, and not by the annealing or reorientation of individual defects.

Systematic investigations on the dependence of the annealing behaviour of damaged layers, and that of the lattice location of the implanted ions, on the ionic radius have not yet been carried out. However, it is probable that well-fitting ions are incorporated into the crystal lattice during recrystallization without the use of additional energy.

Dependence on Orientation

It has recently been found that the recrystallization can depend on the orientation of the lattice [163], [165], [516]. The annealing behaviour of ⟨100⟩- and ⟨111⟩-oriented silicon is compared in Figure 3.8. In order to avoid additional effects due

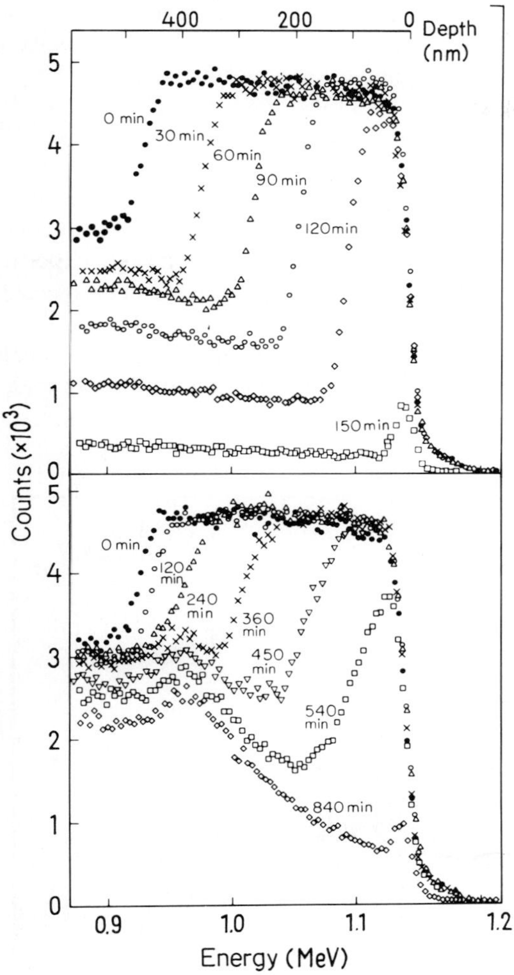

Figure 3.8 Channelling spectra of silicon samples after silicon implantation with 8×10^{15} cm^{-2} at 50 and 250 keV and 77 K, preannealing at 400 °C for 60 min and annealing at 550 °C for different times. The upper section shows the annealing behaviour of a ⟨100⟩-oriented sample, the lower section a ⟨111⟩-oriented sample [165]. Reproduced by permission of the American Institute of Physics

to the implanted ion species, silicon was implanted. The back-scattering measurements clearly show a faster recrystallization of the ⟨100⟩-oriented layers than the ⟨111⟩-oriented ones. Furthermore, there appears in the latter case during annealing a second maximum in the defect distribution at the boundary between the undamaged substrate and the previously amorphous layer. The growth rate at

550 °C is 8 nm/min in ⟨100⟩-oriented crystals; for ⟨110⟩-orientation it is about 2.5 nm/min. The activation energy of the process amounts in both cases to 2.3 ± 0.1 eV. A further example for annealing at 950 °C after implantation of boron is shown in Figure 3.9 [164]. Even though no second maximum in the defect distribution is found in this case, there is significantly less crystallinity in the ⟨111⟩ direction. This could be caused by a slight misorientation of the recrystallized layers in respect to the substrate. The cause of these effects is still mostly unclear, though differences in growth behaviour in semiconductor epitaxy have been previously observed. Corresponding experiments for other semiconductors have not been published, but it is to be expected that very similar effects occur.

For ⟨111⟩-oriented silicon, a high dependence of the annealing behaviour on the previous thermal history has also been found. If arsenic- or boron-implanted samples are annealed at progressively increasing temperatures, a steady decrease

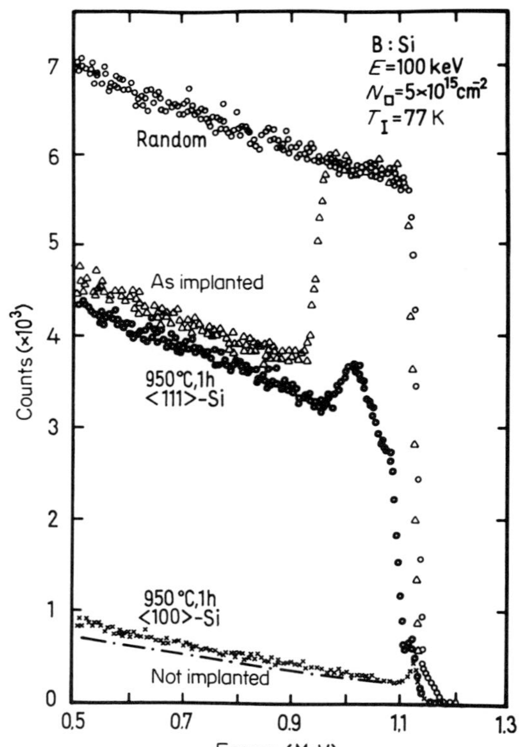

Figure 3.9 Random and channelling spectra of 2 MeV He ions in ⟨100⟩- and ⟨111⟩-oriented silicon after boron implantation (100 keV, 5×10^{15} cm^{-2}) and annealing at 950 °C for 1 h [164]. Reproduced by permission of Gordon and Breach

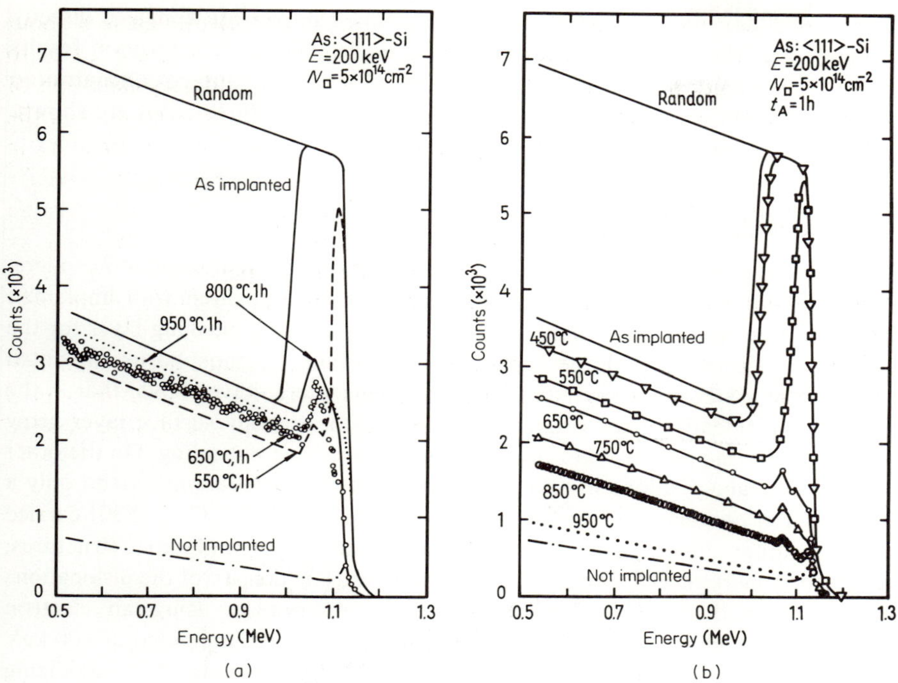

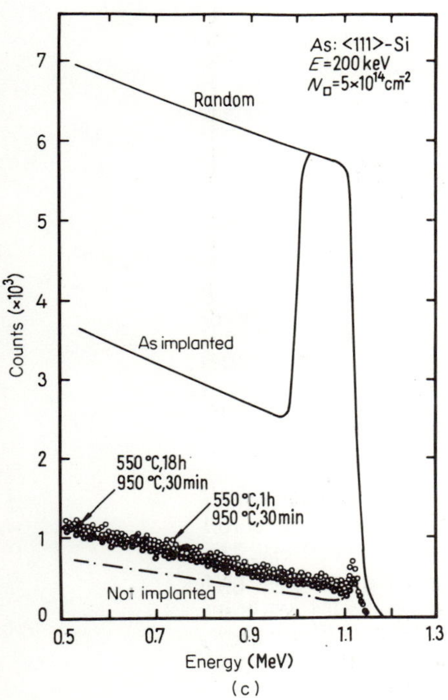

Figure 3.10 Random and channelling spectra for ⟨111⟩-oriented silicon implanted with arsenic at 77 K (200 keV, 5×10^{14} cm^{-2}). The samples were (a) annealed in one step, (b) annealed in temperature steps of 100 °C, (c) preannealed at 550 °C; all annealing was done in a nitrogen atmosphere [164]. Reproduced by permission of Gordon and Breach

in the defect concentration results. However, if the samples are annealed directly at temperatures between 550 and 950 °C, a higher defect concentration results [164]. By pre-annealing at 550 °C, this increased radiation damage concentration can be avoided completely. In Figure 3.10(a) to (c), these three cases are shown, using the example of an arsenic implantation.

Influence of Masking Layers

In semiconductor technology, one often diffuses in an oxidizing atmosphere, following a predeposition step. This process is also used often with implanted layers in order to obtain, simultaneously with annealing, a masking layer for the next technological step. Experiments with annealing of phosphorus-implanted layers, in an oxidizing as well as in an inert atmosphere, have shown that in the first case, dislocation lines originating from the amorphous surface layer grow deeply (up to one micron) into the semiconductor during annealing. On the other hand, if annealing is done in an inert atmosphere, the dislocations extend only a few hundred nanometres into the crystal. Prussin and Fern [579], [580] carried out a similar investigation for boron-implanted semiconductor structures, annealed in inert as well as oxidizing atmospheres. The density of the dislocations was determined by etching and subsequent observation using an electron microscope. A characteristic picture of a silicon structure, implanted at 100 keV with a boron dose of 1.2×10^{15} cm^{-2} followed by annealing in an oxidizing

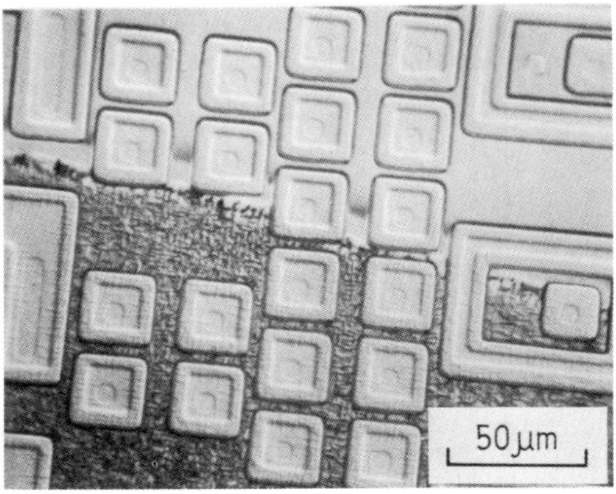

Figure 3.11 Dislocations after annealing in an oxidizing atmosphere (15 min in wet N_2, 30 min in dry N_2 at 1140 °C), following a boron implantation with 1.2×10^{15} cm^{-2} at 30 keV. The sample was Sirtl-etched for 15 s. The upper part of the sample was masked against the ion beam [579]. From Prussin and Fern, *J. Electrochem. Soc.*, **122** (1975) 830, reprinted by permission of the publisher, The Electrochemical Society, Inc.

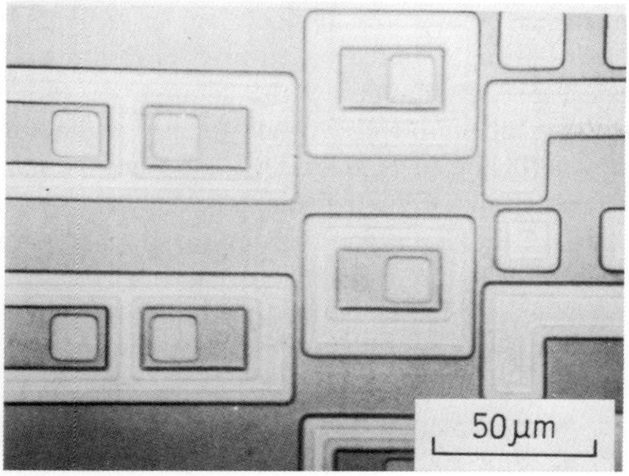

Figure 3.12 Integrated circuit implanted with boron the same as the circuit in Figure 3.11, but first annealed in an inert atmosphere (15 min N_2, 15 min wet N_2, 15 min N_2). No dislocations are visible after Sirtl etching [579]. From Prussin and Fern, *J. Electrochem. Soc.*, **122** (1975) 830, reprinted by permission of the publisher, The Electrochemical Society, Inc.

atmosphere, is shown in Figure 3.11. The upper part of the sample was completely masked against the ion beam. Clearly seen are the many dislocations, which seriously degrade the electrical properties of the *p–n* junctions. A different result is obtained when the samples are first annealed in an inert atmosphere. Figure 3.12 shows an example for this case: no more dislocations are visible. These two examples show clearly that, in the case of silicon, an inert pre-annealing before annealing in an oxidizing atmosphere, or before drive-in diffusion, results in a considerably better crystal quality. The reason for this could be that stress originating during the recrystallization of the disordered layer can only be released in the form of dislocations, due to the oxide layer. The Sirtl etch [689] is suitable for making the dislocations visible. Its composition is 120 cm^3 HF, 100 cm^3 H_2O, and 50 g Cr_2O_3; the etching time is about 15 to 30 s.

Defects in Annealed Layers

In the above paragraph it has been mentioned that, after annealing of implanted layers (though in this context one often speaks of epitaxial recrystallization), many defects remain which in some cases were formed during annealing. These defects consist of various kinds of dislocations and stacking faults. The usual methods for determining radiation damage, such as back-scattering and electron-spin resonance, are generally not able to determine these defects; in the first case because the concentration is too low, and in the second because of spin saturation after annealing at high temperatures. Dislocations in semiconductor devices can manifest themselves through increased reverse currents and low breakdown voltages, and can especially influence the lifetime of the minority carriers. Dislocations are especially intolerable when further technological steps are to follow, for example, diffusion or epitaxy. Detailed studies concerning the growth of epitaxial layers on arsenic-implanted silicon have been carried out by Moline *et al.* [508]. In the case of implantation doses up to 10^{14} cm^{-2}, very good epitaxial layers could be deposited after annealing at 1000 °C for 20 min. In this dose range, no amorphous layer is obtained for implantation at room temperature. At higher doses up to 10^{16} cm^{-2}, good epitaxial layers resulted only if the implantation was carried out at 500 to 600 °C, and if the formation of an amorphous layer was avoided. After implantations at room temperature, it was necessary to anneal in an oxidizing atmosphere to oxidize the damaged layer, and then to etch with HCl before epitaxy. During the first process, 0.15 μm silicon was removed; during the second process, at least 0.2 μm silicon. These conclusions partly contradict the results cited above. One can see from this that the annealing behaviour for each ion species must be considered separately. However, pre-annealing in an inert atmosphere, with subsequent oxidation, is also advantageous in the case of arsenic, since good epitaxial layers can then be deposited too; and at the same time, possible adverse effects based on the 'snow-plough effect' (see Section 3.8.3) at high arsenic concentrations can be avoided. Further examples are given in Chapter 6, for various ion species.

The growth of epitaxial layers on heavily damaged semiconductors, in extreme cases, can take place in a polycrystalline manner. Research is currently in progress on the constructive use of this effect, to achieve a polycrystalline growth on implanted, amorphous areas. The aim would be the use of selective polycrystalline areas for the insulation of integrated circuits, or the use of a high diffusion coefficient in polycrystalline layers, for example, to produce subcollector-contact diffusions simultaneously with the emitter diffusion in integrated circuits. Of course, the amorphous area must not be allowed to recrystallize during the epitaxy process. Therefore, for the implantation, only ions producing permanent radiation damage at the epitaxial temperature can be considered. For example, argon and perhaps silicon appear to be suitable.

3.2 ELECTRICAL ACTIVATION OF IMPLANTED IONS

In addition to bringing about the recrystallization of the lattice, the annealing step should also serve to electrically activate the implanted ions. This process can take place simultaneously with recrystallization, for example, if the ionic radius of the doping atoms is very similar to that of the host lattice atoms; or it can also require additional energy, for example, in the case of implantation into GaAs. In the latter case, back-scattering measurements show that radiation damage is already eliminated after annealing at 500 °C; in general, however, temperatures of more than 700 °C are necessary for electrical activation of the implanted ions.

For the electrical activation of implanted ions, i.e., their incorporation on to lattice sites, it is not possible to state any general rules. Basically, it is necessary to conduct measurements for the desired ion–substrate combination, in order to determine the exact temperature dependence of the activation. Within very broad limits, predictions are possible, for example, that after a sufficient annealing time at 900 °C for silicon, all activatable ions will have been activated; however, the situation at low temperatures can be entirely different.

Besides the temperature, the time is also an important parameter in annealing processes to obtain activation or recrystallization. This is especially important in comparing the results of measurements from different laboratories. The usual annealing times are between 10 and 30 min. They depend mainly on the experimental arrangement, and are limited at the lower end by the time needed to heat the samples. Since electrical activation is a thermodynamic process, it is possible within certain limits to substitute the time for the annealing temperature. In practice, however, one chooses—apart from basic research—a temperature which makes it possible to use the times mentioned above. Longer periods are used when, besides activation of the ions, a drive-in diffusion is desired. The electrical activation is generally determined by Hall effect and sheet resistivity measurements, or more simply by four-point probe measurements (see Chapter 5). In this way, one obtains information on the effective sheet carrier concentration $N_{S,eff}$ and the effective sheet mobility $\mu_{S,eff}$, as well as the sheet resistivity ρ_S. Further important semiconductor parameters such as the carrier lifetime are usually not considered, and their influence is only determined indirectly during device examinations. The annealing behaviour of implanted layers, in most cases, is highly dependent on the implantation dose. With increasing dose, and thus with increasing radiation-damage concentration, higher annealing temperatures are often necessary to achieve a given level of electrical activation. However, at very high doses leading to an amorphous layer (regarding dose, dose rate, and temperature dependence of the dose causing amorphousness, reference is made, for example, to Müller [517]), a relatively low temperature (for silicon, 500–650 °C) is adequate to achieve nearly complete electrical activation, since the ions in the amorphous layer (which recrystallizes at temperatures between 500 and 650 °C) are incorporated into the lattice during this quasi-epitaxial growth without any significant additional energy requirement, which could only be caused by the different ion radii of the doping atoms and the host lattice atoms. Only at temperatures from 900 to 1000 °C (in the case of silicon) will all ions become electrically active independent of the dose, as long as the solubility is not exceeded. In case of implantation with an amorphizing dose, or after a previous implantation causing amorphousness, one finds all the 'dose ranges' discussed above when observing the doping profile from the semiconductor surface to its inner layers. The results of related investigations suggest that different activation mechanisms, with differing activation energies, play a role here.

During the annealing process, various effects can occur. In addition to the simple, temperature-dependent incorporation of the implanted ions with a definite activation energy, complexes of greatly differing shapes can be formed, either as intermediate or as final stages. In the field of implantation, one must almost always depend on assumptions, since the microscopic structure can only be determined directly in a few cases, for example, by electron-spin resonance measurements.

3.2.1 Isochronal Annealing

As an example of a complex annealing process, boron in silicon will now be discussed (Figure 3.13). At low doses, a slow increase in the electrical activation with increasing temperature is seen. If the implantation dose increases above $3 \times 10^{13}\ cm^{-2}$, a retrograde annealing behaviour (reverse annealing) of the sheet carrier concentration results. This has been explained by Blamires [81] in terms of pair formation of boron ions. These pairs dissociate again at higher temperatures, and the atoms are incorporated individually into the crystal lattice. If one implants into an amorphous layer, or with such a high dose that an amorphous layer is formed (in the case of boron, this is practically only possible by implantation at low temperatures), a steep increase in the electrical activation is seen around an annealing temperature of 600 °C, which is usually explained in terms

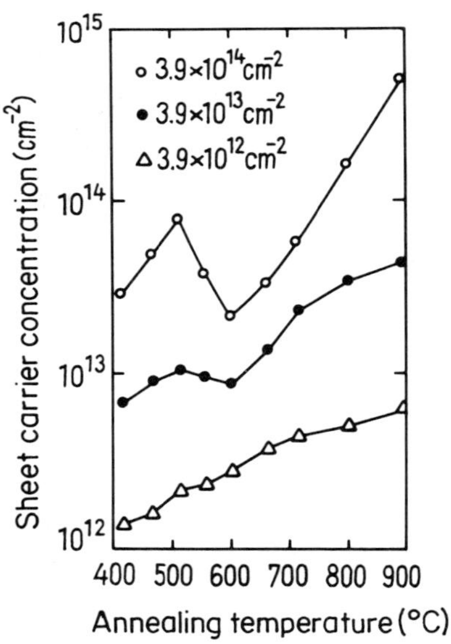

Figure 3.13 Isochronal annealing of boron-implanted layers [762a]. Annealing time 30 min. Reproduced by permission of Taylor & Francis, Ltd.

of recrystallization of the crystal lattice. Often, however, one observes a behaviour involving a steady increase in the electrical activation with increasing temperature.

For semiconductor devices applications, one is generally interested only in the maximum activation obtainable at a certain temperature, which is often predetermined by other processing steps (MOS process). Several corresponding experiments are discussed in Chapter 6.

Basically, it should be possible to express the annealing behaviour of implanted ions by reaction-kinetic equations, which are not yet connected with some physical interpretation, but which can make such interpretations significantly easier. Surprisingly, only very little research in this field has been published up to now [53], [78], [118], [371], [652]; the great majority of the research work on the electrical activation of implanted ions is restricted to phenomenological results.

The equation for a reaction of the nth order is:

$$\frac{dC}{dt} = -KC^{n} \tag{3.1}$$

where C is the concentration of the substance which changes its state, and K is the velocity constant of the reaction. In the general case, with more than one reacting substance (e.g. two), one obtains equations of the following kind:

$$\frac{dC_j}{dt} = -KC_j^{n} C_k^{m} \tag{3.2}$$

where the order of the reaction in this case is $n + m$, and $C_{j,k}$ is the concentration of the j, k substance.

The velocity constant K is independent of the concentration of the reaction partners. Based on Arrhenius, the following is valid for this constant:

$$K = K_0 \exp\left[-\frac{E_a}{kT}\right] \tag{3.3}$$

K_0 is the value of K for $T \to \infty$; E_a is the activation energy of the process. Since one cannot directly measure the concentration C of the reacting particles, but only the reaction product, i.e. the electrically-active ions on lattice sites N_S, equation (3.1) must be rewritten. The concentration, depending on the reaction time t and the annealing temperature T, is determined as follows:

$$\begin{aligned} N_S(t, T) &= N_S(0, 0) + C(0, 0) - C(t, T) \\ N_{S,max} &= N_S(0, 0) + C(0, 0) \end{aligned} \tag{3.4}$$

$N_S(0, 0)$ is the portion of ions which are already electrically active without annealing; $C(0, 0)$ is the quantity of ions which are not active without annealing.

By integration of equation (3.1), one obtains, substituting equation (3.4), for a first order process:

$$\ln\left[\frac{N_{S,max} - N_S(0)}{N_{S,max} - N_S(T)}\right] = Kt \tag{3.5}$$

In most cases, $N_S(0)$ will be negligibly small, and one obtains, by plotting of:

$$\ln\left[\ln\left(\frac{N_{S,max}}{N_{S,max} - N_S(T)}\right)\right]$$

for different samples after annealing at various temperatures versus $1/T$, due to equation (3.3), a straight line of the slope E_a/k, from which the activation energy can be determined. In general, however, the same sample is used for measure-

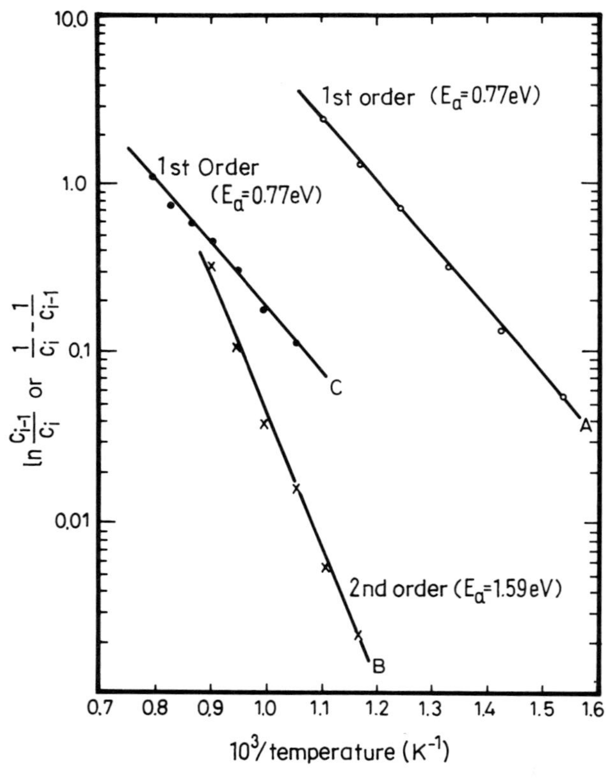

Figure 3.14 Arrhenius plots for isochronal annealing of phosphorus-implanted silicon layers ($E = 20$ keV, $N_{\square} = 3.13 \times 10^{14}$ cm^{-2}), annealing time in all cases 120 min [78]. Three different activation mechanisms with three activation energies are found. Reproduced by permission of Taylor & Francis, Ltd.

ment. Equation (3.5) is then written as:

$$\ln\left[\frac{N_{S,max}-N_{S,i-1}}{N_{S,max}-N_{S,i}}\right]=Kt \tag{3.6}$$

where $N_{S,i}$ is the electrically-active portion after the ith annealing at temperature T_i. Plotting of:

$$\ln\left[\ln\left(\frac{N_{S,max}-N_{S,i-1}}{N_{S,max}-N_{S,i}}\right)\right]$$

versus $1/T_i$ results, analogous to the above, in the activation energy.

For processes of a higher order, one obtains for the case of equation (3.6), after analogous calculation:

$$\frac{1}{n-1}\left[(N_{S,max}-N_{S,i})^{1-n}-(N_{S,max}-N_{S,i-1})^{1-n})\right]=Kt \tag{3.7}$$

By choosing a suitable n and plotting versus $1/T$, one obtains the activation energy and the order of the process. For practical purposes, however, this method should only be applicable up to $n=2$.

N_S is the total concentration of the electrically active ions; for practical purposes, it is approximately represented by $N_{S,eff}$ (see Section 5.5). $N_{S,max}$ is the

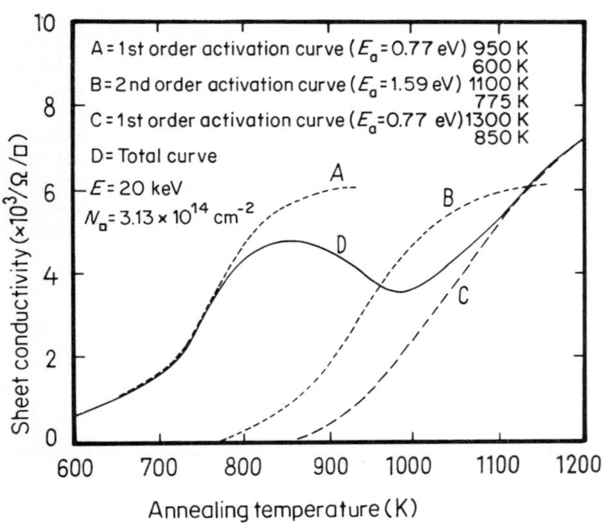

Figure 3.15 Isochronal annealing of the sheet resistivity for phosphorus-implanted silicon layers [78]. Reproduced by permission of Taylor & Francis, Ltd.

maximum portion of the ions which can be activated; i.e., equal in the ideal case to the implanted dose $N_{\square}$; in practice, equal to $N_{S,eff,max}$.

In Figure 3.14, evaluations of annealing curves according to equations (3.6) and (3.7) are shown, as determined by Bicknell [78] by measurements on a phosphorus-implanted sample ($E = 20$ keV; $N_{\square} = 3.13 \cdot 10^{14}$ cm^{-2}; $t = 120$ min). The corresponding annealing behaviour is shown in Figure 3.15. In this concentration range, phosphorus has a reverse-annealing characteristic, in which three different activation energies play a part. In this example, sheet resistivity measurements were used to determine the activation energy. The more correct measurement parameters, however, would actually be N_S as well as $N_{S,eff}$.

3.2.2 Isothermal Annealing

Isothermal annealing is used much less than isochronal annealing, although it can be used to good advantage in some cases where one is limited as to choice of temperature. In the case of isothermal annealing, the time is varied and the temperature is kept constant. An example for such a measurement is shown in Figure 3.16, for an arsenic implantation ($E = 80$ keV; $N_{\square} = 5 \cdot 10^{14}$ cm^{-2}) at annealing temperatures between 390 and 490 °C. The isochronal-annealing curves are so steep for the generally used temperatures that no activation energy

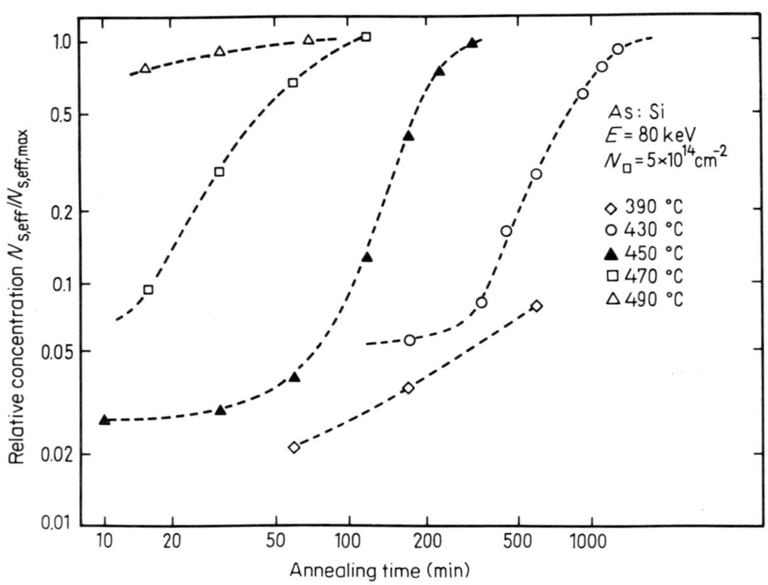

Figure 3.16 Isothermal annealing behaviour of an arsenic implantation at 80 keV and a dose of 5×10^{14} cm^{-2} in silicon, for temperatures between 390 and 490 °C [607]

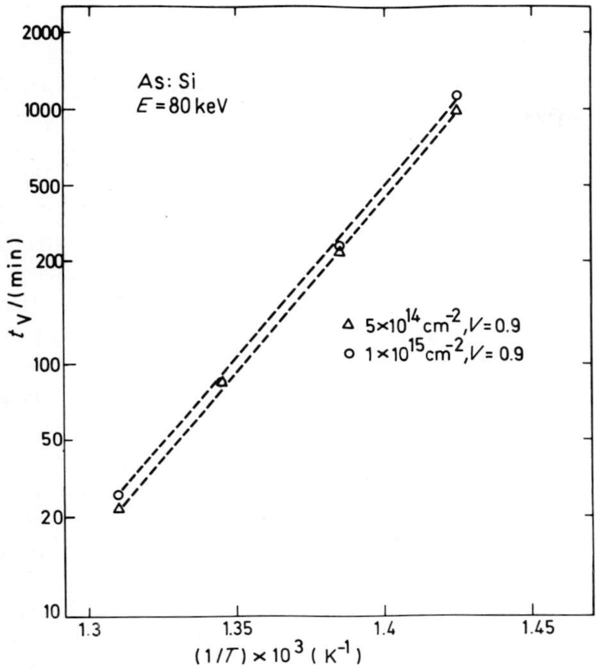

Figure 3.17 t_v (v = 0.9) versus $1/T$ for arsenic implantations in silicon at 80 keV with doses of 5×10^{14} cm^{-2} and 1×10^{15} cm^{-2} [607]

can be determined [610]. The reason for this could be the epitaxial recrystallization of the silicon layer which is made amorphous at the dose used. The arsenic is probably incorporated simultaneously on to electrically active lattice sites, due to its relatively well-suited ion radius. Equations (3.6) and (3.7) are also valid for isothermal annealing. Thus, a simple determination of the activation energy is possible by plotting ln t_ν (t_ν is the time necessary to activate the portion ν) versus $1/T$. The slope of the line then has the value E_a/k. The order of the reaction is determined by plotting:

$$\ln\left[\frac{N_{S,max}}{N_{S,max} - N_S(t)}\right] \quad \text{or} \quad \frac{N_S(t)}{N_{S,max} - N_S(t)}$$

versus different annealing times. One obtains a straight line for a first-order reaction in the first case, and for a second-order reaction in the second case.

An evaluation according to the method outlined above, for the example in Figure 3.16 as well as for a further arsenic implantation, is shown in Figure 3.17. The activation energy is 2.75 eV, and is probably the recrystallization energy of the silicon lattice; the process is a first-order one.

3.3 LASER ANNEALING

Recently, a different method for annealing implantation damage, and for the electrical activation of implanted ions, has been found. By irradiation of an implanted layer for a very short time with laser light, a recrystallization of the lattice and an electrical activation of the implanted ions are possible [393a], [420a], [675a]. Depending on the power density, the irradiation time lies between some nanoseconds and several seconds; the power density is between 10^8 W/cm^2 and 100 W/cm^2. In principle, this annealing too is a thermal annealing, but there are some differences. In the case of thermal annealing, the wafer is brought to a constant temperature below the melting point for a certain time. The temperature is constant throughout the wafer. The annealing time is usually chosen so that the heating phase plays no important role ($t \geqslant 30$ min). In the case of laser annealing, two different types have to be distinguished: pulsed laser annealing and continuous laser annealing. In the first case, a melting of the surface layer takes place, and the rest of the wafer remains cool [45a], [463a], [755a]. Subsequently, after a short time, the liquid layer recrystallizes epitaxially. In the second case, which is more similar to thermal annealing, a surface layer or the whole wafer is heated up, depending on the irradiation time and power, and is recrystallized without melting [279a, 776a].

The main parameters important in laser annealing are the power density of the laser beam, the energy deposition per pulse and the pulse length, the wavelength and the absorption coefficient of the light, the thermal properties of the semiconductor, and the diameter of the laser beam.

As a practical point, one has to consider that furnace annealing is a batch process where up to several hundred wafers can be annealed at the same time, whereas laser annealing is basically a single-wafer process similar to ion implantation, if no special target chamber is used. The diameter of the annealed spot is between a few dozen micrometres and several millimetres, so that a scanning of the laser beam or a movement of the wafer are mandatory for large area annealing. Therefore, the combination of implantation and annealing seems to be possible.

Many results of research on laser annealing were presented in three recent conferences: 'Laser Effects in Ion-Implanted Semiconductors' [591a], 'Laser-Solid Interactions and Laser Processing—1978' [240a], and 'Laser and Electron Beam Processing of Electronic Materials' [40a].

There are several related short-time annealing methods. Pulsed and scanned electron-beam annealing are very similar to the two types of laser annealing [289a, 305a]. The main difference is the relatively free choice of the penetration depth of the electron beam, which is set by the energy of the electron beam used. Since there are no basic differences between these methods, but only practical differences (e.g., a vacuum is required for electron-beam annealing), only a few results will be given in Chapter 6 without further discussion of this latter method.

In addition, the concentrated light of a flash lamp [94a], concentrated solar light [424a], and perhaps still other high-energy sources, can be used to anneal wafers. The only problems are power density and the absorption coefficient of the radiation. To our present knowledge, all these methods are basically thermal. The very often proposed influence of ionization, however, cannot be ruled out completely.

3.3.1 Pulsed Laser Annealing

Pulsed laser annealing is usually done with ruby, Nd:glass and Nd:YAG lasers. The respective wavelengths are 0.69 μm, 1.06 μm and 0.53 μm, with frequency doubling using a KDP crystal [45a], [124a], [246a], [393a], [420a], [463a], [675a], [755a]. Other lasers with shorter wavelengths could also be used. The lasers are Q-switched, with pulse lengths between 10 and 100 ns; longer pulses have also sometimes been used [45b].

Pulsed laser annealing always results in a melting of the surface of the irradiated semiconductor, followed by an epitaxial regrowth. The time the layer stays molten depends on the energy of the pulse and the thermal properties of the crystal. The calculation of the heating, melting and cooling process is difficult, since most material parameters are not well known at high temperatures.

A very important parameter is the wavelength of the light. Whereas, in the cases of the ruby light ($\lambda = 0.69$ μm) and the doubled Nd light ($\lambda = 0.53$ μm), a very strong absorption takes place in Si and GaAs close to the surface over a distance which is small compared to the thermal-diffusion length, the 1.06 μm light of Nd lasers is only very weakly absorbed. The main mechanism in this case is probably free carrier absorption. The threshold energy for melting can be approximately calculated [84a], for 0.53 and 0.69 μm radiation, as follows:

$$E_{\mathrm{th}} = \frac{1}{1-r} C_{\mathrm{p}}(T_{\mathrm{m}} - T_{\mathrm{o}})\sqrt{(D\tau_{\mathrm{p}})} \tag{3.8}$$

where R is the solid phase reflectivity, C_{p} the heat capacity, and T_{m} and T_{o} the melting and ambient temperatures, respectively. D is the thermal diffusivity, and τ_{p} the pulse duration. For the melting of silicon, approximately 0.3 J/cm^2 are required in the case of short wavelength lasers, and 6 J/cm^2 in the case of Nd lasers.

During the melting, a very fast diffusion of the dopants, and also for specific elements a very pronounced solid–liquid segregation, take place. The latter is only valid for elements which are of no interest for doping applications, such as copper and lead [51a], [246b]. The former, however, is a very significant restriction upon applications where very steep profiles close to the surface are required. Two examples for these two effects are given in Figures 3.18 and 3.19, where arsenic and copper, respectively, have been implanted into silicon. Whereas in Figure 3.18 the diffusion during the melting phase is clearly seen, the segregation of copper (liquid–solid segregation coefficient 4×10^{-4}) is very pronounced in Figure 3.19. On the other hand, the melting and subsequent epitaxial recrystallization result in an absolutely perfect crystal, without any dislocations

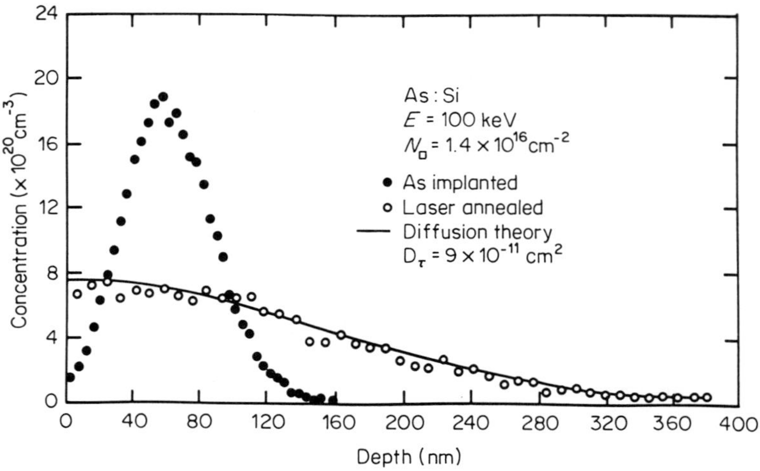

Figure 3.18 Arsenic profiles in silicon before and after laser annealing with a ruby laser, compared with a theoretical diffusion profile (solid line) [41c]; $E = 100$ keV, $N_{\square} = 1.4 \times 10^{16}$ cm^{-2}. Reproduced by permission

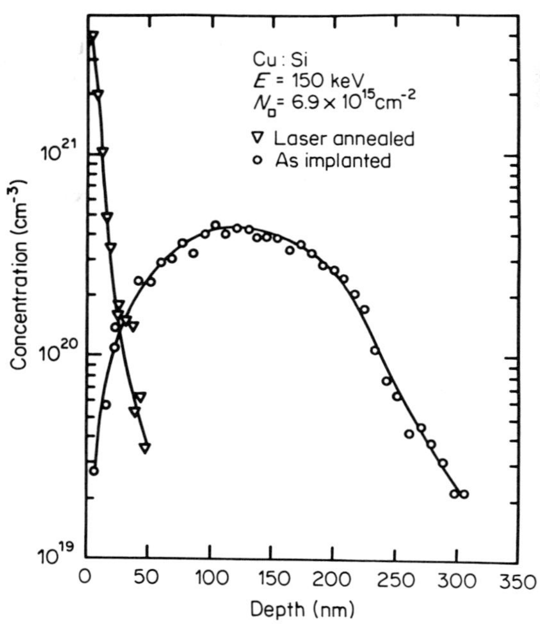

Figure 3.19 Profiles of Cu implanted into ⟨111⟩-oriented Si ($E = 150$ keV, $N_{\square} = 6.9 \times 10^{15}$ cm^{-2}), before and after laser annealing (ruby laser, 1.4 J/cm^2, 20 ms) [41d]. Reproduced by permission

or other crystalline defects [529a]. Moreover, due to the extremely fast process of recrystallization, which takes place in a few hundred nanoseconds, more dopant atoms than theoretically possible on account of the solubility can be incorporated into the crystal lattice, and are electrically active there. This has been found up to now for antimony in silicon, and for tellurium in GaAs, with a difference amounting to up to a factor of 10 [296a, 586, 586a, 647a]. Upon subsequent thermal treatment, however, the solubility equilibrium is again established [586]. The melting can also dissolve dislocations which form during diffusion [340a].

One very important problem with this type of laser annealing, and also to a lesser extent with CW-laser annealing, is the stability of masking and passivating layers. During the melting phase, for example, SiO_2 and Si_3N_4 layers are still solid, and float on top of the molten semiconductor. During solidification, or perhaps during melting, these layers often lift off due to acoustical or mechanical effects. There is still, therefore, much work to be done for the development of this new and promising annealing method. Only for solar cells does there seem to be an immediate application possibility.

3.3.2 Continuous Laser Annealing

For continuous or CW-laser annealing, argon, krypton and CO_2 lasers have been used up to now [167c], [279a], [715a], [776a], [787a]. This type of laser annealing does not result in a melting of the irradiated layer, but rather, in a heating to approximately 1000 to 1300 °C, which is higher than the temperatures usually used for thermal annealing. The results are very similar to furnace annealing; only the time required for annealing is shorter, because of the higher temperature. Defects such as those found in thermally annealed semiconductors are again found, and no defect free material is obtained.

The recrystallization takes place via solid-phase epitaxy. The regrowth velocity is of the form [45b]:

$$v = v_o \exp(-E_a/kT) \qquad (3.9)$$

where $v_o = 2.9 \times 10^9$ cm/s and $E_a = 2.7$ eV, for ⟨100⟩-oriented silicon. For example, at 1100 °C, the regrowth of a 1000 Å thick layer should take place in about 390 ms. Since such a long time is required for solid-state epitaxy, it is unlikely that this can occur using a Q-switched laser.

For lasers with short wavelengths, the optical absorption length is again small compared to the thermal diffusion length, and the average temperature rise can be calculated [84a] as follows:

$$\Delta T = (1 - R) I \tau_p / C_v \rho \, (2D\tau_p)^{1/2} \qquad (3.10)$$

where R is the reflectivity, C_v the heat capacity, τ_p the pulse duration, D the thermal diffusivity, I the pulse energy and ρ the density of the semiconductor. The thickness of the heated layer is $\sqrt{(2D\tau_p)}$. The cooling time is of the same order of magnitude as the heating time.

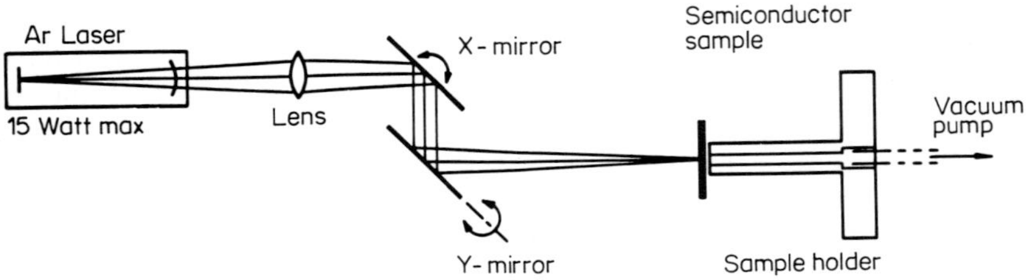

Figure 3.20 Illustration of a scanning system for CW laser annealing [279a]. Reproduced by permission of the American Institute of Physics

Since no melting takes place, the profile shape is not changed. Therefore, a profile tailoring using multiple implants seems to be possible.

The dwell time is usually in the order of several milliseconds; however, much longer times have been used [715a].

In Figure 3.20, a typical arrangement for laser annealing, using an argon laser and an optical scanning system, is shown. Other approaches involve mechanical movement of the wafer.

In contrast to furnace annealing, it is possible with both types of laser annealing to work selectively. By this means, new types of semiconductor devices seem to be possible.

Recently, data on some experiments using CO_2 lasers have been published. The wavelength of this laser is well below the band-gap energy of most semiconductors. Therefore, absorption of energy can only take place via free carrier absorption [124b, 167c, 715a]. The experiments show that this annealing is more or less identical to thermal annealing. The times involved are up to several seconds. Therefore, the entire thickness of the wafer is heated. This method is obviously independent of the band-gap energy. Nevertheless, it seems to be very promising, especially for compound semiconductors, since no protective layers are required for these short-time anneals.

3.4 RANGE DISTRIBUTION IN TWO-LAYER STRUCTURES

For implantation into a two-layer structure, in general, one cannot use the calculated range distributions directly. Only in special cases is the stopping power the same in both substances, or the difference small enough so that no corrections are necessary.

Some combinations for which no corrections are necessary are, for example: ZnSe/Ge; ZnSe/GaAs; Ge/GaAs; ZnTe/CdTe (deviation 1%); InAs/CdSe; GaSb/InAs (deviation 2%); ZnS/GaP (deviation 2%). As can be seen, these combinations are not especially interesting for devices. Important for practical use are combinations such as SiO_2 on silicon or Si_3N_4 on silicon, which are employed for integrated circuits. In the case of boron implantation through SiO_2 into silicon, one can use the values for silicon as acceptable approximations (error $\leqslant 5\%$); for other combinations, however, a more exact treatment of the problem is to be recommended.

Furukawa and Ishiwara [266], using quasi-Monte Carlo calculation, obtained corresponding distributions. For practical purposes, however, this procedure is too complicated. On the other hand, a simple model by Ishiwara *et al.* [375] makes an estimate possible. The profile in the first and second layer is constructed from two Gaussian profiles. In Figure 3.21, this is shown using the example of SiO_2/Si. The following equations are valid for the profile:

$$N_1(x) = \frac{N_\square}{\sqrt{(2\pi)}\Delta R_{p1}} \exp\left[-\frac{(R_{p1}-x)^2}{2\Delta R_{p1}^2}\right] \qquad 0 \leqslant x \leqslant t \qquad (3.11)$$

$$N_2(x) = \frac{N_\square}{\sqrt{(2\pi)}\Delta R_{p2}} \exp\left[-\frac{[t+((R_{p1}-t)\Delta R_{p2}/\Delta R_{p1}-x)]^2}{2\Delta R_{p2}^2}\right] \qquad x \geqslant t \quad (3.12)$$

The model is based on the assumption that, for the combinations used, the mean atomic numbers and masses are relatively similar.

The applicability of equations (3.11) and (3.12) can be estimated by calculating the error according to the following equation:

$$\eta = \frac{R_{p2} - R'_{p2}}{R_{p2}} \times 100\% \qquad \text{with} \qquad R'_{p2} = R_{p1}\frac{\Delta R_{p2}}{\Delta R_{p1}} \qquad (3.13)$$

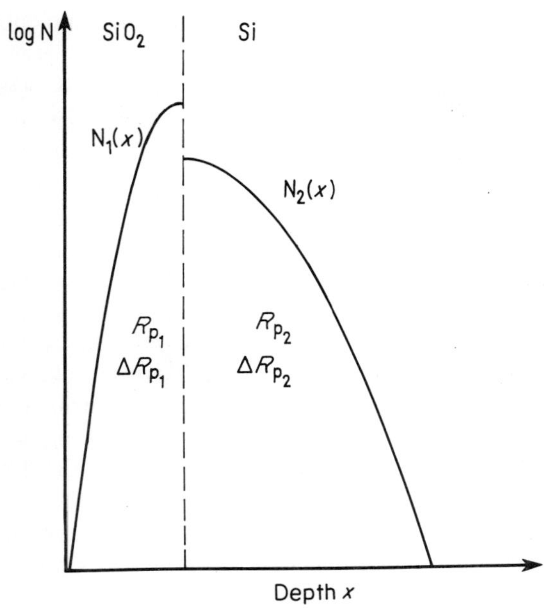

Figure 3.21 Range distribution of ions in a two-layer structure (schematic)

A similar model has been published by Satya and Palanki [621]. Better results are obtained by calculation of energy distribution functions and approximation of the range distribution by an asymmetrical joint half-Gaussian profile [287].

3.5 MASKING LAYERS

In most applications, a local doping of the solid is desired, for example, to fabricate a transistor by implanting into silicon, or a light guide by implanting into glass. Basically, there are two possibilities for masking: by use of a mask which is in contact with the substrate, or using a mask placed in front of the substrate (projection masking). Both methods have advantages as well as disadvantages.

3.5.1 Contact Masking

This technique is usual in the fabrication of semiconductor devices. The masking material is in direct contact with the substrate. Metals, insulators (SiO_2, Si_3N_4), or photoresist can be used. The photolithographic processes developed in the last few years for use in silicon planar technology are especially well suited for fabrication of the desired structures in these layers.

The requirements placed upon the masking layers are not as stringent as those in planar technology regarding temperature stability (diffusion temperatures for silicon are up to 1300 °C); on the other hand, however, they must be thick enough to be able to stop the ions, the sputtering yield should be low, and the layers must be capable of being stripped off following implantation.

Frequently used layers include silicon dioxide (SiO_2), silicon nitride (Si_3N_4), aluminium oxide (Al_2O_3), various photoresists such as Kodak KPR, KTFR, Shipley AZ 1350 J, AZ 111, Waycoat IC and PMMA, as well as metal layers deposited by evaporation.

Masking through Insulating Layers

SiO_2 is the most important masking layer, due to its significance as the natural oxide of silicon. It can be formed by thermal and anodic oxidation of silicon, by pyrolytic decomposition of gaseous silicon compounds, or by ion sputtering. Thermal oxides are formed by dry or wet oxidation of silicon at temperatures above 900 °C. The ratio between the thickness of the consumed silicon layer and the oxide thickness amounts to 0.44. The advantage of pyrolytic oxide is found in the low reaction temperatures (according to the process, 400–800 °C), which are especially important when temperature-sensitive semiconductors such as GaAs are used. It should be noted that the density of pyrolytic oxide without annealing is always lower than that of thermal oxide. The masking layer can become insoluble or difficult to dissolve after high-dose implantation. This has been found, for example, by Schmid [623a] for Al implantation in SiO_2, through formation of mixed or SiO_2–Al_2O_3 layers. In some cases, there can be a significant increase in

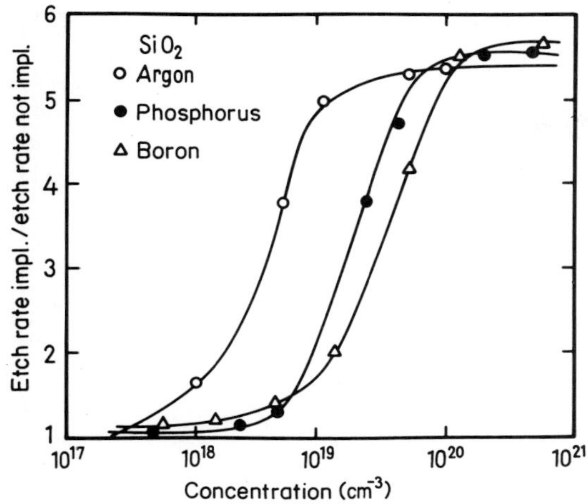

Figure 3.22 Increase of the etch rate in SiO_2 due to irradiation with Ar, P and B ions. Implantation at different energies to obtain a constant distribution; according to [511]. Reproduced by permission of Springer–Verlag, Heidelberg

the etch rate, as shown in Figure 3.22, depending on the dose of the implanted boron. Si_3N_4 layers can be produced pyrolytically at about 700 to 800 °C, and in a plasma discharge at as low as 300 °C. Al_2O_3 can also be produced pyrolytically at temperatures between 200 and 400 °C. If even lower temperatures are necessary during deposition, sputtered SiO_2, Si_3N_4, or Al_2O_3 layers are well suited. An increase in the etch rate due to ion bombardment has also been found in the case of Si_3N_4 [34], i.e., for B, N, P, Ar, Zn and Sb ions. The etch rate is increased by up to a factor of ten, according to the energy and the dose (Figure 3.23); after annealing at 600 °C, however, the original rate is again reached. This leads to the assumption that radiation damage is the cause of the increase.

In Table 3.2, a few suitable etchants are listed for SiO_2, Al_2O_3 and Si_3N_4. The etch rates are based on the non-implanted case.

Masking with Photoresist

If the implantation temperature is not too high (<100 °C), photoresists can be used for masking. Polymerization occurs at ion doses above 10^{14} cm^{-2}, and the resists are then no longer dissolvable in the usual solvents, but only by plasma stripping or in hot sulphuric acid. The range of ions is significantly greater in resist than in SiO_2, so that thicker layers are necessary for masking. Photoresists are especially well suited for masking against light ions, since they can be applied with suitable viscosity in thicknesses up to 5 μm. In Figure 3.24, the thickness of Waycoat SC resist, depending on the rotation speed during spinning-on of the resist, is shown. The viscosity can be varied as needed by dilution, between 0.45

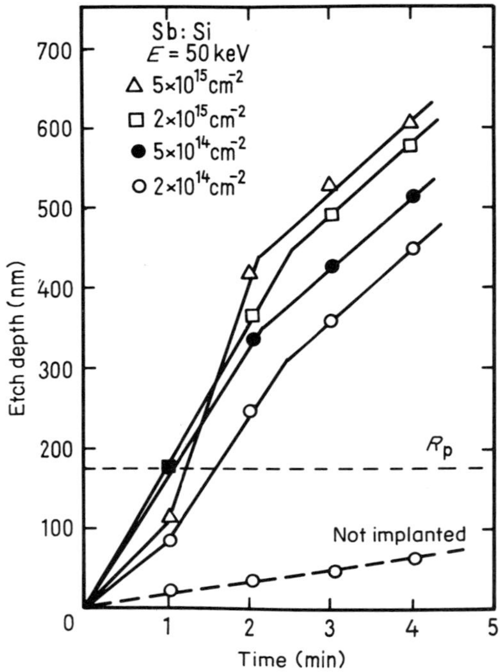

Figure 3.23 Thickness of etched Si_3N_4 layers depending on ion dose and etch time, after implantation with 50 keV antimony ions [34]. Reproduced by permission

Ns/m² (no dilution) and 0.08 Ns/m² (2 parts SC: 1 part thinner). In mass production, photoresist is generally used for masking in the fabrication of depletion-load MOS transistors (cf. Section 7.1.3).

Metal Masking

In planar technology, metal layers made of aluminium or gold, as well as multilayer contacts made of platinum, titanium or certain other metals, are used. Since semiconductors are very sensitive to metal contamination, special care is needed in the use of metals as masking layers, especially for metals with a high sputtering yield. In order to avoid contamination of the semiconductor, a thin additional insulating layer is generally used, i.e., SiO_2 in the case of silicon. After implantation, it is possible to remove the metal contamination by means of etching. This process is only possible for implantations using ions which can be accelerated high enough so that they can penetrate the insulator, i.e., using light ions. Generally, the choice of the metal is determined more from the standpoint of

Table 3.2 Etchants for masking and passivating layers

Layer	Etchant	Etch temperature (°C)	Etch rate (nm/min)	Remarks	References
SiO_2	p-etch	25	10–12	Thermal oxide	[573]
	2 HNO_3 (70%) : HF (48%) : 60 H_2O		20–120	Sputtered and pyrolytic oxide	
	75 NH_4F (40%) : 9 HF (40%) : 16 H_2O	20	50	Thermal oxide	
	10 NH_4F (40%) : 1 HF (48%)	RT	≈100		[109], [723]
		26	85		
PSG	10 NH_4F (40%) : HF (48%)	26	140	Phosphorus silica glass	[723]
Si_3N_4	H_3PO_4 (85%)	180	6–10	SiO_2 masking is necessary (SiO_2 is etched at 1 nm/min)	[109]
	HF (48%)	RT	14–100	Only for etching of large areas	[143]
$Si_xO_yN_z$	HF (48%)	RT	1–500	Strongly dependent on N content; suitable only for large area etching	[109], [583]
Al_2O_3	H_3PO_4 (85%)	150	15	Pyrolytic SiO_2-masking necessary	[31]
AlN	NaOH (10%)	20	$1–10^3$	Etching rate strongly dependent on growth temperature	[140]
	H_3PO_4	100	≈30	SiO_2 masking necessary	[140]

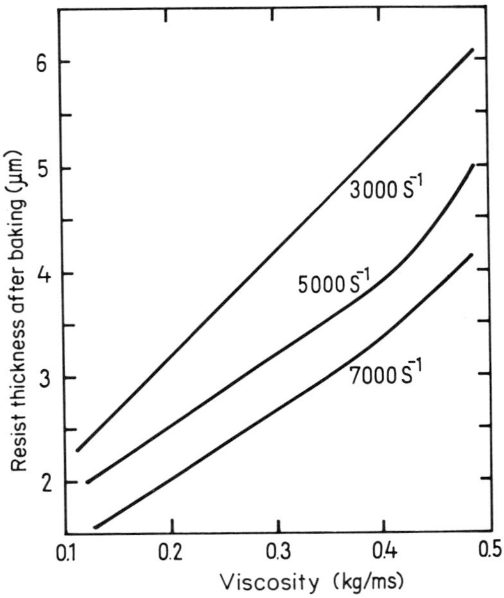

Figure 3.24 Film thickness of Waycoat SC resist (negative resist) after baking, depending on viscosity and rotation speed during resist spinning [761]

semiconductor technology than the requirements of ion implantation. If adhesion on SiO_2 is important, aluminium and molybdenum are to be preferred.

In Table 3.3, a few etchants for metals are shown. All of these metal layers can only be applied without problems up to approximate thicknesses of 0.5 to 1 μm.

Table 3.3 Etchants for metal masking layers

Layer	Etchant	Etch temperature	Etch rate (nm/min)	Remarks	References
Al	HNO_3 (65%): 5 CH_3COOH (conc.): :25 H_3PO_4 (85%)	70 °C	1000	Photoresist usable	
Mo	92 g $K_3[Fe(CN)_6]$, 20 g KOH in 300 ml H_2O or Al-etchant			Photoresist usable	[109]
Au	KCN (10%)			Photoresist unstable	
Ni	Acid iron chloride			Photoresist usable	

Electroplating to obtain thicker layers is possible, but can lead to an undesirable blurring at the edges.

Thickness of the Masking Layer

An important problem is the thickness of the masking layer. It should not be any thicker than necessary, in order to allow the resolution of small structures. For a determination of the minimum required thickness, one must know the range data in the masking layer and the permissible doping in the substrate. The corresponding calculations must generally be carried out as described in Section 2.4. If the stopping power in the masking layer and the substrate is approximately the same (i.e., aluminium or SiO_2 on silicon), the calculation of the number of ions stopped in the masking layer is simple. In general, the number of implanted ions stopped in the masking layer is:

$$N_{\square,\text{oxid}} = \frac{N_{\square}}{2}\left(1 + \text{erf}\,\frac{t - R_p}{\sqrt{(2)}\Delta R_p}\right) \tag{3.14}$$

where the values for the masking layer must be used for R_p and ΔR_p. The simplest procedure is to determine the thickness of the masking layer experi-

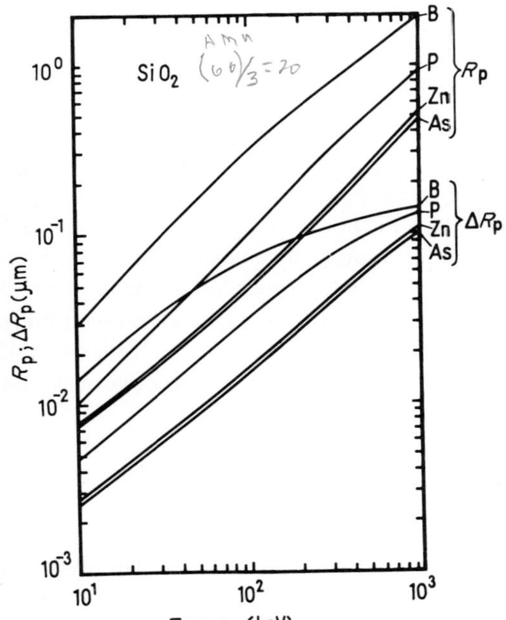

Figure 3.25 Range and range straggling of arsenic, boron, phosphorus and zinc in SiO_2, according to Reference [5]

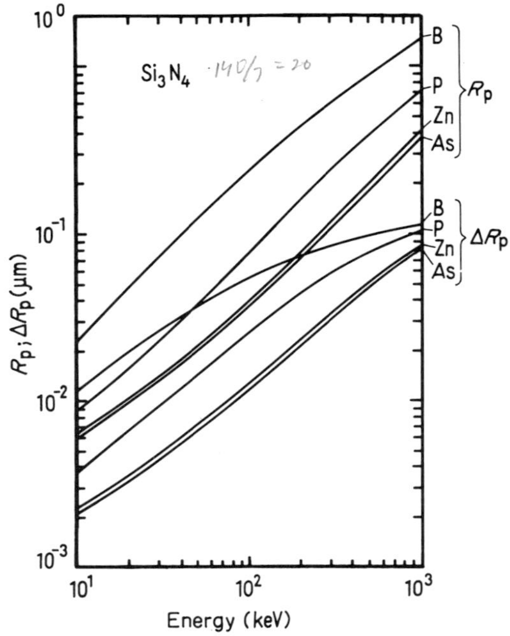

Figure 3.26 Range and range straggling of arsenic, boron, phosphorus and zinc in Si_3N_4, according to Reference [5]

mentally, for example, using a desk calculator and drawing the range distributions, again using the range parameters of the masking layer. If an exact treatment is necessary, the methods discussed in Section 3.4 are appropriate. Unfortunately, the range parameters in insulators are not precisely known in some cases. It is therefore always advisable to make some allowance for error. Range and range straggling in SiO_2, Si_3N_4 and aluminium are plotted in Figures 3.25 to 3.27, as determined by Gibbons *et al.* [5] for some of the more frequently used ions; due to their varying mass, they provide a basis to make estimates for other ion species.

3.5.2 Projection Masking

Masks which are not in contact with the substrate have the advantage that they can be used repeatedly. Due to their thickness, which is necessary for mechanical reasons, they can also mask against high-energy ions without difficulty. The disadvantage of such methods is the low adjustment accuracy, which excludes their use in connection with integrated circuits; furthermore, the fabrication of isolated structures (e.g., rings) is not possible, since the masks must be self-supporting. The masks must be connected with the target, in order to avoid charg-

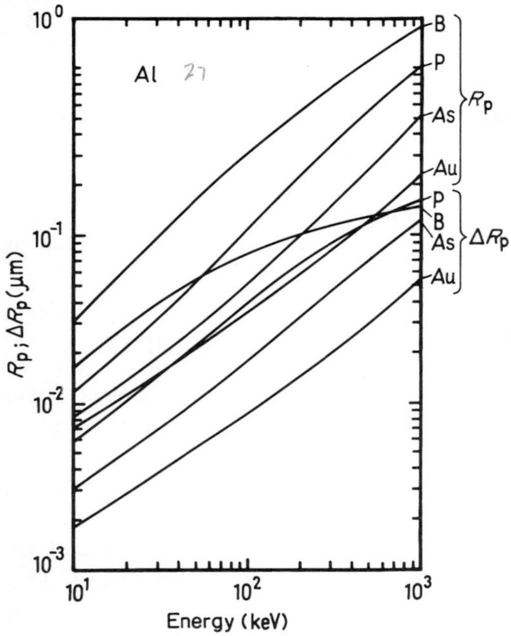

Figure 3.27 Range and range straggling of arsenic, boron, gold and phosphorus in aluminium, according to Reference [5]

ing effects and current measurement errors. For this reason, ceramic masks are also not suitable.

Metal foils used for this purpose must have a sufficiently high melting point, a low vapour pressure, good mechanical stability, and a low sputtering yield (in order to avoid contamination). Stainless steel sheets are well suited for large structures, while thin foils of molybdenum, tantalum or tungsten can be chosen according to need for small structures. Molybdenum masks are especially suitable, due to their easy structuring by photolithographic methods. Gold and copper are poorly suited because of their high sputtering yield. Metal masks are used, for example, for simple measuring structures and for the masking of relatively large area devices, such as nuclear detectors.

3.5.3 Edge Effects

In the following, the lateral scattering of ions at edges will be considered theoretically. Real masks only correspond to the assumption of abrupt edges, which is made below, more or less approximately. Especially for thick masking layers, one must expect the occurrence of tapered slopes. Figure 3.28 shows diagrams of etched slopes, as obtained for SiO_2 or metals. Tapered edges are

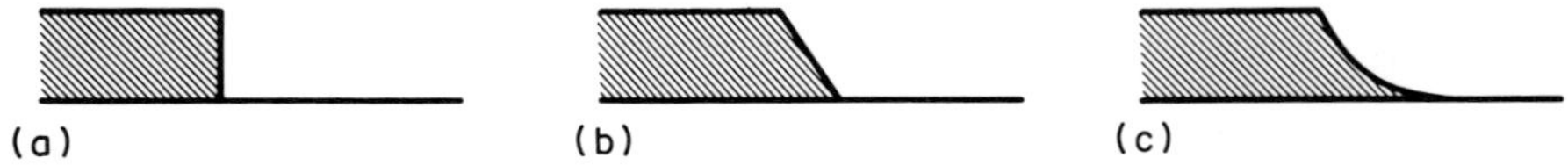

Figure 3.28 Schematic illustrations of etched masking-layer slopes: (a) ideal case, (b) tapered slope (e.g. SiO_2), (c) parabolic slope (e.g. aluminium)

often necessary with SiO_2 layers to avoid interruptions in the metallization, and are sometimes produced by special double-layer masking or ion implantation |504|.

3.6 LATERAL SPREAD

In addition to the projected range R_p and the standard deviation ΔR_p, there is (neglecting the higher moments) a further parameter which is important for the practical application of ion implantation: the lateral spread, ΔR_{pL}. This term refers to the spread to which ions are subjected away from the vertical direction within the crystal. Extensive calculations for this have been carried out by Furukawa *et al.* |268|, |269|. Data on the lateral spread are also given in the series of tables by Gibbons *et al.* |5|. The lateral spread, ΔR_{pL}, is always of the same order as ΔR_p. These values of the lateral spread are small in comparison to the lateral diffusion, which is of the same order of magnitude as the diffusion perpendicular to the surface, and in some cases can be significantly larger than this (e.g., Zn in GaAs). The structures being used nowadays in semiconductor technology, however, are so small that this effect too must be taken into account. If one implants through an opening in the mask into the semiconductor, one obtains an effective broadening of the structure. The distribution of the ions beneath this opening is (according to |268|):

$$N(x, y, z) = \frac{N_\square}{(2\pi)^{3/2}\Delta R_p \Delta R_{pL}^2}\left(\exp\left[-\frac{(x-R_p)^2}{2\Delta R_p^2}\right]\right). \tag{3.15}$$

$$\times \int_{-\infty}^{\infty}\int_{-\infty}^{\infty} \exp\left[-\frac{(y-\xi)^2-(z-\eta)^2}{2\Delta R_{pL}^2}\right] d\xi\, d\eta$$

For the case of implantation through an infinitely long mask with a width of $2a$, the following is valid for the implantation profile:

$$N(x, y, z) = \frac{N_\square}{\sqrt{(2\pi)}\Delta R_p}\left(\exp\left[-\frac{(x-R_p)^2}{2\Delta R_p^2}\right]\right). \tag{3.16}$$

$$\times \frac{1}{2}\left(\operatorname{erfc}\frac{y-a}{\sqrt{(2)}\Delta R_{pL}} - \operatorname{erfc}\frac{y+a}{\sqrt{(2)}\Delta R_{pL}}\right)$$

Values of the complementary error function, erfc, are given in the Appendix. A

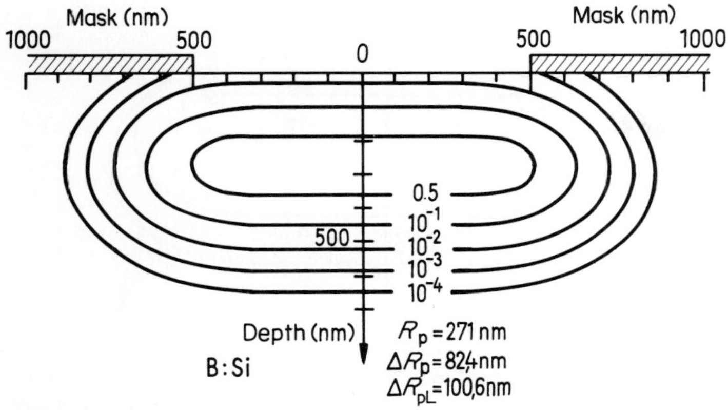

Figure 3.29 Contour lines for constant boron concentration (normalized to 1) after implantation at 70 keV through a 1μ-wide mask, according to [268]. Reproduced by permission of the Japanese Journal of Applied Physics

very interesting result of the calculations is that the doping already decreases before the edge of the mask; under the assumption that $a \gg R_{pL}$, $N = N_{max}/2$ is already valid directly below the edge of the mask. In Figure 3.29, normalized values of the concentration are shown, depending on the depth, for a boron implantation with 70 keV through a 1 μm wide mask. If we assume, for this case, a maximum doping of 10^{20} cm^{-3} and a substrate doping of 10^{15} cm^{-3}, then the *p–n* junction is located approximately 350 nm from the edge of the mask. Figure 3.30 shows a cross-section at the depth of R_p for the same case. In Figure 3.31, contour lines are shown for a concentration of 10^{-3}, relative to the maximum concentrations of boron, gallium, nitrogen, antimony and phosphorus implantations, in all cases at 70 keV.

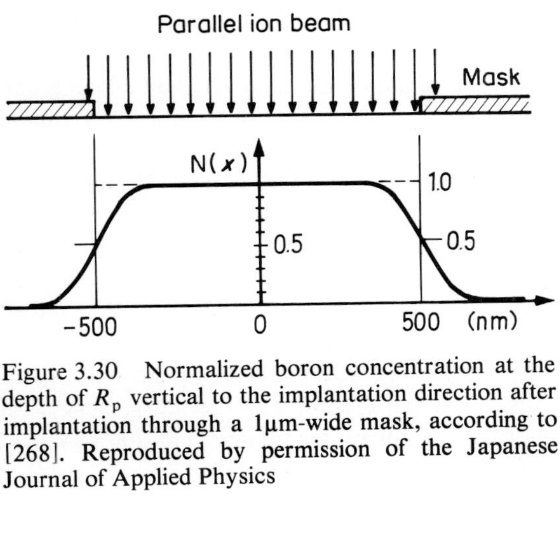

Figure 3.30 Normalized boron concentration at the depth of R_p vertical to the implantation direction after implantation through a 1μm-wide mask, according to [268]. Reproduced by permission of the Japanese Journal of Applied Physics

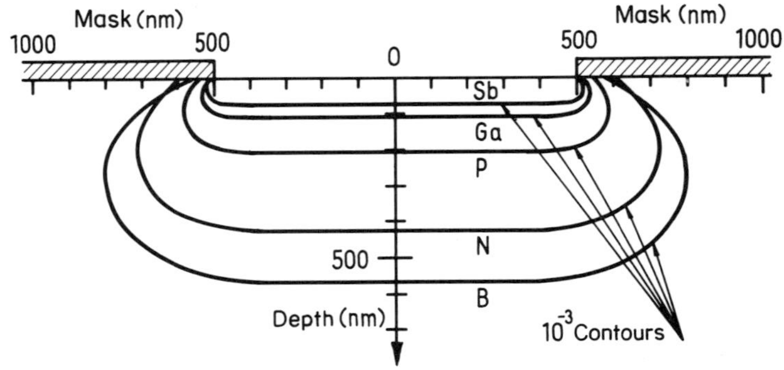

Figure 3.31 10^{-3} contour lines of boron, gallium, nitrogen, antimony, and phosphorus for a 70 keV implantation through a 1μm-wide mask into silicon, according to [268]. Reproduced by permission of the Japanese Journal of Applied Physics

Real masks and layers do not have walls as steep as those assumed in these calculations. Typical are walls similar to those shown in Figure 3.28. An additional contribution is thus made to the lateral spread.

A discussion of the lateral spread for arbitrarily shaped masks has been published by Runge [602a]. He found a universal expression for the profile obtained by implanting through an infinitely long window with an arbitrarily shaped mask:

$$N(x, y) = \frac{N_{\square}}{2\pi\Delta R_{\mathrm{pL}}\,\Delta R_{\mathrm{p}}} \int_{-\infty}^{\infty} \left[\exp\left(-\frac{(x-\xi)^2}{2\Delta R_{\mathrm{pL}}^2} - \frac{(x - t_{\mathrm{ox}}(\xi) - R_{\mathrm{p}})^2}{2\Delta R_{\mathrm{p}}^2}\right)\right] \mathrm{d}\xi \tag{3.16a}$$

with t_{ox} being the thickness of the masking layer. For non-Gaussian profiles, the two-dimensional profile is obtained by convolution of the one-dimensional profile with a Gaussian distribution.

The distribution of the radiation damage also has a lateral spread. Matsumura and Furukawa [479] calculated the lateral distribution of radiation damage, based on the work of Brice [104] and Tsurushima [739]. Figure 3.32 shows an example of their results, for 200 keV antimony in silicon. The dotted lines represent the doping distribution; the continuous curves are the contour lines of the radiation-damage distribution. The actual radiation-damage distribution per volume unit is obtained from these curves by multiplication with the dose $N_{\square}$. It can be seen from the curves that the lateral spread of the radiation damage becomes smaller in the direction of the surface than that of the ions, since the ions which come to rest underneath the mask were scattered out of the bulk of the semiconductor towards the surface, and have thereby lost the most energy. Further curves for antimony, phosphorus, and boron, for 50 keV and 200 keV, can be found in the above-mentioned study.

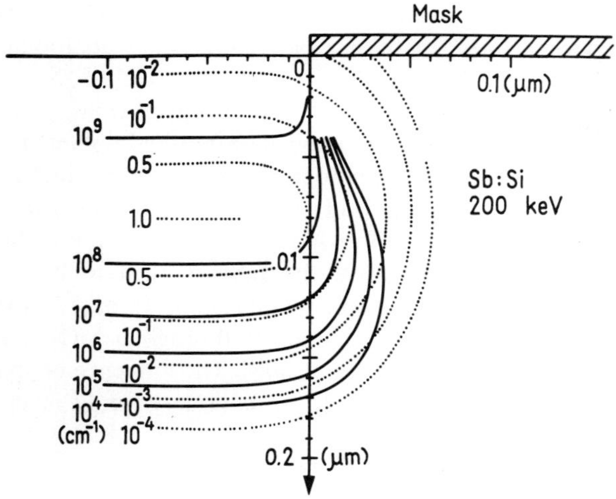

Figure 3.32 Radiation-damage distribution at a mask edge for antimony implantation into silicon at 200 keV. The dotted lines indicate equal-doping concentration contours [479]. Reproduced by permission of the Japanese Journal of Applied Physics

3.7 PASSIVATING LAYERS

A number of semiconductors are so sensitive that they have to be protected by a passivating layer during implantation, or during the annealing which follows. An example of a semiconductor which has to be protected during implantation is the III–V compound semiconductor InSb; semiconductors which have to be passivated during annealing, in order to prevent vaporization of a semiconductor component, include practically all III–V semiconductors as well as silicon and germanium, for prevention of the out-diffusion of implanted ions. For this reason, annealing often takes place in an oxidizing atmosphere for silicon. The resulting doping profile modifications will be discussed in detail in Section 3.9.

3.7.1 Passivation During Implantation

In order to protect the bare surface of the semiconductor, implantation is often done through a thin covering layer. This can be necessary for semiconductors which easily dissociate, or can be applied in general to prevent contamination of the semiconductor surface during sample manipulation and implantation. Sputtered metal atoms, which originate from masks or beam lines and generally diffuse very quickly, can have especially negative effects upon device characteristics. In this case, one can obtain an uncontaminated surface after implantation by slightly etching, or completely etching off, the covering layer. This procedure is routinely used, for example, with MOS transistors (cf. Section

7.1) for adjustment of the threshold voltage, or for self-alignment of the gate and source or drain electrodes; in these cases there is no additional etching step needed.

For several sensitive semiconductors, for example InSb, passivating layers must be used to prevent surface decomposition by the implanted ions. A scanning electron microscope (SEM) photograph of an InSb surface is shown in Figure 3.33. The upper half of the wafer was implanted with a dose of 10^{16} cm^{-2} zinc ions at 150 keV, while the lower half was covered by a metal mask during implantation. In the implanted area, the InSb is seen to have a sponge-like structure, which is not even removable by annealing. In many experiments, even with silicon, it can be necessary to implant through a SiO_2 layer to prevent surface decomposition due to sputtering [608]. The influence of sputtering on the profile shape, which normally needs to be considered only in connection with high-dose implantations, will be discussed in Section 3.8.2.

All layers with good adhesion which can be applied thinly and do not react with the substrate, for example, SiO_2, Si_3N_4, Al_2O_3 and evaporated metal layers, are suitable for use as passivating layers, depending on the substrate. The passivating layers must be thin enough so that a sufficient quantity of ions can penetrate them. For thicker layers, calculation of the range distribution is analogous to the

Figure 3.33 SEM photograph of InSb after implantation of 10^{16} cm^{-2} zinc at 150 keV through SiO_2 and directly into bare material [77]

methods described in Section 3.4; for thin layers one can assume, for the sake of simplicity, a decrease in the primary energy, which is easily calculated if the stopping power and the thickness of the passivating layer at the primary energy E are known:

$$E' = E - \Delta E = E - \left[\left(\frac{dE}{dx}\right)_n + \left(\frac{dE}{dx}\right)_e\right] \Delta x \qquad (3.17)$$

The range parameters in the substrate are then obtained from the tables, using this energy E'. In the case of somewhat thicker layers, it is better to calculate only the range with the help of this energy; for the range straggling, one should use the value corresponding to E, since the ions are already scattered in the passivating layer. A further effect which can occur during implantation through thin layers will be discussed in the next section.

3.7.2 Secondary Implantation

The following effect can occur during implantation through passivating layers, as well as at the edge of masking layers: If the mass of the implanted ions is comparable to that of the atoms in the covering layer, then the implanted ions can transfer, by collisions, a relatively large amount of energy to these atoms (cf. Section 2.1), and can thereby 'implant' them into the substrate (knock-on or recoil implantation). Figure 3.34 [123] shows the result of an experiment in which

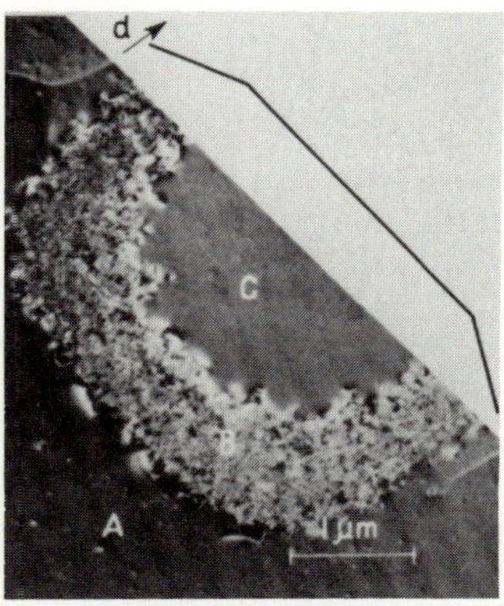

Figure 3.34 SEM photograph of silicon implanted through an SiO_2 layer of varying thickness. By secondary implantation of oxygen, defects were produced in a certain region [123]. Reproduced by permission of the American Institute of Physics

arsenic ions were implanted into silicon through a SiO_2 layer of varying thickness. In the areas where the SiO_2 layer was thick enough to stop all arsenic ions as well as all silicon and oxygen atoms affected by collisions with these ions, or in the areas where no SiO_2 layer was present, an undamaged surface of the implanted sample is found after etching. However, where silicon or oxygen atoms were implanted into the silicon by secondary implantation, many dislocations have been formed. Detailed investigations by Sigmon *et al.* [678], using simulating implantations of oxygen, have clearly shown that this effect can be attributed to a knock-on implantation of oxygen. Theoretical calculations have been carried out by Moline *et al.* [510]. Recently, there have been several reports of experimental results on determination of the profiles of secondary implanted oxygen ions [141], [294], [456], [505]. A secondary oxygen profile obtained by Goetzberger *et al.* [294], by means of capacitance-voltage measurements after arsenic implantation in SiO_2, is shown in Figure 3.35. The experimental conditions in this case were so arranged that the arsenic ions could not reach the sub-

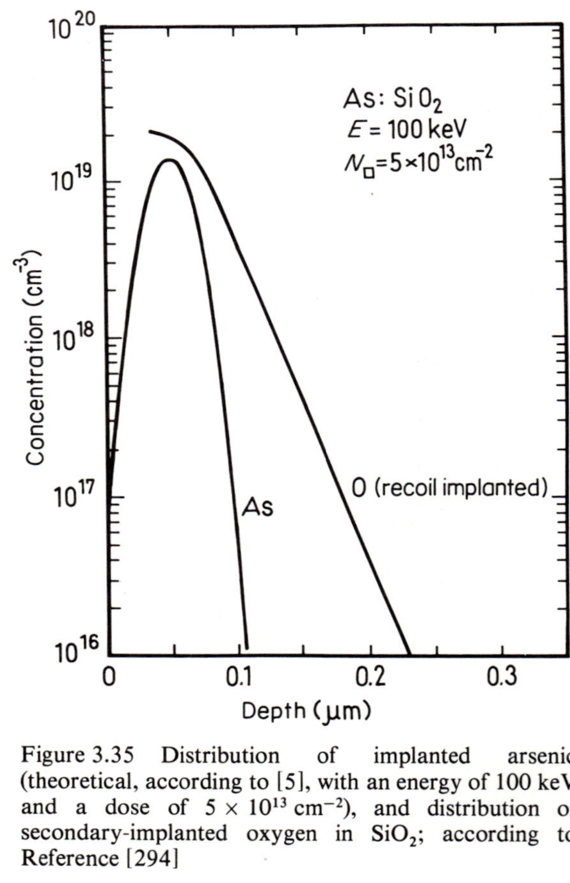

Figure 3.35 Distribution of implanted arsenic (theoretical, according to [5], with an energy of 100 keV and a dose of $5 \times 10^{13}\ cm^{-2}$), and distribution of secondary-implanted oxygen in SiO_2; according to Reference [294]

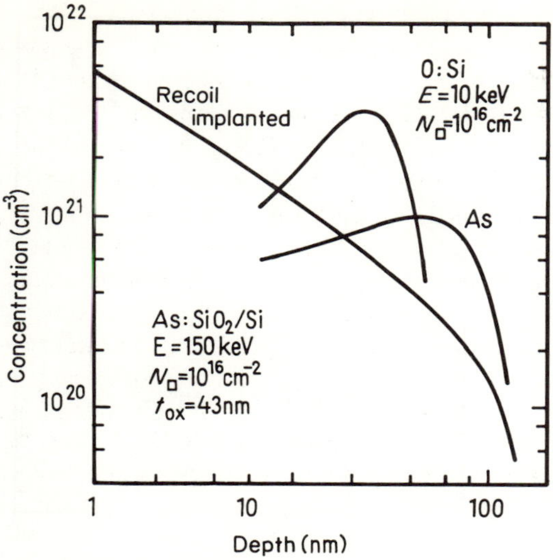

Figure 3.36 Theoretical concentration profiles of recoil-implanted oxygen and arsenic in silicon after implantation of $10^{16}\ cm^{-2}$ arsenic at 150 keV through 43 nm SiO_2. A $10^{16}\ cm^{-2}$ oxygen implantation at 10 keV into bare silicon is shown for comparison [505]. Reproduced by permission of the American Institute of Physics

strate. The number of oxygen atoms affected by collisions with ions, as well as their range, are significantly greater. This distribution can be converted into an oxygen profile in silicon under a SiO_2 layer, if one takes into account that the range of oxygen is approximately 20 per cent higher in silicon than in SiO_2.

At greater depths, an approximately exponential profile was measured. Moline and Cullis [505], using comparative measurements with a transmission-electron microscope, showed that practically the same defects are produced in the case of arsenic implantation through SiO_2 as those in the case of simulated oxygen implantation. Experimental and theoretical work has also shown that the concentration of the secondary implanted ions can be considerable. A calculation for an arsenic implantation at 150 keV with a dose of $10^{16}\ cm^{-2}$, through 43 nm of SiO_2, is shown in Figure 3.36 [505]. An oxygen implantation at 10 keV with the same dose, into bare silicon, is shown for comparison. The completely different shapes of the oxygen profile of the recoiled atoms, and of the Gaussian implantation profile of the arsenic and the oxygen, can be seen clearly. An experimental determination of the quantity of oxygen ions recoil implanted into the silicon substrate by krypton ions at 24 keV and 48 keV, from SiO_2 layers of varying thickness, is shown in Figure 3.37 [510].

The negative effects of oxygen in many semiconductors are well known. Thus,

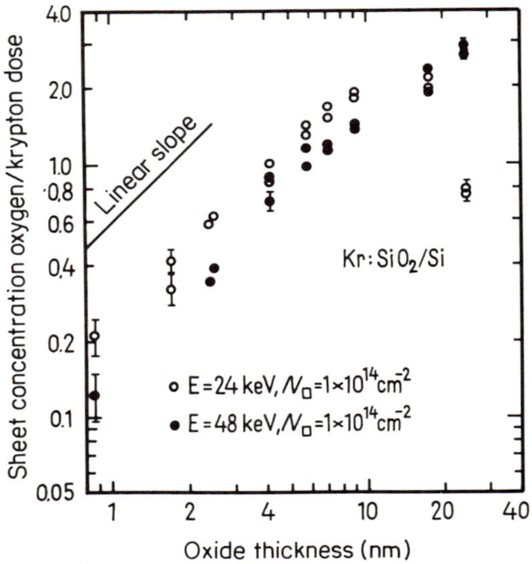

Figure 3.37 Number of ^{18}O atoms recoil-implanted by krypton at 24 and 48 keV from SiO_2 layers of varying thickness into silicon substrate [505]. Reproduced by permission of the American Institute of Physics

for example, GaAs becomes semi-insulating due to oxygen; in silicon, very poor *p–n* functions are obtained. In the case of silicon, such undesired effects can be eliminated by suitable annealing and diffusion steps. If only the quality of the *p–n* junction is important, a drive-in diffusion of the implanted ions a few nanometres into the bulk silicon is adequate. In this way, for example, one can fabricate arsenic emitters of excellent quality [58], [303], [562]. If it is necessary to grow an epitaxial layer on the implanted layer, a drive-in diffusion is no longer sufficient. However, good results can be obtained, for example for arsenic, by a drive-in diffusion in an inert atmosphere, followed by oxidation of the region which was heavily damaged by the secondary implantation of oxygen [508]. Whenever implantation is to be done through masking layers, it is therefore always necessary to estimate or investigate whether the desired passivation effect might be negated, or even reversed, due to an undesired secondary implantation.

Non-semiconductors have been investigated by Perkins [565], who produced Cermet films by secondary implantation of aluminium with argon ions. Figure 3.38 shows the relation which he determined between the argon dose and the sheet concentration of the aluminium secondary-implanted into the film, at an aluminium layer thickness of 50 nm and an argon energy of 50 keV. The aluminium was measured by activation analysis; the substrate was a SiO_x layer ($x = 1.5$) produced by evaporation. Approximately three aluminium atoms were implanted into the substrate per argon ion. Changes in the surface due to

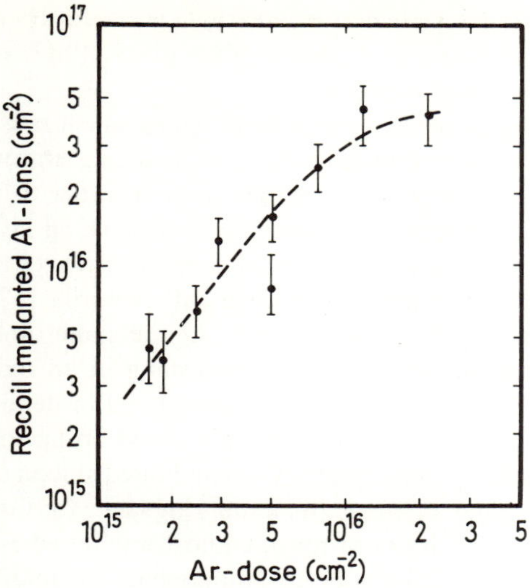

Figure 3.38 Concentration of Al atoms secondary-implanted from a 50 nm thick Al layer into glass, depending on Ar dose at 50 keV [565], as measured by activation analysis. Reproduced by permission of North-Holland Physics Publishing

secondary implantations, for various oxides, have been investigated by Kelly and Sanders [403].

Basic theoretical investigations on the energy and range distribution of recoil-implanted atoms, independent of two-layer structures, have been carried out by Ishitani and Shimuzu with Monte Carlo calculations [372], as well as by Tsurushima and Tanoue [738].

3.7.3 Passivation Against Out-diffusion

In some cases, passivation layers must be used during annealing. Depending on the semiconductor in question, such layers mask against out-diffusion of a semiconductor component (e.g., arsenic in the case of GaAs), or against out-diffusion of implanted ions (e.g., Al in Ge), or both. In the case of the single-element semiconductors silicon and germanium, only the second point is of importance; in cases of compound semiconductors, the covering layers are mostly necessary for both reasons, whereby the first is almost always the more important one. For this purpose, SiO_2, Si_3N_4, and Al_2O_3 layers are often used. Diffusion coefficients of several elements in SiO_2 are given in the Appendix. The diffusion coefficients of the standard doping elements in SiO_2 (except for gallium) are considerably lower than in silicon; therefore, SiO_2 is very well suited as a diffusion mask. Si_3N_4 can be used as a passivating layer for gallium, which is only of minor importance in silicon technology. These layers are also well suited for germanium; in addition, phosphorus-silica glass (PSG) is used for accommodation of the expansion coefficient.

Since silicon is easy to oxidize, and due to the excellent masking effect of SiO_2 against most dopants, one often anneals in an oxidizing atmosphere or performs an oxidizing drive-in diffusion after implantation, in order to avoid out-diffusion of the ions while simultaneously obtaining a new masking layer. This method is very simple, but can result in poor electrical characteristics of the implanted layer, since the radiation damage may not anneal satisfactorily [579]. Details on this have already been discussed in Section 1.3. At the same time, the implantation profile is changed considerably more by annealing in an oxidizing atmosphere than in an inert atmosphere. This will be considered in detail in Section 8.2. A related problem which has received very little attention up to now is that of the different oxidation rates of implanted and unimplanted silicon (cf. Section 6.1.10).

SiO_2 only has a limited suitability as a masking layer for GaAs and GaP since, among other problems, it does not mask against implanted zinc ions (*p*-doping), or against gallium as a semiconductor component. Therefore, sputtered Si_3N_4 layers [219] and AlN are often used instead of SiO_2. In the case of other compound semiconductors (e.g., III–V, II–VI and IV–VI semiconductors, as well as the corresponding ternary compounds), an experimental investigation of the passivation characteristics is necessary in most cases, since data on the diffusion coefficients of dopant substances, or of semiconductor components in masking layers, are practically non-existent. In addition to the above-mentioned SiO_2, Si_3N_4 and AlN, usable layers also include Al_2O_3, SiO_2–Si_3N_4 mixed layers, phosphorus-silica glass, and metal layers such as evaporated aluminium (e.g., for GaAs [647]) or molybdenum, as well as natural oxides (e.g., for GaAs [646]). In addition to the degree of passivation provided by the layers, important factors are their adhesion, especially at higher annealing temperatures, and the ease with which they can be etched without simultaneously etching the semiconductor.

The adhesion in the implanted region is often poor, since a change in the volume can be caused by the ion bombardment, with a subsequent return to the original volume during annealing, thus weakening the adhesion of the passivating layer. A preannealing step at temperatures high enough to reduce radiation damage and volume changes, but at which out-diffusion does not yet take place, can be helpful in this case. The adhesion also depends to a very high degree upon the pretreatment (etching, cleaning) of the wafers.

The procedure needed to obtain suitable masking layers is also highly dependent upon the deposition temperature. Low-temperature processes include evaporation and sputtering using appropriate precautions (cooling), whereby the latter process, under some circumstances, can lead to undesired radiation damage. Pyrolytic layers are deposited at temperatures between 200 and 800 °C. These layers can have different growth rates in implanted and unimplanted regions. Table 3.4 lists some pyrolytic masking layers, together with their deposition temperatures, which are frequently used for III–V semiconductors and other semiconductors which easily dissociate. A further possibility for the production of

Table 3.4 Dielectric passivation and masking layers

Layer	Substrate temperature (°C)	Process	Growth rate (nm/min)	Remarks	References
SiO_2	280–480	Pyrolytic	20–250	From silane	
	900–1200	Pyrolytic	10–100	From $SiCl_4$, SiF_4, etc.	[109], [142]
	RT–500	Sputtered		Quality dependent on substrate temperature	[331], [332]
PSG	350–400 900–1150	Pyrolytic	100–400	Phosphorus silica glass	[524], [627], [723]
BSG	400–450	Pyrolytic	100–400	Boron silica glass	[627], [723]
Si_3N_4	700–1200	Pyrolytic	1–100	At 800 °C	[65], [142], [143], [152], [331]
	260–350	Pyrolytic	4–18	With glow discharge [589]	[421], [589]
	RT–250	Sputtered	1–12		[407]
$Si_xO_yN_z$	800–1000	Pyrolytic	3–70		[109]
Al_2O_3	200–450	Pyrolytic	6		[31], [142]
	RT–350	Electron beam evaporation	10–300		[152], [524]
AlN	800–1200	Pyrolytic	9–16		[140]

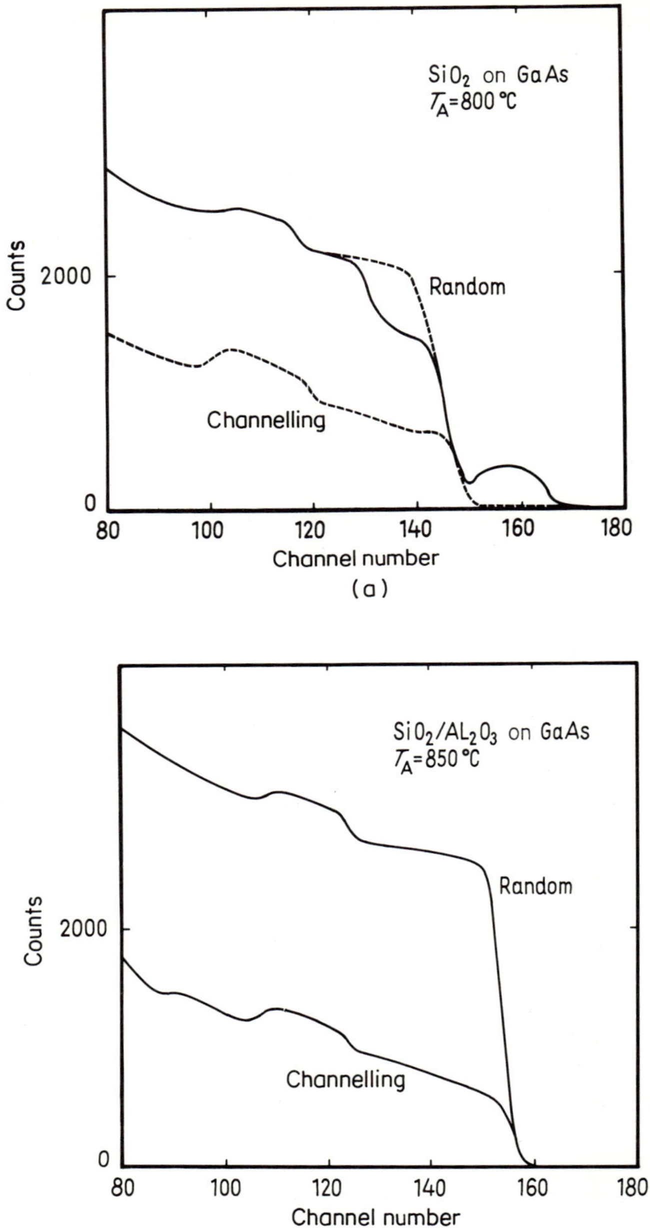

Figure 3.39 Comparison of passivation provided by SiO_2 (a) and SiO_2/Al_2O_3 (b) to reduce GaAs decomposition during annealing at 800 or 850 °C

masking layers is the use of silica films (obtainable from Emulsitone, Merck, etc.), which are applied by spinning as is done with photoresist.

The most important method for the investigation of masking effects is the backscattering technique. For example, in Figure 3.39 one observes a difference in the out-diffusion from GaAs, masked either with SiO_2 or with a mixed layer of Al_2O_3 and SiO_2 (about 1:1). For annealing above 800 °C, an out-diffusion of gallium into the SiO_2 is seen, while the Al_2O_3–SiO_2 layer prevents dissociation of the semiconductor up to a temperature of 950 °C. If changes in the carrier concentration of the semiconductor beneath the masking layer are to be investigated, capacitance-voltage measurements are especially suitable.

3.8 SPUTTERING DURING IMPLANTATION

An effect which is often neglected in the context of ion implantation is the sputtering of substrate material by the implanted ions. This effect can be especially noticeable in cases of heavy ions and high implantation doses, and can lead to alteration of the implantation profile up to the profile saturation due to the achievement of equilibrium between implanted and sputtered ions. The most important parameter in dealing with this problem is the sputtering yield S, which represents the quantity of atoms displaced by an implanted ion.

3.8.1 Sputtering Yield

At sufficiently high energy, the sputtering yield is proportional to the energy deposited at the surface of the sample by impact cascades, and is given according to Sigmund [683], [684], for the case of vertical incidence of the ions, by:

$$S = \frac{3}{4} \frac{S_n(E)\alpha(M_2/M_1)}{\pi^2 C_0 U_0} \tag{3.18}$$

C_0 is a constant ($C_0 = 1/2\pi\lambda_0 a^2$; $\lambda_0 = 24$; $a = 0.0219$ nm); $\alpha(M_2/M_1)$ is a numerically calculable function of the relationship of the ion mass M_1 to the mass of the target atoms M_2; U_0 is the surface binding energy (7.81 eV for silicon [683]). The function $\alpha(M_2/M_1)$ is shown in Figure 3.40 for two different power potentials. Though the influence of the power is relatively small, the mass ratio has a very significant effect on the sputtering yield. Theoretical sputtering yield curves, for bombardment of aluminium with various ions, are shown in Figure 3.41.

According to a further theory on ion sputtering by Thomson [726], the sputtering yield can be calculated as follows:

$$S = \frac{\pi^2 a_0^2 N^{2/3}}{8q} \frac{E_R}{E_B} \frac{M_1 (Z_1 Z_2)^{5/6}}{(M_1 + M_2)} \frac{1}{\cos\theta} \tag{3.18a}$$

E_R is the Rydberg energy, E_B the surface binding energy, N the atomic density,

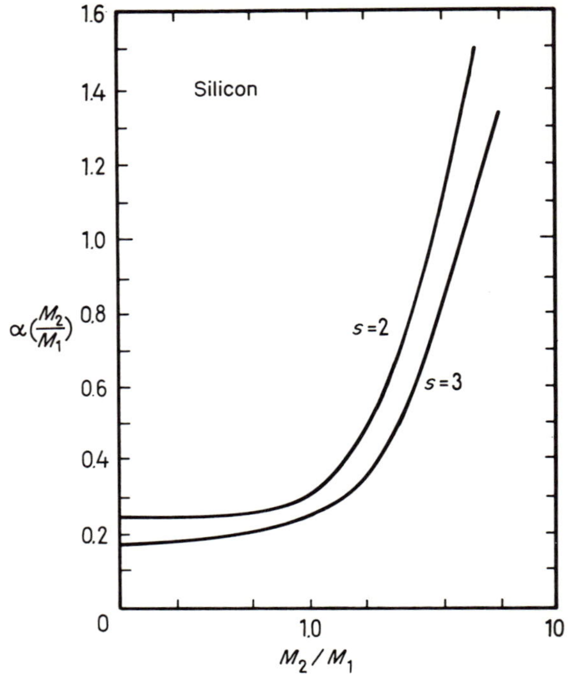

Figure 3.40 Values of α (M_2/M_1) for power potentials with $s = 2$ and $s = 3$, according to Reference [683]. Reproduced by permission of the American Physical Society

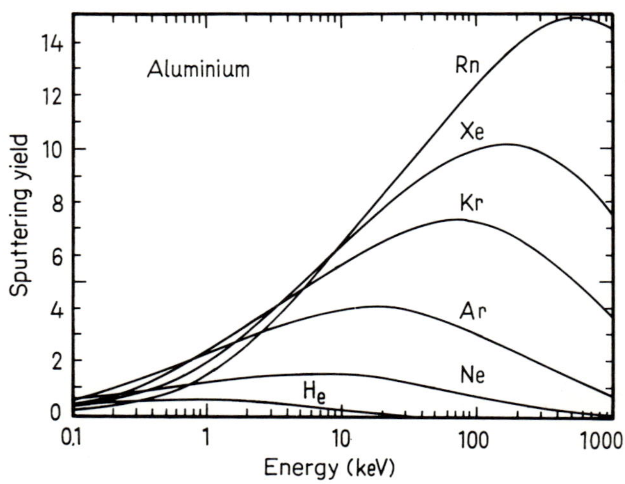

Figure 3.41 Sputtering yields of He, Ne, Ar, Kr, Xe and Rn ions in aluminium as a function of energy [683]. Reproduced by permission of the American Physical Society

Table 3.5 Sputtering yield for several metals [146]

Ions:	He			Ne				Ar					Kr	Xe	Hg					
Energy: (keV)	0.4	2	5	0.4	0.6	1	5	0.4	0.6	1	3	5	0.4	0.8	0.2	0.4	0.8	2	3	10
Tungsten				0.34	0.32				0.62						0.3	0.54	1.3	2.0	2.75	3.0
Molybdenum				0.44	0.54				0.93						0.1	0.44	1.0	2.0	2.75	5.0
Platinum					0.7				1.56						0.5					10.0
Aluminium				0.68	0.83				1.24						0.14	0.45		1.5		
Silver		0.36	0.48	1.6	1.98	2.3	5.5		3.4	5.1	7.0	9.2			0.85					22.5
Germanium	0.08			0.63	0.82			0.92	1.27				0.91	1.41	0.21	0.63	1.23			
Nickel					1.34				1.52						0.5		1.4		3.6	7.0
Iron				0.76	0.97				1.26						0.13	0.5	1.1	2.0	2.75	4.5
Copper				1.53											0.54					12.0
Gold				1.0											1.0			8.0		21.0

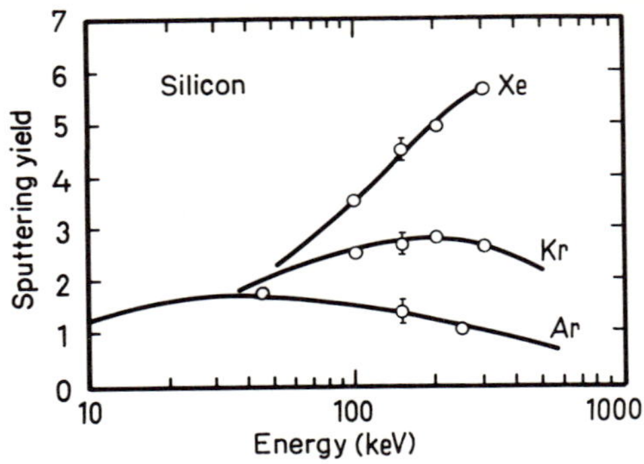

Figure 3.42 Sputtering yields of Ar, Kr and Xe in silicon depending on energy. The Ar curve is based on the theory of Sigmund [683]; the Ar point at 45 keV is according to Reference [39]

and θ the incidence angle of the particles. This dependence is valid for $\theta < 60°$ if $M_2 \gg M_1$. Thomson has made use of values for E_B taken from Honig [345].

Although experimental sputtering rates in the case of metals are known for many ion species (several values are given in Table 3.5 [146]), only a small amount of data has been published up to now for semiconductors (with the exception of silicon). EerNisse has measured the sputtering yields of silicon, depending on the energy, for argon, krypton and xenon [213]. His results are shown in Figure 3.42. The typical shape indicating a flat maximum of the yield is observable, which is shifted to higher energies with increasing mass of the implanted ions. Experiments regarding the sputtering yields of silicon for various ion species have been carried out by Andersen *et al.* [39]. Their measurements were made at an energy of 45 keV, and are shown in Figure 3.43, together with a theoretical curve according to Sigmund [683].

The thickness of the removed layer is calculated from the sputtering yield S, at a given ion dose $N_\square$, as follows:

$$t = \frac{S}{N} N_\square \tag{3.19}$$

Table 3.6 Removal of silicon by sputtering with 45 keV; S according to Reference [39]

		Dose (cm^{-2})	
Ion	10^{15}	10^{16}	10^{17}
B	6×10^{-2} nm	0.6 nm	6 nm
Ar	0.35 nm	3.5 nm	35 nm
As	0.6 nm	6 nm	60 nm
Sb	0.78 nm	7.8 nm	78 nm

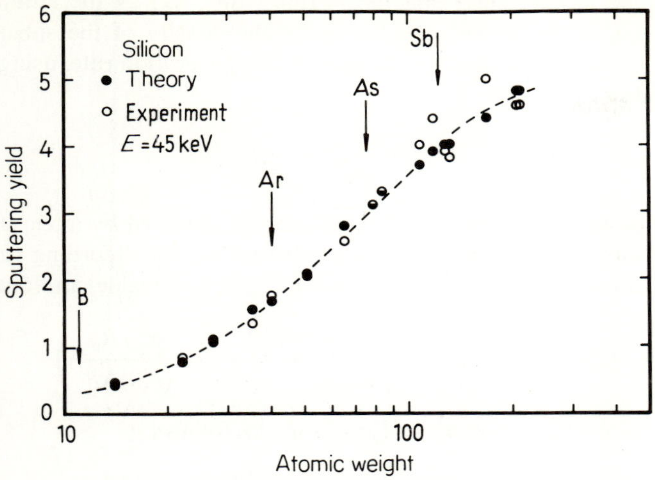

Figure 3.43 Sputtering yields of ions having different masses at 45 keV in silicon [39]. The theoretical values are according to Sigmund [683]

Table 3.6 lists removed layer thicknesses for doses between 10^{15} cm^{-2} and 10^{17} cm^{-2} at 45 keV, for several frequently implanted ion species in silicon, according to Figure 3.43 and equation (3.19). Depending on the ion species, saturation can already be reached with implantation doses of 10^{16} cm^{-2} in the case of GaAs [729].

3.8.2 Profile Changes Due to Ion Sputtering

The modification of the implantation profile due to sputtering can be calculated if a few simplifying assumptions are made: (a) the sputtering yield is constant, and is the same for the substrate and the implanted ions; (b) there is no knock-on effect; (c) the volume change due to radiation damage can be neglected; and (d) the implantation of ions results in a Gaussian profile.

The number of ions implanted per unit time (generation rate) is given (cf. equation (2.23)) as follows:

$$g(x, t) = \frac{j}{q\sqrt{(2\pi)}\Delta R_{\mathrm{p}}} \exp\left[-\frac{(x - R_{\mathrm{p}})^2}{2\Delta R_{\mathrm{p}}^2}\right] \tag{3.20}$$

The implantation profile can be calculated, if ion sputtering is not taken into account, by integrating equation (3.20). If the semiconductor surface is eroded due to sputtering, then the origin of the x coordinate changes its position during implantation. The following is valid for the removal rate:

$$v = m_{\mathrm{a}} \cdot \frac{AS}{Q}\frac{j}{\rho} = \frac{S}{N}\frac{j}{q} \tag{3.21}$$

where m_a is the atomic mass unit ($m_a = 1.66 \cdot 10^{-27}$ kg), N the atomic density ($N = \rho/M \cdot m_a$), A the atomic weight, and ρ the density of the substrate. The implantation profile is obtained by integrating the generation rate, using equation (3.21), as follows:

$$N(x) = \int_0^t g(x + vt', t')\,dt' \tag{3.22}$$

The number of atoms remaining in the substrate is obtained by integrating equation (3.22) from 0 to ∞. For a Gaussian generation rate according to equation (3.20) with constant current density, equation (3.22) is solvable, and one obtains:

$$N(x) = \frac{N}{2S}\left(\operatorname{erf}\frac{x - R_p + N_\square(S/N)}{\sqrt{(2)}\Delta R_p} - \operatorname{erf}\frac{x - R_p}{\sqrt{(2)}\Delta R_p}\right) \tag{3.23}$$

The saturation profile is obtained, for $t \to \infty$, as follows:

$$N(x) = \frac{N}{2S}\operatorname{erfc}\frac{x - R_p}{\sqrt{(2)}\Delta R_p} \tag{3.23a}$$

i.e., the maximum concentration, contrary to the implantation profile, is at the surface, and is given by:

$$N_{max} = \frac{N}{2S}\operatorname{erfc}\left(\frac{-R_p}{\sqrt{(2)}\Delta R_p}\right) \approx N/S \text{ for } R_p \geqslant 3\Delta R_p \tag{3.24}$$

This maximum concentration is independent of the implantation dose, and depends mainly on the relationship between the atomic density of the substrate and the sputtering yield, since the value of the complementary error function in equation (3.24) is between 1 and 2. In silicon, this maximum concentration for arsenic, which is relatively heavy ($S \approx 3$ at 45 keV), is found at approximately 2×10^{22} cm^{-3}. However, in the case of semiconductors with higher sputtering yields (e.g. GaAs), the effect can be considerable, and can lead to significantly lower saturation values.

In Figure 3.44, profiles according to equation (3.22) are given for different implantation doses, in the cases of Si and GaAs. In the first case, a value of 3 has been chosen for S; in the second, a value of 15. Although sputtering in the case of silicon can only be observed at very high doses, this effect already takes place for GaAs at a dose of 10^{15} cm^{-2}. Carter *et al.* |119| have measured values of S as high as 40 for krypton implantations into GaAs.

The maximum achievable sheet concentration is obtained, by integrating equation (3.23a), as follows:

$$N_{S,max} = R_p \cdot N_{max} - \frac{N\Delta R_p}{S\sqrt{(2\pi)}}\exp\left[-\frac{R_p^2}{2\Delta R_p^2}\right] \tag{3.25}$$

If no Gaussian profiles are present, the integration must generally be done

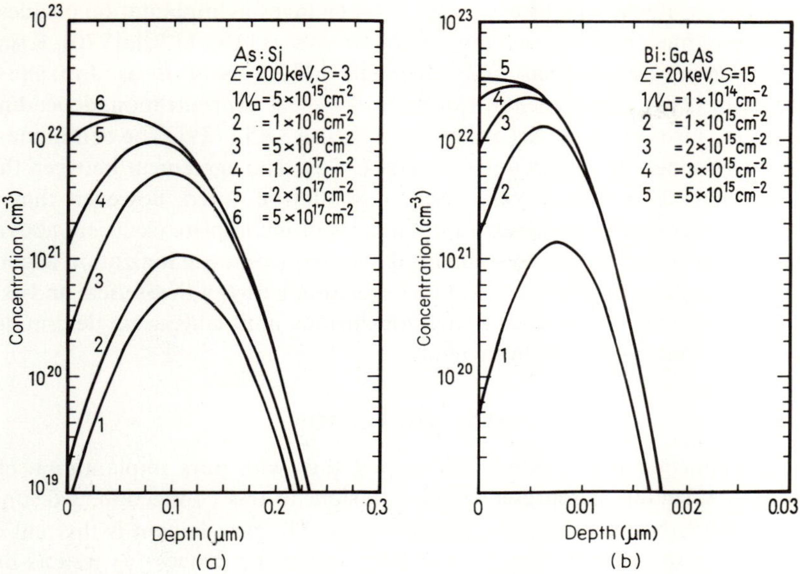

Figure 3.44 Doping profiles depending on implantation dose and sputtering yield for As in silicon ($E = 200$ keV; $s = 3$) and Bi in GaAs ($E = 20$ keV; $s = 15$)

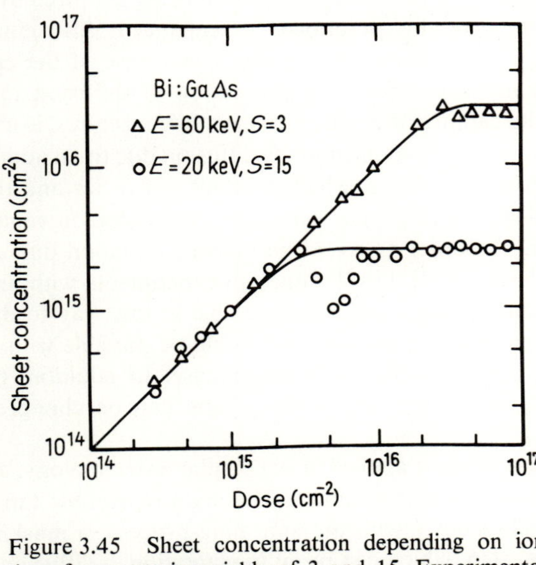

Figure 3.45 Sheet concentration depending on ion dose for sputtering yields of 3 and 15. Experimental data according to Reference [729]

numerically. Calculations or experiments on changes in implantation profiles due to sputtering have been carried out by Carter *et al.* [119], [120], [770], Krimmel and Pfleiderer [419], Tsai and Morabito [732], as well as Fritsche and Rothemund [261]. Experimental values for the sheet concentration, depending on the dose for bismuth in GaAs, are given in Figure 3.45 [729]. The continuous line has been obtained by integrating equation (3.23). The agreement between theory and experimental results is very good; it should be noted, however, that such results are also partly explainable on the basis of incomplete electrical activation at high concentrations, due to solubility limits or incomplete ionization according to Fermi statistics. In other words, the correlation between theoretical and experimental curves cannot yet lead to any conclusions until a separate determination of the sputtering yield has been made.

3.9 DIFFUSION

A very important effect, which occurs together with pure implantation effects such as channelling, is represented by diffusion. In this connection, a distinction must be made between several different types. The best known is thermal diffusion, which can occur during the necessary annealing process as well as during implantation, if the semiconductor sample is heated by the beam to high temperatures. Many vacancies are produced by the ion bombardment, which can lead to diffusion during implantation even at relatively low temperatures, since many doping elements in semiconductors diffuse via vacancies. Interstitial diffusion can also play a significant role, since some of the implanted ions come to rest on irregular positions in the lattice, and can diffuse quickly on interstitial lattice sites before being captured by a vacancy. The latter effects are generally described as enhanced diffusion, without one's generally having a clear knowledge in any given case of the exact mechanism involved. As an opposite effect, the prevention of diffusion through radiation damage, due to the capture of doping atoms at defect centres, is also a possibility.

A further effect which bears mention is diffusion due to an additional bombardment with non-doping particles, which produce vacancies and thus increase the diffusion coefficient. In this case, one refers to 'radiation-enhanced diffusion', abbreviated RED. In order to avoid permanent radiation damage, protons are generally used [532], [608], [547], although experiments with argon and helium have also been carried out [735]. It is possible in this way to diffuse implanted distributions deeper into the crystal than would be possible with direct implantation, due to the limited range in the latter case. In addition, by choosing the proton energy appropriately, the profile shape can be changed within certain boundaries.

When ion implantation is applied in silicon planar technology, annealing is very often done in an oxidizing atmosphere. The main reason for this is that one can thus obtain, simultaneously with the annealing process, a masking layer for the next technological step, which can be an implantation, a diffusion, or a metallization. As a result of annealing in an oxidizing atmosphere, the implantation profile is changed considerably more than when annealing in an inert atmosphere, regardless of whether or not a passivation layer is used, due to segregation at the SiO_2–Si interface.

The following sections will deal in more detail with thermal diffusion, diffusion during annealing in an oxidizing atmosphere, and radiation-enhanced diffusion: several other effects which can lead to increased diffusion coefficients will also be touched upon briefly.

3.9.1 Thermal Diffusion

Following implantation, the implanted ions are usually not electrically active, or are only active to a small extent. Their electrical activation must be attained by an appropriate annealing process. The required temperatures, in the case of silicon, can range up to 1000 °C, and can be considerably higher for other semiconductors (e.g. silicon carbide).

Constant Diffusion Coefficient

During annealing (usual times are between 10 minutes and 1 hour), thermal diffusion can occur. The theoretical treatment of this problem is very simple if one assumes a semiconductor extending from $-\infty$ to $+\infty$. The diffusion equation (Fick's second law), in the one dimensional case, is written as:

$$\frac{\partial N}{\partial t} = D\frac{\partial^2 N}{\partial x^2} \tag{3.26}$$

where N is the concentration of the diffusing material and D is the diffusion coefficient.

For Gaussian ion distributions according to the LSS theory (equations (2.23) and (2.25)), one obtains $N(x)$ as a special solution of this differential equation as follows:

$$N(x) = \frac{N_{max}}{\sqrt{\left(1 + \dfrac{2Dt}{\Delta R_p^2}\right)}} \exp\left[-\frac{(R_p - x)^2}{2\Delta R_p^2 + 4Dt}\right] \tag{3.27}$$

where t is the annealing time.

If the influence of the surface on the diffusion is also taken into account, an analytical solution is more difficult, but is none the less possible.

With the boundary condition $\partial N(x, t)/\partial x|_{x=0} = 0$, i.e. no out-diffusion at the surface, one obtains for the general solution:

$$N(x, t) = \frac{1}{\sqrt{(4\pi Dt)}} \int_0^{\infty} f(x') \left\{\exp\left[-\frac{(x' - x)^2}{4Dt}\right] + \exp\left[-\frac{(x' + x)^2}{4Dt}\right]\right\} dx' \tag{3.28}$$

where $f(x)$ is the ion distribution for $t = 0$. If in this case a Gaussian distribution is again used, one obtains [606]:

$$N(x, t) = \frac{N_{max}/2}{\sqrt{\left(1 + \frac{2Dt}{\Delta R_p^2}\right)}} \left\{ \exp\left[-\frac{(x - R_p)^2}{2\Delta R_p^2 + 4Dt}\right] \times \left[1 + \operatorname{erf} \frac{\frac{R_p\sqrt{(4Dt)}}{\sqrt{(2)}\Delta R_p} + \frac{x\sqrt{(2)}\Delta R_p}{\sqrt{(4Dt)}}}{\sqrt{(2\Delta R_p^2 + 4Dt)}}\right] + \exp\left[-\frac{(x + R_p)^2}{2\Delta R_p^2 + 4Dt}\right] \left[1 + \operatorname{erf} \frac{\frac{R_p\sqrt{(4Dt)}}{\sqrt{(2)}\Delta R_p} - \frac{x\sqrt{(2)}\Delta R_p}{\sqrt{(4Dt)}}}{\sqrt{(2\Delta R_p^2 + 4Dt)}}\right] \right\} \quad (3.29)$$

For $t = 0$, this solution becomes the Gaussian implantation profile. In Figure 3.46, diffusion profiles calculated according to equation (3.29) are shown, in the case of a 40 keV boron implantation into silicon with an annealing time of 10 min at 800, 900 and 1000 °C. The dashed lines indicate profiles according to equation (3.27). It can be seen that there is not much error in the profiles towards the bulk, but there can be a considerable error towards the surface. For the example shown in Figure 3.46, the error amounts to 96 per cent after annealing at 900 °C for 10 min, and is 100 per cent for 1000 °C and 10 min. A general expression for the error at the surface is given by:

$$\eta = \operatorname{erf} \frac{R_p\sqrt{(4Dt)}}{\sqrt{(2)}\Delta R_p \cdot \sqrt{(2\Delta R_p^2 + 4Dt)}} \quad (3.30)$$

Diffusion during annealing always plays a role when $\sqrt{(2Dt)}$ is of the same order of magnitude as ΔR_p or greater. For estimation purposes, diffusion-coefficient values for several elements in silicon are given in Table 3.7. A more comprehensive table of diffusion coefficients may be found in the Appendix.

The condition $\partial N(x, t)/\partial x|_{x=0} = 0$ is only valid for dopants with a negligible vapour pressure at the corresponding annealing temperature. If evaporation of the implanted ion species is to be expected, the condition $N(0, t) = 0$ provides a closer approximation to the practical case. Under this condition, the general solution is given by:

$$N(x, t) = \frac{1}{\sqrt{(4\pi Dt)}} \int_0^\infty f(x') \left\{ \exp\left[-\frac{(x' - x)^2}{4Dt}\right] - \exp\left[-\frac{(x' + x)^2}{4Dt}\right] \right\} dx' \quad (3.31)$$

Assuming a Gaussian profile, one obtains a solution similar to equation (3.29),

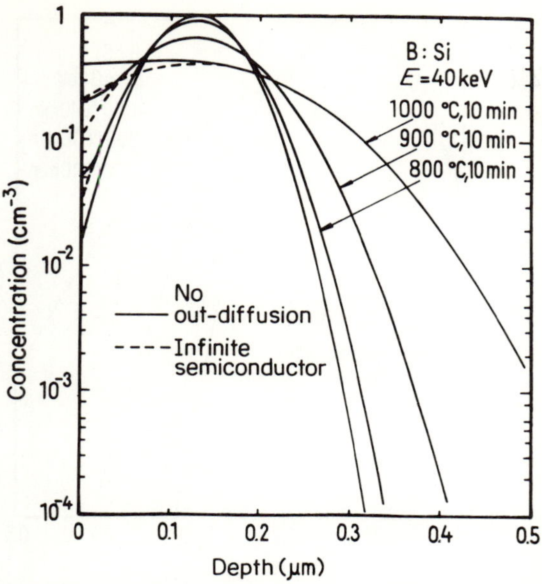

Figure 3.46 Alteration of an implantation profile due to thermal diffusion, normalized to the maximum concentration according to equation (3.29). $R_p = 0.13$ μm; $\Delta R_p = 0.044$ μm (40 keV boron into silicon). The diffusion coefficients were chosen according to Table 3.7; the diffusion time is 10 min in all cases. For comparison, profiles for an ∞-extending semiconductor according to equation (3.27) are shown as dashed curves

the only difference being that a minus appears between the two expressions in the braces. In Figure 3.47, calculations are shown with the same parameters as in Figure 3.46, and are compared with the result of equation (3.29). Between the two related curves the profile must appear, if other effects (non-constant diffusion coefficient, no Gaussian profile) do not play a role. The treatment of diffusion if a masking layer (thermal or pyrolytic oxide) is present cannot be represented by comparably simple expressions; it depends on the segregation coefficient between the oxide and the semiconductor as well as the diffusion coefficients in both. The

Table 3.7 Diffusion coefficients of several elements in silicon; $D = D_0 \exp(-E_a/kT)$

Element	D_0 (cm^2/s)	E_a (eV)	D (800 °C) ($\times 10^{-16}$ cm^2/s)	D (900 °C) ($\times 10^{-16}$ cm^2/s)	D (1000 °C) ($\times 10^{-16}$ cm^2/s)	Reference
Al	8	3.47	4.1	100	1500	[265]
As	24	4.08	0.017	0.72	17.1	[137]
B	0.065	3.19	0.69	12.98	155	
P	10.5	3.69	0.5	15	200	[265]
Sb	5.6	3.95	0.016	0.61	13	[265]

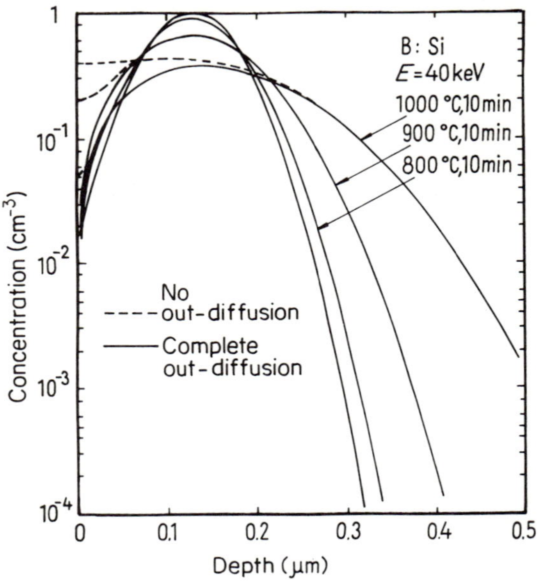

Figure 3.47 Thermal diffusion with out-diffusion using the same parameters as in Figure 3.46, compared to the case of diffusion without out-diffusion according to equation (3.29)

relationship of the concentration in the masking layer (oxide, nitride), N_1, to that in the semiconductor at the interface, N_2, is referred to as the segregation coefficient m, i.e.:

$$m = \frac{N_1(0, t)}{N_2(0, t)} \tag{3.32}$$

An analytical solution is possible for application of a masking layer before annealing, assuming an extension of the masking layer from $-\infty$ to 0 (the semiconductor extends from 0 to $+\infty$) [204]. In the general case, for a finite thickness of the masking layer, numerical solutions are necessary. For implantation through a masking layer followed by diffusion, a solution has been found by Perloff [566] on the same assumption. In the case of a finite masking-layer thickness, only numerical solutions are possible. As a good approximation for both cases, however, equation (3.27) can be used in selecting appropriate range parameters. This is especially relevant because, at present, too little is known about the parameters which are important for diffusion in the temperature range used for annealing of implanted layers. A theoretical example for a boron implantation into silicon through 130 nm of SiO_2, with a segregation coefficient of 3 and a difference of 10^2 between the diffusion coefficients in the masking layer and the substrate, is shown in Figure 3.48. In silicon, the segregation coefficient

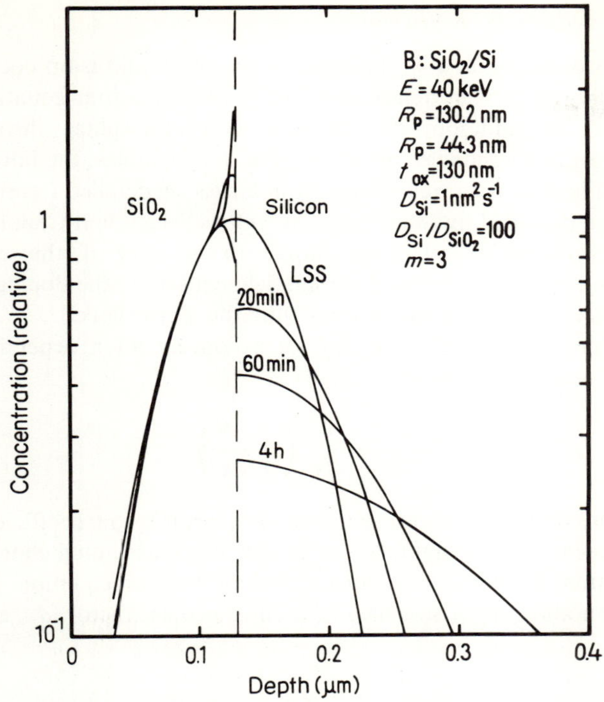

Figure 3.48 Diffusion in an SiO_2–Si structure after implantation through 130 nm SiO_2 at 40 keV; $D_{Si}/D_{SiO_2} = 100$; segregation coefficient 3

ranges between about 100 (boron) and 10^{-2} (antimony) for the usual dopants. Segregation-coefficient values are given in Table 3.8.

The case of annealing in an oxidizing atmosphere will be discussed in the following section.

Table 3.8 Segregation coefficients at the SiO_2–Si interface

Doping	Thermodynamic estimation [727]	Experimental values	Orientation	Reference
B	10^{-3}–10^{3}	100		[307a]
		3.3		[306]
		$2.33 \times 10^{-4} \exp(1.135/kT)$	⟨111⟩	[576]
		$0.05 \exp(0.52/kT)$	⟨111⟩	[145]
		$0.03 \exp(0.52/kT)$	⟨100⟩	[145]
Ga	$<10^{-3}$	20		[307a]
In	$<10^{-3}$	0.1		[306]
P	$<10^{-3}$	0.1		[306]
Sb	$<10^{-3}$	0.1		[306]
As	$<10^{-3}$	0.1		[306]

Non-constant Diffusion Coefficient

For a number of dopants, a strong dependence of the diffusion coefficient upon the concentration has been determined |457|, |458|, so that equation (3.26) no longer applies. In addition, the diffusion can take place through various mechanisms, e.g., by vacancies or interstitials. As examples, the latter is the case for In and Ga in silicon as well as for Zn in GaAs. A detailed treatment of these topics has been provided by Shaw |665| as well as Seeger and Chik |648|, among others. The problem is relatively simple to describe if the concentration dependence is due to the internal electrical field caused by the doping gradient, or due to the vacancy concentration (and thus the Fermi level).

The diffusion equation, for the case of a concentration dependent diffusion coefficient, is written as follows:

$$\frac{\partial N}{\partial t} = \frac{\partial}{\partial x}\left(D\frac{\partial N}{\partial x}\right) \tag{3.33}$$

where N is the dopant concentration and D the diffusion coefficient.

The influence of the internal field can be calculated assuming charge neutrality, from the equation for the particle current and the Poisson equation. The diffusion coefficient is modified by a so-called field enhancement factor, d_F, and is written as follows |431|:

$$D(N) = D_i d_F = D_i\left(1 + \frac{N}{\sqrt{(N^2 + 4n_i^2)}}\right) \tag{3.34}$$

where D_i is the intrinsic diffusion coefficient at low doping concentrations.

In this calculation, influences of semiconductor degeneracy, as well as the possibility of incomplete activation of the implanted ions at high concentrations, have not yet been taken into account. The influence of the field on a theoretical Gaussian arsenic distribution, applying equation (3.34), is shown in Figure 3.49. The much steeper shape of the profile is clearly seen, compared to Figure 3.46.

The influence of the vacancy concentration, C_V, can be calculated if one assumes that the diffusion coefficient is proportional to the vacancy concentration. Shaw [664] and Fair [233a] propose a different diffusion coefficient for every charge state of the vacancies. For the intrinsic diffusion coefficient, the following applies:

$$D_i = D_i^0 + D_i^- + D_i^= + D_i^+ \tag{3.35}$$

where the index 0 refers to the diffusion coefficient for diffusion via neutral, – single-negative, = double-negative, and + positive-charged vacancies. Since the extrinsic vacancy concentrations can be calculated by:

$$C_{V^-} = C_{V,i}^{=}\frac{n}{n_i} \tag{3.36}$$

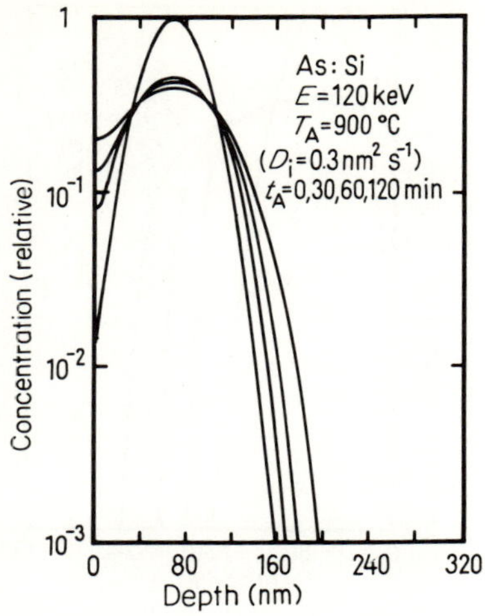

Figure 3.49 Influence of internal electric field on diffusion. $R_p = 69$ nm; $\Delta R_p = 24$ nm (120 keV arsenic in silicon), no out-diffusion, dose 10^{16} cm^{-2}

$$C_{V^=} = C_{V^=,i} \left(\frac{n}{n_i}\right)^2 \tag{3.37}$$

$$C_{V^+} = C_{V^+} \frac{p}{n_i} \tag{3.38}$$

the extrinsic diffusion coefficient is given by:

$$D = D_i^0 + D_i^- \frac{n}{n_i} + D_i^= \left(\frac{n}{n_i}\right)^2 + D_i^+ \frac{p}{n_i} \tag{3.35a}$$

Using the intrinsic diffusion coefficient D_i (which describes the diffusion for $N \ll n_i$) to normalize different vacancy dependent diffusion coefficients, and assuming only the neutral and singly-negative-charged states to be of any importance for the diffusion, one obtains for donors [41f], [352a]:

$$D = D_i \frac{1 + \beta(n/n_i)}{1 + \beta} \tag{3.39}$$

where β is an unknown parameter which in theory is given by $\beta = D_i^- / D_i^0$, this being the relationship of the diffusion coefficient for diffusion via negatively-charged vacancies in the doped case to that in the intrinsic case.

For acceptors, n/n_i in (3.39) has to be replaced by p/n_i, assuming diffusion via neutral and singly-positive-charged vacancies.

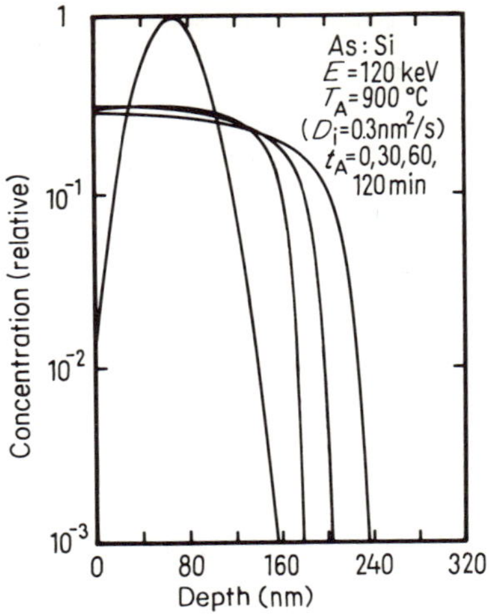

Figure 3.50 Influence of vacancy concentration on diffusion, using the same parameters as in Figure 3.49 ($R_p = 69$ nm, $\Delta R_p = 24$ nm, 120 keV arsenic into silicon, dose: 10^{16} cm^{-2})

In Figure 3.50, this influence of the vacancy concentration on the diffusion coefficient is shown for comparison, using the same diffusion times as in Figure 3.49. It can be seen clearly that the influence of the vacancies is considerably greater that that of the electrical field. Taking into consideration the influences of the field and the vacancies, the diffusion coefficient is obtained as follows:

$$D = D_i d_F d_V \tag{3.40}$$

A further improvement of the model can be obtained if an incomplete electrical activation of the implanted ions at high concentrations |137|, |235|, |351|, as well as the formation of immobile complexes |235|, are taken into consideration.

In Figures 3.51 and 3.52, measured and calculated profiles are compared, with consideration given to the effects discussed above; here, the influences of immobile arsenic complexes and incomplete activation are also taken into account. The correlation between experiment and calculation is quite good, in spite of the relatively simple model. Similar studies have been carried out by van Overstraeten |383|, as well as by research groups at Texas Instruments |633|, IBM |137|, |351|, Stanford |41e|, |41f|, and Bell |232|, |233|, |234|. Such research has become important nowadays as device structures, along with their required tolerances, are continuing to shrink; in the past, on the other hand, the depth of the *p–n* junction alone provided adequate information. The use of ion implantation with its manifold possibilities has made it necessary to consider anew such elementary processes as diffusion, which had long been regarded as fully investigated.

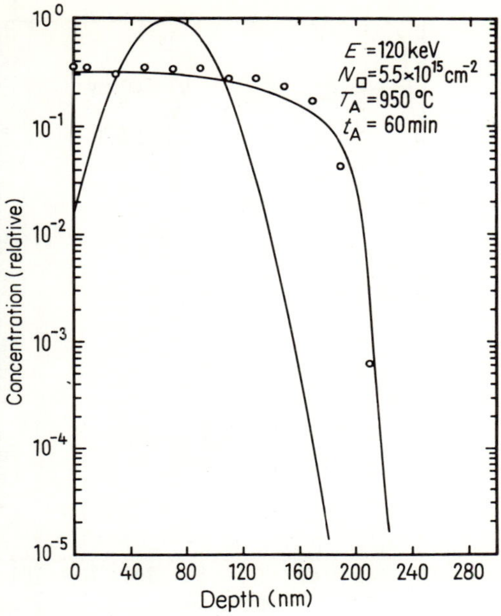

Figure 3.51 Comparison between a measured arsenic doping profile in silicon ($E = 120$ keV, $N_\square = 5.5 \times 10^{15}$ cm^{-2}) and a theoretical profile |628|

Interstitial Diffusion

Interstitial diffusion occurs in the case of atoms with a small atomic radius, for example, gold or copper in silicon. For a treatment of this problem, atoms on regular lattice sites and on interstitial sites, as well as the diffusion of vacancies, must be considered. A detailed discussion for gold in silicon has been provided by Schulz *et al.* |635|, and for GaAs by Zölch |796|. Very deep profile tails after diffusion and implantation are often attributed to this interstitial diffusion |175|, |178|, |186|, |452|. Davies *et al.* |178| have hypothesized ions which are completely stopped within channels, then diffuse via interstitial sites into the crystal, and are finally trapped at vacancies, dislocations, or complex defects. A precondition for a mechanism of this kind is that the ion does not produce any defects while being stopped in the channel. According to Anderson |40|, however, this is only valid for $Z_1 < Z_2$. A further explanation by Bourgoin and Corbett |96| is based on the increase of the diffusion coefficient due to ionization during the ion bombardment. Dearnaley *et al.* |86|, |184| have demonstrated that, in the case of phosphorus in silicon, these profile tails are attributable to the channelling of ions which are scattered into channels while slowing down and are not due to an enhanced diffusion. It is probable that this also applies in many other cases. Interpretations of these so-called 'supertails' are often based on measurement

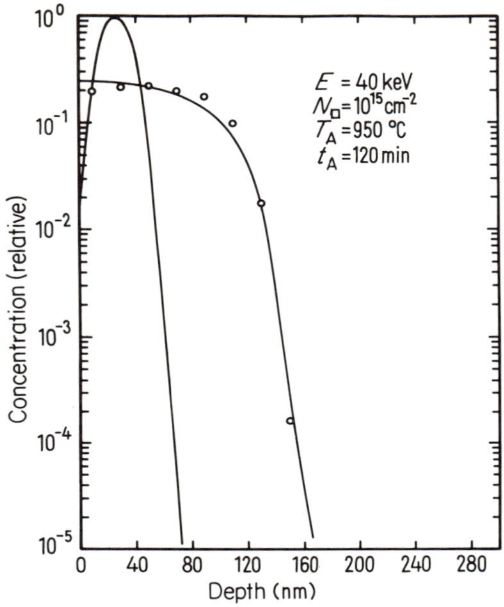

Figure 3.52 Comparison between a measured arsenic doping profile in silicon ($E = 40$ keV, $N_\square = 10^{15}$ cm^{-2}) and a theoretical profile [628]

errors made during electrical profile measurements in semiconductors, as shown by Kennedy [405] (see also Chapter 5).

3.9.2 Annealing in an Oxidizing Atmosphere

Implanted layers are often annealed in an oxidizing atmosphere in silicon planar technology, especially when a masking layer for the next technological step is to be produced simultaneously with the annealing, or when the use of implantation for predeposition makes a subsequent conventional diffusion step necessary.

Although it is possible to oxidize at temperatures so low that no diffusion of the implanted ions takes place (e.g., wet oxidation at high pressure [575]), the processes well known in planar technology are used in practically every case, i.e., the oxidation temperatures are between about 900 and 1100 °C. A pronounced diffusion can already occur at these temperatures. The oxide thicknesses are mostly between 10 nm and 30 nm, which means that about 5 nm to 14 nm of the implanted silicon are consumed. The possible undesired effects which can occur as a result of annealing in an oxidizing atmosphere have been mentioned in the first section of this chapter. Therefore, only the influence of oxidation on the implantation profile will be discussed here. Depending on the segregation coefficient and the diffusion coefficient in oxide and silicon, different doping profiles are obtained. If the segregation coefficient is larger than one, there occurs a depletion of the dopant at the semiconductor surface; in the opposite case, an enrichment (or pile-up, the 'snow-plough effect' [519]) generally occurs. The segregation coefficients of the most important elements, for the SiO_2–Si system, are listed in Table 3.8. Boron is the only element which has a segregation coefficient larger than one. Detailed studies of segregation have only been made for the case of boron; these were carried out by Colby and Katz [145], who examined the temperature and orientation dependencies of this effect. In addition to a knowledge of the diffusion coefficient and the segregation coefficient, knowing the oxidation behaviour of silicon is also important for a mathematical treatment of this problem. In the case of thin oxide layers (⩾20 nm), the growth is linear-dependent on the time, and later obeys a parabolic law [181], [182], [313], [391], [587], [588]. For the total oxide thickness, the following applies [182], [306]:

$$x = \frac{k_p}{2k_c}\left[\sqrt{\left(1 + 4\frac{k_c^2}{k_p}(t+\tau) - 1\right)}\right] \tag{3.41}$$

where k_p is the parabolic growth constant and k_c the linear growth constant; $\tau = 4(d^2/k_p + d/k_c)$; d is the oxide thickness at the beginning of oxidation (natural oxide). For dry oxidation at 900 °C, for example, the corresponding constants are $k_c = 0.6$ nm/min and $k_p = 25$ nm^2/min, and are independent of the doping level [313]. Curves for k_p and k_c, for dry as well as wet oxidation, are given by Grove [306], [307]. At high implantation doses, the oxidation rate can be increased considerably; it can also be reduced greatly by the implantation of ions which form a chemical compound with silicon (e.g., nitrogen); cf. Section 6.1.10.

Several publications have dealt with the mathematical description of the impurity atom redistribution during oxidation [35], [133], [400], [576]. Kato and Nishi [400] have provided exact solutions for error function profiles; Prince and Schwettmann [576] have determined, by computer simulations, diffusion and segregation coefficients for boron between 1000 and 1200 °C, using implanted distributions.

If one places the origin of the coordinate system at the SiO_2–Si interface (cf. Figure 3.53), and if the diffusion coefficient is concentration dependent, the diffusion equation is written as follows [576]:

$$\frac{\partial N}{\partial t} = \frac{\partial}{\partial x}\left(D\frac{\partial N}{\partial x}\right) + k\frac{\mathrm{d}X}{\mathrm{d}t}\frac{\partial N}{\partial x} \tag{3.42}$$

In equation (3.42), k is the relationship between the thickness of the silicon consumed during growth of a SiO_2 layer and the thickness of the SiO_2 layer itself; k has a value of approximately 0.44 [44]. If one further assumes that no out-diffusion takes place through the oxide (this assumption is not entirely correct in the case of elements with high vapour pressures, for example arsenic and phosphorus; however, due to the low diffusion coefficient in SiO_2, the magnitude of the error is very slight), then the following applies, due to the conservation of

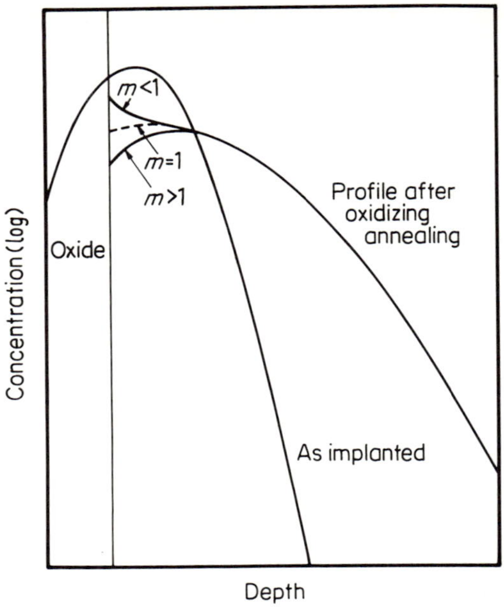

Figure 3.53 Profile change due to annealing in an oxidizing atmosphere (schematic)

the dopant:

$$D \frac{\partial N(x, t)}{\partial x}\bigg|_{x=0} = (m - k)\frac{dX}{dt} N(0, t) \tag{3.43}$$

where m is the segregation coefficient,

$$m = N(0^-, t)/N(0^+, t) \tag{3.44}$$

In some cases the opposite relationship is also thus described |306|.

Prince and Schwettmann |576| have calculated profiles based on implanted boron distributions, assuming a diffusion coefficient as in equation (3.34), slightly modified by an expression $(1 + A \cdot N/n_i)$, $10^{-1} < A < 10^{-2}$, which is intended to represent the influence of vacancies but has practically no influence on the profile shape. A typical example is given in Figure 3.54. The implantation was carried out at room temperature with 80 keV and a dose of 5×10^{14} cm^{-2}; the diffusion was done at 1200 °C for 15 to 45 min. The average range at the energy employed is 0.25 μm; i.e., the boron depletion at the surface results from the segregation, and not from the Gaussian implantation profile or other influences. The diffusion coefficient could be determined, from these experiments, to be $D_i = 0.0322 \exp(-3.02/kT)$ cm^2/s, and the segregation coefficient to be $m = 2.33 \times 10^{-4} \cdot \exp(1.135/kT)$. If the segregation coefficient is smaller than one, i.e., if the dopant

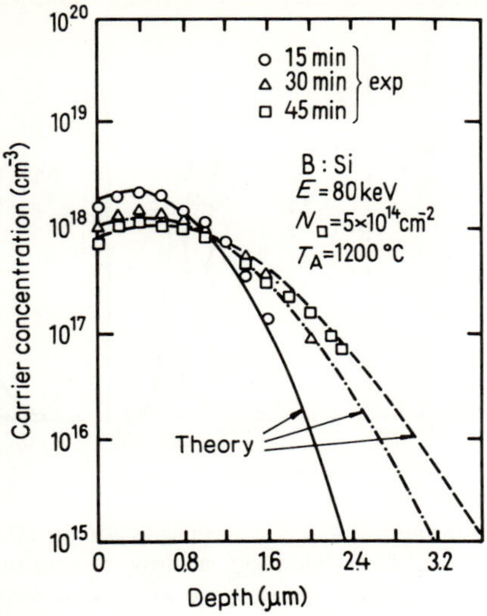

Figure 3.54 Boron profiles after annealing in an oxidizing atmosphere at 1200 °C for 15 to 45 min [576]. From Prince and Schwettmann, *J. Electrochem. Soc.*, **121** (1974) 705, reprinted by permission of the publisher, The Electrochemical Society, Inc.

dissolves less easily in the oxide than in the semiconductor, then the opposite effect occurs. This effect has been measured for arsenic as well as antimony |482|, |519| in silicon. In Figure 3.55, a back-scattering measurement by Müller *et al.* |519| is shown, where the 'snow-plough' effect for arsenic can be seen, depending on the oxidation temperature. If the oxidation temperature is very low, for example wet oxidation at 850 to 900 °C, the implanted ions cannot diffuse into the semiconductor and thus pile up at the interface.

This effect might be utilized, with appropriate ion species, to eliminate by oxidation the layer near the surface which has heavy radiation damage due to implantation. A subsequent relatively high temperature diffusion of the dopant accumulated in the semiconductor at the interface would lead to profiles which, without the concentration-dependent expression, would correspond to a complementary error function. Such experiments have already been made with antimony |311|, |519|. However, it is not yet clear whether dislocations grow into the crystal during oxidation, as has often been observed when annealing in an oxidizing atmosphere (see also Section 3.1).

3.9.3 Radiation-enhanced Diffusion

To conclude this chapter, a relatively new application of ion implantation will be

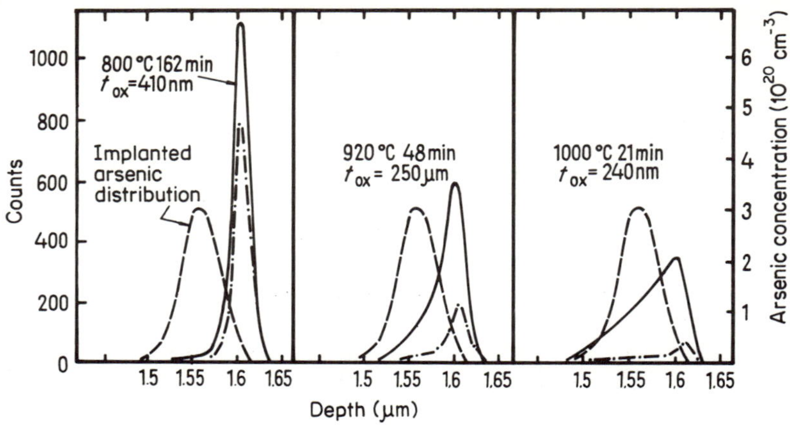

Figure 3.55 'Snow-plough' effect for arsenic in silicon after annealing in an oxidizing atmosphere. After oxidation at low temperatures, considerable steepening of the arsenic profile and incorporation on to irregular sites results; at high temperatures, thermal diffusion is prevalent. Implantation energy 200 keV, implantation dose 10^{16} cm^{-2}; broken line: initial distribution, continuous line: random measurement, dot-and-dash line: channelling measurement [519]. From Muller *et al.*, *J. Electrochem. Soc.*, **122** (1975) 1234, reprinted by permission of the publisher, The Electrochemical Society, Inc.

outlined; that is, radiation-enhanced diffusion. In contrast to thermal diffusion, this is a diffusion effect which is produced by means of bombardment with high energy particles. This effect can be produced during implantation by means of the implanted particles themselves, or after implantation (or diffusion) by irradiation with non-doping particles (α-particles, neutrons, protons, etc.). Defects, and in the ideal case vacancies, are produced by this bombardment. These generally increase the diffusion coefficient, since most doping elements in silicon and germanium diffuse via a vacancy mechanism, and such a mechanism plays an important role in diffusion for other semiconductors too. Relatively high temperatures are necessary during irradiation to avoid the formation of stable defects (in the case of silicon, the temperature must be over 750 °C). Experimental results at lower temperatures have been reported [732]; in these studies, however, crystal damage was not taken into account. The effect of radiation-enhanced diffusion can be calculated approximately by assuming that the diffusion coefficient is proportional to the vacancy concentration (cf. equation (3.23)), and that the interstitial atoms also produced during bombardment are distributed very quickly throughout the semiconductor, due to their low migration energy (0–0.4 eV [648]). If one neglects self-diffusion, the diffusion equations for doping atoms and vacancies can be written as follows:

$$\frac{\partial N}{\partial t} = \frac{\partial}{x}\left(D \cdot \frac{\partial N}{\partial x}\right) + g(x) \tag{3.45}$$

$$\frac{\partial C_V}{\partial t} = D_V \frac{\partial^2 C_V}{\partial x^2} - \frac{C_V}{\tau_V} + g_V(x) \tag{3.46}$$

The diffusion coefficient D of the dopant is assumed as proportional to the vacancy concentration C_V; $g_V(x)$ is the generation rate of the vacancies; τ_V is their lifetime; and D_V is their diffusion coefficient, which is assumed to be constant; $g(x)$ is the generation rate of the dopant ions. Radiation-enhanced diffusion due to the implanted ions themselves ($g(x) \neq 0$) has very little practical importance; however, it can lead to errors in the case of high-temperature implantations, since this effect can lead to profile-shape changes, depending on the ion dose. The dose rate has relatively little influence on the profile shape since when the dose rate is increased, even though the diffusion coefficient then increases accordingly, the implantation time decreases to the same extent. Experimental results on this sort of radiation-enhanced diffusion have been published by Tsuchimoto and Tokuyama [735]. In Figures 3.56 and 3.57, results of their studies, for 100 keV boron and 50 keV phosphorus in silicon at 750 °C, are shown for various doses. Qualitatively seen, these profiles show the broader shape which is to be expected at higher doses. Namba *et al.* [273], [276], [529] have carried out a number of experiments on the enhanced diffusion of antimony, gallium and indium in silicon. Their experiments indicated that the diffusion takes place via an interstitial mechanism at temperatures which are very low in some cases [273]. In Figure 3.58, an example is shown for antimony in silicon. Between

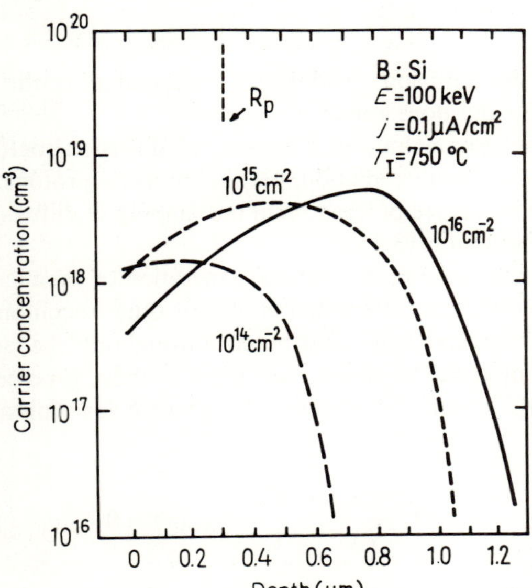

Figure 3.56 Radiation-enhanced diffusion of boron in silicon depending on dose at 750 °C [606]

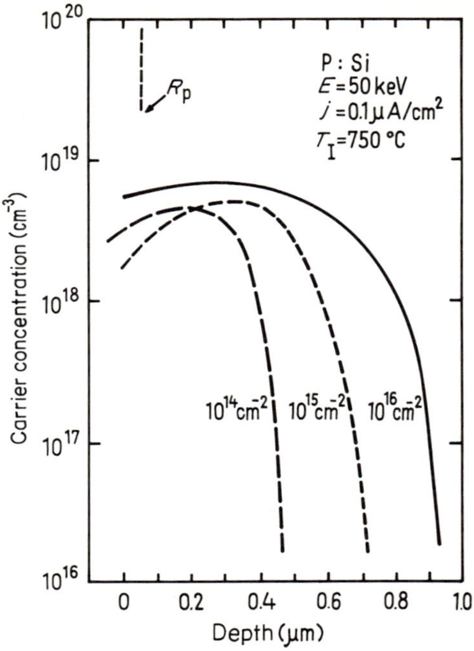

Figure 3.57 Radiation-enhanced diffusion of phosphorus in silicon depending on dose at 750 °C [606]

500 and 700 °C, the diffusion coefficient is independent of the temperature, and depends only upon the dose rate.

Generally, in radiation-enhanced diffusion, the diffusion coefficient is increased by means of an additional irradiation, preferably using protons. In this way, one obtains an additional degree of freedom in the shaping of diffusion profiles. In this case, $g(x)=0$ in equation (3.45).

In deriving equation (3.46), the thermal generation of vacancies was neglected, and it was assumed that the interstitial atoms, or other recombination centres for vacancies, are homogeneously distributed throughout the semiconductor. If $\tau_V \ll t$ (t is the implantation time), then after a short time-period $\partial C_V/\partial t = 0$ applies, and the equilibrium distribution of the vacancies is obtained after solving the equation:

$$D_V \frac{\partial^2 C_V}{\partial x^2} - \frac{C_V}{\tau_V} + g_V(x) = 0 \tag{3.47}$$

An analytical solution for equation (3.47) is possible under the assumption of a Gaussian generation rate for the vacancies, as obtainable for example from the theory by Sigmund and Sanders [685], or as easily derivable from the studies by Brice [103]. If one defines a surface annihilation velocity, s, for vacancies at the

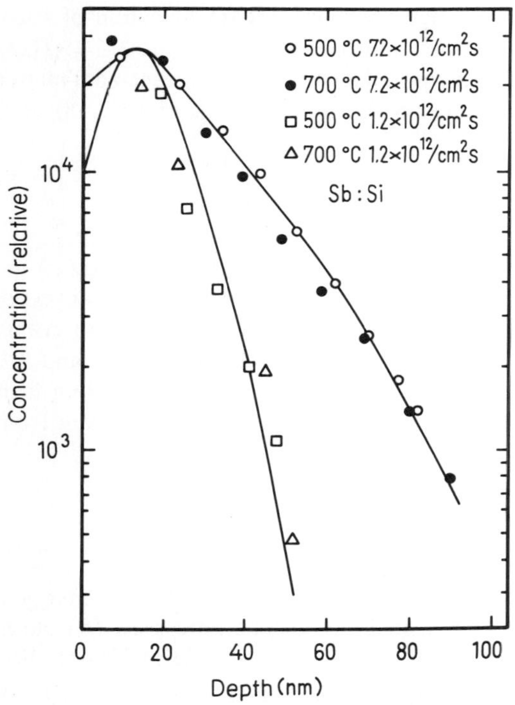

Figure 3.58 Radiation-enhanced diffusion of antimony in silicon depending on dose rate and temperature [276]. Reproduced by permission of the American Institute of Physics

surface:

$$D_V \left.\frac{\partial C_V}{\partial x}\right|_{x=0,t} = sC_V(0,t) \tag{3.48}$$

then the following solution results:

$$C_V(x) = \frac{2\pi \Delta X_D \cdot g_0 L_V}{4D_V} \exp\left(\frac{\Delta X_D^2}{2L_V^2}\right) \left\{ \exp\left(-\frac{x-X_D}{L_V}\right) \left[\frac{1}{\alpha} \exp\left(-\frac{2X_D}{L_V}\right) \right.\right.$$
$$\times \left(1 + \operatorname{erf}\frac{L_V X_D - \Delta X_D^2}{2\Delta X_D L_V}\right) + \operatorname{erf}\frac{L_V(x - X_D) + \Delta X_D^2}{2\Delta X_D L_V}$$
$$\left.\left. + \operatorname{erf}\frac{L_V X_D + \Delta X_D^2}{2\Delta X_D L_V}\right] + \exp\left(\frac{x-X_D}{L_V}\right)\left[1 - \operatorname{erf}\frac{L_V(x-X_D)+\Delta X_D^2}{2\Delta X_D L_V}\right]\right\} \tag{3.49}$$

X_D and ΔX_D are the range and the standard deviation of vacancy generation, respectively; g_0 is the maximum generation rate; and $L_V = \sqrt{(D_V \tau_V)}$ is the diffusion length of the vacancies. α depends on the surface annihilation velocity of the vacancies, and is obtained with the use of equation (3.48) as follows:

$$\alpha = \frac{(D_V/L_V) - s}{(D_V/L_V) + s} \tag{3.50}$$

For high values of X ($X \gg X_D$), the expression $\exp(-x/L_V)$ in equation (3.49) determines the vacancy distribution, and a profile with an exponential decrease into the bulk of the semiconductor is obtained. In order to evaluate the equation, values for X_D, ΔX_D, g_0, D_V, L_V, and s are required. X_D and ΔX_D can be determined experimentally (e.g., with use of the back-scattering technique following implantation at low temperatures); they can also be taken from the theoretical works mentioned above. g_0 is obtained as follows:

$$g_0 = \frac{(\mathrm{d}E/\mathrm{d}x)_{n,max} \cdot j}{qE_d} \tag{3.51}$$

where j is the current density, E_d the vacancy-formation energy, and $(\mathrm{d}E/\mathrm{d}x)_{n,max}$ the maximum energy deposited into atomic processes. D_V can be obtained from experimental values of the self-diffusion coefficient [648]. For L_V in silicon, experimental values are available, ranging between 0.3 μm and 3.5 μm [61], [708].

Figure 3.59 shows vacancy distributions such as are to be expected during bombardment with 10^{14} cm^{-2} s^{-1} protons at 200 keV through 1 μm of SiO_2. The stopping power in SiO_2 and in silicon is assumed to be equal; the SiO_2 layer is necessary in practice to avoid sputtering of the silicon, as well as to adjust the range of the protons to a suitable value. For protons, $X_p \approx R_p$ and $\Delta X_p \approx \Delta R_p$ are approximately valid, so that the data of Gibbons *et al.* [5] could be used.

Examples of experimentally measured profiles are given in Figure 3.60. The samples were irradiated for 10 min at a dose rate of 6.6×10^{14} cm^{-2} s^{-1} (≈100 μA/cm^2), following an arsenic implantation ($E = 80$ keV, $N_\square = 10^{16}$ cm^{-2}). The average range of the arsenic ions was 48 nm; the average range of the protons was between 0.2 and 0.85 nm. The diffusion is especially pronounced near the generation point of the vacancies; a diffusion length of the vacancies amounting to about 0.2 to 0.4 μm at 900 °C can be estimated from this measurement. Further studies on radiation-enhanced diffusion have been carried out by Nelson *et al.* [532], Baruch *et al.* [60], Ohmura *et al.* [547], Minear *et al.* [500], Tsuchimoto and Tokuyama [735], as well as by Zelevinskaya *et al.* [793]. Tarui *et al.* [721] have used radiation-enhanced diffusion to fabricate lateral transistors. Investigations have shown that only hydrogen implantation can be used for radiation-enhanced diffusion. In the case of bombardment with helium or argon, i.e., rare gases which in themselves do not react with silicon or other semiconductors, permanent radiation damage and precipitations of the implanted

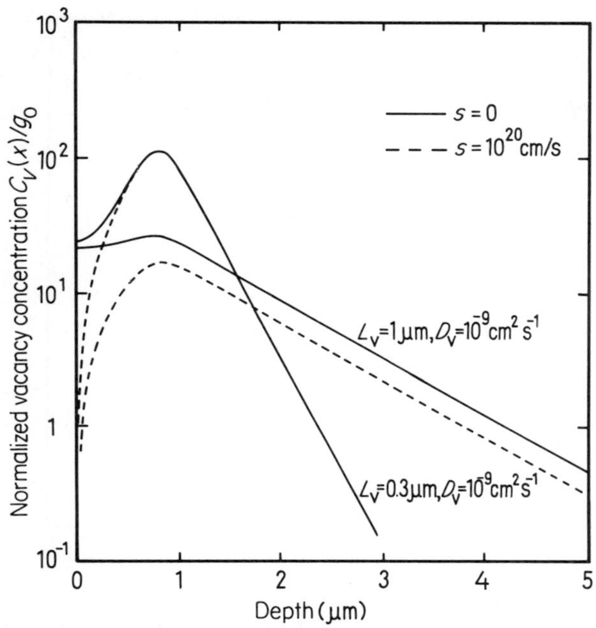

Figure 3.59 Vacancy distribution in silicon during irradiation with 10^{14} cm^{-2} s^{-1} protons at an energy of 200 keV through 1 μm SiO_2

ions remain on account of the low diffusion coefficients (in extreme cases, the formation of 'bubbles' occurs); these factors influence the electrical properties negatively.

The possibilities of this new method for diffusion of implanted or previously diffused impurity atoms into solids at relatively low temperatures (800–900 °C), with the diffusion being done selectively using suitable masking layers and considerably further within short time periods than with thermal diffusion, have only begun to be explored in initial experiments. The considerable range of the protons makes it possible to dope significantly deeper semiconductor regions than previously feasible using implantation.

3.9.4 Other Diffusion Effects

In connection with ion implantation, diffusion can also be influenced by other effects than the mechanisms discussed in the last few sections. The very high degree of ionization during implantation can produce mobile defects, which influence the annealing of radiation damage as well as the diffusion of impurity atoms [96]. Diffusion which is dependent upon radiation damage can also occur. Here, two basically different effects can occur. On the one hand, vacancy forma-

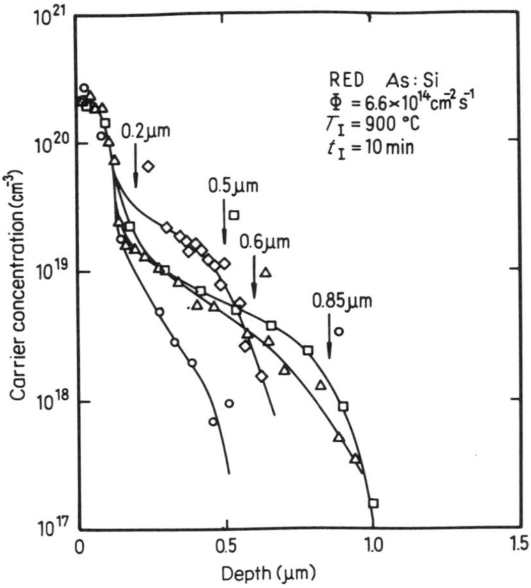

Figure 3.60 Radiation-enhanced diffusion of arsenic in silicon by proton bombardment at 900 °C for 10 min, depending on location of vacancy generation, indicated in each case by an arrow

tion can take place during annealing of the lattice, resulting in an increased diffusion coefficient for the implanted ions [380]. On the other hand, there is the effect of the stable dislocations produced by the implantation, leading to an increased diffusion coefficient as in the case of high concentration boron and phosphorus diffusions. The first problem can be dealt with in a manner analogous to the radiation-enhanced diffusion discussed above, a difference being that the vacancy concentration decreases with time. Research by Itoh [380] suggests that the source of the vacancies is to be found in the transition region between amorphous and undamaged areas, and is not connected with lattice restoration. The vacancies are released at temperatures above 600 °C. The time constant for this vacancy emission is in the range of several minutes. If one assumes equilibrium for the vacancy concentration, equation (3.49) can be used for approximation purposes, multiplying the generation rate by a factor of $\exp(-t/\tau)$, where τ is the time constant for the process of vacancy release.

3.10 SAMPLE HEATING

During high-dose implantation, heating of samples due to the ion beam can become a problem. The power released in the sample is calculated as follows:

$$P = \frac{E}{q} jA = VI \tag{3.52}$$

where E is the acceleration energy, j the current density, A the area of the sample, V the acceleration voltage, and I the ion current. At an implantation energy of 150 keV and a current density of 10 μA/cm², for example, the power deposited at the target amounts to 1.5 W/cm². At an implantation dose of 10^{15} cm^{-2}, this corresponds to an energy of 150 Ws.

If multiply-charged particles are implanted, equation (3.52) remains valid. In general, it is important to keep the implantation time as short as possible for the sake of a high throughput. The wafer manipulation makes a minimum implantation time of a few seconds necessary, generally 5 to 10 s. The implantation time at high doses is limited either by the available current density or by the occurrence of excessive sample heating.

The sample temperature also depends greatly upon heat conduction via the sample holder. Often, the sample is pressed on to a metal holder by metal clamps or springs. These, of course, are not clearly-defined thermal conditions. Therefore, extremely varying results might be obtained in the fabrication of semiconductor devices, especially after low temperature annealing, even in the case that the implantation itself was homogeneous. In calculating sample heating due to the beam, the most important factors are the emissivity of the sample, the heat conduction loss to the sample holder, and the sample thickness. If the sample loses energy by the mechanism of radiation only—in other words, if the heat contact to the sample holder is negligible, and the radiation cooling takes place on both sides of the sample—then the sample's temperature increase is calculated by solving the differential equation, assuming a constant temperature within the sample, as follows:

$$\frac{\partial T}{\partial t} = \frac{P}{Ad\rho C} - \frac{2\varepsilon\sigma}{d\rho C}(T^4 - T_0^4) \tag{3.53}$$

where A is the area, d the thickness, ρ the density, C the heat capacity of the sample (for Si, $C = 0.753$ Ws/gK), P the power supplied by the beam, σ the black-body radiation constant ($\sigma = 5.67 \times 10^{-12}$ Wcm^{-2} K^{-4}), T_0 the environment temperature, and ε the emission factor of the sample surface.

The equilibrium temperature, T_∞, is obtained for the case that $\partial T/\partial t = 0$ as follows:

$$T_\infty = \left(\frac{P}{2A\varepsilon\sigma} + T_0^4\right)^{1/4} \tag{3.54}$$

The heating time up to a temperature T is calculated, by integration of equation (3.53), in the following manner:

$$t = \frac{d\rho C}{8\varepsilon\sigma}\frac{1}{T_\infty^3}\left(\ln\frac{T_\infty + T}{T_\infty - T}\frac{T_\infty - T_0}{T_\infty + T_0} + 2\tan^{-1}\frac{T}{T_\infty} - 2\tan^{-1}\frac{T_0}{T_\infty}\right) \tag{3.55}$$

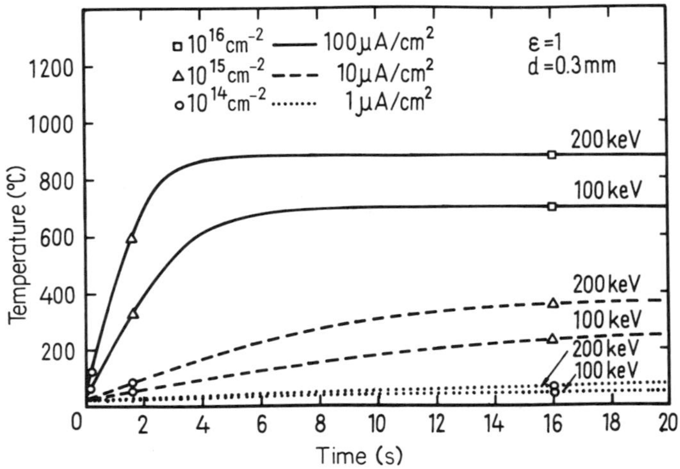

Figure 3.61 Temperature of a silicon wafer depending on implantation time and current density at 100 and 200 keV, $d = 0.3$ mm, $\varepsilon = 1.0$

Only a small amount of experimental data is available for the factor of emissivity at higher temperatures, and these data are subject to modification when working with semiconductors, especially if masking layers of SiO_2, Si_3N_4, etc. or metallic contact layers are applied. Such values, however, should amount to between 0.5 and 1. In the case of silicon, ε (300 K) ≈ 0.65; at 900 °C, it approaches the value of 1.

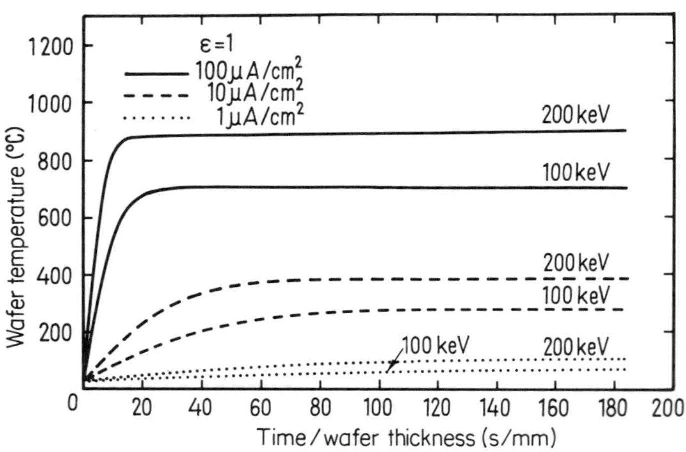

Figure 3.62 Normalized temperature increase of a silicon wafer depending on implantation time and current density

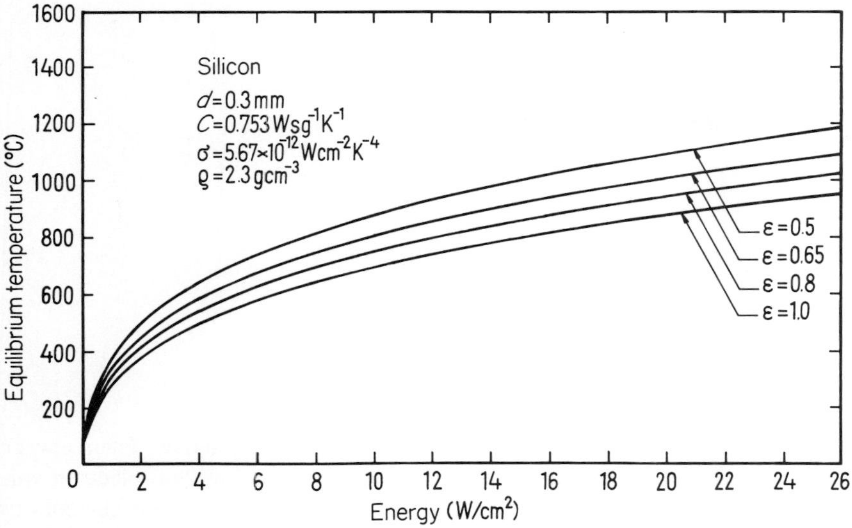

Figure 3.63 Equilibrium temperature of a silicon sample during implantation depending on beam power, with the emission factor as parameter

The temperature change in a silicon sample having a thickness of 0.3 mm, depending on the implantation time and the current density at 100 keV and 200 keV, is shown in Figure 3.61. Temperature values for various implantation doses are also shown. The temperature curves shift to the right or to the left depending on whether thicker or thinner samples are used. Figure 3.62 shows temperature curves standardized in regard to wafer thickness. The heating time is obtained by multiplying the standard values by the wafer thickness in centimetres. An evaluation of equation (3.54) is shown in Figure 3.63. The equilibrium temperature, which is independent of the wafer thickness, is plotted here as a function of the beam power with the emission factor as a parameter.

Excessive heating during implantation can be avoided by good heat contact with a sample holder possessing a high heat conductivity. For example, indium or vacuum grease are usable [2]. These procedures are not suitable for routine implantations, especially for reasons of sample contamination. Therefore, a better method is to implant several wafers simultaneously, thus keeping the effective current density per wafer at a lower level. This is possible, for example, using an implantation chamber with a carousel-type rotating sample holder. Details on such equipment will be presented in the next chapter.

4
Ion-implantation Machines

Many different kinds of ion accelerators are now in existence. Their energies range in some cases between about 10 keV and several million electron-volts; most, however, are between 20 keV and 300 keV, and their beam currents can vary between a few microamps and several milliamps. The reasons for the general limitation to not more than a few hundred kilo-electron-volts have to do with apparatus cost factors as well as the complexity of ion production and analysis. Accelerators developed for use in nuclear physics offer energies above 300 keV for specialized applications.

The principle of all ion accelerators is the same. Gaseous, liquid or solid substances are ionized in an ion source, the ions are then accelerated by an electrostatic field, and the resulting ion beam is separated according to ion mass, whereby the latter process can also take place before acceleration, directly after extraction from the source. The separated ion beam is directed towards the sample; in order to obtain a homogeneous doping, a raster scanning is generally used. The implanted dose is determined by time integration of the current. Further differences between accelerator types have to do with the production of the acceleration voltage, with that of the voltages for extraction, focus, and source (e.g., by means of a transformer, a Van de Graaff generator or a motor-generator system), and with the ground potential.

Problems involved with high-current machines (the accelerator types of the future) include sample heating during implantation, manipulation of a considerable number of wafers within a short time, and implantation homogeneity. Two books deal with the technical equipment aspects of ion implantation: the book by Dearnaley *et al.* [2] and the work by Wilson and Brewer [13], whose main topics are ion sources, beam transport and discussions of various systems.

4.1 ION SOURCES

The beam currents necessary for implantation amount to between a few microamps and several milliamps. Although low currents are adequate for MOS devices (e.g., threshold-voltage adjust), high beam currents are needed for applications involving bipolar transistors or an alteration of chemical material properties, if relatively short implantation times are to be achieved.

Of the various ion source types, the best known are the hot-cathode, high frequency, cold-cathode (Penning), duoplasmatron and sputter sources. The choice of the most suitable source depends on the specific application. Basically, one must distinguish between sources for gaseous substances and those for solids, as well as between high-current sources and those which provide low ion currents but have a relatively good lifetime. The various source types have varying energy spreads. This can be important in regard to mass separation. Values below 10 eV are necessary as a rule for research systems, while the values of industrial systems (e.g., dedicated boron systems) can be considerably higher.

The greatest difference between research and production implantation systems has to do with the spectrum of the ions delivered by the source. In the first case, as many elements and their compounds as possible should be ionizable (gaseous, liquid, and solid, with melting points over 2000 °C); for production purposes in silicon technology, on the other hand, one to four ion species (boron, phosphorus, arsenic and antimony) are adequate in most cases. Gaseous or liquid compounds are often used to facilitate production of a desired ion species. A simple adaptation of the ion source to other species is important, for example, by the use of several gas-inlet valves, or using air locks for solids.

Simple maintenance, as well as a long operating time before the source must be cleaned or overhauled, are especially important in the case of production systems. One hundred hours represent a good period of operation, but very often, much shorter times must be accepted.

The production of double-charged or triple-charged ions is desirable, in order to extend the range of the accelerator to higher energy levels. Hot-cathode ion sources or cold-cathode sources provide up to 10 per cent double-charged and 1 per cent triple-charged ions. The separation of multiply-charged ions is no problem. Because of their higher velocity, they can be separated in a lower field than single-charged ions. Thus, for example, it is possible to produce ions with a maximum energy of 600 or 900 keV by means of a relatively simple 300 keV system.

Ion sources should not be considered alone; their adaptation to the accelerator is always of critical importance in determining their behaviour. In addition to the energy spread, the current density achievable at the target is the decisive factor. A very detailed discussion of the present state of the art in ion sources has been presented by Dearnaley *et al.* [2]. The book by Wilson and Brewer [13] is also very comprehensive, with numerous quotes from original works. Fundamental research on ion sources can be found, among others, in the work by von Ardenne [43].

4.1.1 Hot-cathode Sources

At present, this type of source is probably the most widely used. The reasons have to do with its flexibility regarding the ion species which can be ionized, as well as the high beam currents which can be achieved. The principle of this source is relatively simple. From a directly heated tungsten cathode, an ionizing discharge burns towards the anode. The path of the electrons is generally lengthened by a magnetic field, in order to obtain a higher ionization rate. Because of the high gas pressure (typically 1–10^{-2} Pa), a stable arc burns between cathode and anode. According to the construction, the energy straggling amounts to between 1 and 50 eV, but in most cases it is less than 10 eV; the working pressure can be 10^{-4} Pa to 1 Pa.

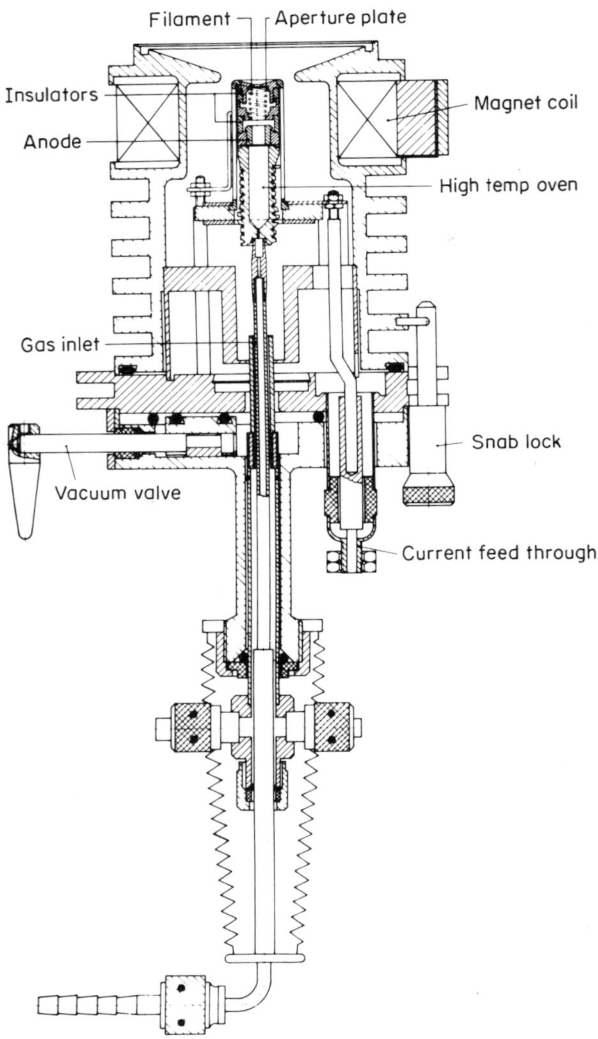

(a)

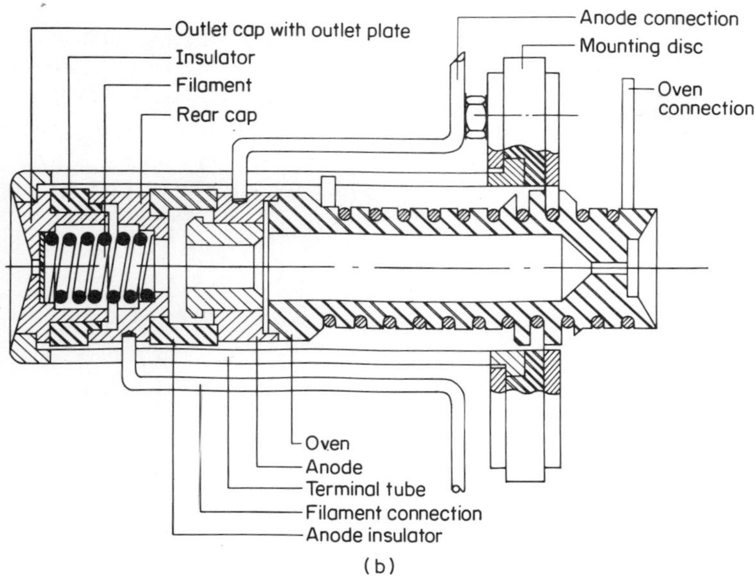

Figure 4.1 Cross-section through a Sidenius source produced by Danfysik: (a) general view, (b) discharge chamber with heater

Two different kinds of ion extraction are usual: either axially or vertically to the axis of the plasma. In the first case, a beam with a circular cross-section is extracted at currents of several microamps; in the second case, the beam is extracted from a slit-shaped aperture in the shape of a wedge. Because of the large extraction slit, currents up to several milliamps are possible with the latter geometry. In fact, currents as high as several 100 mA are possible (Calutron source [310], [413]) in the case of very large sources for isotope separators. Most sources of this type are equipped with ovens, so that solids can be vaporized. Therefore, they are among the most universally applicable types of sources.

A cross-section through an ion source for low to medium currents is shown in Figure 4.1. This so-called Sidenius source [676] is especially advantageous because of its small size and low energy requirement (≈200 W). Furthermore, it is equipped with an oven which permits temperatures up to about 1800 °C. The cathode and the anode are axially positioned, and the ion beam is extracted through an aperture of approximately 0.5 mm diameter.

A typical example of a slit-extraction source is the so-called Freeman source [257], Figure 4.2. It delivers currents up to 5 mA; the cathode has a lifetime of approximately 23 mAh. The cathode is parallel to and extremely close to the extraction slit, in order to achieve a strong discharge in the most favourable position. Therefore, only a very weak magnetic field ($\leqslant 10^{-2}$ T) is necessary. The oven reaches temperatures up to 1000 °C. A sputter electrode can be included for the production of ions from elements with high melting points.

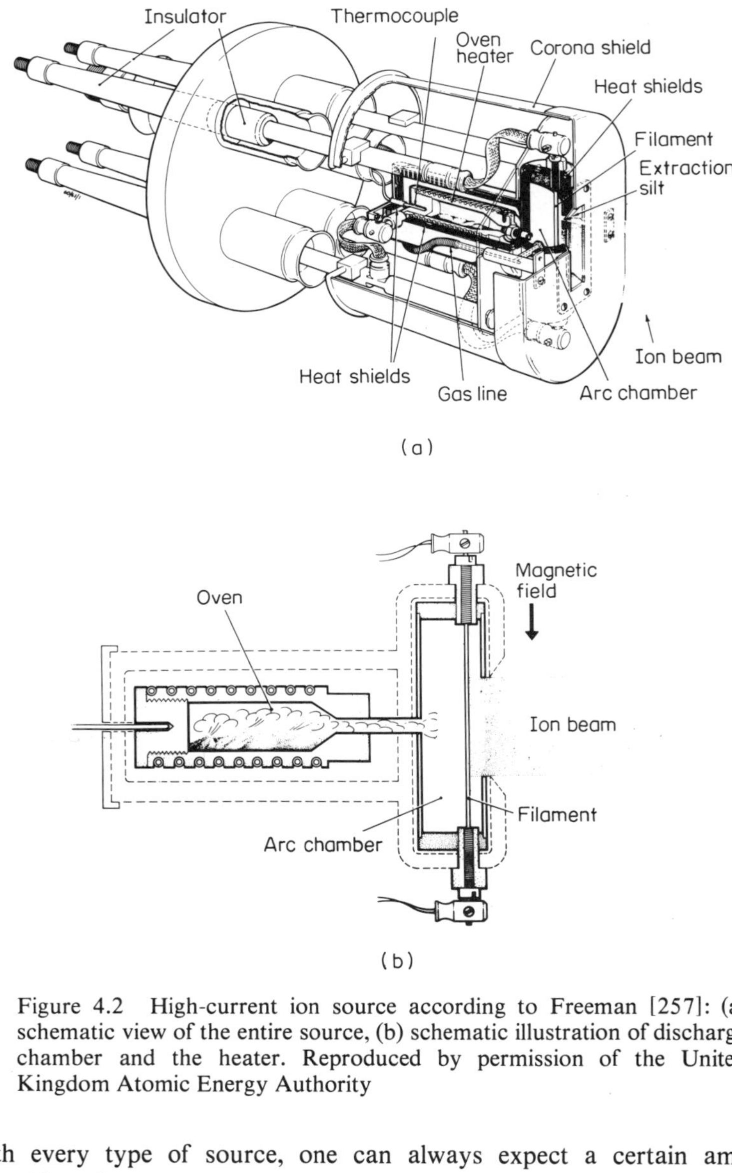

Figure 4.2 High-current ion source according to Freeman [257]: (a) schematic view of the entire source, (b) schematic illustration of discharge chamber and the heater. Reproduced by permission of the United Kingdom Atomic Energy Authority

With every type of source, one can always expect a certain amount of sputtered ions from the source materials, mainly from the filament (i.e., tungsten, tantalum or molybdenum). Special sputtering sources have a negatively biased sputter electrode which accelerates ions from the plasma on to its surface, whereby the atoms are sputtered. The efficiency of such sources is relatively low, although maximum currents in the mA range have been extracted. The deposition of conducting layers on insulators in the source is an especially serious problem, due to the high proportion of neutral sputtered particles. The energy straggling of sputtering sources is generally somewhat greater than that of simple hot cathode sources, about 10 eV to 50 eV; the operating pressure ranges from 10^{-2} Pa to several pascal.

Duoplasmatron sources, a variation of the hot-cathode source (von Ardenne [43]), are used relatively seldom for the production of heavy ions, although they are capable of delivering pulsed currents up to several amperes of hydrogen or deuterium. In order to produce heavy ions, gases or vapours fed into the plasma expansion chamber are ionized by primarily produced ions, usually helium, by means of a charge–exchange reaction [475]. In this way, the corrosion of the cathode can be reduced. The energy straggling is approximately 10 eV; the operating pressure is 1–10 Pa.

Deposition of conducting layers on insulators, as well as corrosion and sputtering of the cathode, are problems affecting all hot-cathode sources. An important advantage of these sources is their low energy straggling, amounting to only a few eV. This is especially important for elements with many isotopes, for example, zinc, for an adequate mass separation.

4.1.2 Radio-frequency Ion Sources

Radio-frequency ion sources [83] (RF sources) have only a limited applicability, in spite of their low power input and their simplicity. A RF source, as shown in Figure 4.3, consists of a glass or quartz cylinder with two electrodes, an anode and a cathode. The ionizing radio-frequency field can be coupled capacitively (see Figure 4.3) or inductively. The frequency is in the megahertz range; the power consumption is a few hundred watts. A magnetic field serves to stabilize the discharge; for inductive coupling it should be applied vertically, and for capacitive

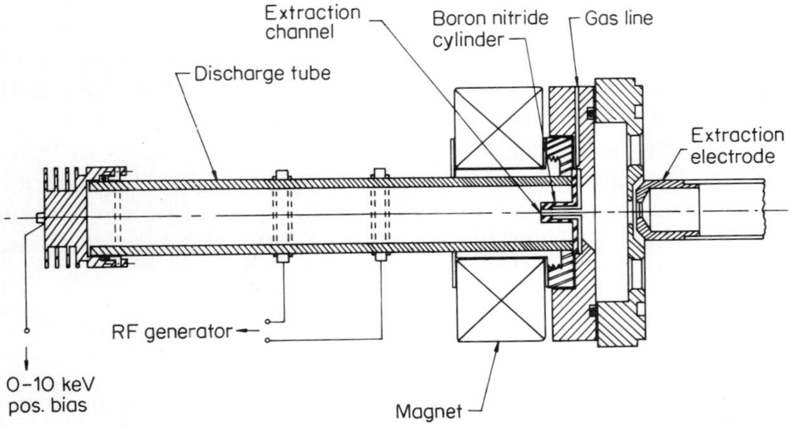

Figure 4.3 Radio-frequency ion source, with capacitative coupling of the RF field

coupling in a parallel manner [270]. An anode voltage of a few kilovolts is necessary to produce a stable plasma and to facilitate ion extraction. For the cathode, a metal channel with an insulating covering is generally used. The gas consumption is thus reduced greatly; in the narrow channel, however, considerable quantities of ions are also neutralized. The energy straggling of RF sources amounts to 10–500 eV, so that the mass resolution is less than for hot-cathode sources; the operating pressure is between 10^{-2} Pa and 1 Pa. With sources of this kind, it is possible to produce currents up to several mA from rare gases and nitrogen. However, problems arise when a production of other ions is desired. Layers which are deposited on the quartz or glass tube influence the efficiency of the source. The field distribution is thus altered, and a stable operation is made impossible. This is especially serious in the case of compounds such as BF_3 and PF_5, for production of boron and phosphorus ions, or in the case of metals. None the less, ions of Na, Ni, Si, Ge, B, P, Si and Ag have been produced under laboratory conditions, in some cases with currents of several milliamps [415]. Further research work is necessary to attain universal characteristics such as those of hot-cathode sources.

However, RF sources have an important area of application in the case that a very long period of uninterrupted operation with inert gases is required. For this reason, practically all Van de Graaff accelerators are equipped with such sources.

4.1.3 Penning Sources

Penning or cold-cathode ion sources are based on the Penning principle [69], well known in connection with vacuum gauges. By applying a high voltage, electrons are emitted from the cathode; these are then made to oscillate by a magnetic field, and ionize the gas in the source.

Penning sources are extremely simple, and can be constructed very ruggedly to achieve a long lifetime. In general, they are exclusively used for ionization of gases, and deliver currents up to several hundred microamps. Rare gases can only be ionized in Penning sources with great difficulty. Mainly, these sources are used for routine implantations of boron, arsenic, and phosphorus from BF_3, AsH_3 or PF_5 gas. Here, their somewhat high energy requirement, limited flexibility, and considerable energy straggling (about 100 eV) are compensated for by their ease of operation. They are also used in cases where their ability to produce large quantities of multiply-charged ions is advantageous. Their operating pressure is generally very high, 10 Pa to 100 Pa; a pump in the vicinity of the source is therefore necessary.

4.1.4 Other Ion Sources

Several other ion sources used in special individual cases may also be mentioned for the sake of completeness. Surface ionization sources [424] enable the production of relatively high currents for elements with a suitable ionization potential. Their energy straggling is only 0.2 to 0.5 eV. Laser sources [225] use a finely focused light beam, vaporizing the source substance in a vacuum which is high compared to the other sources mentioned. Surprisingly, no attempt has yet been made to use an electron-beam gun, such as that used for vacuum evaporation, as an ion source. The spark-ion source [778] operates according to the Wagner hammer principle; here, one electrode consists of the material to be ionized, and the other can be graphite or another appropriate substance. The energy spread is about 100 eV, in extreme cases up to 10^4 eV. Large quantities of multiply-charged ions are produced as is the case with the Penning source. The spark-ion source too can operate in a high vacuum, but its disadvantage is the high fluctuation of the current.

The use of a glow discharge should be discussed as a final source principle. Glow discharges were frequently employed in the past for sample cleaning prior to evaporation, and are still in use for evaporation on glass substrates. In this way, very high currents can be produced; the acceleration energy of the ions, however, is poorly defined, and ranges for practical purposes between zero and the maximum voltage. None the less, this is an extremely simple method for the implantation of ions. The first successful experiments were carried out as early as 1963 by Strack [708]. It must be said, though, that only relatively few ion species are suitable to maintain discharges. Besides this, the penetration depth is relatively small. Therefore, this method would mainly be suitable for implanting ohmic contacts or extremely shallow layers. Surprisingly, though, very little research has been done regarding this simple implantation procedure, which requires no additional acceleration or separation of the ions. Applications should especially be possible in the field of solar cells.

4.1.5 Ion-source Operation

For successful operation of ion sources, the choice of the appropriate substance as the source feed material is very important. In this respect, vapour pressure tables are useful, for instance those by Honig [345]. The necessary pressure, depending on the type of source and the ionization cross-section, can be between 10^{-2} and 10 Pa. Table 4.1 lists source feed materials which are well suited for the most important dopants in semiconductor technology, as well as for a few other elements. These recommendations are based mainly upon our experience with a hot-cathode source (Danfysik type 911 A); the data agree well, in general, with the values given by Dearnaley and Freeman [29], [259]. Further data on appropriate source materials have been compiled by Ryding [605], who quotes research done in Oak Ridge between 1942 and 1960, and also by Axmann [46]. In the Appendix, vapour pressure tables by Honig and Kramer [345] are reproduced; these make it very easy to determine whether a desired ion species can be ionized directly from the element, or whether it is necessary to make use of a compound.

Problems which all ion sources have in common, and which can only temporarily be suppressed or eliminated by using appropriate measures, include deposition of conductive films on insulators, chemical reactions between ions and

Mercury	Hg	202, 200, 199	Hg	–39		Poisonous
			HgF_2		[2]	
Nickel	Ni	58, 61	Ni	1453		
			$NiCl_2$	1001	[2], [261]	
Nitrogen	N	14	N_2	–210		
Oxygen	O	16	O_2	–219		
Phosphorus	P	31	P	44		Red phosphorus
			PCl_3	–28	[2], [261]	
Platinum	Pt	194, 195, 196	Pt	1773		
Selenium	Se	80, 78	Se	220		Poisonous
			CdSe	1350	[2], [261]	
			SeO_2	345	[261]	
Sulphur	S	32	S	118	[261]	Operating temperature of many sources is too high to permit a continuous vaporization of sulphur
			H_2S			Problem-free. SO_2 [261], [2]: due to the equal masses of S and O_2, separation by magnet is not possible
			CdS		[2]	
Tellurium	Te	130, 128, 126	Te	450		Poisonous
			CdTe		[2]	
Tin	Sn	120, 118, 116	Sn	232		
			$SnCl_4$		[2], [261]	
Zinc	Zn	64, 66, 68	Zn	47		ZnS [261], [2]: due to the equal masses of Zn and S_2, separation by magnet is not possible

* In the case of solid substances, an auxiliary gas, generally argon, is necessary during operation or to ignite the source. With solid substances it is generally better to use coarse powder or granulated materials, because of the controlled evaporation, rather than a fine powder or large pieces.

Table 4.1 Materials for ion-source operation

Element	Symbol	Atomic mass of the most frequent isotope	Recommended source material*	Melting point (°C)	References	Remarks
Aluminium	Al	27	Al	659		
Antimony	Sb	121, 123	Sb	630		
Argon	Ar	40	Ar	−189		
Arsenic	As	75	As	817		Poisonous
			GaAs	1237	[2]	
Beryllium	Be	9	Be	1284		Poisonous (carcinogenic)
Boron	B	10, 11	BF_3	−128		High ion current with relatively low corrosion in the source
			BCl_3	−107	[2]	
			B	2300	[261]	
Cadmium	Cd	114, 112	Cd	321		Poisonous
			CdS	1750	[2]	
Fluorine	F	19	BF_3	−128		Poisonous (corrosive)
			NaF		[2]	
Gallium	Ga	69, 71	Ga	30		
Germanium	Ge	74, 72, 70	Ge	936		
Helium	He	4	He	−272		
Hydrogen	H	1	H	−262		
			H_2O		[2]	
Krypton	Kr	86, 84	Kr	−157		
Lithium	Li	7	Li	179		
			LiCl	614	[2], [261]	
Magnesium	Mg	24	Mg	657		
Manganese	Mn	55	Mn	1244		

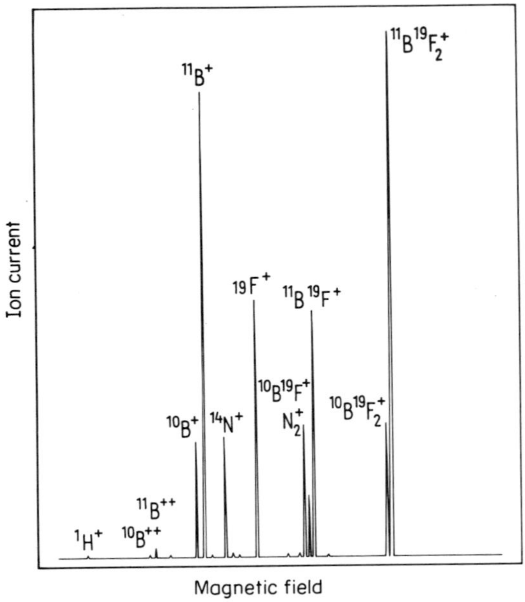

Figure 4.4 Ion spectrum of boron, with BF_3 as source material

metal parts (corrosion) or source insulators, and sputtering. Especially affected by these problems are the negatively-biased parts of the source. These factors too can be decisive in choosing source feed materials.

Because of the many different kinds of ions in an ion beam, it is often difficult to identify a desired isotope. This is especially the case when molecules form the source feed material. For example, if BF_3 is used to produce boron ions, there can be found in the ion beam, in addition to the boron isotopes $^{10}B^+$ and $^{11}B^+$, all combinations with fluorine too, i.e., $^{11}BF^+$, $^{10}BF^+$, $^{10}BF_2^+$, $^{11}BF_2^+$, $^{10}BF_3^+$, $^{11}BF_3^+$, and multiply-charged molecules as well. In Figure 4.4, a typical spectrum for this case is shown.

A description of all isotopes in terms of their mass in respect to their abundance is very helpful for identification purposes. They can then be found in the ion current with approximately the same distribution. In the Appendix, the abundance of the isotopes of all elements is given to scale.

4.2 ACCELERATION AND FOCUSING

The high voltages necessary for acceleration are generally produced by a transformer followed by a rectifier. Cockroft–Walton and Van de Graaff generators are also used at higher voltages (500 keV to 2 MeV). The necessary operating voltage for the ion source is transmitted via an insulation transformer or a motor-generator system to high potential.

4.2.1 Acceleration

The required maximum energy or voltage depends upon the desired penetration depth of the ions which are used; the same applies for the minimum energy. Low energy systems (up to about 50 keV) use a single acceleration electrode. At higher energies, an acceleration tube with several electrodes (up to 20) is used. The former system is considerably simpler, and is more advantageous from the vacuum viewpoint; the latter has the advantage of better ion-optical properties.

For research systems, a maximum energy of 300 to 400 keV is suitable; this can be achieved with relative ease by air-insulated systems. For special applications, energies up to 2 MeV are sometimes used. A pressure vessel is then needed, and a Van de Graaff generator is used in most cases. Systems with pressure vessels involve the major disadvantage that access to the source is difficult. It is not possible to construct an accelerator for all energy ranges. Typical ranges are about 2 to 15 keV, 10 to 100 keV, 50 to 300 keV or 0.5 to 2 MeV. Production systems for present-day applications usually require a maximum energy of 200 keV. The lower limit of the acceleration voltage is determined by reduction of beam current and focusability. A very important question is the speed with which the acceleration voltage can be changed. The ion beam is often separated after acceleration, so that a voltage change is only possible in connection with a new adjustment of the separation parameters. Special attention also has to be given to the stability and ripple-free quality of the acceleration voltage.

Various concepts have been developed for the arrangement of the acceleration electrodes in regard to mass separation and ground potential. The most frequently used configurations are shown diagrammatically in Figure 4.5. In addition, mixed types are also possible. The advantage of the design shown in Figure 4.5(a) is that only the parameters for source and extraction have to be changed at high potential. All other elements of beam manipulation and measurement are on ground potential. The separation magnet, however, must be relatively large. If separation takes place before acceleration (Figure 4.5(b)), a very small magnet is adequate, and an alteration of the acceleration voltage during implantation is possible without difficulty. This makes it possible to produce nearly arbitrarily shaped profiles. Due to modern technology (light guides, opto-electronic couplers), this accelerator type is no longer associated with any technical problems; therefore, it can be considered as the most advantageous configuration. In Figure 4.5(c), the possibility of post-acceleration is also shown for the sake of completeness. This method is only used to achieve higher acceleration voltages when no sufficient voltage can be applied to the acceleration lens. However, access to the target (heating, cooling, current measurement) is difficult here, due to the application of the post-acceleration voltage. This is the case, for example, with units having a single acceleration electrode, as is often used for slit-shaped extraction of the ion beam (e.g., Reference [255]). A new machine produced by Extrion therefore uses a separation magnet which focuses the wedge-shaped ion beam approximately to a circle, so that higher acceleration voltages can be reached by means of a multiple-electrode acceleration tube.

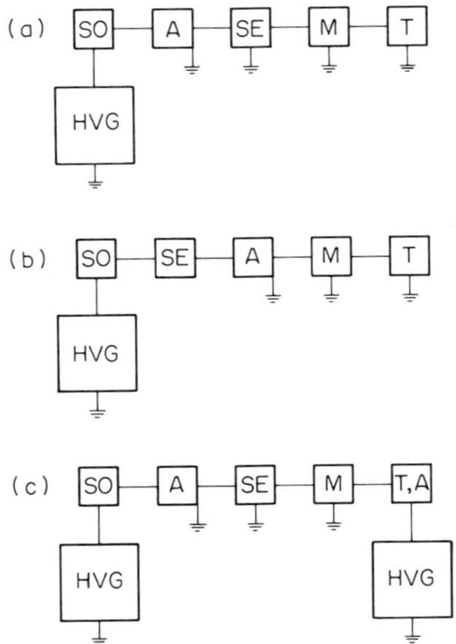

Figure 4.5 Schematic illustrations of ion-accelerator concepts. HVG = high-voltage generator, SO = ion source, A = acceleration, SE = ion separation, M = beam manipulation, T = target. (a) Standard type: only the ion source is at high voltage. Large analysing magnet is necessary. Energy variation requires adjustment of the magnet. (b) Analysis of the beam at high voltage: Only a small analysing magnet is necessary. Energy variation is accomplished by adjustment of the high voltage. (c) Post-acceleration of the ions at the target: the energy can be varied by adjusting the post-acceleration voltage. Relatively small analysing magnet. Access to target is difficult (current and temperature measurement)

4.2.2 Focusing

Focusing is accomplished by means of a focus electrode and an extraction electrode, directly following the ion source and prior to acceleration. Low-energy and medium-energy systems sometimes have combinations of acceleration and focusing. In order to improve the low-energy properties of high-energy systems, a focusing einzel lens or a quadrupole is sometimes used following separation. A defocusing of the beam and a deterioration of the focusing quality always result at low energies, or at very high currents due to space-charge effects. Wilson and Brewer [13] have treated this problem in detail. In Figure 4.6, a diagram from their book is reproduced, allowing an easy estimation of the possible space-charge influence. In the field-free area following the acceleration and focusing electrodes, the ion beam can trap electrons or negative ions until it becomes neutral. In this state, the beam forms a plasma with the same number of positive and negative charges. Such a compensation is especially important for the beam in low-energy, high-current implanters. The relaxation-time constant up to achievement of neutrality has been measured, for example, by Bernas *et al.* [71], as a function of pressure and energy; this time is in the range of microseconds.

Aberrations produced by the optical components have further influences upon the shape of the beam. Since mainly electrostatic elements are used, in contrast to electron-optical systems where magnetic lenses are employed, the former can influence the beam quality very considerably. A detailed discussion on this subject is also provided by Wilson and Brewer [13].

In order to observe the ion beam, apertures can be inserted into it; still better is the use of beam-profile monitors, of which many different types are commercially available. The latter are especially useful for focusing in the case of long systems. In addition to rotating and oscillating wires, arrangements consisting of several Faraday cups are well-suited. However, metal apertures (of tantalum, etc.) are

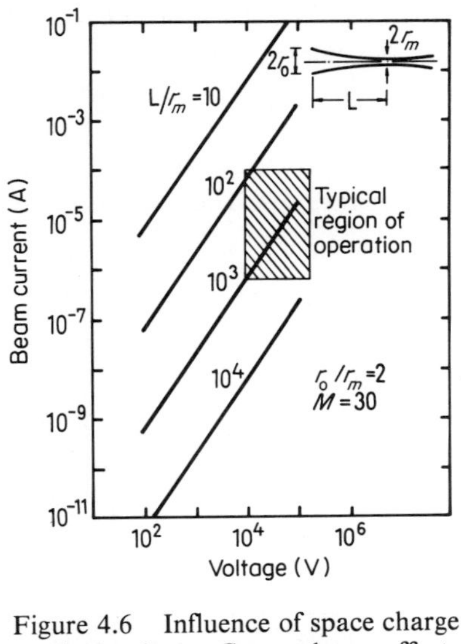

Figure 4.6 Influence of space charge on the ion beam. Space-charge effects can become important above the lines; below the lines, they probably have no influence [13]

simpler; these become luminescent at currents above 1 μA, and begin to glow at high currents. At low voltages and currents, a quartz plate with a fine wire mesh in front of it or a luminescent screen can be used.

4.2.3 Beam Writing

In the usual ion implantation, the whole areas of the samples are implanted, either by defocusing the beam or (more generally) by scanning the beam or moving the sample. A new possibility for applying ion implantation (sometimes called 'beam writing') is based on the controlled movement of a finely focused ion beam upon the area to be implanted, without the use of masking layers. An alternative to this method is the use of imaging systems [72]; these, however, do not seem to hold much promise, due to the low achievable dose rates. An important condition for the attainment of realistic writing speeds is the development of a high-intensity ion source, as well as that of efficient focusing optics which do not cause high intensity losses.

Possibilities for applying this new method include the selective activation of 'read only memories' (ROM) [655] by sputtering, mask repair, the implantation of entire integrated circuits using various ion species for *p*-doping and *n*-doping, and the implantation of light guides [293]. While the third application still requires much development work, the others can probably be realized with less effort, since they involve lower beam currents. Recently, ion beam lithography has emerged as a new application of ion beam writing [612a], [656a].

Brewer [100] has analysed the smallest beam diameters obtainable at the sample. He was able to demonstrate that space-charge effects, chromatic aberration, and diffraction can be neglected. Seliger and Fleming [655] have experimented with a simple construction using an einzel lens and two apertures, producing minimum beam diameters of about 5 μm at a beam energy of 60 keV and a total maximum current of 20 nA (calculated from the data given). This system is shown diagrammatically in Figure 4.7. The beam is reduced to the desired size by

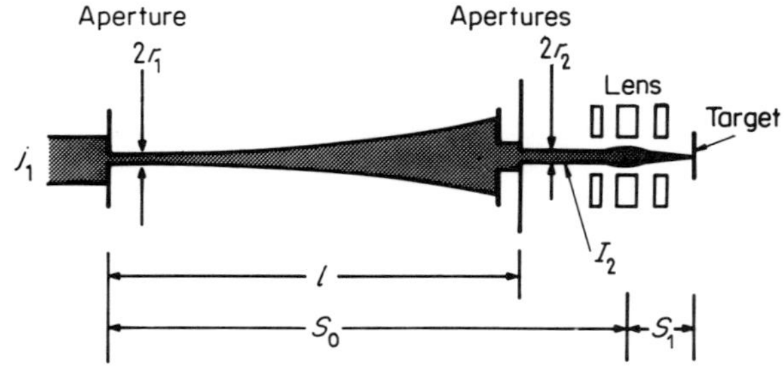

Figure 4.7 Beam-writing system, according to Reference [655]. Reproduced by permission of the American Institute of Physics

two apertures at a distance from each other of 1, and is focused by an einzel lens of the Liebmann type [436]. For the beam diameter, the following is valid:

$$D = \frac{4I_2 f^2/S_0^2}{\pi j_1 [1 - \exp(-\theta^2 f^2/2\sigma^2)]} + \frac{1}{4} C_s^2 (\Theta^6)^{1/2} \quad (4.1)$$

In equation (4.1), j_1 is the current density prior to the first aperture, I_2 the current at the sample, Θ the half-angle of the beam at the sample ($\Theta \approx r_2/f$), f the focal length ($f = 1.92\,D \approx S_i$), σ the standard deviation of the beam profile at the second aperture, C_s the coefficient of the spherical aberration of the lens, and D the lens diameter. For S_i and S_0, see Figure 4.7.

Diameters of the ion beam calculated according to equation (4.1), dependent on Θ as well as the relationship of I_2/j_1, are shown in Figure 4.8. The triangle represents an experimental point with a beam diameter of approximately 5 µm [655]. In recent years, field ion sources with a very high brightness have been developed [314a], [552a]. Up to 10^7 A cm^{-2} sr^{-1} for hydrogen [314a] and

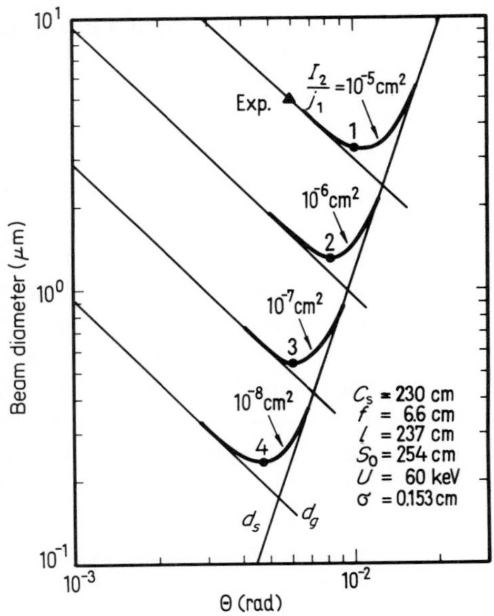

Figure 4.8 Ion-beam diameter, depending on the half-angle of the beam (M) at the target as well as relationship between target current I_2 and primary current density j_1. The symbols d_g and d_s represent beam diameters produced by Gaussian reduction and spherical aberration, respectively [655]. Reproduced by permission of the American Institute of Physics

Table 4.2 Implantation times depending on dose and beam diameter for a micro beam system. Current density at the target: 4 A/cm^2; area: 2.5 mm^2

Beam			
Diameter (μm)	Half angle (mrad)	Dose (cm^{-2})	Implantation time (s)
0.5	10	10^{15}	400
0.5	10	10^{13}	4
0.1	2	10^{15}	10 000
0.1	2	10^{13}	100

10^4 A cm^{-2} sr^{-1} for dopant ions such as boron or arsenic [277a], [419b] have been obtained. These sources are field-emission gas or liquid-metal sources with an emission spot of several ångstrom to several hundred ångstrom. Beam diameters below 500 Å have been obtained [656a]. Such beams can be used for implantation or micromachining purposes. For boron, the best current density at the target has been 4 A cm^{-2} up to now. In Table 4.2, implantation times are shown for ion doses of 10^{13} cm^{-2} and 10^{15} cm^{-2}, and for beam diameters between 0.1 μm and 0.5 μm at the above current density, calculated for a 5 × 5 mm^2 semiconductor chip under the assumption that 10 per cent of the area is to be implanted. With the current state of ion-source technology, in cases of small beam diameters and high doses, these times can become unacceptably long. Additional treatment of this topic can be found in the work by Wilson *et al.* [13].

Closer to practical application is the technique of ion-beam lithography. Lithography is a very important technological step for the definition of patterns in device processing. For VLSI (very large scale integration) circuits, dimensions of 1 to 2 μm are used. Electron-beam writing is used for mask making and, to a lesser extent, for direct resist exposure on the wafer as well [327a], [745c], [761a]. It is limited by fast back-scattered and secondary electrons to dimensions of about 1–0.3 μm under practical conditions in semiconductor fabrication. Ion-beam lithography is a new technique which shows great promise for the fabrication of circuits with submicron details. The reasons for this are the high resolution possible with ions as well as the high sensitivity of resists to ions. The sensitivity of resists to ions is higher than to electrons or X-rays, since the energy deposition per unit path length is higher for ions. Consequently, a lower exposure dose is required. The reason for the high resolution is that ions do not suffer from the so-called proximity effect of electrons, and do not produce photoelectrons (200 eV to 2 keV) as in the case of X-rays. Ions produce only low-energy secondary electrons (5 eV to 50 eV), and no high-energy back-scattered and secondary electrons, as electrons do. Therefore, the structures are defined by the primary ion beam, and are limited only by the straggling of the ions, which is less than the range of the electrons produced by electron beams. Typical required doses are

10^{11} to 10^{13} cm^{-2}, thus greatly reducing the demand for sources with high brightness.

4.2.4 Examples of Ion Accelerators

In Figures 4.9 to 4.13, several types of ion accelerators are depicted, illustrating some of the different construction forms which now exist. Figure 4.9 shows one of the first commercial accelerators made by Accelerators Inc. (now Veeco). The high voltage is produced by means of a transformer. RF sources, as well as Penning and hot-cathode sources, can be used for ion production; their operating voltage is superimposed on the high voltage by an insulation transformer, and the mass separation is carried out at ground potential using a Wien filter (see Section 4.3). A diagram of this accelerator, including target chambers, can be seen in Figure 4.10. An accelerator made by High Voltage Engineering and Danfysik suitable for general laboratory use, is shown as a further example in Figure 4.11. The high voltage, in this case too, is produced by a transformer. The necessary voltages for operation of the source, the pumps (which are at high potential), and other components are provided by a motor-generator system. A hot-cathode source is used. This accelerator is currently in wide operation as an implantation system for research purposes.

In addition to Veeco and High Voltage Engineering, there are three other important manufacturers of implantation equipment. Applied Materials produces a high-current accelerator (the former Lintott machine) originally developed at Harwell, equipped with a Freeman source (see Figure 4.2). This unit is suitable for production purposes, and thanks to its universal source, is equally well suited for

Figure 4.9 Implantation machine made by Accelerators Inc. (now Veeco) with separation after acceleration. Maximum voltage: 300 keV

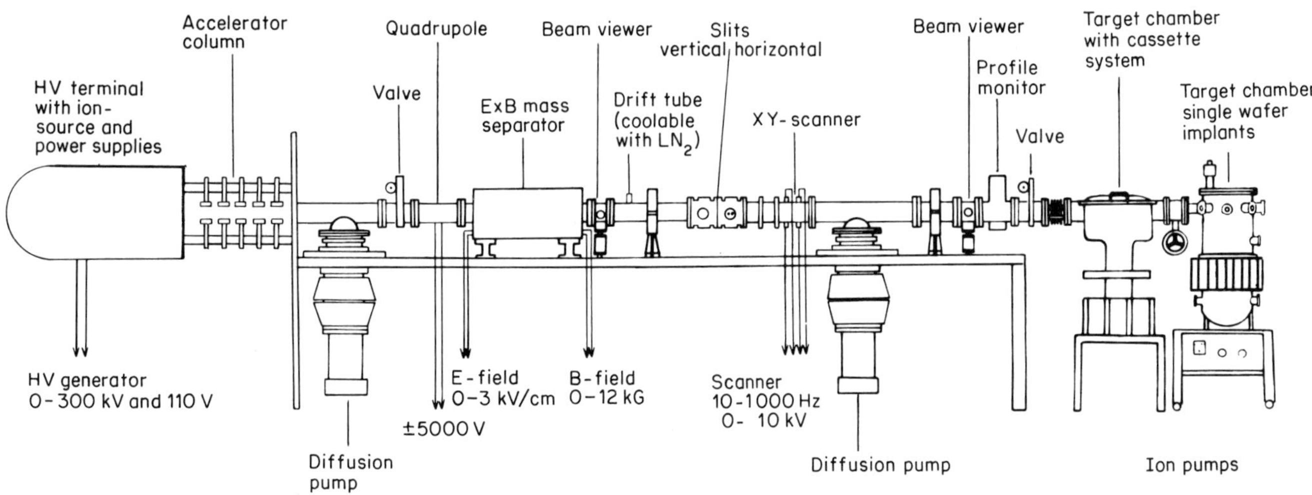

Figure 4.10 300 keV accelerator made by Accelerators Inc.

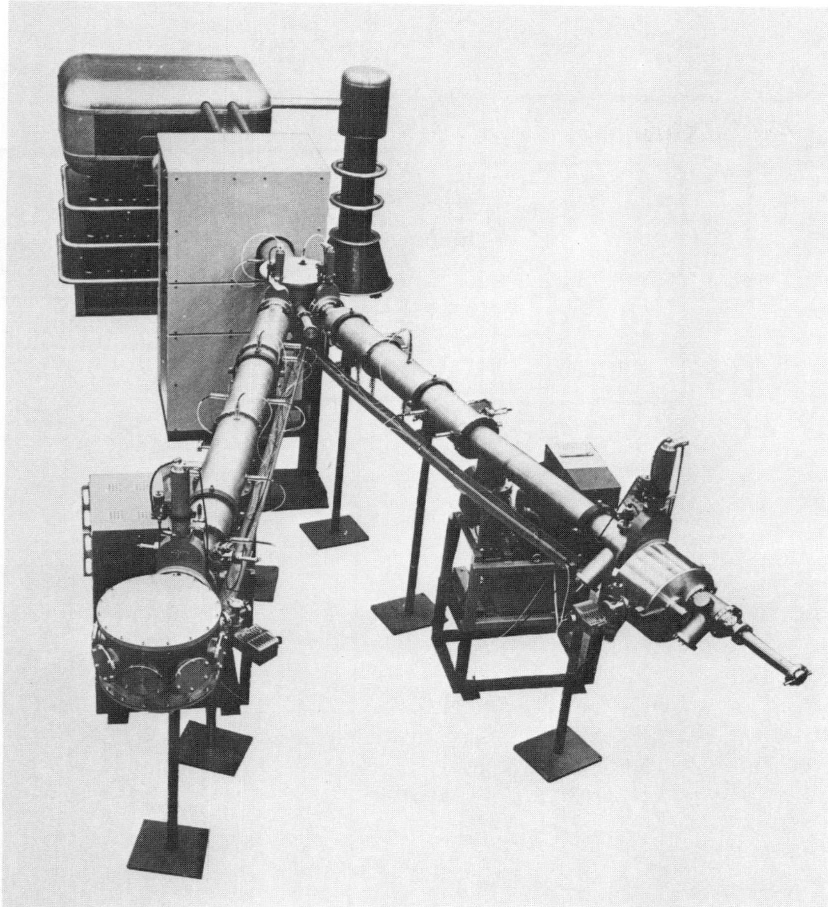

Figure 4.11 Ion accelerator with separation before acceleration made by High Voltage Engineering. Maximum voltage: 350 kV; two beam lines

research. The largest programme of accelerators is offered by Varian-Extrion. All systems made by this firm are intended for semiconductor production, and are relatively easy to operate. A medium current implanter from this company is shown in Figure 4.12. This model is a high-throughput multipurpose ion implanter. The system has an energy range from 10 to 200 keV, and utilizes a hot-cathode type ion source to provide stable operation over a wide range of beam currents. An electrostatic deflection system scans the beam across individual wafers. Microprocessor control of the implanted dose ensures precise dose accuracy. The Wayflow end station (see Figure 4.26) is designed for high speed atuomatic processing of wafers up to 125 mm in diameter. Varian-Extrion also produces high current implanters with mechanical (spinning-disc) and magnetic scanning.

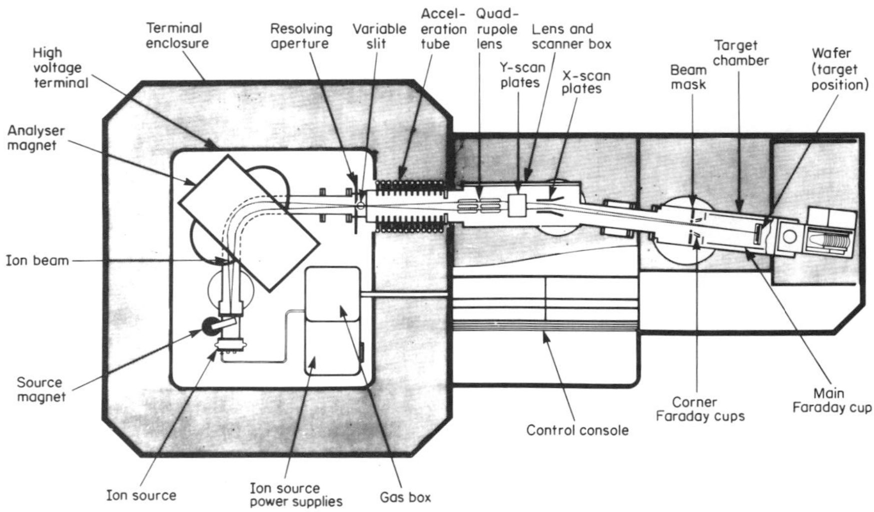

(a)

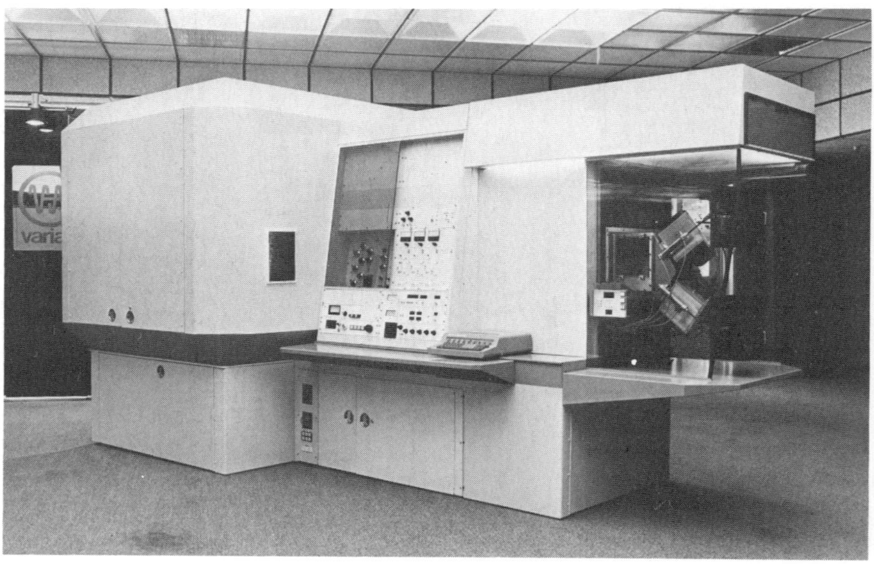

(b)

Figure 4.12 (a) Schematic illustration of an Extrion medium-current implanter; maximum energy 200 keV, maximum current 1.5 mA. (b) Photograph of this implanter

A high current implanter from Eaton-Nova is shown in Figure 4.13. This implanter is designed for high dose applications, such as pre-deposition for source and drain. Ions are generated in the ion source from either gaseous or solid materials, and are extracted in a beam at an energy between 20 and 80 keV. An analysing magnet then separates the components of the beam according to their

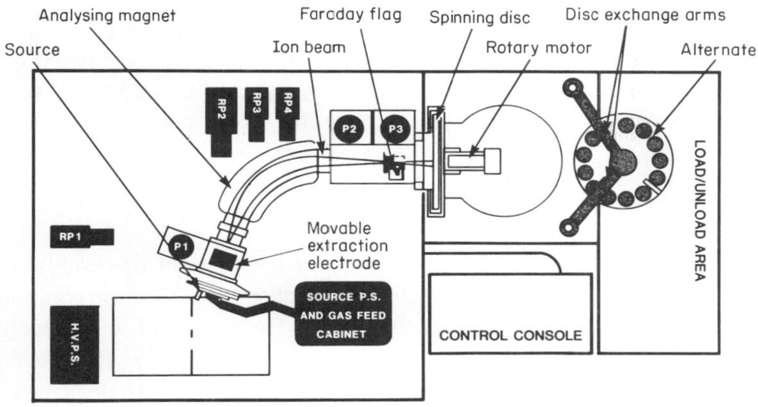

(a)

(b)

Figure 4.13 High-current predeposition ion implanter made by Eaton-Nova. Maximum energy: 80 keV. The accelerator provides currents up to 10 mA. (a) Schematic illustration, (b) photograph

atomic mass, and focuses a beam of the selected ion species through a resolving aperture. A retractable Faraday flag is located just beyond the aperture. This flag is moved into the beam path when not implanting, and is used during set-up to measure the beam current on target. The implanter uses a 'spinning disc' system for wafer processing. Here, the wafers are mounted around the perimeter of a disc. During the implantation, the spinning motion of the disc provides high-speed scanning of the wafers across the stationary ion beam. A second scanning motion is provided by moving the process chamber up and down, so that the wafers are slowly scanned from top to bottom as they are swept across the beam.

4.3 BEAM ANALYSIS

Since a vacuum system always contains a certain amount of residual gas and because, in addition, substances which are not absolutely pure or compounds of the desired ion species are used as the source feed material, it is necessary to separate the individual components of the ion beam according to their mass. In principle, there are several possibilities to accomplish this. At present, the only methods are magnetic separation and using a Wien filter ($E \times B$ filter). An important point in evaluating beam analysis systems is the available mass resolution, which depends on the characteristics of the separator as well as the energy dispersion of the source.

As mentioned above, there are two points at which separation is possible: either directly after the ion source and prior to acceleration, or following acceleration. Both possibilities have advantages as well as disadvantages. In the first case, the mass filter is at high potential, which makes operation more difficult; however, the filter has the advantage of being very small. The main advantage of separation before acceleration is that in the case of an energy change, the separation parameters do not need to be changed too, since an already separated beam is being accelerated. It is thus possible to change the energy during implantation, which makes it quite simple to obtain a doping profile deviating from the Gaussian shape, the profile being determinable practically at will. In the second case, the mass filter is at ground potential and must therefore be relatively large, since it has to separate ions which have already been accelerated. The magnetic field must be adjusted whenever the acceleration voltage is changed. The only advantage is a simple current supply to the magnet.

For ease of operation, the important factors are good reproducibility and adjustment stability. The magnetic field should be field-regulated (not only current-regulated). For certain production applications, however, a universal adjustability of the separation would in fact be superfluous. Here, one or a few fixed settings which could be applied for certain ion species and energies would be adequate.

The quality of the separation is determined by the resolution, $M/\delta M$ (δM is the minimum resolvable mass difference). Values of about 70 are adequate for simple implantation systems; in some production applications (e.g., dedicated boron systems), the resolution can be even lower. Research systems, however, require a better resolution to make possible the necessary separation in cases of higher mass, as well as for the exact identification of ions. Values of up to 1000 are obtained with units having qualities approaching those of isotope separators.

The resolution is not determined exclusively by the magnet, but rather (in addition to the energy dispersion of the ions) mainly by the focusing of the ion beam. The better the focusing, the better the separation of neighbouring isotopes. Therefore, slit apertures are used after separation to stop the undesired ion species.

4.3.1 Magnetic Separation

Charged particles in a homogeneous magnetic field move in a circular path of radius R:

$$R = \frac{Mv}{qB} = \frac{1}{B}\left(\frac{2M}{q}V\right)^{1/2} \tag{4.2}$$

where q is the charge, M the mass of the ions, B the magnetic induction, V the acceleration voltage and v the velocity of the particles. As can be seen from equation (4.2), particles with different mass and identical energy are deflected differently. The separation, at a deflection of 180° and a mass difference of ΔM, is approximately:

$$D = \frac{\Delta M}{M} R \tag{4.3}$$

The magnet also acts as a lens. If the beam has a divergence of 2α, the width of the beam at the target is given by [2]:

$$W = 2R(1 - \cos\alpha) = R\alpha^2 \qquad \text{(for small values of } \alpha\text{)} \tag{4.4}$$

The influences of non-stabilized acceleration voltage and magnetic-field variations can be determined easily by differentiation of equation (4.2).

Usually, homogeneous sector magnets are used for separation. Their calculation is quite complicated; it has been carried out, for example, by Herzog [329]. Further analyses have been made by Steffen [702] and Enge [226].

If the ion beam enters the magnet perpendicularly and leaves it again perpendicularly (*cf.* Figure 4.14), then the analysis is simpler [226] and one obtains for the separation at a mass difference of ΔM:

$$D = \frac{R\,\Delta M}{2M}\left(1 + \frac{q}{p}\right) \tag{4.5}$$

and the size of the image is obtained as follows:

$$W = \frac{\alpha^2}{2} R \left(\frac{p^2}{q^2} + \frac{q}{p}\right) \tag{4.6}$$

For a symmetrical magnet, i.e., $p = q$ and $\gamma = \lambda$, the same expressions are

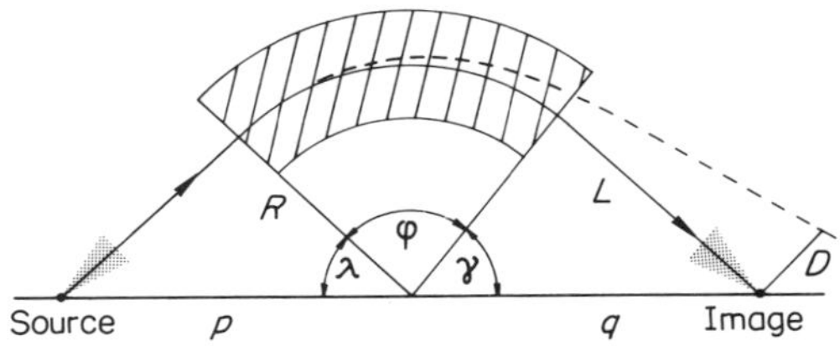

Figure 4.14 Homogeneous sector magnet

obtained as for the homogeneous 180° magnet. If object and image are not on the same plane, then instead of equation (4.5), the following applies [2], [13]:

$$D = \frac{1}{2}\frac{R\,\Delta M}{M}\left[1 - \cos\varphi + \frac{L}{R}\sin\varphi\right] \tag{4.7}$$

In the above, effects caused by stray fields have been neglected. A detailed treatment of this subject can be found in References [227] and [786]. It is possible to suppress these effects in the analysis direction by means of specially shaped pole pieces; perpendicularly to this direction, the stray field can be used to reduce the divergence of the beam.

4.3.2 Wien Filter

The use of Wien (or E × B) filters [656], [775] is limited to low-current implanters ($\leqslant 100\,\mu A$), since space-charge effects worsen the separation effect at higher currents. In the Wien filter, a magnetic field and an electrical field are perpendicular to each other. If the influences of both fields upon a charged particle are the same, i.e.:

$$q\mathbf{E} = q(\mathbf{v} \times \mathbf{B}) \tag{4.8}$$

(where **B** is the magnetic field strength, **v** the ion velocity, and **E** the electrical field strength), then this particle will not be influenced; all particles having a larger or smaller mass, however, will be deflected either to the left or to the right.

If the magnetic field is perpendicular to the incidence direction of the ion beam, the following is valid for the separated mass:

$$M = 2qV(B/E)^2 \tag{4.9}$$

Particles of mass $M + \Delta M$ traverse circular paths having the following radius:

$$R = 2\frac{V}{E}\cdot\left[\left(\frac{M}{M + \Delta M}\right)^{1/2} - 1\right]^{-1} \tag{4.10}$$

A diagram of a Wien filter is shown in Figure 4.15. For small angles, the separation at the target (which is at a distance L from the centre of the filter) is

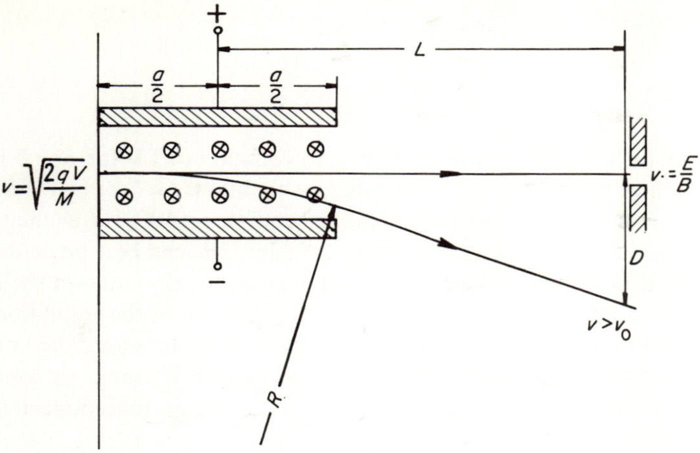

Figure 4.15 Wien filter

calculated as follows:

$$D = \frac{1}{2} La \frac{E}{V} \left[\left(\frac{M}{M + \Delta M} \right)^{1/2} - 1 \right] \approx \frac{1}{4} La \frac{E}{V} \frac{\Delta M}{M} \qquad (4.11)$$

The separation, and thus the resolving power of the filter, can be changed by altering the electric or magnetic field. A disadvantage of this type of separator is that neutral particles can strike the target without being deflected by the filter. For this reason, the use of an additional electrostatic deflector (neutral-beam trap) is absolutely necessary. In addition, unless special measures are taken [13], [656], this filter produces distortions; it is therefore useful only for special applications at low currents.

In the case of multiply-charged particles, nq should be written instead of q in equations (4.2) and (4.9); n is the number of elementary charges. The particle is then separated as if it had the mass M/n. In the ion spectrum, therefore, double-charged ions appear with one-half of their mass, and triple-charged ions with a third of their mass.

4.3.3 Resolving Power

The resolving power of an acceleration system depends (in addition to the separator) upon the energy spread of the ion source, the stability of the acceleration voltage, the beam diameter, and the slit width of the ion detector. Two ion beams are resolvable if the separation of the beams is greater than the sum of the beam diameter ($2R$) and the slit width (S):

$$D \geqslant S + 2R \qquad (4.12)$$

The resolving power can thus be calculated, for a mass difference of 1, as follows:

$$\frac{M}{\delta M} = \frac{D \cdot M}{2R + S} \qquad (4.13)$$

For magnets, a value of $M/\delta M$ according to equation (4.13) for $S = 0$ is always given, as well as a separation, D, according to equations (4.7) or (4.11) for a mass difference of one. For a complete system, equation (4.13) is evaluated using values measured at the target. The beam diameter, $2R$, can best be defined as the full width half maximum (FWHM), or as the standard deviation in the case of a Gaussian beam ($R = \sigma$). For implantation, the definition of the resolution according to equation (4.13) is not adequate. From case to case, one must take undesired neighbouring masses into account, as well as the shape of the ion beam too. Caution is especially recommendable in cases of inhomogeneous beam profiles (astigmatism) with extensive tails.

A problem which has recently come to light, especially in connection with high current machines, is that of particles sputtered from slits and apertures, which are partly implanted into the sample. This problem can be countered by use of suitable materials, for example, apertures and slits made of silicon, aluminium, or graphite in the case of an implanter for silicon doping.

4.4 BEAM DEFLECTION AND HOMOGENEITY

Since it is desirable for ion-optical reasons to use a narrow beam (several millimetres to 1 cm diameter), the beam must be deflected, the sample must be moved, or the beam must be defocused in order to obtain a homogeneous implantation. Defocusing results in very inhomogeneous implantations, and is only suitable for primitive orientational investigations; it was therefore only used in the beginnings of ion implantation. Particularly for semiconductor applications, a very homogeneous implantation is necessary (inhomogeneity $\leqslant \pm 1$ per cent) over samples with diameters of 3 in. to 5 in.

Many factors are important in determining the design of a beam-deflection system. Among the key points are the size of the sample, the beam diameter, the channelling angle, distortion factors, the length of the machine, vacuum-technological factors, and beam heating effects.

The simplest ion-beam deflection method is an electrostatic deflection in the X and Y directions. It is especially important, by means of a suitable analysis system prior to the deflection plates, to ensure that only the desired ion species is scanned. For this reason, beam scanning before mass separation is not practicable.

The deflection of the ion beam by a parallel-plate system on to the target (*cf.* Figure 4.16) is given for small deflection angles by:

$$D = L \tan \alpha \approx L \frac{a}{2d} \frac{V_a}{V} \qquad (4.14)$$

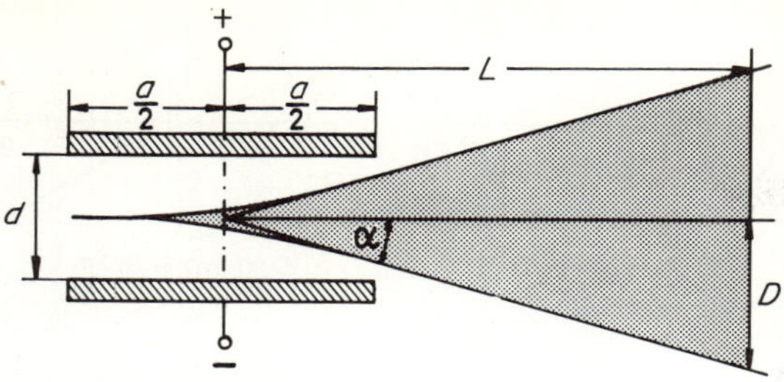

Figure 4.16 Beam scanning by means of deflection plates

(a is the length of the plates, d the distance between the plates, L the distance between the plate system and the target, V_a the deflection voltage, and V the acceleration voltage).

The deflection voltages generally have a triangular shape; their frequencies are between 1 Hz and 10 kHz. It is important to be sure that no Lissajou figures are produced, since these would cause an inhomogeneous implantation. Either a very large frequency difference, for example, about 10 Hz and 1 kHz, or a very small frequency difference, for example, 999 Hz and 1 kHz, have proven effective. It must in all cases be ensured that the sample is scanned an adequate number of times, and that the lines are close enough to each other. A further condition for a homogeneous implantation is an adequate degree of over-scanning, in order to eliminate the possibility of peripheral effects. The distance between the lines should be about $\sqrt{(2)}\sigma$, assuming a Gaussian-shaped beam with a standard deviation of σ. The over-scanning should amount to at least three beam diameters. Under these conditions, the implantation homogeneity is better than 1 per cent, if no additional effects due to the deflection voltages occur. The current loss caused by this over-scanning can be very large. In the case of a 5×5 cm^2 aperture, such as is used, for example, in front of 2 in. wafers, this means that only 20 per cent of the total current is effectively utilized assuming a 1 cm beam diameter, or 40 per cent for a 0.5 cm beam diameter.

As Figure 4.16 shows, the deflection (as seen from an integral point of view) produces a divergent beam. The deflection system must be far enough away from the target so that the maximum deflection angle necessary to implant the entire sample is smaller than the critical angle for channelling. It is generally chosen to be smaller than 1°. For implantations in the channelling direction, this can be too much. In this case, a parallel deflection system is used having a second set of deflection plates, which serve to rebend the deflected beam so that it is again parallel. However, this can cause distortions and thus lead to inhomogeneous implantations. In the case of 300 keV boron, for example, the critical angle for channelling in the ⟨100⟩ direction is 2.03°, and the distance between the deflection plates and the target, assuming a 2 in. wafer, would have to be 70 cm in order to make channelling experiments possible. The deflection angle must be as small

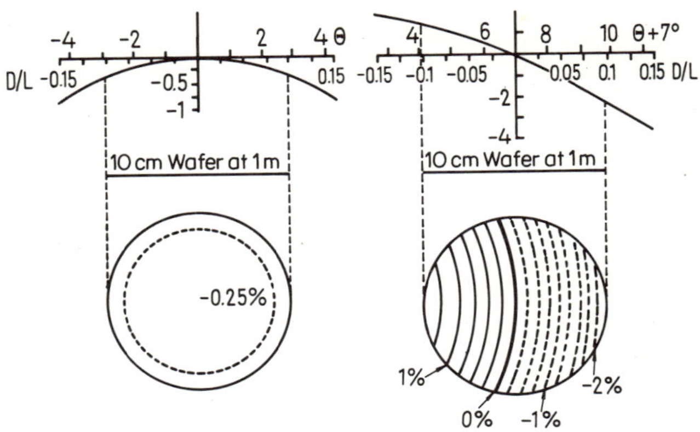

Figure 4.17 Inhomogeneity resulting from electrostatic x–y deflection. (a) Sample vertical to beam direction; (b) sample tilted 7° from vertical [254]

as possible for an additional reason too: the tilting which is necessary to avoid channelling results (together with the deflection angle of the beam) in an inhomogeneous ion deposition. In Figure 4.17, an example is shown for a 7° tilting of a 100 mm wafer, with a distance of 1 m between the sample and the deflection plates [254].

For currents above several milliamps, electrostatic deflection becomes difficult because of space-charge effects. The electrons which neutralize the charge of the beam are stripped by the deflection plates, and a space-charge blow-up occurs. Therefore, mechanical or magnetic deflection is used for high-current implanters. Magnetic deflection is more complicated than electrostatic deflection, since it is not only energy dependent, but also mass dependent. The deflection for small angles is given by:

$$D = \frac{La}{R} = \sqrt{\left(\frac{q}{2M}\right)} \frac{LaB}{\sqrt{V}} \tag{4.15}$$

where L is the distance between the centre of the magnet and the target, a the length of the magnet, R the deflection radius, B the magnetic field, and V the acceleration voltage.

With an electrostatic or magnetic deflection system, considerable inhomogeneities due to neutralized ions can occur, depending on the vacuum. Some of the ions are neutralized after acceleration and mass separation, due to collisions with atoms of the residual gas as well as electron capture. Since these particles are not deflected in the electrostatic or magnetic field, they form a 'hot' spot on the sample with a greatly increased local dose. This additional component, since it is neutral, cannot be determined by current measurement.

The inhomogeneity, $\Delta N_\square / N_\square$, which is produced by this beam component can be calculated [605] as follows:

$$\frac{\Delta N_\square}{N_\square} = 2.48 \cdot 10^{14} P L \sigma_{10} \frac{A}{a} \tag{4.16}$$

(P is the pressure (Pa), L the path length of the particles (cm), σ_{10} the effective cross-section for the formation of neutral particles (cm^2) from single-charged particles, A the area over which the beam is scanned, and a the area of the non-deflected (focused) beam.)

The cross-section for electron capture by ions of velocity v can be calculated, according to Bohr and Lindhard [94], as follows:

$$\sigma_{10} = \pi a_0^2 Z_1^2 Z_2^{1/3} \left(\frac{v_0}{v}\right)^3 \tag{4.17}$$

where a_0 is the Bohr radius ($a_0 = \varepsilon_0 h^2/\pi m q^2 = 0.0529$ nm) and v_0 is the Bohr velocity ($v_0 = q^2/2\varepsilon_0 h$). The values of σ_{10} amount to approximately 10^{-16} cm^2.

For a pressure of 10^{-4} Pa, a path length of 8 m, a beam cross-section of 1 cm^2 and a scanned area of 100 cm^2, assuming for σ_{10} the value of 2×10^{-16} cm^2, the inhomogeneity ($\Delta N_\square / N_\square$) amounts to 40 per cent.

The neutral beam can be reduced by a better vacuum or a 'neutral-beam trap', which additionally deflects the charged particles. Either a third pair of deflection electrodes or preferably, a superimposed dc voltage on the last pair of plates are usable. Both of these possibilities are shown diagrammatically in Figure 4.18. The

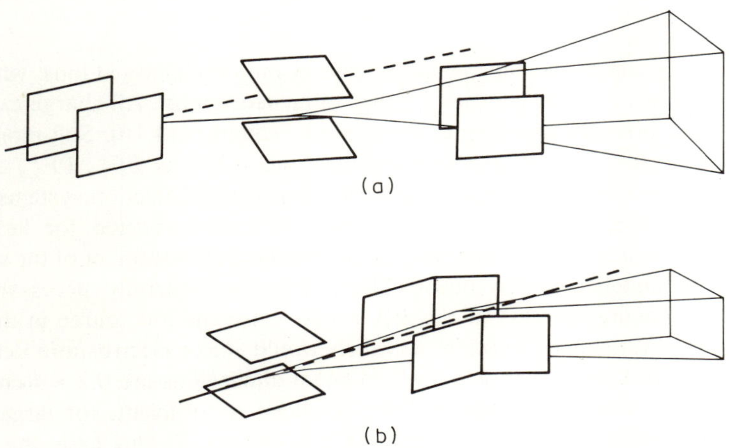

Figure 4.18 Suppression of neutral beam by electrostatic deflection, using (a) separate deflection plates, or (b) a superimposed voltage on one pair of plates

cross-section σ_{12} for the production of double-charged ions has the same order of magnitude as σ_{10} [507]. This effect, however, only causes problems under very poor vacuum conditions, due to falsification of the current measurement. A simple method to investigate the implantation homogeneity is the use of transparent plastic foils (for example, Kalle Hostaphan BN 250 (transparent) or Mylar foil are well suited), which become coloured when ions are implanted. Depending on the ion species and the intensity, brownish to red colours are obtained. Dose differences below 5 per cent can be seen clearly. This method is especially useful for neutral beam detection. For implantations at high energy densities ($\geqslant 0.1$ W/cm^2), caution is advised due to heating of the foil, which can lead to the release of gases.

A further source of possible error is the implantation of other ion species with the same effective molecular weight, for example, as in the case of a $^{32}S^+$ implantation of oxygen molecules (O_2^+) or double-charged zinc ions ($^{64}Zn^{++}$). In order to avoid such effects, careful attention must be given to the substance used for ionization as well as the materials of the ion source. ZnS, for example, would be a practical material to produce zinc and sulphur ions, but since $^{64}Zn^{++}$ has the same effective mass as $^{32}S^+$, this substance is only suitable for zinc.

The implantation homogeneity can also be influenced greatly by charge-exchange reactions, depending on the pressure in the vacuum system. If ions change their charge state, or if the number of atoms in the molecule is changed:

$$A_p^{n+} \rightarrow A_q^{m+} \tag{4.18}$$

(n and m are the number of charges, p and q the number of atoms), then a new effective mass, M', is obtained:

$$M' = \frac{q^2}{p}\frac{n}{m^2}M \tag{4.19}$$

This effect especially plays a role in the case of multiply-charged ions, which are frequently used to extend the energy range of an accelerator. All charge-exchange reactions are pressure dependent, according to equation (4.16). Still more complicated reactions can occur with molecular ions, such as BF_2^+, PF_3^+, etc.

For high currents, difficulties occur with electrostatic deflection systems due to space-charge effects. The above-mentioned alternative method for large-area implantation, which helps in this case, is the mechanical movement of the samples through the undeflected ion beam. This method is generally necessary with accelerators having currents in the mA range, since the ion source in this case usually has a slit-shaped extraction, which would make electrostatic deflection relatively difficult and expensive. Typical beam dimensions are 0.2×4 cm^2. For small wafers, movement in the X direction alone is sufficient; for larger ones, movement in both the X and Y directions is necessary. In this case, the above-mentioned problems concerning the influence of the neutral-beam component on the homogeneity no longer occur. Mechanical scanning is sometimes combined with magnetic scanning in the case of spinning-disc systems. An advantage of

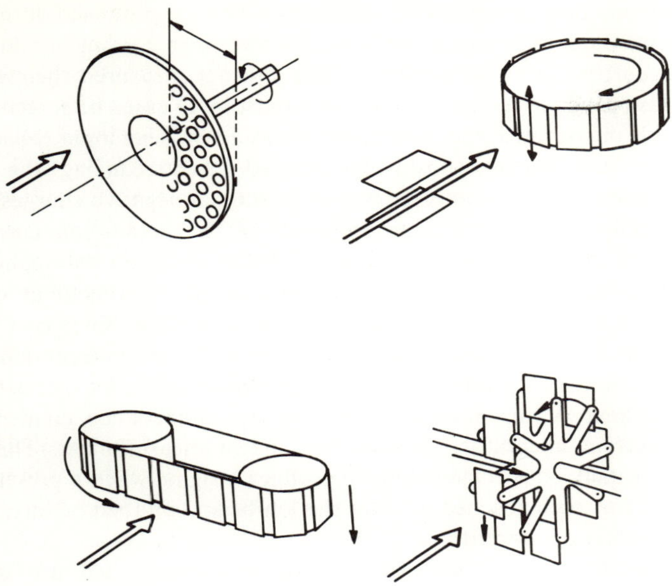

Figure 4.19 Examples of mechanical beam-scanning systems; from References [254] and [593]. Thick arrows: ion beam; thin arrows: movement direction of the wafer holder

mechanical deflection is that the ion beam is not subjected to any additional divergence due to the deflection. On the other hand, in the case of rotating-drum systems, the definition of the angle can present a problem. Examples of various mechanical beam-scanning systems are shown diagrammatically in Figure 4.19.

4.5 SAMPLE CHAMBERS

The last topic in this chapter's discussion of ion-accelerator components is the vacuum chamber in which the samples are implanted. Depending on the intended use, this component has to meet varying requirements; these differ greatly according to whether a unit for universal experimentation purposes is being considered, or one for industrial production. In the first case, flexibility regarding the sample's temperature, its orientation, and its dimensions is essential; in the second case, the critical points are the wafer throughput, the implantation, homogeneity, and the operational dependability. The construction of implantation chambers, accordingly, is different from case to case.

4.5.1 Current Measurement

A knowledge of the exact dose, i.e., the time integral of the implanted ion-current density, is necessary for all ion-implantation experiments. A major problem in this connection is posed by the secondary electrons which are produced during the ion bombardment. Since they are negatively charged, their current is added to the positive ion current. Neutral particles which are not measured themselves, but which can also produce secondary electrons, must in all cases be screened out by taking suitable measures. If the current meter has a very high input resistance, the target holder can become positively charged, thus reducing the effective secondary-electron yield; it can even attract additional negative charges and thus reduce the indicated current below the level of the actual ion current. The measurement of ion beams itself is not new. During ion implantation, however, a relatively large area is usually irradiated, the sample is sometimes cooled or heated, and the latter should be visible through a window. Since one is always interested in the implanted dose, which is determined by the current density only, there must be an aperture in front of the sample, permitting an exact definition of the implanted area. A current meter with a low input impedance, connected to an integrator (current integrator), is used for current measurement. This current integrator is usually constructed so as to produce a signal when a preset ion dose or ion charge has been reached, so that the ion beam can then be interrupted or the next implantation started.

The suppression of secondary electrons can be accomplished most simply by using a negatively-charged electrode (suppression electrode) mounted in front of the sample holder. Constructing the target holder as a Faraday cup improves the secondary-electron suppression still further. It is also possible to design the negatively-charged electrode as a cylinder. The ion beam must not be allowed to strike the suppression electrode, since it would then produce secondary electrons which could be measured at the target. The sputtering of particles from the target holder, as well as secondary electrons produced by these particles, cause further errors (which will usually be slight, however). A further method for suppressing secondary electrons is the use of weak magnetic fields ($\approx 10^{-2}$ T), which direct the electrons back to the target. However, this is usually too complicated if wafer transport systems are used. Figure 4.20 shows three examples: a Faraday cup, a suppression cylinder, and a simple suppression electrode. The necessary voltage depends mainly upon the geometry of the system, and must be determined experimentally.

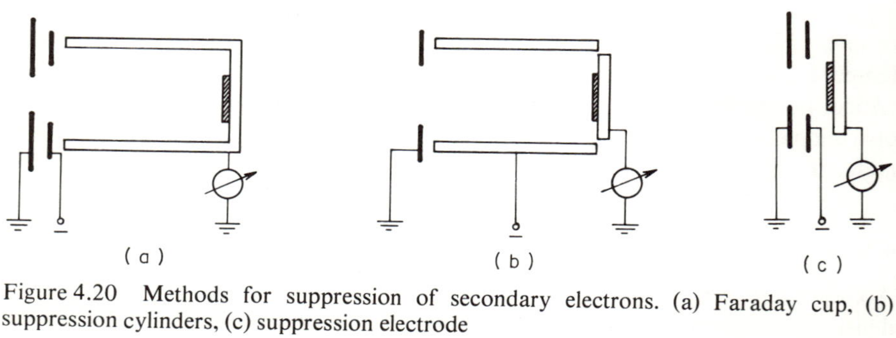

Figure 4.20 Methods for suppression of secondary electrons. (a) Faraday cup, (b) suppression cylinders, (c) suppression electrode

4.5.2 Sample Orientation

In making fundamental investigations, it is often necessary to implant into exactly oriented samples. Optical or X-ray methods (Laue cameras), permitting an accuracy of ±0.5°, are used for this purpose; still better are measurements of the back-scattering yield of ions. For the latter method, helium or protons are generally used (see Section 5.8). The achievable accuracy of this orientation is approximately ±0.02°. Channelling and other experiments can then be made with the oriented samples, or the implantation can be monitored *in situ* by back-scattering measurements. For such experiments, the samples are mounted on goniometers which allow precision tilting of the samples in all directions with an accuracy of ±0.01°. Only a suppression of the channelling effect is of interest for most experiments, as well as for device fabrication. The semiconductor samples, which are oriented by the manufacturer to a precision of ±0.5°, are therefore tilted between 7° and 10° away from the vertical, whereby attention must be given to the correct direction of rotation. Tilting to a greater degree is not done because one would again come into the direction of low-indexed channels or planes. A tilting optimized by back-scattering measurements would lead to best

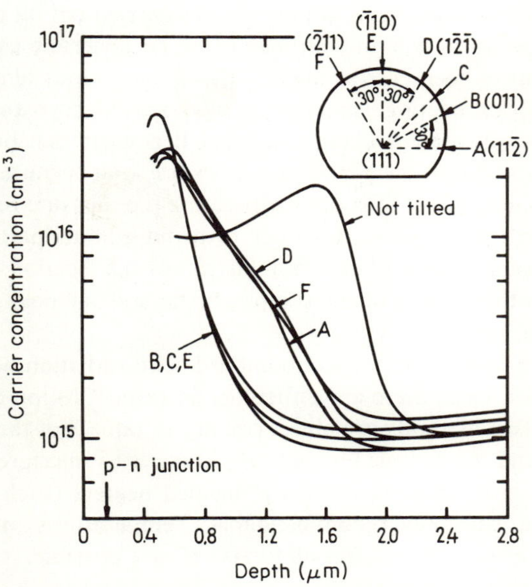

Figure 4.21 Dependence of the profile shape on tilting and rotation for a 450 keV phosphorus implanted silicon sample tilting 10° away from ⟨111⟩ orientation, according to Reddi [585]. Reproduced by permission of Solid State Technology, Technical Publishing, a Company of Dun & Bradstreet

results; this procedure, however, involves too much effort for practical use. Figure 4.21 shows the influences of sample orientation and rotation angle for ⟨111⟩-oriented silicon wafers implanted with phosphorus, based on a study by Reddi [585]. It can be seen clearly that channelling plays a role even in cases of perfectly misoriented samples.

4.5.3 Heating and Cooling

Heating or cooling of samples during implantation, in the case of certain semiconductors, can lead to a better incorporation of the implanted ions (e.g., for GaAs and InSb), or can make fundamental investigations possible regarding the production and annealing of radiation damage. Most research equipment, therefore, includes provisions for heating and cooling. Equipment used for production, on the other hand, seldom includes these features, since the additional expense is prohibitive. If, in the case of production units, sample heating caused by the beam itself becomes a problem (*cf.* Section 3.6), this undesired effect can be countered by simultaneous implantation of several wafers (*cf.* Section 5.4) as well as cooling the wafer holder by Freon or water. Recently, it has been shown that the heat contact between wafer and target holder can be improved considerably by filling the space between wafer and holder with gas [411a] (gas cooling), thus also reducing the wafer temperature.

Cooling to low temperatures is almost always carried out by means of contact cooling. The temperature of liquid nitrogen (77 K) is generally used; a few experiments, however, have been carried out at 4 K. A continuous temperature regulation by simultaneous heating, such as in evaporation cryostats, is practically never used and is not commercially available. It is clear that only single wafers can be implanted, and that the processing times are long because of the necessary heating of the sample holder to room temperature (condensate formation!) before the vacuum chamber is opened. In order to avoid contamination of the cooled sample due to condensation of residual-gas atoms (*cf.* Section 4.6), a cold trap should be used in the vicinity of the sample; better still is a cold wall surrounding the entire sample holder.

Samples can be heated by electron bombardment, radiation heating or contact heating. In the first case, there are difficulties in respect to ion-current measurement. The other two methods are about equally suitable, but the latter is simpler in construction, and the sample temperature is easier to measure and control. As heating elements in a vacuum, coaxially-shielded heaters (such as Thermocoax from Philips) are especially recommendable. Temperatures up to 500 °C are usual; in special cases, temperatures up to 900 °C are possible.

4.5.4 Examples of Implantation Chambers

A universally applicable chamber for research purposes has been developed at Harwell [300]; this is shown in Figure 4.22. The chamber can hold up to 36 samples. Wafer changing takes place in a manner similar to a slide projector. A

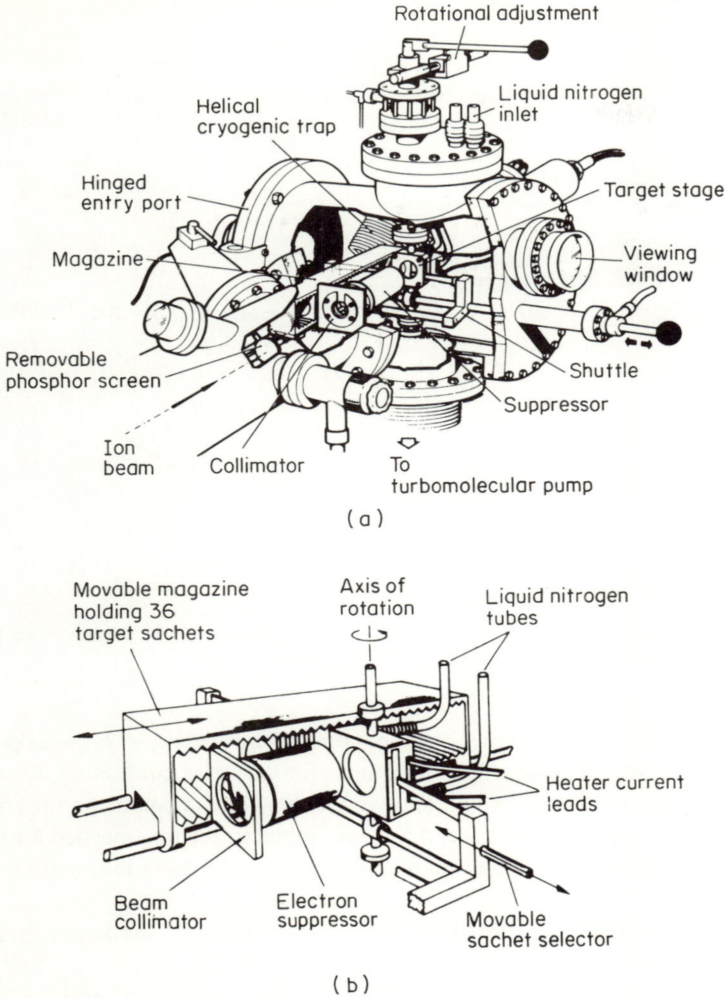

Figure 4.22 Target chamber for universal research purposes developed at Harwell [300]. (a) General view, (b) details of the cassette system. Reproduced by permission of the United Kingdom Atomic Energy Authority

goniometer permits the exact alignment of pre-oriented samples. The samples can be cooled by liquid nitrogen or radiation heated. Target chambers for industrial application have also been developed at Harwell. A cassette system for an accelerator with a wedge-shaped ion beam is shown in Figure 4.23; it has a capacity of 20 sample holders with up to 120 2 in. wafers. In this system the beam is not deflected, but rather, the wafer carriers are moved through the beam, as indicated by the arrows. The linear cassette system shown in Figure 4.24,

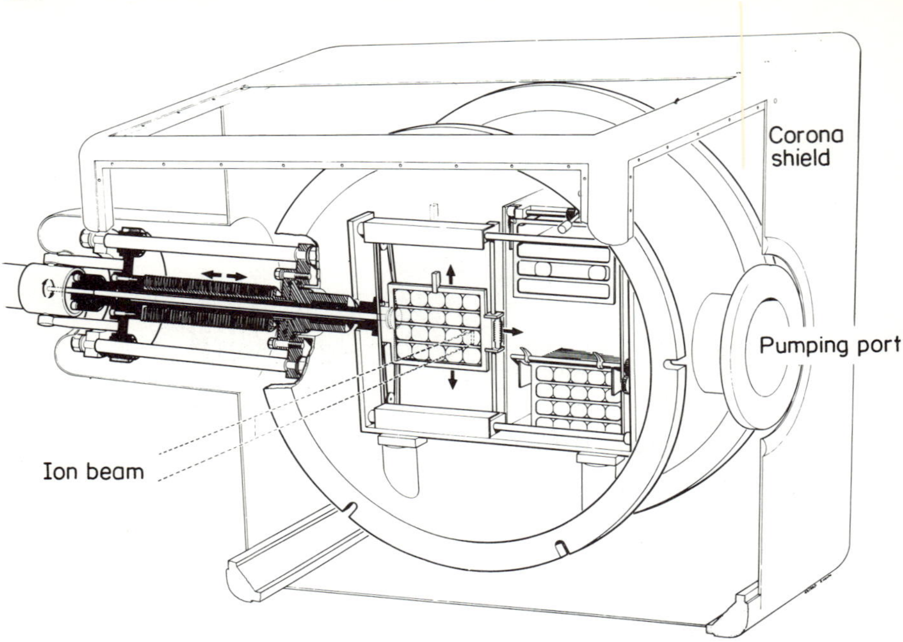

Figure 4.23 Industrial target chamber developed at Harwell [255]. Reproduced by permission of the United Kingdom Atomic Energy Authority

manufactured by Extrion, allows quasi-continuous operation, with the help of two small vacuum chambers which serve as locks for cassettes containing 20 samples each. The locks are located behind the two valves. The ion beam can thus be used much more effectively, and the 'dead time' is limited to the time needed for sample changing, i.e., the insertion and removal of the sample holders into and out of the cassette.

All units described up to now have been so-called 'batch' units, i.e., the wafers are brought into the target chamber in groups, and are then implanted. In the case of the Harwell machines, several wafers are always implanted simultaneously; with the Extrion unit, each wafer is implanted separately.

Continuous systems, in which individual wafers, sometimes mounted on a holder, are transported into the target chamber by means of differential pumping or using vacuum locks, fit in well with the current industrial trend towards fully-automated and continuous production. Such systems are mostly in the development stage at present [722], [780], but some are already on the market. Figure 4.25 shows such a chamber made by Varian-Extrion. A similar chamber is used by the Eaton/Nova medium-current implanter. The samples are transported from the wafer carriers into the entry lock, the upper valve closes, and the small chamber volume is quickly evacuated. The lower valve opens and the sample moves on to the sample holder, which lifts up for the implantation. After implantation, the sample leaves the high-vacuum area via the exit lock and is

Figure 4.24 Target chamber made by Varian-Extrion, with cassette system and vacuum locks for quasi-continuous operation

transported into the lower wafer carrier. A rotatable Faraday cup allows centring of the ion beam; four additional Faraday cups at the entrance aperture are used to adjust the deflection voltage and measure the scanning homogeneity. Figure 4.25(b) shows this continuously operating chamber with the first Faraday cup, the suppression electrode, the entrance aperture, and the additional Faraday cups. Instead of wafer carriers, this chamber can also be used with continuous transport systems.

Sample heating is an extremely important problem when using high-current machines. Figure 4.26(a) shows equilibrium temperatures of silicon samples depending on current density, for implantation energies of 50 to 300 keV (calculated according to equation (3.51)). The temperature rise as a function of

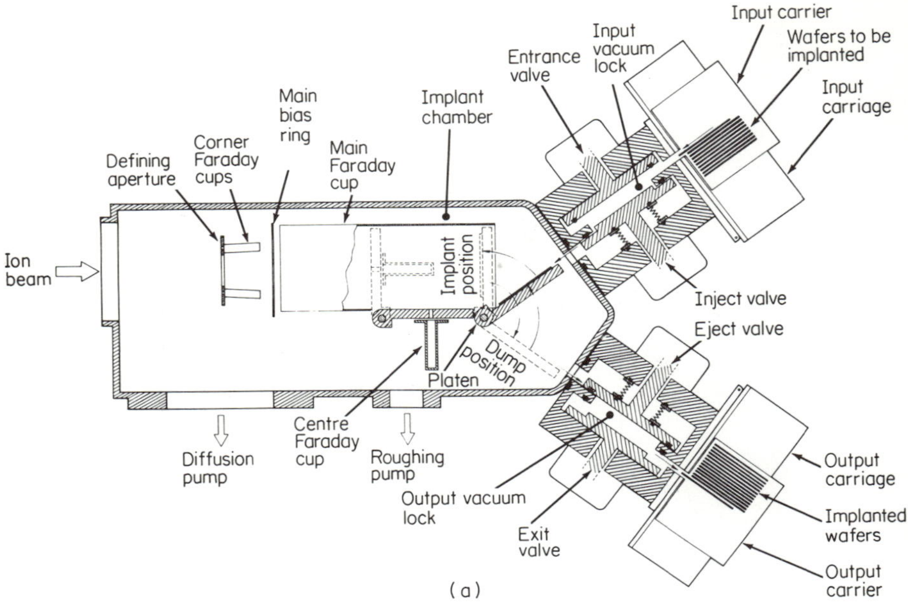

(a)

(b)

Figure 4.25 Target chamber for continuous operation [780]: (a) Schematic diagram, (b) photograph of sample holder with suppression cylinder (partly opened)

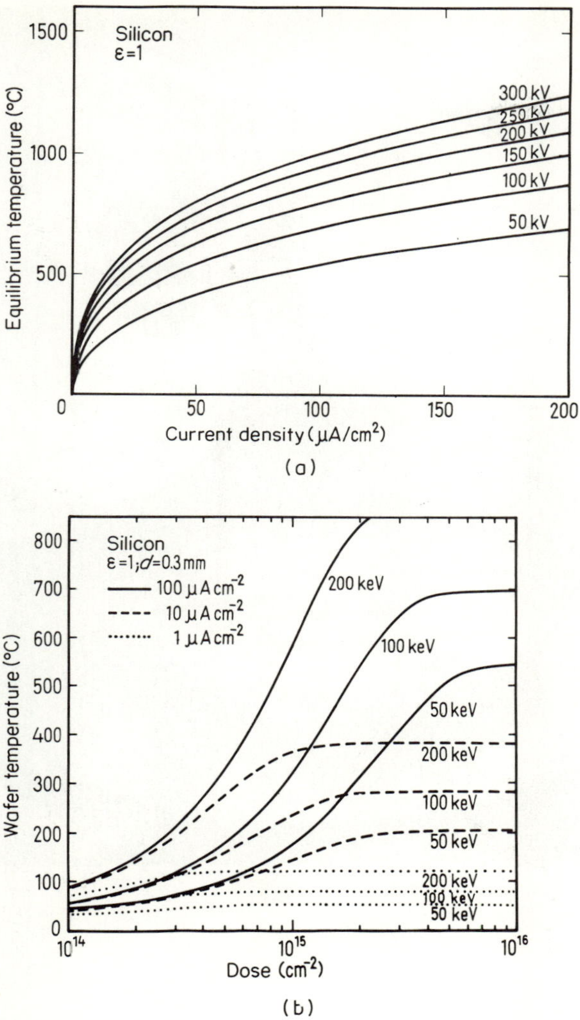

Figure 4.26 (a) Equilibrium temperature of silicon, depending on the current density, at 50 to 300 keV. (b) Temperature rise depending on implantation dose for 50 to 200 keV, with the current density as parameter

implantation dose, with applied current density as a parameter, is shown in Figure 4.26(b) (according to equation (3.52)). From this diagram, it can be seen clearly that temperature rise becomes a serious problem at doses above 10^{14} cm^{-2}. The solution is to implant several wafers simultaneously, as with the unit from Harwell shown in Figure 4.23, or to use disc [593] or drum-shaped sample carriers to

(a)

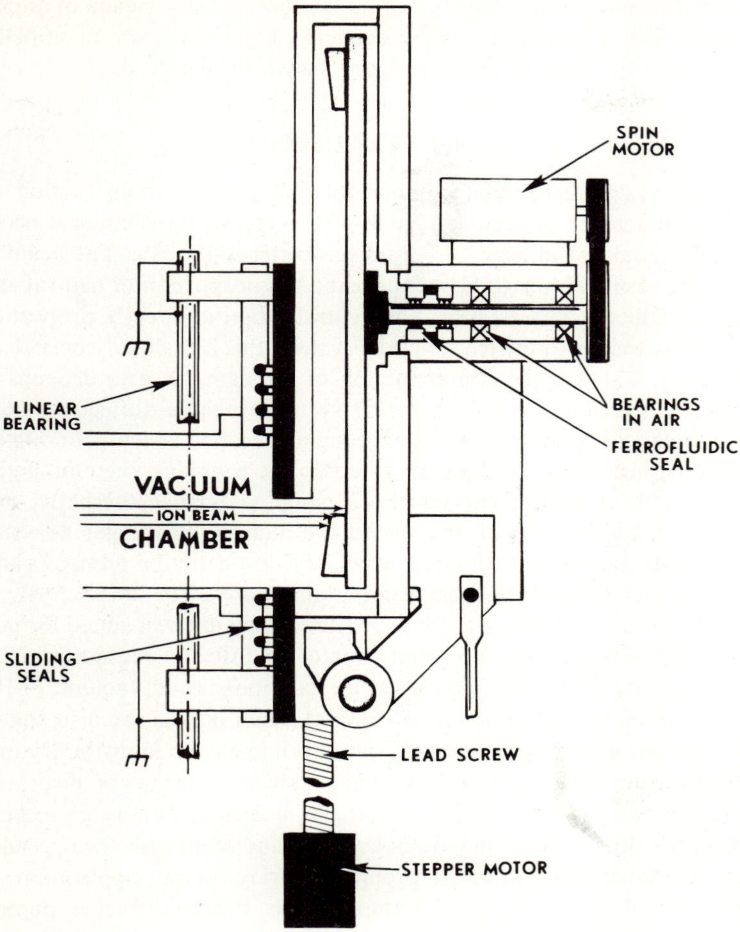

(b)

Figure 4.27 Disc-shaped target chambers of high current accelerators. (a) Photograph of the open chamber of the Extrion 80-10 implanter partly loaded with wafers. In the rear a second chamber is seen in which an implantation is performed for maintenance. (b) Schematic view of the target chamber of the Nova high current implanter

reduce the effective current density during implantation by means of an increase in the area. The disc-shaped sample carriers of the high-current implantation machines made by Extrion and Nova are shown in Figure 4.27.

4.6 VACUUM

The vacuum system is a highly important factor in the construction of ion-implantation machines. A reliable system with a very good vacuum is necessary. The vacuum should be at least 10^{-4} Pa; even better is 10^{-5} Pa. The beam can be focused more easily with a good vacuum, and the proportion of neutral atoms in the beam is reduced considerably. The neutralization is directly proportional to the pressure as well as to the length of the system; this has already been discussed in Section 4.4. The surface contamination of the samples also depends on the quality of the vacuum. Suitable pumping systems include diffusion pumps and turbo-molecular pumps with mechanical roughing pumps or ion-gettering pumps, which only require absorption pumps to attain the roughing vacuum, perhaps in conjunction with an oil-free carbon piston pump. A further alternative are cryo pumps. The choice depends on the nature and quantity of the gas as well as the degree to which an oil-free vacuum is necessary. Ion-gettering pumps produce an absolutely oil-free vacuum, but they cannot pump rare gases satisfactorily, have a low capacity, and are sensitive to air incursions. They are best suited for positions in the vacuum system which are rarely vented and do not require a high pump capacity (e.g., the beam lines), or when an extremely good vacuum ($<10^{-5}$ Pa) and absolutely oil-free conditions are required. Such pumps are also suitable for sample chambers if certain precautions are taken (e.g., venting with dry nitrogen, pressure gradient to the rest of the system, roughing by means of absorption and carbon-piston pumps down to 10^{-2} Pa). Turbo-molecular pumps are more robust and are practically oil-free. They do not require liquid nitrogen for operation, but are very expensive and can cause problems during certain applications due to vibration. The most widely used pump type is the oil-diffusion pump. It is relatively inexpensive, is reliable, and has a high pumping speed. However, in order to reduce hydrocarbon contamination of the samples, one must equip diffusion pumps with cold traps for liquid nitrogen, refrigerant coolers, or thermoelectric cold traps. Even if this is done, though, a complete avoidance of such contamination is not possible. In the vicinity of the ion source, the use of this pump type is generally necessary due to the large quantities of gas to be evacuated, but ion pumps and especially turbo-molecular pumps can also be used for the beam lines and the sample chamber, depending on the technical requirements. The best oil for the operation of diffusion pumps is one that is insensitive to air incursions. Polyphenyl ethers have proven to be better in practice than silicon oils for this purpose, since the former lead to no contamination due to creep effects and have a lower vapour pressure. The recently introduced cryo pumps have an extremely high pumping speed. The gases and vapours used in operating the source condense in the pump and are released during regeneration. Since they are poisonous, this type of pump can only be used in special cases.

In order to keep the leakage rate low, metal seals would be preferable, especially for production systems which are to be operated without alteration after having been installed. In the case of research equipment, a combination system allowing the use of metal or Viton seals is better for reasons of flexibility. Under some circumstances in the case of long systems, it is advisable to equip the beam lines with cold traps. The employment of cold traps in the vicinity of the target is necessary if implantation is to be carried out at low temperatures, in order to avoid contamination of the cooled samples due to condensation of residual gases. The main impurities are hydrocarbons deriving from the pump fluid. The deposition rate of the impurities is higher under high partial pressure, low acceleration voltage, and low acceleration temperature; it can also depend on the current density [788a].

5
Measurement Techniques to Investigate Ion-implanted Layers

During ion implantation, various undesirable effects generally occur together with the desired electrical doping or the alteration of the physical and chemical properties; for this reason, precise measurements to investigate the implanted layers are necessary. The most important problem areas in the context of semiconductor doping are the influences of radiation damage and the electrical activation of the implanted ions.

The following sections describe the most important methods to measure radiation damage, the distribution of the implanted ions, and their electrical activation; the limits of these methods will also be discussed.

Ion implantation involves higher requirements for measurement techniques than other doping methods, due to the low penetration depth of the ions and the resulting steep profile shapes. A number of methods for implantation-profile measurement have been developed from new or adapted from previous techniques. Basically, the profile shape of the implanted distributions is of interest, but often, integral information on the implanted distribution can be adequate—or must be adequate, for reasons of detection sensitivity or measurement speed.

The various measurement techniques can be divided roughly into three groups: (a) chemical methods, (b) electrical methods, and (c) other physical methods. The first group includes the chemical etching of *p–n* junctions, while the second comprises Hall effect, sheet resistivity, capacitance voltage, spreading resistance, and current voltage measurements, among others. The third group includes several very different measurement methods. The most important of these are the Rutherford back-scattering technique, activation analysis, nuclear reactions through bombardment with high-energy atoms, and the excitation of characteristic X-rays.

When semiconductors are doped, one is generally interested in the distribution of the electrically active ions, or more specifically, the distribution of the majority carriers. Non-electrical techniques can be better suited for fundamental investigations, especially since electrical methods can involve a high degree of error. Therefore, non-electrical techniques will be treated below in relatively great detail, though only a few research laboratories—and practically no semiconductor production facilities—have all, or even some, of the corresponding equipment on hand.

It is not easy to categorize these different measurement techniques. According to the problem being investigated, for example, they could be described as profile measurement methods, radiation-damage measurement methods, etc., or according to the nature of the measurement principle, as chemical, electrical or physical measurement methods. Since a certain amount of overlap is inevitable, it is better not to attempt a strict classification, but rather to simply discuss the individual methods one by one. A comparison of their special capabilities will be given in the last section.

5.1 STAINING OF *p–n* JUNCTIONS

A long-known method to determine the depth of diffused *p–n* junctions is the staining or decorating technique. This method can also be used effectively in connection with ion-implanted junctions. However, bevelling is always necessary to obtain sufficient resolution, since the *p–n* junctions are relatively close to the surface (unless a 'drive-in' diffusion has been carried out). The necessary angles are in the range of 1° ($34'40'' = 1{:}100$; $1°9' = 1{:}50$). A mixture of 0.3 µm Al_2O_3 powder in water has proven very well suited for bevelling in the case of silicon. In the case of ⟨100⟩-oriented III–V semiconductors, one generally uses ⟨110⟩-oriented cleaving planes or bevels made with a fine diamond paste. Suitable etchants to make *p–n* junctions visible are listed in Table 5.1. The *p*-areas and *n*-areas become differently coloured, or a fine line is obtained at the *p–n* junction, depending on the etchant used.

By means of this method, it is only possible to determine the depth at which the substrate doping is equal to the implanted doping. If one implants simultaneously into materials with different background dopings, it is possible after annealing to construct a depth profile from the different positions of the *p–n* junction [289], [412]. Annealing is necessary because the etching depends on the carrier concentration. The principle of this method is shown in Figure 5.1.

The method just described, however, requires considerable time and effort, and its results can be unprecise since at high concentrations, under some circumstances, several lines can appear, while it is uncertain at low concentrations whether the line really corresponds to the metallurgical *p–n* junction. Therefore, it is used only seldom, and its use is generally restricted to orientational investigations on the obtainable *p–n* depth.

5.2 DETERMINATION OF THE CONDUCTIVITY TYPE

The determination of the conductivity type is possible by several means: the use of Hall effect measurements, observation of a *p–n* junction, or use of a thermoprobe. The first two methods require contacts or special measurement structures; they will be discussed in more detail later on, since they provide much more informa-

Table 5.1 Staining etches for p–n junctions

Semiconductor	Composition of etch	Etching time	Remarks	Reference
GaAs	3 g KOH dissolved in 25 ml H_2O 2 g $K_3[Fe(CN)_6]$ dissolved in 25 ml H_2O	A few seconds	Also usable for GaAlAs, GaAsP and GaP; should be mixed before etching; suitable also for n^+–n and p^+–p; a line is obtained in these cases.	
	3 CP4 : CH_3OH : H_2O CP4[10 HNO_3(conc.) : 5 HF (48%) : 11 CH_3COOH]	A few seconds	p becomes lighter in colour, n becomes grey.	[317]
Si	2 g H_6JO_6, 1 ml HF (48%) 20 ml H_2O	About 30 seconds	Under UV light; also for conductivity differences; discoloration.	[317]
	5 HNO_3(conc.) : 95 HF (48%)	A few seconds	Under UV light	[265]
	0.5 ml HNO_3(conc.) : 99.5 ml HF (48%)	A few minutes	Under UV light; discoloration.	[90]
Ge	2 HNO_3(conc.) : 1 HF (48%)	A few seconds		[90]

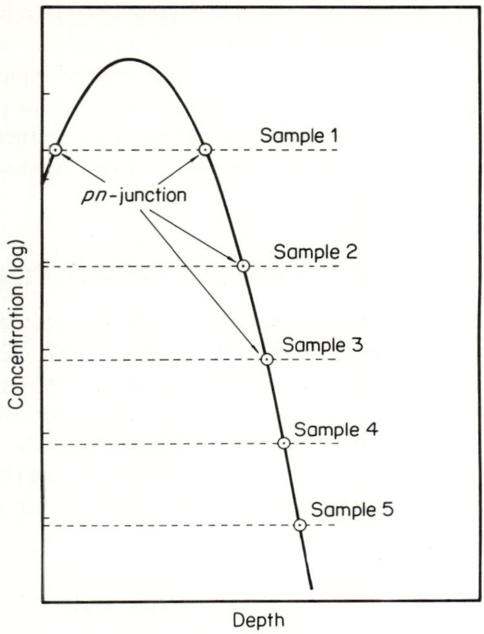

Figure 5.1 Measurement of profiles by simultaneous implantation in differently doped materials and staining of the *p–n* junctions

tion than only the conductivity type. The use of a thermoprobe is a very simple procedure to investigate whether an implantation has led to a change in the semiconductor doping. If one heats an area of the semiconductor locally, the charge carriers diffuse in the direction of the temperature gradient (Seebeck effect) and a potential difference thus results. For example, it is possible to use two point contacts, one of which is heated, and to press them next to each other on to the semiconductor surface. If the semiconductor is *n*-type, a positive voltage results between the hot contact and the cold one; if it is *p*-type, a negative voltage results.

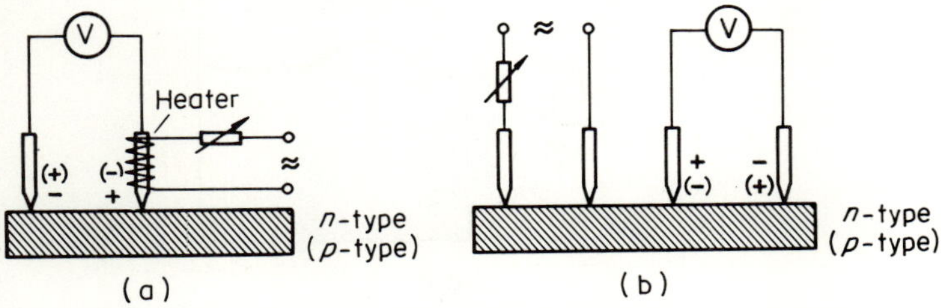

Figure 5.2 Thermoprobe principles: (a) heating with the use of a heating element (e.g., a soldering iron), which heats a measurement contact; (b) use of a four-point probe, with direct heating of the semiconductor

In general, the signal amounts to several millivolts and decreases with increased heating of the crystal. Two thermoprobe types are shown diagrammatically in Figure 5.2. The type shown in Figure 5.2(a) is universally applicable; the contacts can be rectifying. For the type shown in Figure 5.2(b), a four-point probe is used and the heating current passes directly through the semiconductor. This method is only suitable for relatively high-doped materials (otherwise, there can be problems with rectifying metal semiconductor contacts).

5.3 CAPACITANCE–VOLTAGE MEASUREMENTS

The measurement of doping profiles by differential capacitance–voltage measurements (C–V measurements) is a widely used method [334], [724]. An abrupt *p–n* junction or a Schottky contact is made at the semiconductor surface and the capacitance is measured as a function of reverse bias. For evaluation purposes, the approximation by Schottky is used [630], [631]. The semiconductor is depleted of mobile charge carriers up to a depth of x; beyond this, there is charge neutrality (Figure 5.3). The relationship between the capacitance $C(V)$ and the bias V is given by:

$$\frac{1}{C^2(V)} = \frac{2}{q\varepsilon_0\varepsilon_r N(x) A^2}(V_D + V) \qquad (5.1)$$

The doping is obtained by differentiation of equation (5.1):

$$N(x) = \frac{C^3(V)}{q\varepsilon_0\varepsilon_r A^2}\left(\frac{dC(V)}{dV}\right)^{-1} \qquad (5.2)$$

(A is the area of the Schottky contact, q is the elementary charge, ε_0 and ε_r are the permittivity and the dielectric constant, and V_D is the diffusion voltage.)

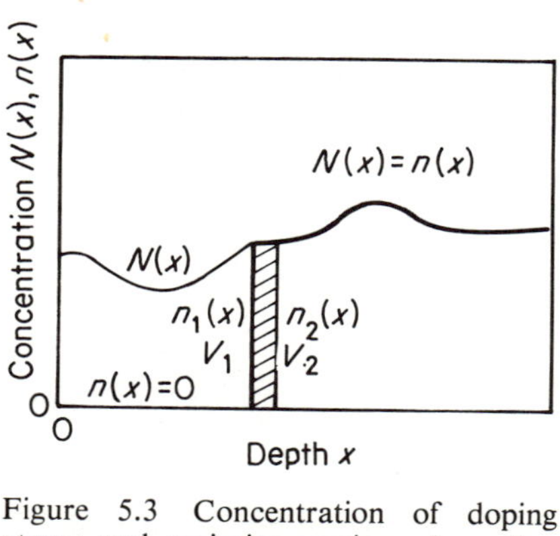

Figure 5.3 Concentration of doping atoms and majority carriers, depending on the bias of a Schottky contact in the case of an abrupt junction

The width of the space-charge (and thereby the related x-coordinate) is obtained from the measured capacitance, according to the capacitor formula, as follows:

$$x = \frac{\varepsilon_0 \varepsilon_r A}{C(V)} \tag{5.3}$$

This approximation assumes that the Debye length L_D is 0.

The following sections will examine the questions: (a) To what extent is this approximation applicable for steep implantation profiles? (b) Which additional errors can occur? and (c) Which different measurement techniques can be applied?

5.3.1 Limits of the Method

Debye Length

A criterion for quasi-neutrality [390], and thus for the validity of equations (5.2) and (5.3), is obtained by considering the Poisson equation and the current equations at the Schottky contact. All derivations below are made for n-type semiconductors, but are also applicable without difficulty for p-type ones.

If the minority carriers are neglected, the Poisson equation is written as:

$$\frac{d^2\Phi}{dx^2} = -\frac{\rho}{\varepsilon_0 \varepsilon_r} = -\frac{q}{\varepsilon_0 \varepsilon_r}[N(x) - n(x)] \tag{5.4}$$

For the current across the Schottky contact, the following applies:

$$I_n = qD_n \frac{dn(x)}{dx} - q\mu_n n \frac{d\Phi(x)}{dx} \tag{5.5}$$

Φ is the potential; $n(x)$ is the majority carrier concentration.

If the current at the contact is 0, then using the Einstein formula $D_n = \mu kT/q$, one obtains for the relationship between the doping concentration $N(x)$ and the majority carrier concentration $n(x)$, from equations (5.4) and (5.5):

$$N(x) = n(x) - \frac{kT}{q}\frac{\varepsilon_0 \varepsilon_r}{q}\frac{d}{dx}\left(\frac{1}{n(x)}\frac{dn(x)}{dx}\right) \tag{5.6}$$

$N(x) \approx n(x)$ must be valid for quasi-neutrality, so that the following is obtained:

$$\left| N(x)\frac{d^2N(x)}{dx^2} - \left(\frac{dN(x)}{dx}\right)^2 \right| \ll \left(\frac{N(x)}{L_D}\right)^2 \tag{5.7}$$

where L_D is the Debye length, which is defined as follows:

$$L_{\mathrm{D}} = \sqrt{\left(\frac{kT\varepsilon_0\varepsilon_r}{q^2N(x)}\right)} \tag{5.8}$$

The Debye length, depending on doping as well as temperature for Si and GaAs, is plotted in Figure 5.4. For example, the Debye length is 0.13 μm at 300 K and a doping concentration of 10^{15} $\mathrm{cm^{-3}}$; for a boron implantation at 30 keV, $R_p = 0.099$ μm and $\Delta R_p = 0.037$ μm. This shows that implantation profiles change significantly within the range of several Debye lengths, depending on the doping concentration.

A criterion for the doping concentration below which the majority carrier deviates from the impurity concentration, so that errors can occur when using equation (5.2), is obtained under assumption of a Gaussian profile:

$$N(x) = N_{\max} \exp\left[-\frac{(R_p - x)^2}{2\Delta R_p^2}\right]$$

If the error is to be <1 per cent, one obtains, by insertion into equation (5.7):

$$\frac{L_{\mathrm{D}}}{\Delta R_p} \leqslant 0.1$$

A critical dose can thus be defined, for which the following applies:

$$N_c = 100\,\frac{kT\varepsilon_0\varepsilon_r}{q^2\Delta R_p^2} \tag{5.9}$$

Curves for this critical doping concentration depending on energy, for boron and

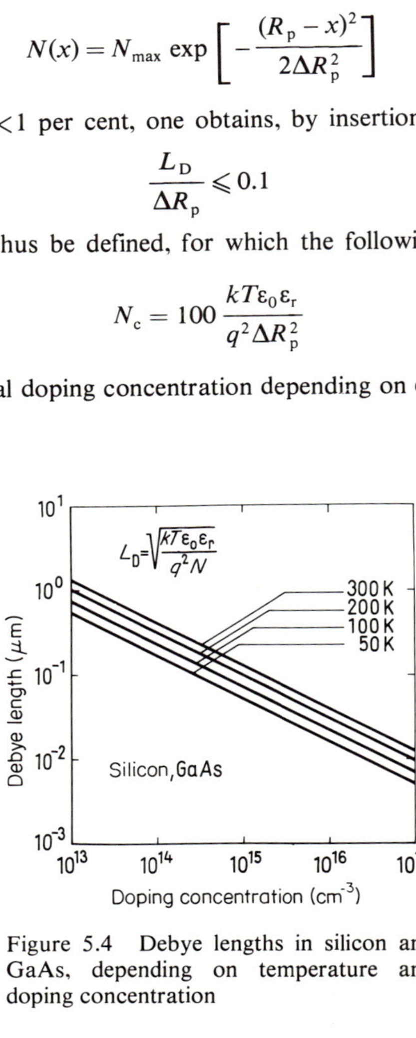

Figure 5.4 Debye lengths in silicon and GaAs, depending on temperature and doping concentration

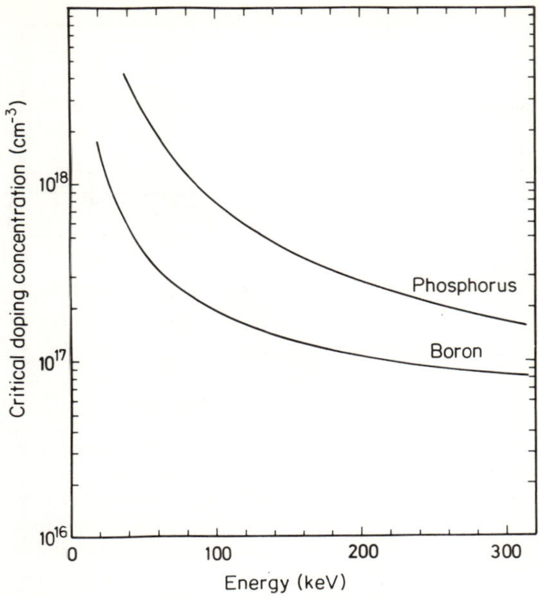

Figure 5.5 Critical doping concentration N_c, for implantations of boron and phosphorus into silicon. Below N_c, the deviation between the doping concentration and the majority-carrier concentration can exceed 1 per cent

phosphorus in silicon, are shown in Figure 5.5. Values of the critical dose under other conditions can be calculated in a similar manner.

In order to obtain more exact information on the deviation of the majority-carrier concentration from the doping concentration, a detailed numerical calculation is necessary. Majority-carrier curves in the case of an abrupt doping change are shown in Figure 5.6. $N(x)$ is the doping concentration, n_0 is the equilibrium concentration of the majority carriers, n_1 the concentration when a negative bias V_1 is applied, etc. The x-coordinate is normalized to L_D. The space-charge increase when the voltage is changed from V_1 to V_2 is given as follows:

$$\Delta Q = q \int_0^\infty (n_1 - n_2)\, dx \qquad (5.10)$$

As can be seen, ΔQ (and therefore $C = \Delta Q/\Delta V$, as well) are blurred over several Debye lengths.

Kennedy [405], [406] has made the derivation that equation (5.3) is the majority-carrier concentration $n(x)$, and applies this in equation (5.6) to thus calculate the doping concentration $N(x)$. In this derivation, though, he uses the expression $dQ = qn\, dx$, which (see equation (5.10)) is not correct and again means

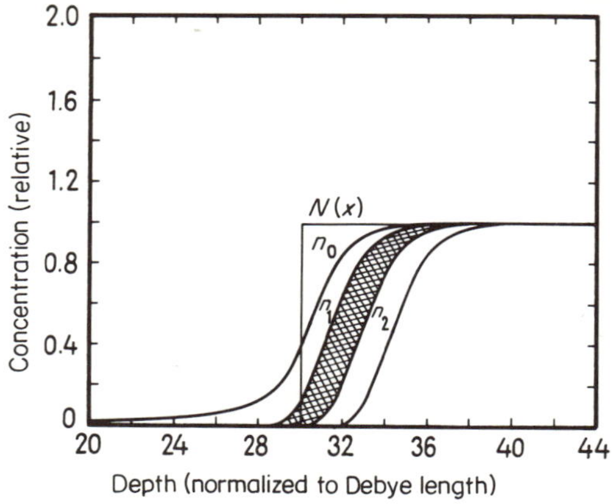

Figure 5.6 Majority-carrier concentration, depending on the bias, in the case of a doping change by a factor of 100. Depth coordinate normalized to the Debye length of the high-doped side. Schottky contact at $x/L_D = 0$. $N(x)$ is the doping concentration, n_0 the majority-carrier concentration without bias, $n_{1,2}$ the majority-carrier concentration with bias $V_{1,2}$ ($V_2 > V_1$) [606]

$L_D = 0$. Therefore, to obtain reliable information on the deviation between $N(x)$ according to equation (5.2), the actual doping concentration, and the majority-carrier concentration $n(x)$, the differential equations (5.4) and (5.5) have to be solved.

The doping concentration compared to the majority-carrier distribution n_0 obtained by an exact solution of the Poisson equation, and also compared to the apparent doping according to equation (5.2), is shown in Figure 5.7 for a 40 keV boron implantation into silicon with a background doping of 10^{15} cm^{-3}. The Debye length is 0.13 μm at a concentration of 10^{15} cm^{-3}; the deviation of the majority-carrier concentration from the doping concentration only begins to decrease after several Debye lengths. The dots in Figure 5.7 indicate the values calculated for $N(x)$ according to equation (5.2). As can be seen, they provide a surprisingly good approximation of $n(x)$. If material with a lower background doping than that in Figure 5.7 is used, the deviation between $N(x)$ and $n(x)$ becomes smaller at the same doping concentration. However, it is not possible with this measurement technique to determine the doping profile at low doping concentrations or to interpret deep measured profile tails. As can be seen from Figure 5.6, a maximum doping gradient of 0.3 $N_{max}/L_{D,max}$ can be measured for an abrupt doping change of 1:100. This limit cannot be exceeded with the measurement technique in question, even if the Kennedy correction is used. One can also conclude from these results that attempts to obtain steep implantation profiles are only purposeful up to a certain limit. Similar considerations have been made by Wu *et al.* [788], among others.

Besides this fundamental limitation of the C–V method, there are several

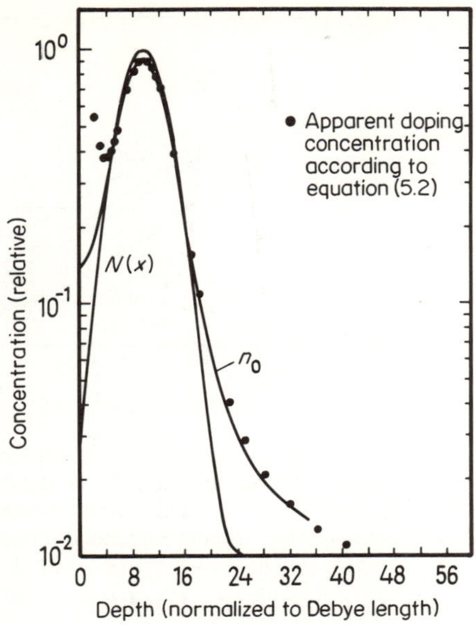

Figure 5.7 Comparison between doping distribution $N(x)$, equilibrium distribution of the majority carriers n_0, and apparent doping concentration distribution according to equation (5.2); dots apply for an implantation of boron at 40 keV. Maximum doping 10^{17} cm^{-3}; background doping 10^{15} cm^{-3}; depth coordinate normalized to the Debye length at 10^{17} cm^{-3} [606]

additional limitations of its applicability. At high doping concentrations and great measurement depths, one limit is the junction breakdown. Additional errors can be due to leakage currents and boundary effects.

Breakdown

The maximum usable bias is limited by the breakdown of the semiconductor junction, which in turn depends on the doping. The maximum measurement depth depends on the space-charge width at breakdown. Values for the breakdown voltage V_B and the maximum space-charge width w_{max} are shown in Figures 5.8 and 5.9. The breakdown voltage is obtained as follows, neglecting diode edge effects (which often cause problems, especially in the case of Schottky diodes):

$$V_B = \frac{E_{max} w_{max}}{2} = \frac{\varepsilon_r \varepsilon_0 E_{max}^2}{2qN_B} \tag{5.11}$$

(E_{max} is the breakdown field strength; N_B is the substrate doping.)

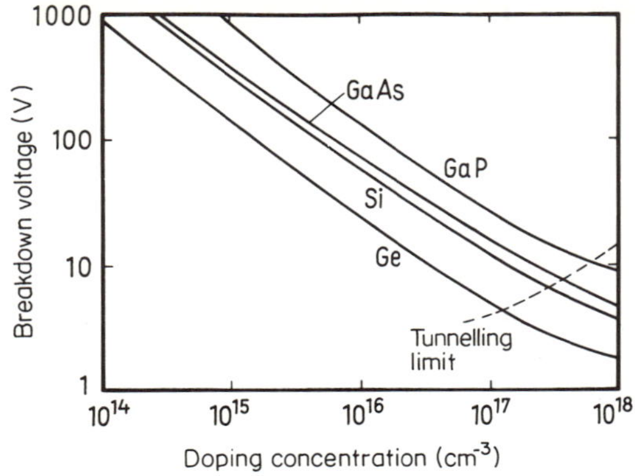

Figure 5.8 Breakdown voltages for abrupt *p–n* junctions in silicon, germanium and GaAs, depending on the doping [713]. The broken line indicates the doping beyond which tunnel currents dominate

An approximate expression according to Sze [713] is:

$$V_B \cong 60 \left(\frac{E_G}{1.1}\right)^{3/2} \left(\frac{N_B}{10^{16}}\right)^{-3/4} \quad \text{(V)} \tag{5.12}$$

where E_G is the bandgap in electronvolts and N_B is the substrate doping in cubic centimetres. The maximum doping which can be measured is given approximately

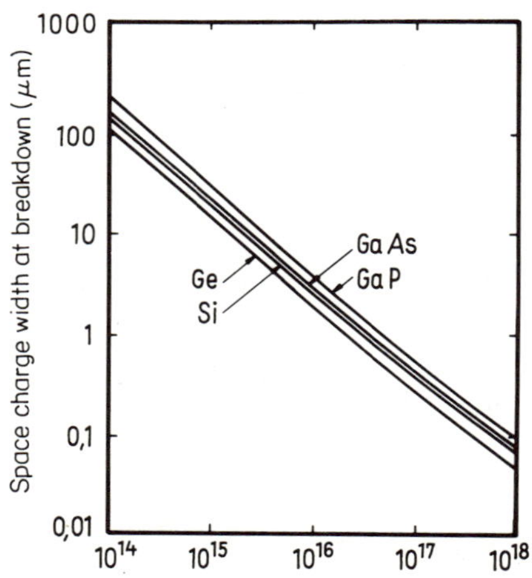

Figure 5.9 Width of the space charge at the breakdown for silicon, germanium, GaAs and GaP, depending on the doping [713]

by $N_B w_{max}$; it amounts to about 10^{13} cm^{-2}. These values can be reduced still further by edge effects of the junction. Equations (5.11) and (5.12) are valid for Si, Ge and GaAs. Other relations can apply for narrow-gap semiconductors, e.g., InSb [433].

Current

The influence which the current across the *p–n* junction or across the Schottky diode has on the measurement was neglected in deriving equation (5.2). The main portion here results from surface-leakage currents, especially in the case of Schottky diodes. The quality factor Q is given by:

$$Q = \omega C R_p \tag{5.13}$$

(R_p is the parallel resistance to the capacitance of the junction.)

Capacitance meters allow measurement up to a Q value of 3, which means that the parallel resistance can be approximately 50 kΩ at a capacitance of 10 pF and a measurement frequency of 1 MHz.

Edge Effects

Especially with small-area diodes, a relatively important source of possible error is the neglect of edge effects at the Schottky or *p–n* junction. A correction by Copeland [151] provides improved values as follows:

$$x = \frac{x_{\text{Meas}} R}{(1 - K x_{\text{Meas}})} \qquad N(x) = \frac{N_{\text{Meas}}}{(1 + K x_{\text{Meas}}/R)^3} \tag{5.14}$$

where R is the contact radius and K is a constant (≈1.5 for silicon).

An additional area correction is needed due to lateral diffusion when *p–n* junctions are used.

Traps

Profile measurements can also be distorted by traps in the forbidden gap [330], [613]. Such traps can occur in the case of incompletely annealed implanted layers. On the other hand, this effect can be utilized to determine the density and energy level of the traps [302], [657]. In general, however, these traps are not distributed homogeneously in implanted layers, so that a detailed discussion would go beyond the scope of this book. With very narrow distributions of traps at low concentration levels, though, capacitance-voltage measurements can be used for their study [634].

5.3.2 Measurement Techniques

In the generally used method, the capacitance of a Schottky or *p–n* diode is measured using a radio-frequency measurement bridge, as a function of a bias amounting to a few millivolts, up to a voltage just below breakdown [334], [724]; the capacitance-voltage data are then evaluated according to equation (5.2). The amplitude of the radio-frequency voltage must be low (10 to 20 mV) to ensure detection of weak signals.

Meyer [493], Copeland [150], and Spiwak [699] have described techniques based on the fact that in the *p–n* junction or the Schottky diode, harmonics of the measurement signal are produced due to the non-linearity of the diode characteristic. If the diode is operated with alternating current $I(t) = I_0 \sin \omega t$, the following applies for the voltage at the diode:

$$V = \frac{I_0 \cos \omega t}{\varepsilon_0 \varepsilon_r A} x + \frac{I_0^2(\cos 2\,\omega t + 1)}{4\omega^2 q \varepsilon_0 \varepsilon_r A^2} \frac{1}{N(x)} \tag{5.15}$$

The signal of the fundamental wave in this case is proportional to the depth x, while the second harmonic is proportional to $1/N(x)$. This method helps to avoid problems in differentiating measured capacitance data according to equation (5.2). The space-charge width is varied by changing the equivalent bias, as in the standard technique.

A further method described by Miller [498] uses a feedback procedure to control the space-charge width. Two operation modes are possible: with constant modulation of the electrical field or with constant modulation of the space-charge width. Very long connection cables can be used with this method, and it is possible, for example, to measure profiles during implantation. A further advantage is that capacitance depending on temperature can be measured at a constant space-charge width.

The use of MOS structures for profile measurement is also possible; in this case, one measures oxide capacitance and space-charge capacitance in series. This method is especially useful in determining trap levels [231].

5.4 SHEET RESISTIVITY MEASUREMENTS

Measurement of the sheet resistivity is an important factor in semiconductor technology; several different techniques have been developed for this purpose. Frequently used approaches include:

(a) Four-point measurement (Valdes [743], Smits [692], Severin [660]);
(b) Spreading resistance measurement (Holm [341], Schumann [636], Mazur [488]);
(c) Sheet resistivity measurement using Hall effect samples (e.g., van der Pauw [744], [745]);
(d) Sheet resistivity measurement using resistor structures;
(e) Breakdown methods (Gardner [278], Brownson [108], Frank [252], Schumann [637]).

The first method is rapid, requires no special structures, is more or less non-destructive, and yields the sheet resistivity directly. Spreading resistance measurement provides a resistivity value which is correlated with the sheet resistivity, but requires very considerable correction; therefore this method is only well suited for profile measurement. It will be discussed in Section 5.6. Sheet resistivity measurement using van der Pauw samples involves much effort and requires a special structure. Thus, it is only purposeful in connection with Hall effect measurement, and will be discussed in the corresponding section. Sheet resistivity measurement using resistor structures is also a structure dependent method, but due to the small dimensions required, it is well suited for implantation homogeneity examination. The last method, sheet resistivity determination by measurement of the breakdown voltage (three-point method), is only of interest for low-doped layers on highly conductive substrates, and is therefore of no relevance for ion implantation.

5.4.1 Four-point Measurements

The four-point measurement procedure uses four metal probes at an equal distance from each other, which are pressed on to the semiconductor surface (see Figure 5.10). The two outer contacts 1 and 4 conduct the current; the potential difference occurring between contacts 2 and 3 is measured using a digital voltmeter. The distance s between the probes (which consist of tungsten, application weight 50 to 100 g) is generally between 0.5 and 1.5 mm. The current is kept as low as possible to avoid sample heating. The measured signals are in the millivolt range.

The sheet resistivity in a thin layer of infinite dimensions is to be calculated, according to Smits [692], as follows:

$$\rho_S = \frac{\pi}{\ln 2} \frac{V}{I} \tag{5.16}$$

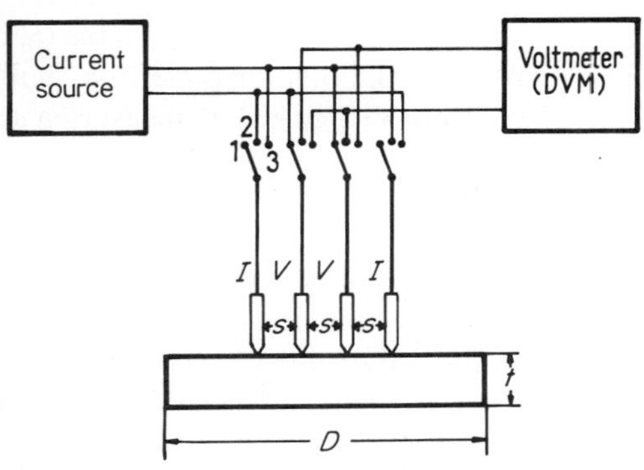

Figure 5.10 Diagram of a four-point probe. The voltage measurement probes are 'formed' in position 1; measurements are made in positions 2 and 3

If the relationship between the sample thickness t and the distance between the probes s is smaller than 0.6, the specific resistivity can be calculated with an error of less than 1 per cent in the following manner:

$$\rho = \rho_S \cdot t = \frac{\pi}{\ln 2} \frac{V}{I} t \tag{5.16a}$$

One generally implants into substrates having the opposite conductivity type; t corresponds to the depth x_j of the p–n junction, which is generally not known; in addition, ρ_S is a weighted average in the case of a doping profile (*cf.* Section 5.5).

Errors in measurement can be caused by:

(a) Surface effects,
(b) Leakage currents,
(c) Sample heating,
(d) Measurement-probe pressure,
(e) Geometrical effects.

Surface Effects

These are related, for example, to inversion layers or conducting layers produced by the influence of light (photo effect). Such effects are especially likely to occur with high resistivity layers. To avoid them, measurement can be done in darkness; and in addition, an appropriate surface treatment can be applied. For n-type silicon layers, storage in air for two hours following an etching treatment (e.g., in HF) is useful, while for p-type layers, etching in hot H_2O_2 is recommendable according to Bader and Kalbitzer [50].

Leakage Currents

In order to avoid leakage currents via the substrate, the measurement currents for implanted p–n junctions should be kept as low as possible (about 1 μA to 1 mA, depending on the resistivity), and the p–n characteristic should naturally be as good as possible. This effect can be recognized by measuring at various currents.

Sample Heating

Heating effects can occur even at very low currents, especially for shallow implanted samples. In the case of incomplete ionization of the dopant atoms, the sheet resistivity changes about exponentially with the temperature. For this reason, the temperature of the surroundings must be kept constant and sample heating by the measurement current must be avoided.

Pressure

The pressure applied to the measurement probes is almost always so high as to cause damage to the implanted layers. The impression caused by a measurement

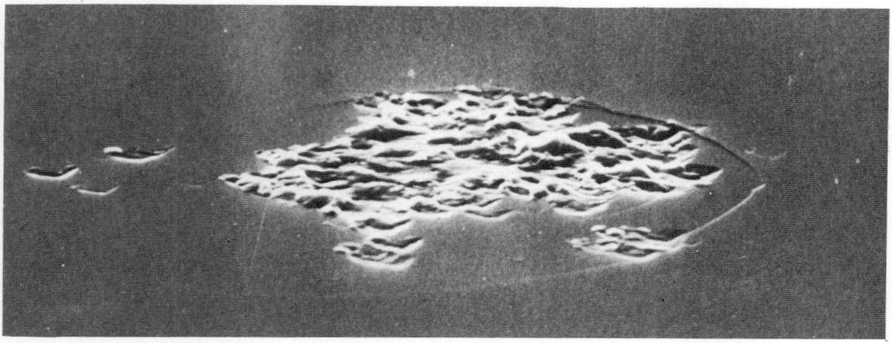

Figure 5.11 Indentation caused by a measurement probe applied at a pressure of 200 g, magnified about 2000× [705]

probe applied at a pressure of 200 g [705] is shown in Figure 5.11. To lessen such damage, one can employ greatly reduced pressure (about 10 to 20 g), larger-area probes, or mercury contacts (Severin [661]). The latter method seems especially well-suited for implanted samples; Figure 5.12 is a diagram of this measurement set-up. After placing the sample on to the four contact holes, the block is tilted and compressed air is applied until the channels are filled. The sample is then returned to the horizontal again and the measurement can be made. Before measurement, the samples should be carefully etched in HF, rinsed in water, and dried. This method is only suitable for *n*-type silicon layers.

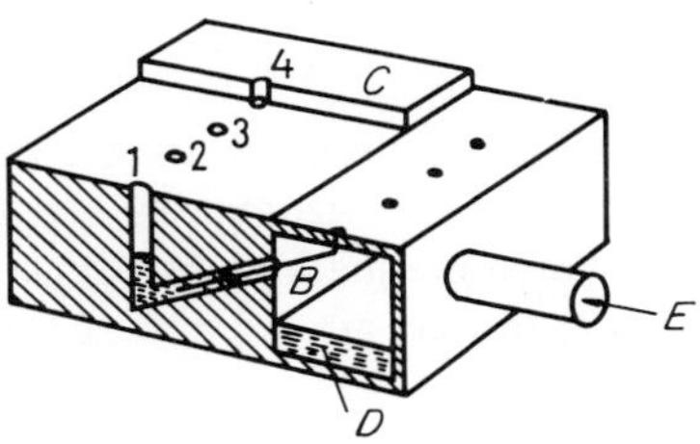

Figure 5.12 Four-point probe with mercury contacts; cross-section through mercury channel 1. Mercury is made to flow from the reservoir D into the channels by means of tilting, pressure is applied through E, and the sample holder is then tilted back. The contact to the mercury is made via a platinum wire, B. The contact radius can be altered by the use of various apertures, C [661]. From Severin and Bulle, *J. Electrochem. Soc.*; **122** (1975), 134, reprinted by permission of the publisher, The Electrochemical Society, Inc.

Table 5.2 Correction factors $C(D/s)$ for four-point probe measurements with finite wafer dimensions [692]

D/s	$C(D/s)$	Error without C (%)
3	2.2662	50
4	2.9289	35.4
5	3.3625	25.8
7.5	3.9273	13.4
10	4.1716	7.9
15	4.3646	3.7
20	4.4364	2.1
25	4.4724	1.3
31.75	4.4934	0.9
40	4.5076	0.5
50.8	4.5172	0.3
76.2	4.5256	0.1
∞	4.5324	0

Geometrical Effects

In deriving equation (5.16), the assumption was made that the measurement probes are on a wafer with infinite dimensions. This assumption is valid with an error of less than 1 per cent when the relationship between wafer diameter D and distance between probes s is larger than 40. If the relationship D/s is smaller, correction factors have to be used [743], [692], and ρ_s is then calculated as follows:

$$\rho_s = \frac{V}{I} C\left(\frac{D}{s}\right) \tag{5.17}$$

The necessary correction factors, depending on the relationship D/s, are given in Table 5.2. Correction factors for other geometries can be found in References [305], [526] and [692]. In deriving equations (5.16) and (5.17), the additional assumption was made that the probes are located at the centre of the wafer. If their location is displaced by a factor of a, corrections for wafer size as well as displacement are necessary. In Table 5.3, corresponding values are given for 1 in., 2 in., 3 in. and 100 mm wafers. For other relationships of D/s or a/D, the correction factors can be calculated as follows [445]:

$$C\left(\frac{D}{s}, \frac{a}{D}\right) = \pi \left\{ \ln 2 + \frac{1}{2} \ln \frac{\left[1 - \left(\frac{2a}{D} + \frac{s}{D}\right)\left(\frac{2a}{D} - \frac{3s}{D}\right)\right]}{\left[1 - \left(\frac{2a}{D} - \frac{s}{D}\right)\left(\frac{2a}{D} - \frac{3s}{D}\right)\right]} \times \frac{\left[1 - \left(\frac{2a}{D} - \frac{s}{D}\right)\left(\frac{2a}{D} + \frac{3s}{D}\right)\right]}{\left[1 - \left(\frac{2a}{D} + \frac{s}{D}\right)\left(\frac{2a}{D} + \frac{3s}{D}\right)\right]} \right\}^{-1} \tag{5.18}$$

Table 5.3 Correction factors $C(D/s, a/D)$ for four-point probe measurements in case the probes are not at the centre of the sample. Probe distance $s = 1.0$ mm (according to equation (5.18) [445])

a/D	1 in. wafer ($D/s = 25.4$)	Error without C (%)	2 in. wafer ($D/s = 50.8$)	Error without C (%)	3 in. wafer ($D/s = 76.2$)	Error without C (%)
0	4.4724	0	4.5172	0	4.5256	0
0.1	4.4673	0.11	4.5159	0.03	4.5250	0.013
0.15	4.4600	0.28	4.5141	0.07	4.5242	0.03
0.2	4.4475	0.56	4.5109	0.14	4.5228	0.06
0.25	4.4260	1.04	4.5054	0.26	4.5203	0.12
0.3	4.3867	1.92	4.4954	0.48	4.5158	0.22
0.35	4.3036	3.78	4.4742	0.95	4.5064	0.42
0.40	4.0747	8.89	4.4158	2.25	4.4803	1
0.41	3.9797	11.0	4.3917	2.78	4.4695	1.2
0.42	3.8475	14.0	4.3582	3.52	4.4546	1.6
0.43	3.6544	18.3	4.3097	4.59	4.4330	2
0.44	3.3534	25.0	4.2355	6.24	4.3999	2.8
0.45	—	—	4.1131	8.95	4.3456	4

a/D	100 mm wafer ($D/s = 100$)	Error without C (%)	150 mm wafer ($D/s = 150$)	Error without C (%)
0	4.5284	0	4.5306	0
0.1	4.5281	0.007	4.5305	0.003
0.15	4.5276	0.02	4.5303	0.008
0.2	4.5268	0.04	4.5299	0.02
0.25	4.5254	0.07	4.5293	0.03
0.3	4.5228	0.12	4.5281	0.06
0.35	4.5173	0.25	4.5257	0.11
0.40	4.5021	0.58	4.5189	0.26
0.41	4.4959	0.72	4.5161	0.32
0.42	4.4872	0.91	4.5123	0.41
0.43	4.4746	1.19	4.5067	0.53
0.44	4.4554	1.61	4.4981	0.72
0.45	4.4238	2.3	4.4841	1.03

Table 5.4 Correction factors for four-point probe measurements of thin samples [692]

d/s	$F(d/s)$	Error without F (%)
0.4	0.995	0.05
0.5	0.9974	0.26
0.5555	0.9948	0.52
0.6250	0.9898	1
0.7143	0.9798	2
0.8333	0.9600	4
1.0	0.9214	7.8
1.1111	0.8907	11
1.25	0.8490	15.1
1.4286	0.7938	20.6
1.6666	0.7225	27.7
2.0	0.6336	36.6

For homogeneously doped thin samples too, correction factors have to be used for determination of the specific resistivity, as provided by Smits [692]. This is important, for example, for characterization of the substrate prior to implantation. The following applies:

$$\rho = \rho_s t = \frac{V}{I} t \frac{\pi}{\ln 2} F\left(\frac{t}{s}\right) \tag{5.19}$$

The correction factors are necessary (error $\geqslant 1$ per cent) for $t/s > 0.6$; they are given in Table 5.4.

If the distances between the probes are not constant, an additional correction factor is to be used [531]:

$$F_{sp} = 1 + 1.082\left(1 - \frac{s_2}{s}\right) \tag{5.20}$$

s_2 is the distance between the two inner probes; s is the average distance between probe and sample. The various geometrical correction factors are multiplicative for $t/s < 4$.

Under certain conditions, it is also possible to measure doping profiles with the use of sheet-resistivity measurements. This possibility will be discussed in Section 5.5.2.

5.4.2 Resistor Structures

A simple method to measure the sheet resistivity in integrated circuits is offered by resistor structures. A further advantage is that resistors can be small compared to the distance between probes in four-point measurement. Investigations on the homogeneity of implanted layers are thus possible without geometrical corrections being needed. In the case of test circuits, special resistor structures are generally used for this purpose.

Two typical structures are shown in Figure 5.13. In most cases, there is a high-doped area under the contact to avoid contact resistances. The resistor is isolated from the substrate by a *p–n* junction. For the sheet resistivity, the following applies:

$$\rho_S = R\frac{W}{L} \tag{5.21}$$

(L is the sample length; W its width; R the measured resistance.)

Possible sources of error are:

(a) Irregularities in the dimensions;
(b) Contact resistances;
(c) Space-charge effects.

In order to avoid geometrical errors in the order of several micrometres, it is best to use the structure in Figure 5.13(b). Influences of contact resistances can be determined by measuring at various currents; they can be reduced by using diffused contacts. The influence of the space charge between substrate and resistor is also expressed by a non-linear current-voltage characteristic.

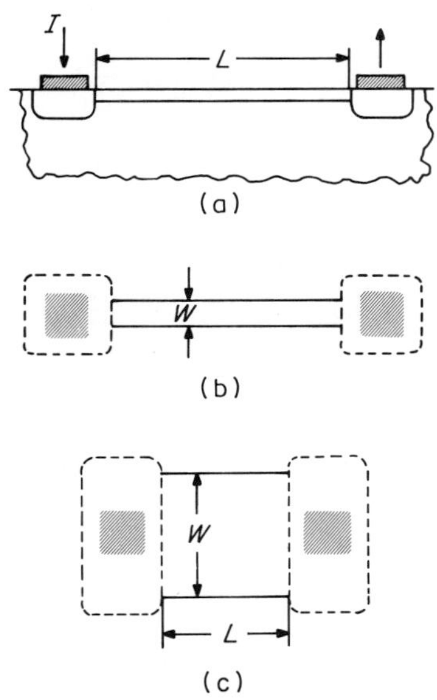

Figure 5.13 Resistor structures with diffused contacts; shaded areas: metallization

5.5 HALL EFFECT MEASUREMENTS

As opposed to sheet resistivity measurement alone, one can obtain additional information by using the latter combined with Hall effect measurement; it is thus possible to determine the carrier concentration as well as its mobility. These parameters are extremely important to be able to understand the processes that take place during annealing of implantation damage and electrical activation. If these parameters are measured not only integrally, but also in relation to the depth, profiles of the electrically active portion of the implanted ions are obtained. These can then be compared with profiles of the total implanted ions, as measured by the back-scattering technique, SIMS, the radiotracer method, or other approaches.

5.5.1 Van der Pauw Structures

Basically, there are several different methods that can be used to measure the Hall effect [113], [144], [291]. The one by van der Pauw [745] has proven the most advantageous. Using this method, four contacts are applied at the circumference of a sample having any desired shape; see Figure 5.14(a). The magnetic field is vertical to the sample. Symmetrical samples are generally used for practical reasons; two such samples are shown in Figures 5.14(b) and (c). According to the theory of van der Pauw, the contacts must be infinitely small and located at the circumference; one therefore uses structures with separate contacts at the circumference, connected only by narrow strips with the area to be measured.

According to van der Pauw [745], the following applies for the sheet resistivity with symmetrical structures:

$$\rho_S = \frac{\pi}{\ln 2} \frac{V_{34}}{I_{12}} \tag{5.22}$$

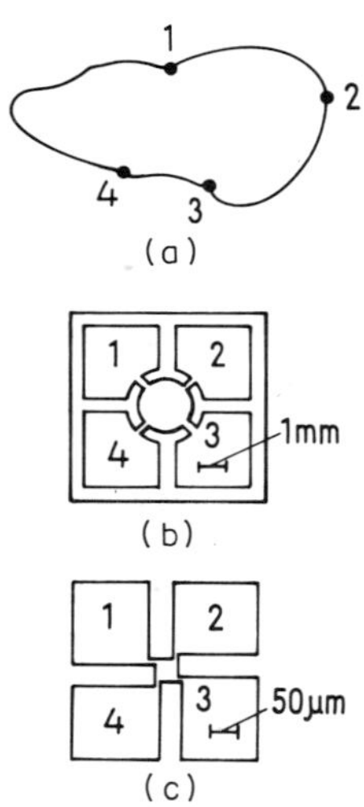

Figure 5.14 Measuring structures according to van der Pauw: (a) asymmetrical structure, corrections necessary; (b) symmetrical structure, suitable for profile measurements [745]; (c) symmetrical structure for high spatial resolution [530]

and for the sheet Hall coefficient $R_{H,S}$:

$$R_{H,S} = \frac{\Delta V_{H,13}}{2I_{24}B} \tag{5.23}$$

(V_H is the Hall voltage; B is the magnetic field.)

In this equation, the sheet Hall coefficient is used (in contrast to the Hall coefficient in classical Hall effect measurements), since the thickness of the layer is generally unknown. The Hall voltage is the voltage difference between contact 1 and contact 3, for positive as well as negative magnetic fields.

In practice, the current electrodes as well as the voltage electrodes are periodically exchanged during both types of measurements to eliminate structural asymmetries and interfering magneto-electrical effects (Nernst, Ettinghausen, Righi–Leduc, etc.) [582], and to reduce measurement errors at the low voltages in the millivolt range. ρ_S and R_S are obtained from these measured values by averaging. The sheet carrier concentration $N_{S,eff}$ and the sheet Hall mobility $\mu_{S,eff}$ are then obtained as follows:

$$N_{S,eff} = \frac{r}{qR_{H,S}} \tag{5.24}$$

and

$$\mu_{S,eff} = \frac{R_{H,S}}{r\rho_S} = \frac{1}{N_{S,eff}q\rho_S} \tag{5.25}$$

where r is the scattering factor, which is the ratio of the Hall mobility and drift mobility.

Using this method, the main possible sources of error are: (a) the scattering factor, (b) averaging, and (c) geometry effects.

Scattering Factor

The scattering factor r, as stated, is the quotient of Hall mobility and drift mobility. The following applies:

$$r = \frac{\mu_H}{\mu_C} = \frac{\overline{\tau_R^2}}{\bar{\tau}_R^2} \tag{5.26}$$

τ_R is the relaxation time of the carriers.

Theoretical considerations result in $r = 1$ for degenerate semiconductors and $3\tau/8$ for non-degenerate ones, assuming that lattice scattering is the dominant scattering process. For practical purposes, the values are between 1 and 1.3 for n-type silicon or between 0.7 and 1 for p-type silicon [243], [603]. Curves for r based on Runyan [603] are shown in Figure 5.15. Since these values are not especially well established, though, it is usual to apply the value $r = 1$ in spite of the relatively large resulting error.

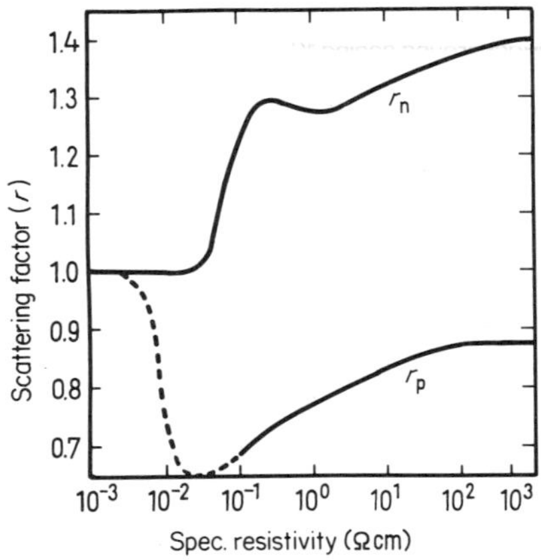

Figure 5.15 Scattering factor for p-type and n-type silicon, according to Runyan [603]. Courtesy of Texas Instruments Incorporated, from data supplied by W. M. Bullis and R. Nasby

Averaging

$N_{S,eff}$ and $\mu_{S,eff}$ are the average values of a profile [59], [567]. The following equations apply:

$$\mu_{S,eff} = \frac{\int_0^t n(x)\mu_H(x)\mu_C(x)\,dx}{\int_0^t n(x)\mu_C(x)\,dx} \tag{5.27}$$

$$N_{S,eff} = \frac{[\int_0^t n(x)\mu_C(x\,dx]^2}{\int_0^t n(x)\mu_H(x)\mu_C(x)\,dx} \tag{5.28}$$

where t is the thickness of the implanted layer up to the p–n junction, or for implantation into material having the same doping type, the wafer thickness. These effective average values are not the exact ones, and can therefore cause a considerable degree of error, especially in the case of a high local mobility. The following apply for the exact average values:

$$\mu_S = \frac{\int_0^t n(x)\mu_C(x)\,dx}{\int_0^t n(x)\,dx} \tag{5.29}$$

$$N_S = \int_0^t n(x)\,dx \tag{5.30}$$

Figure 5.16 shows error curves of $\mu_S/\mu_{S,eff}$ and $N_S/N_{S,eff}$ for boron implantation into silicon at various energies for the case of $r = 1$. A Gaussian profile and a mobility curve according to Reference [243] have been assumed for the computer

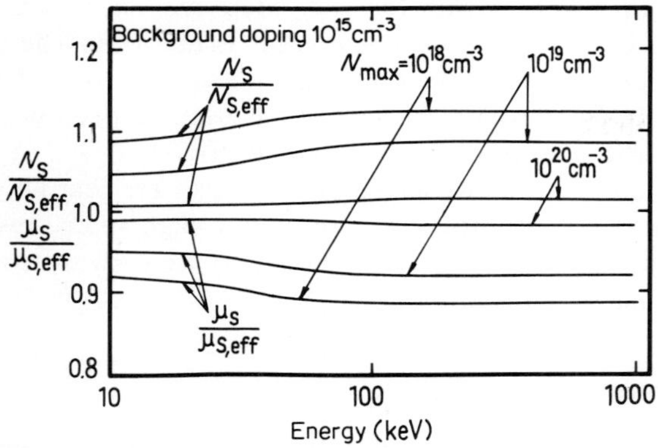

Figure 5.16 $N_S/N_{S,eff}$ and $\mu_S/\mu_{S,eff}$, depending on the energy, for maximum doping of 10^{18} cm^{-3}, 10^{19} cm^{-3}, and 10^{20} cm^{-3}, for implantations of boron into silicon

program. As one can see, the errors when using equations (5.26) and (5.27) can be relatively large, especially for samples which have not been annealed completely. This has been shown in Reference [59] based on a two-layer structure.

Geometry Effects

If the contacts are not at the periphery of the sample and of finite dimensions, the error can be estimated according to van der Pauw [745]. Figure 5.17 shows

	$\Delta\varrho/\varrho$	$\Delta R_H/R_H$
(a) 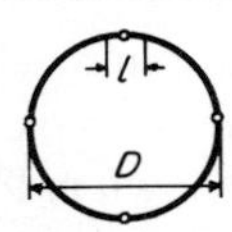	$\approx \frac{-l^2}{16D^2 \ln 2}$	$\approx \frac{-2l}{\pi^2 D}$
(b) 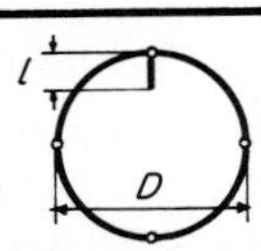	$\approx \frac{-l^2}{4D^2 \ln 2}$	$\approx \frac{-4l}{\pi^2 D}$
(c) 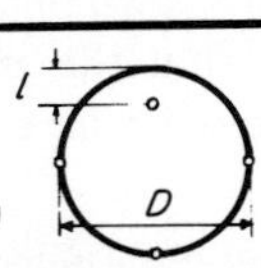	$\approx \frac{l^2}{2D^2 \ln 2}$	$\approx \frac{-2l}{\pi^2 D}$

Figure 5.17 Errors in cases of non-ideal van der Pauw structures [745]. Reproduced by permission of Philips Research Laboratories

approximate relative error values for Hall effect and sheet resistivity measurements, assuming in each case that one contact is not ideal. The total error, in a first approximation, is equal to the sum of the individual errors. An exact calculation for contacts located within the sample has been carried out by Buehler and Pearson [113], who also provide correction factors.

The main advantages of van der Pauw structures are that no corrections are needed and that, if suitable structures are used, a high spatial resolution can be obtained. For these reasons, it can be desirable to use such structures for routine sheet resistivity measurements too.

5.5.2 Profile Measurement

To determine doping and mobility profiles, thin layers of the semiconductor are etched off one after the other, and the Hall coefficient and sheet resistivity are then measured. Using the measured values before and after the ith layer has been etched away, the mean values for the doping as well as the mobility of this layer are yielded by the following equations:

$$\mu_{\mathrm{eff},i} = \frac{R_{\mathrm{H,S},i}\sigma_{\mathrm{S},i}^2 - R_{\mathrm{H,S},i+1}\sigma_{\mathrm{S},i+1}^2}{\sigma_{\mathrm{S},i} - \sigma_{\mathrm{S},i+1}} = \frac{\int_{x_i}^{x_{i+1}} \mu_{\mathrm{C}}(x)\mu_{\mathrm{H}}(x)n(x)\,\mathrm{d}x}{\int_{x_i}^{x_{i+1}} \mu_{\mathrm{C}}(x)n(x)\,\mathrm{d}x} \tag{5.31}$$

$$n_{\mathrm{eff},i} = \frac{\sigma_{\mathrm{S},i} - \sigma_{\mathrm{S},i+1}}{q(x_{i+1} - x_i)\mu_{\mathrm{eff},i}} = \frac{\left[\int_{x_i}^{x_{i+1}} n(x)\mu_{\mathrm{C}}(x)\,\mathrm{d}x\right]^2}{(x_{i+1} - x_i)\int_{x_i}^{x_{i+1}} n(x)\mu_{\mathrm{C}}(x)\mu_{\mathrm{H}}(x)\,\mathrm{d}x} \tag{5.32}$$

The above remarks concerning averaging apply in this case too. If one assumes $r \approx 1$ as an approximation, then $\mu_{\mathrm{eff},i} \approx \mu_{\mathrm{H},i} = \mu_{\mathrm{C},i}$, whereby the error due to inexact averaging will generally be less than that resulting from the assumption $r = 1$.

The influence which the thickness of the removed layer has on the values of $\mu_{\mathrm{eff},i}$ and $n_{\mathrm{eff},i}$ can only be estimated from case to case. Values of $n_{\mathrm{eff},i}$ depending on the thickness of the etched-off layer are shown in Figure 5.18, together with the actual profile.

Although the error in the lower parts of the profile can be considerable, it remains below the error limit of the measurement procedure, assuming a suitable choice of $t = x_{i+1} - x_i$.

As can be seen from equations (5.31) and (5.32), this method is only practicable (due to difference formation) in the case that the measured values differ sufficiently, i.e., that the thickness of the removed layer is not too small in comparison to the total layer thickness.

If the measurement error in σ is of the same magnitude as the difference in the measured value from σ_i to σ_{i+1}, the error can become ∞. In all cases, therefore, one must make certain that the difference in the measured value is significantly greater than the maximum possible error. An exact calculation of the relative error is not possible, due to the differences in equations (5.31) and (5.32). Figure

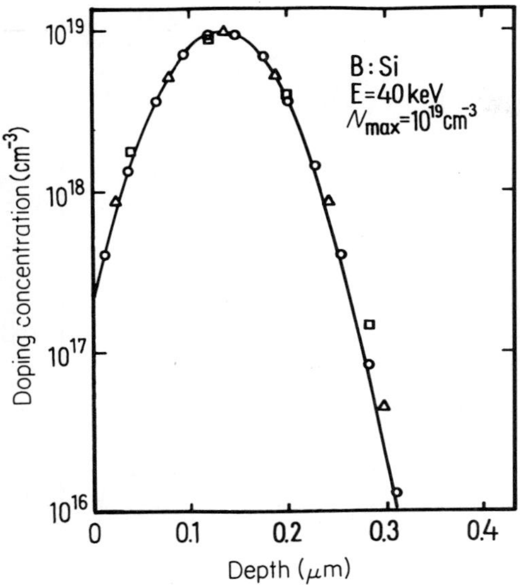

Figure 5.18 Effective doping $n_{eff,i}$ according to equation (5.23) for the case of a boron implantation at 40 keV into silicon (maximum concentration 10^{19} cm^{-3}) depending on the relationship of the thickness of the etched-off layer, t, to the range, R_p. $t/R_p = 0.2$ (○); 0.4 (△); 0.6 (□)

5.19 shows the situation in the case of a boron implantation at 40 keV, with an error alternating between plus and minus 3 per cent in $R_{H,S}$. The first pair of values for $n_{eff,i}$ and $\mu_{eff,i}$ is negative. The decrease in the error together with decreasing residual thickness can be seen clearly. Thus, in general, this method can only yield results which are exact enough in the case of either a steep doping gradient or an adequate thickness of the removed layer relative to the total layer thickness.

To determine doping profiles, it is not absolutely necessary to make Hall effect measurements. If, based on other experiments, one is sure that the relationship between doping concentration and mobility in the implanted layer is the same as in conventionally doped crystals, then sheet resistivity measurements together with layer removal are adequate for profile determination. To calculate the carrier concentration in the case of silicon, one can use either the relationship between specific resistivity and mobility according to Irvin [370], or improved values [48], [753], most preferably those in the analytical approximation by Caughey and Thomas [124].

The specific resistivity ρ in the ith layer can be calculated from the measured values as follows:

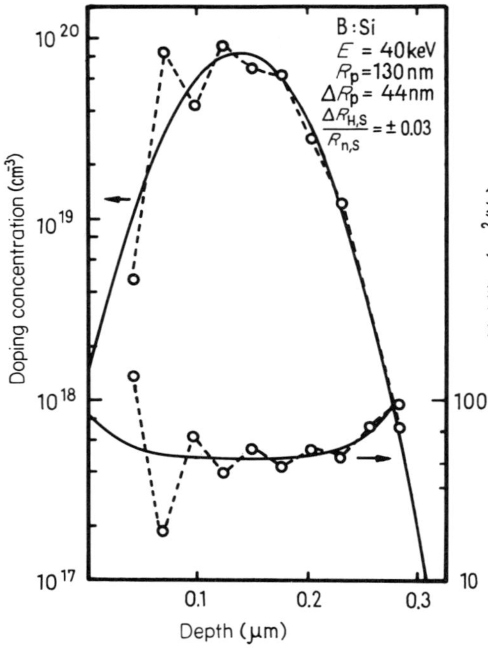

Figure 5.19 Doping and mobility distributions $n_{\text{eff},i}$ and $\mu_{\text{eff},i}$ according to equations (5.31) and (5.32), with an alternating measurement error of 3 per cent in $R_{\text{H,S}}$, for a boron implantation at 40 keV and a dose of 10^{15} cm^{-2} (background doping 10^{15} cm^{-3})

$$\rho_i = \frac{x_{i+1} - x_i}{\sigma_{\text{S},i} - \sigma_{\text{S},i+1}} \tag{5.33}$$

For the relationship between mobility and doping, the following applies [124]:

$$\mu = \mu_{\text{min}} + \frac{\mu_{\text{max}} - \mu_{\text{min}}}{1 + (N/N_{\text{ref}})^{\alpha}} \tag{5.34}$$

Values of μ_{min}, μ_{max}, N_{ref}, and α for n-type and p-type silicon are given in Table

Table 5.5 Mobility parameters according to the model by Caughey and Thomas [124]

	μ_{min}	μ_{max}	N_{ref}	α	Reference
p-type silicon	47.7	495	6.3×10^{16}	0.76	[46]
	47.7	495	1.9×10^{17}	0.76	[53]
n-type silicon	65	1330	8.5×10^{16}	0.72	[46]
	92	1360	1.3×10^{17}	0.91	[52]

5.5. The calculation of the doping concentration with the help of equations (5.33) and (5.34) must be performed iteratively.

5.5.3 Layer Removal Techniques

Depending on the type of solid being investigated, different methods must be used for the removal of thin layers.

Anodic Oxidation

The simplest and most exact method is the removal of thin layers by anodic oxidation. In this procedure, the sample is anodically oxidized in a suitable electrolyte, whereby the desired oxide thickness is determined by the applied voltage. The oxide is then removed by means of a suitable etchant. This method has been used successfully with aluminium [177], tungsten [453], tantalum [746], gold [772], silicon [115], [173], [207], [461], GaAs [321], [520], GaP [638] and InSb [188]. A list of electrolytes for the most usual substrates is given in Table 5.6. The removable layer thickness amounts to between approximately 0.2 and 3 nm/V for all electrolytes. This technique is limited to the removal of relatively thin layers, since only about 10 to 100 nm (according to the dielectric strength of the oxide) can be removed per etching step.

A cross-section through a sample holder permitting measurement as well as anodic oxidation is shown in Figure 5.20. Within the same holder, the removal of thin semiconductor layers by anodic oxidation, the subsequent dissolution of the oxide in hydrofluoric acid, and the measurement of the sample can all be carried out. In addition, the current for the anodic oxidation can be connected via the implanted layer, which is necessary for p-type implantations in n-type material, since otherwise the p–n junction (which is necessary for the insulation of the implanted p-area from the substrate) would be reverse biased. The implanted area is separated from the contacts by placing the etching cover (6) in position. As opposed to the construction generally used in the past, where the seals were Viton or rubber rings (under which the electrolyte could penetrate when the anodizing voltage was applied), Teflon has been employed here. At a constant current, the sample is oxidized up to a preset voltage drop due to the oxide layer formed, and is then densified until the current has been reduced to about 10 per cent of its original value. The van der Pauw structure developed for these measurements with four diffused contacts, as well as a 3 in. silicon wafer having such structures, are shown in Figure 5.21. Diffused or deeply implanted contacts are absolutely necessary for layer removal measurements, in order to avoid a separation of the area to be measured due to the occurrence of a deeper etching at the periphery.

Etching

Etching represents another possible method; however, the constancy of the removal rate, which is dependent on the temperature and the radiation damage as

Table 5.6 Electrolytes for the anodic oxidation of silicon, GaAs, InSb and GaP

Material	Electrolytes	Remarks	References
Si	0.1 M H_3BO_3 in $Na_2B_4O_7$		[461]
	0.04 M KNO_3 in N-Methylacetamide (NMA)	Addition of 5–10% H_2O is necessary for effective oxidation ~0.2 nm/V	[461], [626]
	4 g KNO_3, 900 ml ethylene glycol, 100 ml H_2O	0.7 – 1.3 nm/V	[461], [581]
GaAs	1 part 3% tartaric acid (or citric acid) + 2–4 parts ethylene or propylene glycol	Adjust pH value to 6 by means of NH_4OH	[324], [777]
	30% H_2O_2 solution	Adjust pH value to 2 by means of H_3PO_4 or NH_4OH	[697], [777]
	N-Methylacetamide	Adjust pH value to 8.3 by means of NH_4OH; 1.3 nm/V	[518]
GaP	30% H_2O_2 solution	Adjust pH $\leqslant$ 2.5 by means of H_3PO_4; 1.2 nm/V	[230]
InSb	0.1 N KOH	Growth properties very critically dependent on oxidation current	[188], [777]

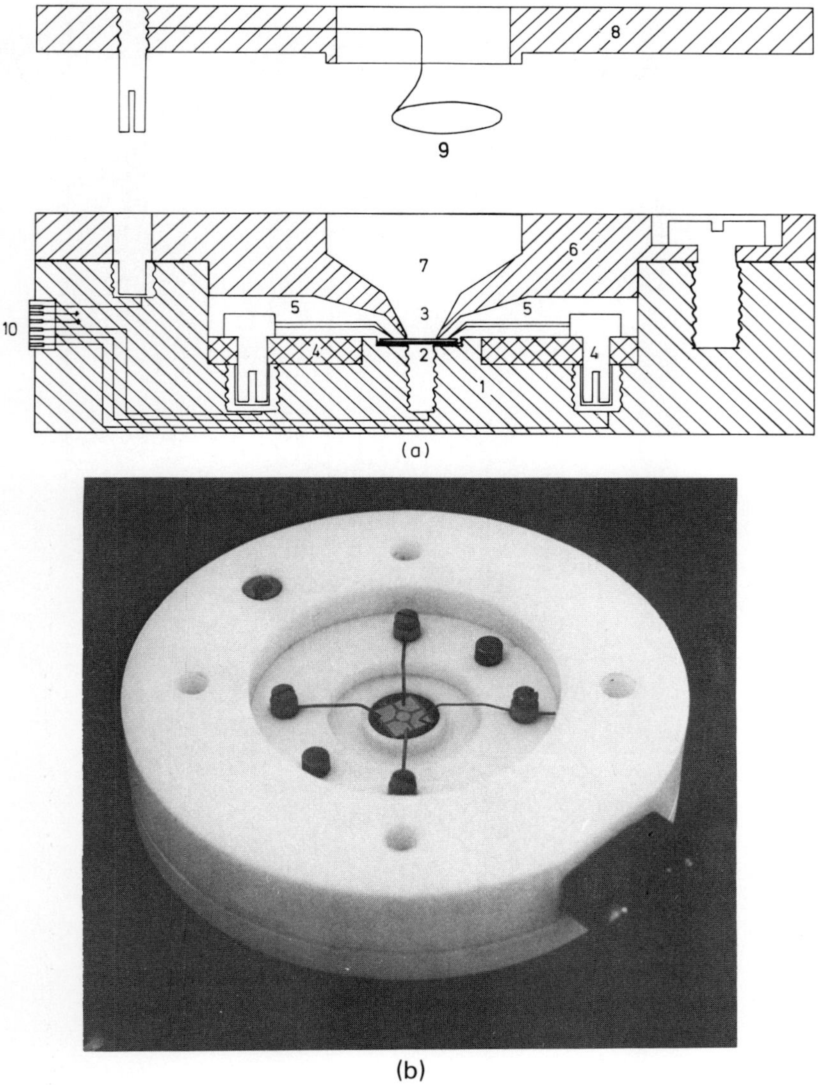

Figure 5.20 Sample holder for layer removal. (a) Cross-section through the holder: (1) body, (2) adjustable support, (3) sample, (4) contact plug, (5) contact probes, (6) etching mask, (7) space for electrolyte, (8) cathode holder (is attached for oxidation), (9) platinum electrode, (10) plug for attachment of measurement equipment or constant current source. (b) Photograph of the sample holder, with sample in place [611]

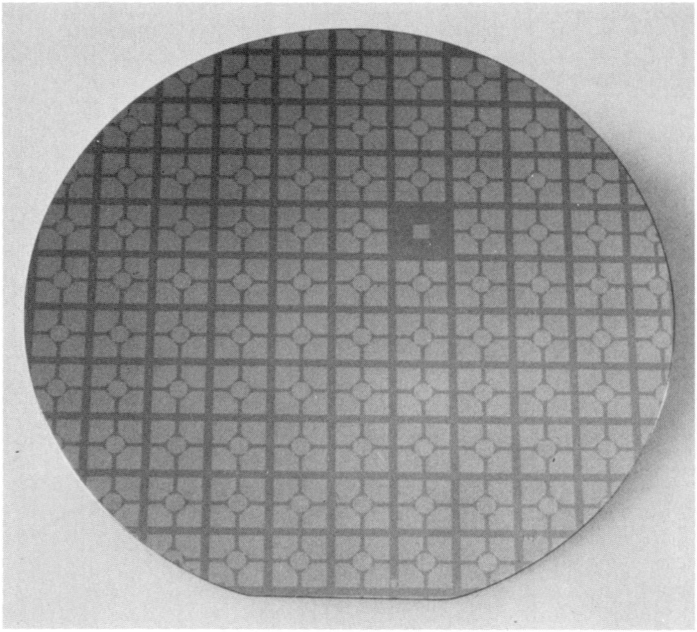

(a)

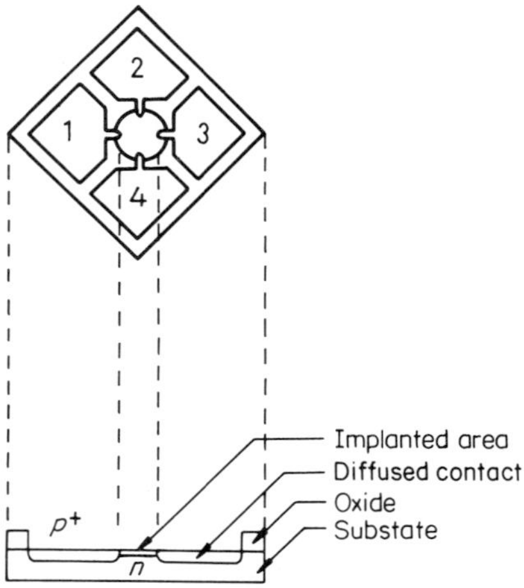

(b)

Figure 5.21 (a) Photograph of a silicon wafer with van der Pauw structures. (b) Diagram of a measurement structure

Table 5.7 Slow etchants for layer removal

Material	Composition of etch	Remarks	Reference
Si	CP4 [HF (40%) : CH_3COOH (conc.) : 1.7 HNO_3 (conc.) Br 0.5 ml per 200 ml etch solution	Etch rate at 23 °C : 0.13 μm/min	[190]
GaAs	H_2O_2 (30%) : 100 H_2O : H_2SO_4 (conc.)	Mix in this order; etch rate at 23 °C : 55 nm/min	[796]
$GaAs_{1-x}P_x$ ($x \approx 0.4$)	H_2O_2 (30%) : 10 H_2O : H_2SO_4 (conc.)	23 °C : 75 nm/min	[796]
InSb	0.1% Br in Methanol	Etch rate at RT: 50 nm/min	

well as on the doping concentration, is a problem. A precise calibration or a measurement after each etching step is necessary. Several etches which are suitable for semiconductors are described in Table 5.7.

Further Methods

Vibration polishing (Whitton [771], [773]) is universally applicable, but is relatively complicated. Ion milling is also a universally applicable method; however, it can alter the layer properties due to radiation damage and due to implantation of the sputtering particles. Furthermore, it is only seldom possible to remove thick layers evenly. This is also a serious problem when using the SIMS method (see Section 5.10), which is also based on the sputtering of the sample surface to be analysed.

5.6 SPREADING-RESISTANCE MEASUREMENTS

A further possibility for determining of doping profiles is offered by spreading resistance measurements in combination with an oblique bevel across the sample. This technique was originally developed for sheet resistivity measurements. Because of its relatively great inaccuracy and the need for corrections and calibrations, however, it is inferior to the four-point method for this purpose. None the less, in combination with an oblique bevel, it offers a rapid measurement method which can be applied over a large range of doping concentrations.

For the spreading resistance in a homogeneous, semi-infinite sample with a contact of diameter D, the following is theoretically valid:

$$R_S = \frac{\rho}{2D} \tag{5.35}$$

In practice, however, one does not obtain this relationship between the specific

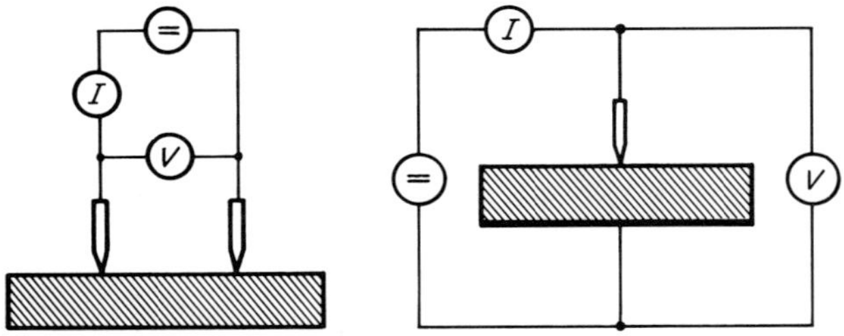

Figure 5.22 Diagrams of spreading-resistance probes

resistance and the spreading resistance, and the following is written:

$$R_S = k(\rho)\frac{\rho}{2D} \tag{5.36}$$

The correction factor $k(\rho)$ must be determined experimentally. Two-point or one-point probes are used as the measurement structure. Both of these possibilities are shown diagrammatically in Figure 5.22. Numerous articles have dealt with the necessary correction factors, especially when dealing with a doping profile. Complicated correction formulas have been developed for two-layer and multiple-layer

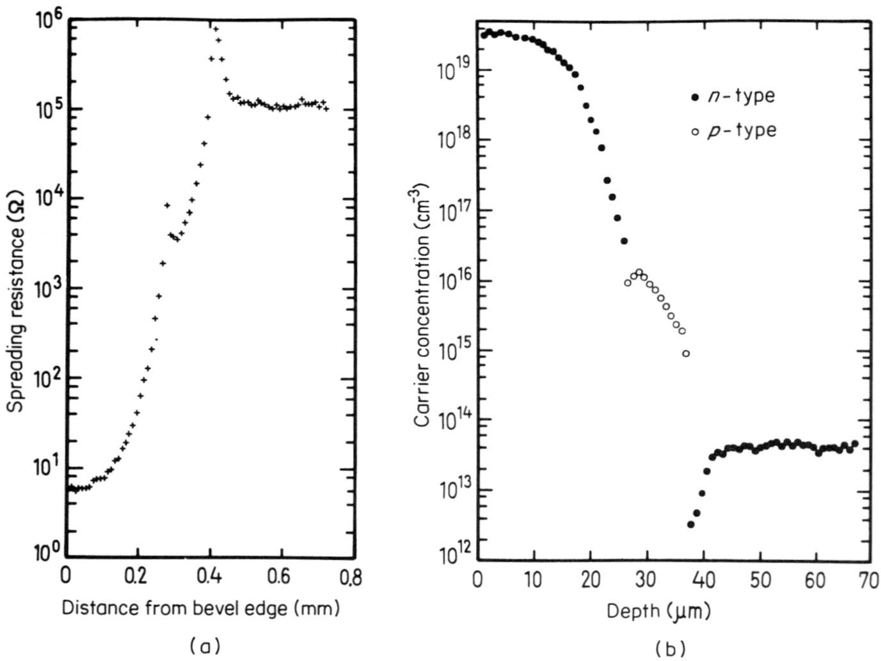

Figure 5.23 Doping profile of a transistor, obtained with a spreading-resistance probe by measuring across a bevel of 5°50′ (1:10): (a) measured spreading resistance; (b) calculated doping profile

structures [352], [429], [515], [636]. These formulas are so extensive that, within the framework of this book, it is not possible to deal with them in detail. A two-point probe (made by Solid State Measurements, Inc.) has been on the market for some time. A general survey on the state of the art can be found in the *Proceedings of the Spreading Resistance Symposium, Gaithersburg, USA, 1974* [216].

A typical example, for measurement of the doping profile of a transistor, is shown in Figure 5.23. The great advantage of the spreading resistance method can be seen clearly: the profile can be measured across *p–n* junctions. This advantage compensates for the method's relatively great inaccuracy during routine examinations. The making of bevels has already been discussed, for example, in Section 5.1. The time needed for a measurement is considerably less than with other methods, even if sample preparation is taken into consideration.

5.7 CURRENT-VOLTAGE MEASUREMENTS

Significant characteristics of semiconductor devices are related to the electrical parameters of *p–n* junctions. Up to now, however, relatively few investigations on the properties of implanted *p–n* junctions have been carried out. Most of the works which exist deal with electrical activation using Hall effect and sheet resistivity measurements, or with the annealing of radiation damage using back-scattering measurements. In the first case, only majority carrier properties are investigated; in the second, only crystallographic properties are dealt with. The characteristics of *p–n* junctions depend greatly on the minority carrier lifetime, while the latter depends on the concentration of recombination centres.

In examining the electrical properties of implanted layers, it is seen that the implanted ions first become electrically active ($\triangleq$ majority carriers); the mobility then reaches the value corresponding to the doping concentration—this is always true at least for silicon, but is not always true for III–V and other types of semiconductors—and during further annealing, often only at significantly higher temperatures, the minority-carrier lifetime is restored. The latter, however, is generally not measured, but is only determined indirectly by device properties. The main reasons for avoiding measurement of the minority-carrier lifetime are related to the facts that this parameter already varies in the starting material, that it is altered by almost every technological process (oxidation, annealing, gettering), and that its direct measurement is difficult, especially in inhomogeneously doped layers.

Investigating the properties of implanted *p–n* junctions—although this method is an indirect one—is none the less the best way to study the influence of implantation upon lifetime. This procedure will therefore be discussed in more detail below.

5.7.1 *p–n* Characteristics

The current flow across a *p–n* junction has been treated in the literature in detail, for example, by Shockley [674], Sah *et al.* [614], Spenke [695], and Sze [713].

The following mechanisms can be responsible for the current flow across a p–n junction:

(a) Diffusion of minority carriers across the space charge;
(b) Recombination or generation of carriers in the space charge;
(c) Tunnelling effect (only important with heavily-doped samples);
(d) Surface-leakage currents;
(e) Series resistance.

Diffusion Current

For the diffusion component of the current density, according to Shockley [674], the following applies (with the limitations already discussed):

$$j_{\text{Diff}} = qn_i^2 \left(\frac{D_n}{L_n N_A} + \frac{D_p}{L_p N_D} \right) \left(\exp \frac{qV}{kT} - 1 \right) \tag{5.37}$$

where n_i is the intrinsic carrier concentration, $D_{n,p}$ are the minority-carrier diffusion coefficients for electrons and for holes, $L_{n,p}$ are the diffusion lengths for electrons and for holes, k is the Boltzmann constant ($k = 1.38 \times 10^{-23}$, $J/K = 8.625 \times 10^{-5}$ eV/K), V is the voltage across the p–n junction, and $N_{A,D}$ are the acceptor concentration and the donor concentration.

The diffusion length is related to the lifetime ($\tau_{n,p}$) of the minority carriers as follows:

$$L_{n,p} = \sqrt{(D_{n,p}\tau_{n,p})} \tag{5.38}$$

From equation (5.37), due to the fact that $N_A \gg N_D$, the following is obtained for p^+n junctions:

$$j_{\text{Diff}} = j_S \left(\exp \frac{qV}{kT} - 1 \right) = q \frac{D_p}{L_p} \frac{n_i^2}{N_D} \left(\exp \frac{qV}{kT} - 1 \right) \tag{5.39}$$

j_S is the saturation current density which flows at reverse voltages significantly above kT/q. A similar result is obtained for n^+p junctions.

Generation-recombination Current

According to the Shockley–Read–Hall model [314], [614], [675] with a single recombination-generation centre (R-G centre) in the middle of the band gap, the reverse current density due to generation is obtained as follows:

$$j_{\text{R-G}} = \frac{qn_i w}{\tau_{\text{eff}}} \tag{5.40}$$

where w is the width of the space charge; τ_{eff} is an effective lifetime which is given by:

$$\tau_{\text{eff}} = \frac{\sigma_n \exp\left(\frac{E_t - E_i}{kT}\right) + \sigma_p \exp\left(\frac{E_i - E_t}{kT}\right)}{\sigma_p \sigma_n v_{\text{th}} N_t} \tag{5.41}$$

σ_n, σ_p are the cross-sections for the capture of electrons and holes, v_{th} is the thermal velocity ($v_{\text{th}} = 3kT/m^* \approx 10^7$ cm/s at room temperature for silicon, m^* is the effective mass), N_t is the concentration of R-G centres, E_t the energy level of the R-G centres, and E_i the intrinsic Fermi level.

For $E_t \approx E_i$ (i.e., the R-G centre is approximately in the middle of the energy gap) and for $\sigma_n \approx \sigma_p \approx \sigma$, the following applies:

$$\tau_{\text{eff}} = \frac{1}{\sigma v_{\text{th}} N_t} \tag{5.42}$$

In this case, it follows that $\tau_{\text{eff}} \approx \tau_n \approx_p$.

The reverse current is proportional to the width of the space charge (see equation (5.40)). The following is valid for abrupt *p–n* junctions:

$$j_{\text{R-G}} \sim w \sim (V_D - V)^{1/2} \tag{5.43}$$

For *p–n* junctions with a linear-graded doping profile, the following applies:

$$j_{\text{R-G}} \sim w \sim (V_D - V)^{1/3} \tag{5.44}$$

For the total reverse current, the following is valid in the case of a p^+n junction ($|V| \geqslant 3\ kT/q$):

$$j_{\text{Tot}} = q \frac{D_p}{L_p} \frac{n_i^2}{N_D} + \frac{q n_i w}{\tau_{\text{eff}}} \tag{5.45}$$

The diffusion component dominates for semiconductors with large n_i (germanium); the generation current dominates for semiconductors with small n_i and relatively low lifetime (GaAs); for silicon, both components are of importance. If the generation current dominates, the shape of the profile at the *p–n* junction can be estimated by measuring the reverse characteristic.

In the forward direction, the recombination current for $V \geqslant kT/q$ is obtained by:

$$j_{\text{R-G}} = \frac{qw}{2} \sigma v_{\text{th}} N_t n_i \exp\left(\frac{qV}{2kT}\right) \tag{5.46}$$

and the total current ($V \geqslant kT/q$) is given by:

$$j_{\text{Tot}} = q \frac{D_p}{L_p} \frac{n_i^2}{N_D} \left(\exp\frac{qV}{kT}\right) + \frac{qw}{2} \sigma v_{\text{th}} N_t n_i \left(\exp\frac{qV}{2kT}\right) \tag{5.47}$$

Thus, the following approximation is written very often for the current in the forward direction:

$$j_{\text{Tot}} \sim \exp \frac{qV}{nkT} \qquad 1 \leqslant n \leqslant 2 \tag{5.48}$$

Leakage Current and Series Resistance

In the case of real *p–n* junctions, surface-leakage currents represent an undesirable effect. These can be caused by inversion layers, stoichiometry changes at the surface, and contaminations such as moisture or solvents. These currents are generally linear, but can also be voltage-dependent. The semiconductor surface must usually be passivated in order to keep this influence minimal. The use of a field electrode above the passivation is often desirable to control the surface states.

At high current densities, the series resistance of the substrate material, the contacts, and the unannealed damaged layer constitutes an ohmic limit. This additional voltage drop results in a sublinear characteristic of the I–V curve in a semi-logarithmic scale. The value of the resistance can be calculated easily by:

$$R_s = \frac{\Delta V}{j_{\text{Tot}}} A \tag{5.49}$$

where ΔV is the deviation from the linear characteristic in a semi-logarithmic plot and A is the area of the diode.

An I–V characteristic in the forward direction, showing all the current components discussed above, is to be seen in Figure 5.24(a). At low voltages in the

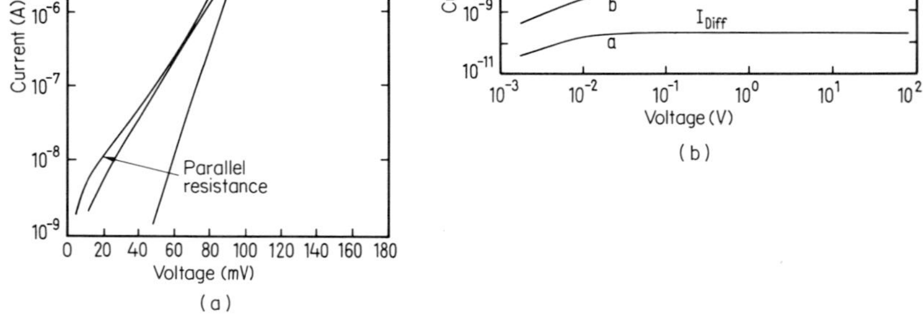

Figure 5.24 Theoretical current-voltage characteristic for InSb at 77 K: (a) forward current (semi-logarithmic scale), (b) reverse current (double-logarithmic scale)

forward direction, surface leakage currents dominate, until the generation recombination current (slope 2 of the semi-logarithmic plot) becomes noticeable. At higher currents, finally, the diffusion current (slope 1 of the semi-logarithmic plot) dominates, until the ohmic limit of the series resistance comes into effect. The reverse characteristics of *p–n* junctions are shown in double-logarithmic scale in Figure 5.24(b). Curve (a) represents the ideal characteristic curve according to Shockley (equation (5.37)); curve (b) indicates total current for an abrupt *p–n* junction, consisting of diffusion and generation current. The breakdown voltage is dependent upon the doping (see e.g., Sze [713]), but this will not be treated further in this context, since no additional information can be obtained from it.

From the reverse characteristic, or from its change resulting from annealing, one can therefore draw direct conclusions regarding the influence of radiation damage produced recombination centres upon the lifetime.

5.7.2 Determination of the Minority Carrier Lifetime

The minority carrier lifetime provides the most sensitive indication of the perfection of the crystal structure, as well as of impurities with levels at the centre of the band. It is generally assumed for ion implanted layers that the effect of an implantation ends at the metallurgical *p–n* junction. However, the radiation damage—as well as dislocations caused by it—can extend very deep into the semiconductor, although the maximum of its distribution is located closer to the surface than that of the doping atoms (see Section 3.1). Bogardus and Poponiak |89| have found damage effects at up to four times the projected range. In addition, dislocations and other defects can diffuse into the semiconductor at the temperatures necessary for annealing. All these defects can reduce the minority carrier lifetime by a production of recombination generation centres. Equation (5.42) is applicable for the lifetime.

If a p^+n junction is abruptly switched from forward to reverse, there initially results a current flow in the reverse direction, which then decreases again down to

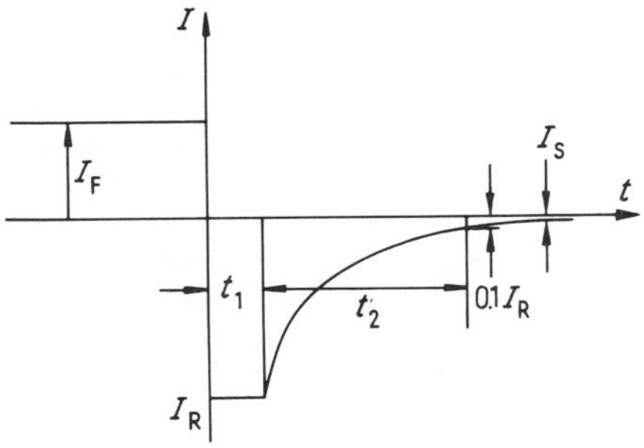

Figure 5.25 Current change in the case that the current flow of a diode is switched from forward to reverse in order to measure the minority-carrier lifetime

the reverse saturation current. This change in current relative to time is shown in Figure 5.25. For the minority carrier lifetime, the following applies (Sze [713]):

$$\operatorname{erf}\sqrt{\left(\frac{t_1}{\tau_p}\right)} = \frac{1}{1 + I_R/I_F} \tag{5.50}$$

The current, which at first is constant in the reverse direction, is adjusted by means of a series resistance. For $I_F/I_R = 1$, one obtains $t_1 = 0.3\,\tau_p$; for $I_F/I_R = 5$, $t_1 = \tau_p$. A further method reported upon by Takusagawa *et al.* [716] makes it possible to measure very short lifetimes by using impedance measurements at the *p–n* junction.

5.8 ANALYSIS OF IMPLANTED LAYERS USING HIGH-ENERGY LIGHT IONS

Interactions which are caused when high-energy light ions strike crystalline solids can be put to good use for implanted-layer analysis, since the probing depth of these ions can be adjusted closely to the thickness of the layer in which implantation effects are to be investigated. Hydrogen, deuterium, and helium ions are frequently used in the energy range between 0.1 MeV and 5 MeV; heavier ions such as carbon or nitrogen are also used occasionally. The interactions which are utilized for layer-analysis purposes (see Figure 5.26) are the following:

(a) Elastic collisions with atomic nuclei (changes in the direction and energy of the ions);
(b) Non-elastic collisions with atomic nuclei (nuclear reactions resulting in the emission of characteristic reaction products);
(c) Elastic and non-elastic collisions with electrons (continuous slowing-down of the ions, emission of characteristic X-rays).

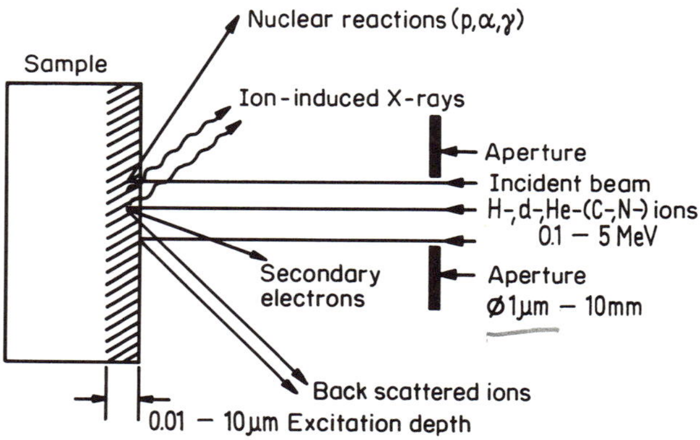

Figure 5.26 Interactions between high-energy light ions and solids

Analysis based on elastic interactions with atomic nuclei is generally referred to as the Rutherford back-scattering technique (based on the Rutherford formula concerning the effective cross-section for elastic collisions between α-particles and atomic nuclei); in addition, there are the techniques based on ion-induced X-ray emission spectroscopy and ion induced nuclear reactions. The advantages of these various methods are that they are non-destructive, do not require any preparatory work, and make it possible to determine the total number of the implanted ions. In this context, as well as in connection with activation analysis (Section 5.9), the expression 'non-destructive' means only that the crystal is not destroyed as a result of the measurement; however, a degradation of the electrical properties, especially the mobility and the lifetime, is always to be expected.

The apparatus needed to carry out such investigations is relatively complicated. An electrostatic van de Graaff generator is used to accelerate the ions up to energies in the MeV range. Ion sources and beam transport systems are comparable to those used in implantation machines; in contrast to the latter machine type, however, expensive shielding is required in this case to protect against the radiation which is produced, due to the much higher energies involved. Because of this fact, only a few of the laboratories dealing with ion implantation are equipped to carry out analysis using high-energy light ions. None the less, though, a great many fundamental experiments regarding implantation effects have been carried out using these techniques. This chapter is intended to contribute to a better understanding of this research; for a more thorough treatment, reference is made to the book by Chu, Mayer and Nicolet [139], as well as to that by Ziegler [793a].

5.8.1 Rutherford Back-scattering

The back-scattering technique offers the possibility of obtaining, without layer removal, depth profiles of implanted ions as well as of damage (see Section 5.8.2). This method is quantitative and relatively fast; in some cases, though, its sensitivity is not adequate, so that other measurement methods have to be used. This technique measures the energy distribution of the ions which are back-scattered from the implanted layer at a defined angle. The particles which strike the sample, in almost all cases, are monoenergetic (1 to 5 MeV) He ions. The energy of a back-scattered ion is determined by two parameters:

(a) The mass of the atomic nucleus with which the eleastic collision takes place (the heavier the target nucleus, the less energy is lost by the colliding light particle);
(b) The depth at which the collision takes place (the light ion loses energy through interaction with the electrons).

In general, therefore, the energy spectrum of the back-scattered ions does not permit any clear-cut determination, either for elements or for depths. However, in many important cases—specifically, when the masses of the different nuclei are far enough apart, and when the distribution of the element in question only extends across a limited depth range—it is possible to correlate various parts of

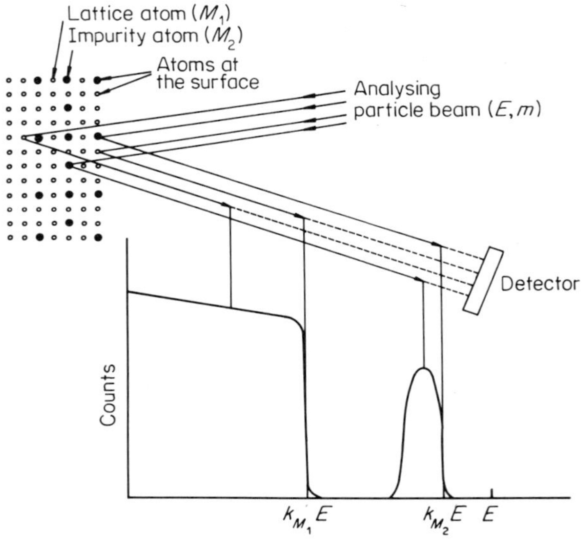

Figure 5.27 Diagram of the back-scattering technique for the determination of impurity-atom distributions; $M_1 < M_2$

the energy spectrum with specific chemical elements, so that depth profiles can be obtained. This is illustrated in Figure 5.27. In a crystal with atoms of the mass M_1 (in this case, silicon), heavy atoms of mass $M_2 > M_1$ (in this case, antimony) are implanted, whereby the antimony profile extends for the most part over a depth of only 0.1 μm, corresponding to an implantation energy of 100 keV.

The following characteristic energies are important in the energy spectrum of the back-scattering ions (energy of the incident He ions $E = 2$ MeV):

(a) $k_{M_2}E$ is the highest possible energy; it corresponds to a collision of a He^+ ion with an Sb atom at the surface (1.769 MeV);
(b) $k_{M_1}E$ corresponds to a collision with a Si atom at the surface (1.170 MeV).

The mass-dependent k_M factor describes the fraction of the energy which is retained by a light particle of mass m (impacting ion) after a collision with a heavy particle of mass M (target nucleus) at the angle of incidence θ. According to the theory of the classical collision between two hard spheres, k_M is given by:

$$k_M = \left[\frac{m\cos\Theta}{m+M} + \left(\left(\frac{m\cos\Theta}{m+M}\right)^2 + \frac{M-m}{M-m}\right)^{1/2}\right]^2 \tag{5.51}$$

If the angle θ is close to 180°, then

$$k_M \approx \left(\frac{M-m}{M+m}\right)^2$$

If no other atoms are present in the sample except silicon and antimony, then all back-scattered ions with energies of E_2 which are greater than $k_{M_1}E$ can be attributed to collisions with Sb atoms; these collisions take place at increasingly greater depths in the sample as the energy of the back-scattered ions decreases. The energy-depth scale for atoms of the mass M_2 can be calculated from the energy loss of the ion during penetration to a depth x, from the energy transfer during the collision with M_2, and from the energy loss until its return to the sample surface. For depths which are not too great, there is proportionality between the energy difference, $\Delta E = k_{M_2} \cdot E - E_2$, and the depth x:

$$\Delta E = [S]_{M_2}^{M_1} x \tag{5.52}$$

The value $|S|_{M_2}^{M_1}$ is called the 'energy-depth conversion factor' for atoms of mass M_2 in a matrix of mass M_1, and can be calculated approximately as follows:

$$|S|_{M_2}^{M_1} = k_{M_2} \cdot S_{e,M_1}(E) + \frac{1}{\cos \Theta} S_{e,M_1}(k_{M_2} \cdot E) \tag{5.53}$$

$S_{e,M_1}(E)$ represents the electronic stopping power of the matrix for ions of the energy E; preconditions for this formula's validity are a relatively low layer thickness ('surface approximation') and a low concentration of the atom M_2 relative to that of the atom M_1. In the present example, there results $[S]_{Sb}^{Si} = 517$ keV/μm, with an accuracy of better than 5 per cent for an analysis depth of less than 0.3 μm, if the average Sb concentration is under 10^{20} cm^{-3}.

The depth resolution achievable with the back-scattering technique is limited by the energy resolution of the detection system (approximately 12 keV FWHM for silicon surface barrier detectors) as well as by the low energy spread of the analysing ions, which increases with increasing depth ('straggling') due to the statistical nature of the atomic-collision processes. At the silicon surface, a depth resolution of approximately 20 nm is obtained. (With electrostatic detectors, or by tilting the sample, resolutions of below 5 nm can be achieved [544], [554], [555].) The depth at which the depth resolution is limited to 20 nm, due to the energy spread of the particles, is about 0.5 μm for silicon.

The detection sensitivity of the back-scattering technique depends on the mass, the atomic number, and the depth of the element to be detected, relative to the matrix in which the element is incorporated. Two cases are to be distinguished:

(a) The energy spectrum (or a part thereof) of the ions back-scattered from the atoms to be detected is superimposed on the energy spectrum of the ions back-scattered from the matrix atoms. This always applies for light elements in a heavy matrix, and can also apply for heavy elements in a light matrix at high analysis depths, if the masses are not different enough from each other. The detection sensitivity is low in these cases due to the background represented by the matrix; it amounts to about 10 atomic %.

(b) The energy distributions do not overlap (heavy atoms in a light matrix, low analysis depth). In this case, the detection limit—depending on the atomic number—is between approximately 10^{18} cm^{-3} (As in Si) and 10^{17} cm^{-3} (Sb in Si).

For detecting light elements in a heavy matrix, the methods of ion-induced X-ray excitation (Section 5.8.3) and ion-induced nuclear reactions (Section 5.8.4) are more suitable than the back-scattering technique.

5.8.2 Channelling and Lattice Localization

If the analysing particles are injected parallel to a crystal axis into the sample, they penetrate considerably deeper than in the case of a random irradiation direction relative to the crystal lattice (*cf.* Section 2.3), and the yield of the back-scattered ions decreases drastically.

Figure 5.28 shows the yield of ions back-scattered at an angle of 165° to the angle of incidence (integrated over all energies), as a function of the angular deviation between the ⟨111⟩ axis of a silicon crystal and the direction of the incident beam. A comparison of the energy spectra in channelling and random directions is also shown in Figure 5.28. In the channelling spectrum a surface peak can be seen; this is due to the disordered atoms of the native SiO_2 layer (about 2 nm) and the surface atoms of the single crystal. Deeper-lying lattice atoms—relative to the beam direction—are in the shadow of the surface atoms, and contribute only slightly to the back-scattering yield.

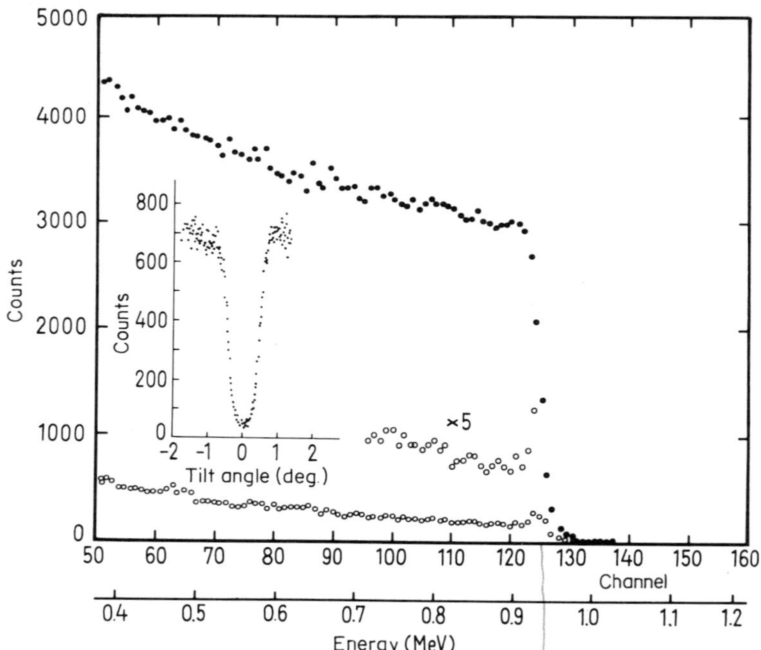

Figure 5.28 Back-scattering spectra of silicon in random (●) and ⟨111⟩ (O) channelling directions. Also shown is a spectrum obtained by tilting the sample around the <111> direction

If within the crystal an atom is displaced from a lattice site, an increased back-scattering yield occurs. The atoms in question can be either host lattice atoms or impurity atoms. It is thus possible, via this increased back-scattering yield, to measure the distribution of displaced (for example, due to radiation damage as a result of ion implantation) atoms. Because of the lower stopping power, the depth scale in the channelling direction is different from that in the random direction. The following is approximately valid [9]:

$$S_{\mathrm{e,channelling}}/S_{\mathrm{e,random}} = 0.7$$

Back-scattering spectra of a 150 keV boron implantation in silicon, depending on the annealing temperature, are shown in Figure 5.29. The annealing process progresses epitaxially from the bulk. A considerable increase in the crystallinity can be observed at 650 °C, but all radiation damage does not disappear until after annealing at 900 °C. One can see from the spectra that the back-scattering yield from layers which lie deeper than the damaged region does not decrease to the value for a totally-undamaged crystal. This can be explained as follows: Ions which, while passing through the damaged layer, are scattered forward by displaced atoms at an angle which is greater than the critical angle, leave the channel ('dechannelling') and then pass through the crystal in a random direction; these ions are therefore back-scattered with a higher yield in comparison to the ions which proceed further in the channelling direction, and are superimposed in the measured energy spectrum upon the ions which are directly back-scattered by displaced atoms. Since the intensity of the dechannelling also depends on the

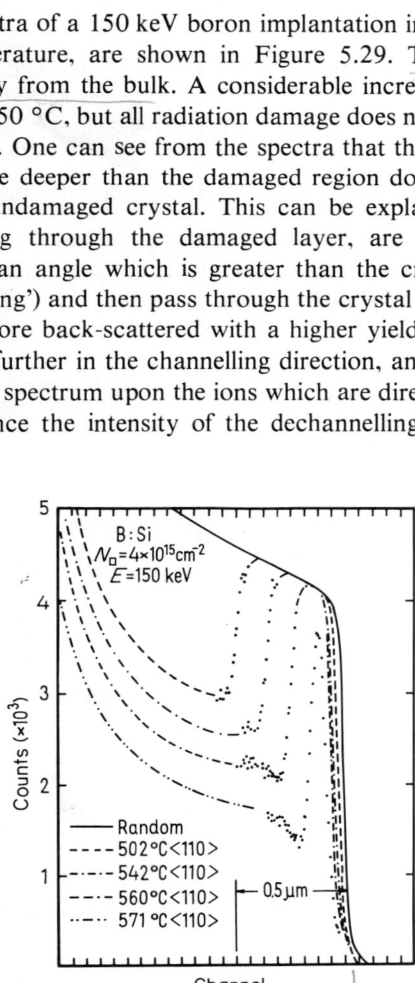

Figure 5.29 Back-scattering spectra of a boron implantation in silicon at 150 keV, after annealing at temperatures between 502 and 571 °C [409]

Surface of Si wafer

distribution of the lattice damage, the calculation of the actual concentration profile of displaced atoms from the channelling spectrum of a damaged crystal is a complicated problem, which must be solved by an iterative method [624], [754]. In practice, however, one is generally interested in the annealing process in a qualitative sense only, i.e., one chooses those annealing conditions under which a maximum recrystallization can be seen from the back-scattering spectra.

By directing the beam in various crystal directions, it is possible in principle to draw conclusions about the position of impurity atoms in the lattice. Figure 5.30 diagrams the methods by which the position of impurity atoms in a silicon lattice can be determined:

(a) If the sample is oriented in the ⟨111⟩ direction, atoms in random positions are measured;
(b) If the sample is oriented in the ⟨110⟩ direction, atoms on random and regular interstitial sites are measured;
(c) All atoms are measured if the sample is randomly oriented.

A spectrum of thallium in silicon, measured in various channelling directions (Figure 5.31), clearly shows the different amounts of implanted ions on regular lattice sites and interstitial sites [528a].

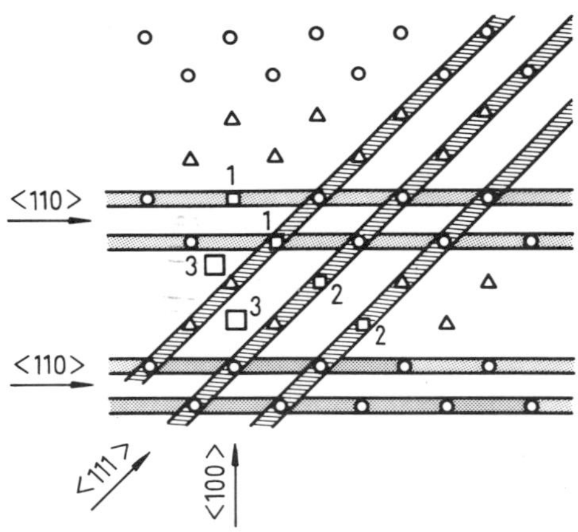

Figure 5.30 Method of lattice-site localization in silicon. The ⟨110⟩ plane is shown. The circles indicate regular lattice sites; the triangles indicate regular interstitial sites in the ⟨110⟩ direction. The lines mark the Thomas–Fermi screening distance around the lattice sites, which apply for MeV particles in the ⟨111⟩ and ⟨110⟩ directions. The squares represent doping atoms on lattice sites (1), regular interstitial sites (2), and random sites (3); according to Reference [485]. Reproduced by permission of Academic Press, Inc.

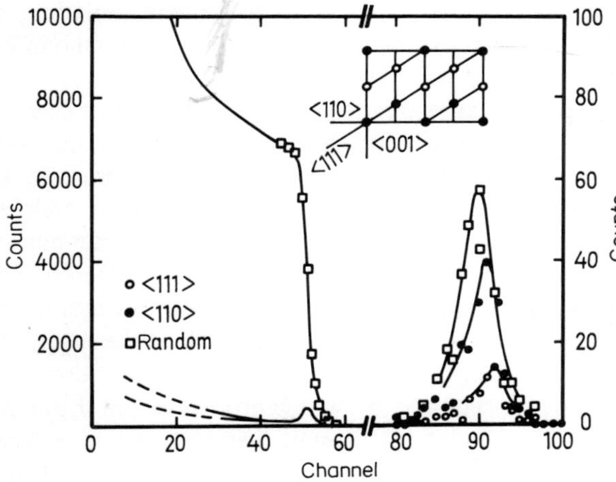

Figure 5.31 Energy spectra of helium ions back-scattered from silicon implanted with thallium. The primary energy was 1 MeV. By comparing the areas under the Tl curves, one can see that approximately 40 per cent of the Tl atoms are on lattice sites, approximately 40 per cent on tetrahedral interstitial sites, and the rest on random sites; according to Reference [528a]. Reproduced by permission of Academic Press, Inc.

The methods just discussed for lattice site localization are also applicable in connection with nuclear reactions, as well as with characteristic X-rays (see Sections 5.8.3 and 5.8.4).

5.8.3 Ion-induced X-Rays

Analysis of the elemental chemical composition of solids by measurement of characteristic X-rays which are emitted from the atoms when an appropriate excitation is applied is one of the oldest methods for non-destructive determination of a material's composition. In addition to the conventional excitation mechanisms using X-rays (X-ray fluorescence) or irradiation with electrons in the energy range from 10 keV to 100 keV in a scanning electron microscope or electron microprobe, the technique of excitation with the use of high energy ions in the energy range from 0.1–5 MeV has become more and more important in recent years. This is because the detection sensitivity in X-ray spectroscopy is not only limited by the properties of the detection system, such as efficiency and noise, but also by the continuous background upon which the characteristic lines are superimposed. This background is generated in the case of electron excitation by the bremsstrahlung; in the case of ion excitation, this bremsstrahlung-induced background is theoretically reduced by the factor (electron mass/ion mass)^2, i.e., by more than 10^6. When using MeV ions, however, there is the possibility of nuclear excitations which decay while emitting gamma radiation, so that under practical conditions one can expect increases in sensitivity amounting to a factor of 10 to 100.

Silicon diodes having a large area (e.g., 1 cm^2) and a large space-charge width (several millimetres, 'Si(Li)' detectors), operated in the reverse direction and cooled by liquid nitrogen, are generally used for X-ray spectrometry. They produce energy-proportional current pulses, so that by means of a multichannel pulse-height analyser, the entire energy spectrum of a sample can be registered simultaneously. These so-called energy-dispersive spectrometers are suitable for measuring the characteristic X-rays of elements with an atomic number of 11 (sodium, energy of K-line 510 eV) or above; at lower energies, the beryllium window generally used for protection of the cooled detector absorbs too greatly. If the beryllium window is replaced by a polypropylene foil 1 μm thick, or if an arrangement without a window is chosen, then the carbon K-line can still be resolved at 280 eV. A crystal spectrometer (so-called wavelength dispersive system) is suitable for measurement of elements with a still lower atomic number; with such equipment, however, only one energy can be measured at a time, corresponding to the angle of the analyser crystal employed according to the Bragg condition.

The cross-section σ_x for the excitation of an atom by light ions, with subsequent emission of characteristic X-rays, increases greatly with an increase in the energy E of the irradiating ions (e.g., $\sigma_x \approx E^4$ for arsenic). Since the ions lose energy through electronic stopping during penetration of the sample, there consequently exists an 'effective depth of excitation' from which characteristic X-rays are emitted. For light MeV ions, this depth is at about 1 to 2 μm; in general, therefore, depth profiles of elements cannot be measured without layer removal. Since implantation profiles generally extend over only a few tenths of a micrometre, the energy variation of the incident ions across this depth range is insignificant, so that the sheet concentration of the implanted ions is measured. The depth profile is obtained by differentiation of the measured sheet concentrations following step-by-step layer removal. Figure 5.32 shows an arsenic profile in silicon, measured from characteristic X-rays using proton excitation (1.6 MeV), compared with electrical measurements and the LSS theory [304]. Anodic oxidation has been used as the layer removal technique. After implantation at 120 keV with a dose of 5×10^{15} cm^{-2}, annealing was carried out in nitrogen at 750 °C for 1 hour; diffusion does not yet occur at this temperature. The low electrical activation at concentrations above 10^{20} cm^{-3} can be seen clearly. The measurement limit is at about 2×10^{13} cm^{-2}, corresponding to a concentration of 10^{19} cm^{-3}, at a 20 nm depth resolution.

During the excitation of characteristic X-rays using heavy ions, in the case of certain elements, threshold energies occur at which the cross-section increases sharply. An example of this is shown in Figure 5.33. For irradiation with krypton ions, there exists a threshold of 65 keV for excitation of the M-line of antimony (0.93 keV); the K-line of silicon (1.7 keV) is only emitted above about 140 keV. Based on these facts, selective excitation is possible; i.e., at a krypton-ion energy

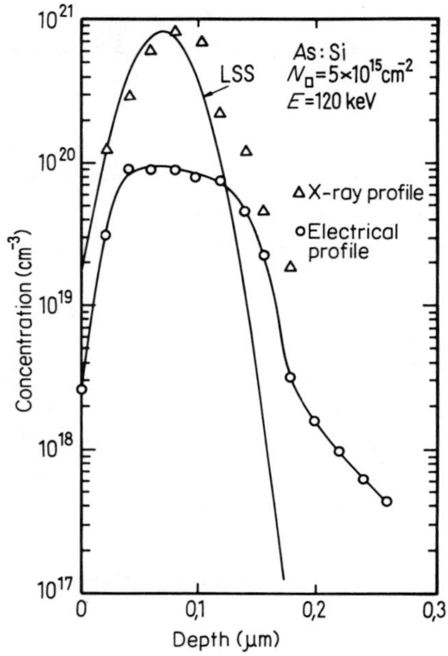

Figure 5.32 Profile of As in silicon, measured by proton-induced X-rays (△), in comparison to a doping profile (○) obtained by differential sheet resistivity measurements; according to [304]. Reproduced by permission of Elsevier Sequoia

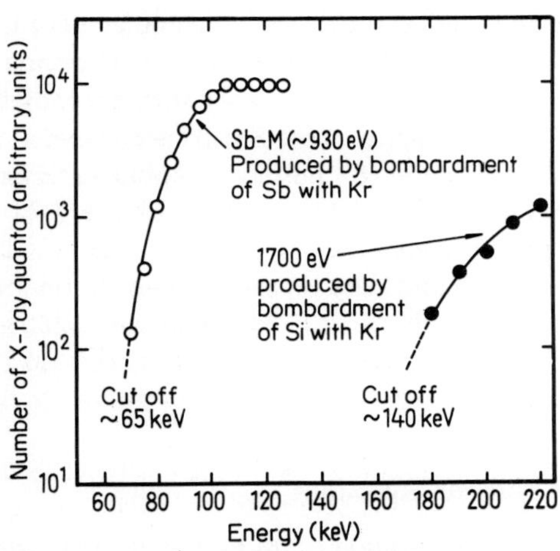

Figure 5.33 Excitation of X-rays using Kr ions at various energies for antimony and silicon [128]. Reproduced by permission of North-Holland Physics Publishing

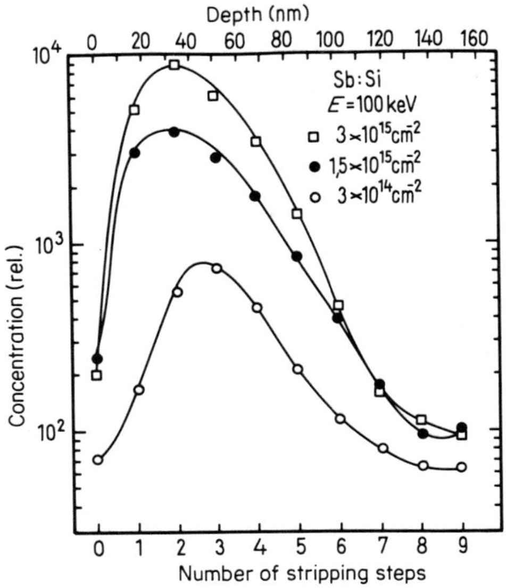

Figure 5.34 Antimony profiles in silicon, measured with Kr-induced X-rays, and using anodic layer removal [128]. Reproduced by permission of North-Holland Physics Publishing

of about 100 keV, in the case of antimony implanted silicon, only the antimony radiation is emitted while the silicon radiation, which (because of its higher energy) cannot be reduced by an absorption foil, is suppressed. In this way, a detection sensitivity of about 4×10^{12} cm^{-2} can be achieved; this corresponds to 2×10^{18} cm^{-3} at a 20 nm depth resolution. In the case of non-selective excitation, for example, using light ions, the considerably stronger radiation from the matrix would obscure the antimony signal in the Si(Li) spectrometer, since it has a limited count-rate capacity (maximum 10^4 pulses/s without deterioration of the energy resolution of approximately 200 eV (FWHM)). Figure 5.34 shows implantation profiles measured by this technique.

For detailed discussions and data on the method of ion-induced X-ray excitation, reference is made to the article by Folkmann [245], and especially for semiconductor applications, to the works by Chairns [127], [128], [129] and Gray [304].

5.8.4 Ion-induced Nuclear Reactions

A prerequisite for the transmutation of an atomic nucleus which is in itself stable is that the particle producing the transmutation approaches the nucleus within a distance in the order of magnitude of 10^{-13} cm, since nuclear forces are not effective beyond these distances. Positively-charged particles (e.g., protons) must, for this purpose, possess an energy with which they can overcome the electrical repulsion of the nuclear charge, the so-called Coulomb wall, which is also positive. This is only possible for the lightest elements in the energy range of the usual van de Graaff accelerators (maximum about 5 MeV). Deuterons, tritons, and ^{3}He particles are especially effective in causing nuclear reactions, since they contain relatively weakly-bound neutrons. These particles split as they approach the nucleus; the positive charges are reflected at the Coulomb wall, while the neutrons can penetrate to the nucleus unhindered by electrical forces. Based on this fact there exist, using these particles as projectiles, a number of reactions with elements of higher atomic number which are important for analytical purposes, and which occur with good yield even at low energies. In comparison to the back-scattering technique, the advantage here is that especially light nuclei can be measured. Nuclei of medium and high atomic numbers make almost no contribution. A disadvantage is that the detection sensitivity is adequate for a few reactions only, due to the small cross-sections.

Table 5.8 Nuclear reactions caused by bombardment with protons and deuterons, for several light isotopes [38]

Isotope A	Q_0 (MeV)	Isotope A	Q_0 (MeV)	Isotope A	Q_0 (MeV)
		(p, α)-Reactions			
^{7}Li	17.347	^{6}Li	4.02	^{9}Be	2.125
^{11}B	8.582	^{18}O	3.970	^{31}P	1.917
^{19}F	8.119	^{37}Cl	3.030	^{27}Al	1.594
^{15}N	4.964	^{23}Na	2.379	^{17}O	1.197
				^{10}B	1.147
		(d, α)-Reactions			
^{10}B	17.819	^{11}B	8.022	^{32}S	4.890
^{6}Li	22.36	^{15}N	7.683	^{18}O	4.237
^{7}Li	14.163	^{9}Be	7.152	^{30}Si	3.121
^{14}N	13.579	^{25}Mg	7.047	^{16}O	3.116
^{19}F	10.038	^{23}Na	6.909	^{26}Mg	2.909
^{17}O	9.812	^{27}Al	6.701	^{24}Mg	1.964
^{14}N	9.146	^{29}Si	6.012	^{29}Si	1.421
^{31}P	8.170	^{13}C	5.167		
		(d, p)-Reactions			
^{10}B	9.237	^{17}O	5.842	^{26}Mg	4.212
^{25}Mg	8.873	^{27}Al	5.499	^{12}C	2.719
^{14}N	8.615	^{24}Mg	5.106	^{16}O	1.919
^{29}Si	8.390	^{6}Li	5.027	^{18}O	1.731
^{32}S	6.418	^{23}Na	4.734	^{14}N	1.305
^{28}Si	6.253	^{9}Be	4.585	^{11}B	1.138
^{13}C	5.947	^{19}F	4.379	^{15}N	0.267
^{31}P	5.712	^{30}Si	4.367		

A listing of significant nuclear reactions is given in Table 5.8 [38]. The formulation A(x, y)B is used for the characterization of nuclear reactions, meaning that the nucleus A is transmuted into the nucleus B when struck by a projectile x, whereby a high-energy particle y is emitted. The Q_0 values in the table indicate the heat difference of the reaction, i.e., the difference before and after the reaction in the rest masses of the particles involved in the reaction. This mass difference is compensated for in the form of kinetic energy of the particles: if the sum of the rest masses of the reaction products is less than the sum of the rest masses before the reaction ($Q_0 \geqslant 0$, exothermal reaction), then the mass difference emerges as kinetic energy of the final nucleus as well as of the emitted particles, including γ-quanta. Otherwise, the particle inducing the reaction must possess at least a kinetic energy corresponding to the mass difference (threshold value) in order to make the reaction possible. The reactions listed in the table are all exothermal, i.e., the particles emitted in the reaction have a greater energy than the particles inducing the reaction, so that the detector can be screened off from the back-scattered primary particles by an absorption foil of suitable thickness.

The detection of oxygen in silicon, using the reaction $^{16}O(d, p)^{17}O$, can serve as a first simple example of this measurement technique. A typical detection system is shown diagrammatically in Figure 5.35(a); a ring-shaped silicon surface-barrier detector ('annular detector') is generally used to increase the counting yield. Figure 5.35(b) shows the particle spectrum which is obtained when a silicon crystal with a thin oxygen containing layer on its surface (e.g., native oxide) is irradiated by deuterium ions at 0.6 MeV. The reaction with ^{16}O produces protons with energies of 1.02 MeV and 1.90 MeV, which are to be correlated with the excited and the ground states of ^{17}O, respectively. Particles from the reaction $^{28}Si(d, \alpha)^{26}Al$ are also recognizable. Because of the exponential decrease of the cross-section with decreasing energy during penetration of the deuterons, only the

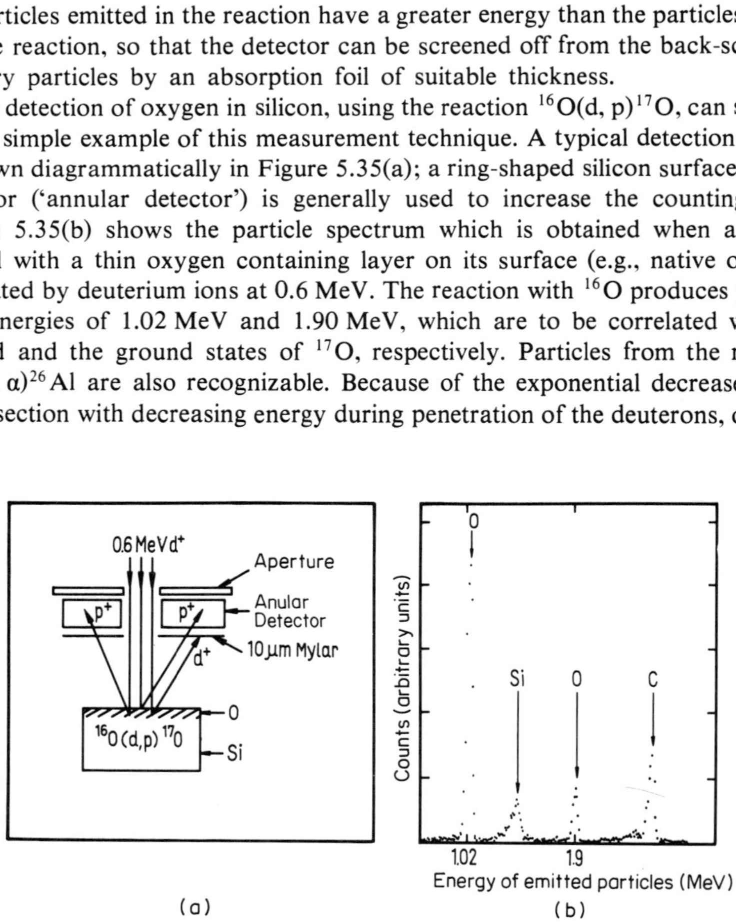

Figure 5.35 Measurement of oxygen in silicon with the ^{16}O (d, p) ^{17}O nuclear reaction: (a) measurement system, (b) particle spectrum

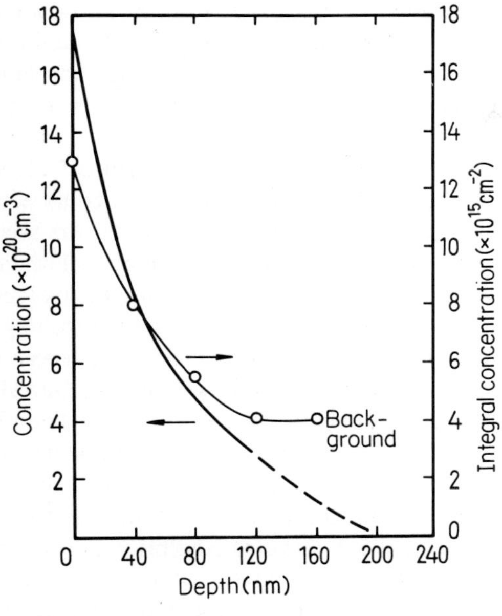

Figure 5.36 Integral oxygen distribution at various depths, following an arsenic implantation at 180 keV and a dose of 10^{16} cm^{-2} through 125 nm of SiO_2, together with the oxygen profile calculated therefrom

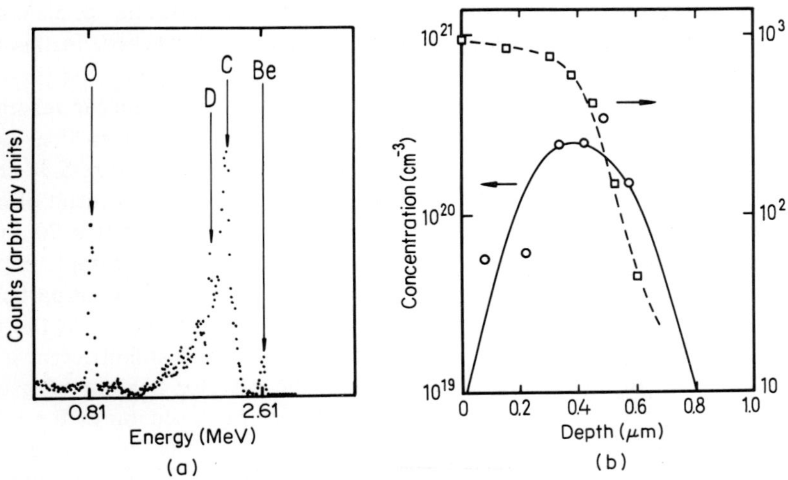

Figure 5.37 Measurement of a Be profile in GaAsP ($E = 150$ keV, $N_{\square} = 10^{16}$ cm^{-2}), using the ^{9}Be (d, α) ^{7}Li nuclear reaction; deuterion energy 0.5 MeV: (a) typical spectrum, (b) determination of the profile by successive etching (etchants: see Table 5.7), integral measurement of the Be concentration, and differentiation of the measurement curve

surface layer is significantly excited; therefore, a considerable widening in the energy distribution of these particles does not occur. In addition, protons from the reaction $^{12}C(d, p)^{13}C$ can be recognized; the thin carbon layer (a few nm) has been formed during the analysis due to hydrocarbon cracking from the residual gas atmosphere.

An application of this technique is shown in Figure 5.36. The oxygen profile produced in silicon as a result of implantation through an SiO_2 layer (knock-on effect) has been obtained by successive etching steps; the sheet concentration of oxygen has been measured in each case. The absolute calibration was made using an SiO_2 layer having a known thickness. The constant oxygen level is produced by the native oxide which forms immediately after the etching process. A further example for the determination of a light element in a heavy matrix is shown in Figure 5.37(a): beryllium implanted as an acceptor in GaAsP can be measured using the reaction $^{9}Be(d, \alpha)^{7}Li$. The decrease in the beryllium signal after successive layer removal, as well as the concentration profile thus derived, are shown in Figure 5.37(b).

By irradiating the crystal using different orientations, according to the principles outlined in Section 8.2, it is also possible with this measurement method to determine the lattice positions of impurity atoms. For example, it is possible to measure the relative amounts of boron atoms implanted into silicon which are on lattice sites versus interstitial sites, as well as to determine the corresponding profiles [244]; this is not possible using the back-scattering technique, due to the low mass of boron.

Some of the reactions which occur are resonant reactions, for example, there is a resonance with a width of 100 eV in the $^{27}Al(p, \gamma)^{28}Si$ reaction at 992 keV. If an ion beam with a higher energy is used, then the reaction does not occur until, due to the stopping of the particles, the resonance energy is reached. In this way, a profile can be measured by variation of the primary energy.

In many cases, the particles emitted as the result of a nuclear reaction are monoenergetic. If this is so, then based on their energy loss from their point of origin to the sample surface, a depth profile can be derived. Figure 5.38 shows a comparison between two particle spectra which are emitted as a result of irradiation with 1.5 MeV deuterium ions: in the first case, the target is a 70 nm thick layer of silicon nitride on GaAsP; in the second, it is a silicon sample implanted with 5×10^{18} cm^{-2} nitrogen atoms. Via the reaction $^{14}N(d, \alpha)^{12}C$, α-particles are emitted having energies of 6.7 MeV and 9.9 MeV; these appear in the energy spectra on the right-hand side. The back-scattered ions are still observable at low energies, since an absorption foil in front of the detector has not been used in order to achieve optimal energy resolution. An enlarged section of the 6.7 MeV nitrogen signal is shown in Figure 5.39. Compared with the thin nitride layer on GaAsP, the implanted layer yields a depth profile; the main line appears at 220 keV, corresponding to a depth of 1.7 μm (implantation energy 1 MeV), and nearby a lower concentration nitrogen distribution decreasing from the surface can be seen. The depth scale is calculated as indicated in Section 5.8.1.

If a focused ion beam is used for the measurement, two-dimensional element

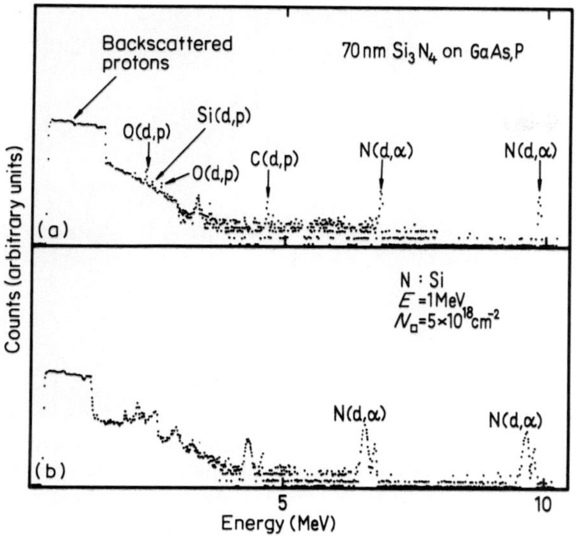

Figure 5.38 Measurement of nitrogen using the ^{14}N (d, α) ^{12}C nuclear reaction: (a) thin Si_3N_4 layer on GaAsP; (b) N-implanted silicon layer ($E = 1$ MeV, $N_□ = 5 \times 10^{18}$ cm^{-2})

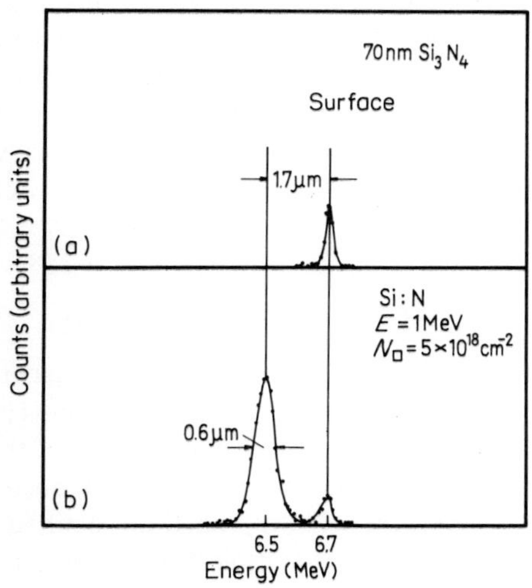

Figure 5.39 Enlarged sections of the spectra from Figure 5.38. (a) Thin Si_3N_4 layer; the width of the distribution indicates the energy and the depth resolution of the spectrum, respectively, (b) nitrogen profile after annealing at 1000 °C for 30 min

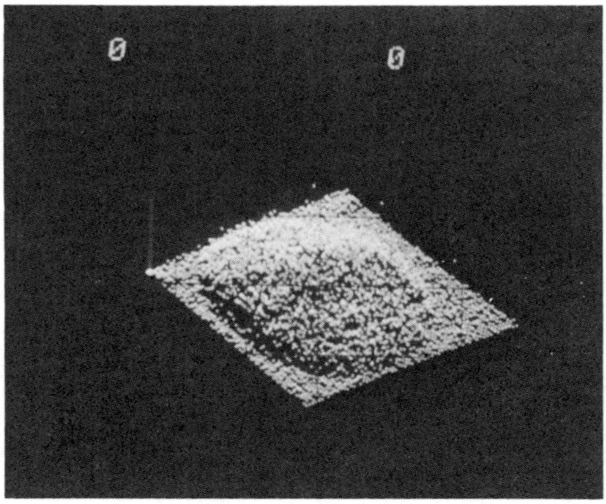

Figure 5.40 Nitrogen concentration at a depth of 1.7 μm in silicon, after a high-energy implantation of nitrogen ($E = 1$ MeV, $N_{\square} = 5 \times 10^{18}$ cm^{-2}, $T_A = 1000$ °C, $t_A = 30$ min). The sample is relatively inhomogeneous (± 20 per cent), since it has been implanted using a defocused beam with an aperture of 4 mm diameter, without deflection

distributions at a predetermined depth can be obtained by scanning the sample. In this way, for example, the lateral homogeneity of an implantation, a diffusion, or an alloying can be determined. As an example, Figure 5.40 shows the nitrogen concentration at a 1.7 μm depth in the nitrogen implanted sample which was measured in Figures 5.38 and 5.39. The distance between the reference points is 100 μm; the diameter of the analysing beam is also 100 μm. A minimum of 1 μm is possible for the distance between reference points as well as for the beam diameter.

5.9 ACTIVATION ANALYSIS

This method is one of the most sensitive measurement techniques for impurity atoms. In activation analysis, the impurity-atom distributions to be examined (and under some circumstances, of course, other components of the solid as well) are activated by irradiation of the sample with neutrons in a nuclear reactor, and are then examined by measurement of the radiation emitted during decay. A modification of activation analysis is the radio-tracer method. In this case, a previously activated element is implanted into the solid to be examined, and its distribution is then measured in the same way as with activated samples. The detection sensitivity of the tracer method is significantly higher, since no other undesired components of the solid are activated. This method is used relatively seldom, however, because of the unavoidable contamination of the ion source and the accelerator. Many fundamental experiments on the channelling effect have been carried out at Harwell using the tracer method. Further details, as well as accelerator requirements, can be found in the work by Dearnaley *et al.* [2].

The activation analysis technique (or the production of source substances in the radio-tracer method) is carried out using thermal neutrons (≈0.025 eV). Matrix effects can usually be neglected. Due to the neutron irradiation, the following reaction takes place:

$$ {}^{A}_{Z}\mathrm{M} + \mathrm{n} \rightarrow {}^{A+1}_{Z}\mathrm{M} + \gamma \tag{5.54}$$

This is generally abbreviated as ${}^{A}\mathrm{M}(\mathrm{n}, \gamma){}^{A+1}\mathrm{M}$, where M is the element, A the atomic weight, and Z the atomic number. Due to the incorporation of neutrons, heavy isotopes are formed; these are generally unstable and decay while emitting α-, β-, or γ-rays, as well as positrons in some cases. This radiation can be measured using suitable detectors. Its energy is characteristic for individual elements, similar to the case described in the previous section.

The activity D in decays per second of a particular isotope at the end of irradiation is obtained as follows:

$$D = \frac{N_L W \Phi_{th} \sigma_{eff}}{A} (1 - e^{-\lambda t}) \tag{5.55}$$

In this equation, W is the mass of the irradiated isotope, A the atomic weight, Φ_{th} the thermal flux of the neutrons, σ_{eff} the activation cross-section, λ the decay constant ($\lambda = \ln 2/T_{1/2}$, $T_{1/2}$ half-lifetime), t the irradiation time, and N_L the Loschmidt number ($N_L = 6.02 \times 10^{23}$ mol^{-1}).

Based on the number of implanted ions, the following applies for the activity per area unit:

$$D_{\square} = N_{\square} \Phi_{th} \sigma_{eff} (1 - e^{-\lambda t}) \tag{5.56}$$

A limitation for the applicability of activation analysis is the non-availability of practicable isotopes for a number of interesting dopants, i.e., isotopes with a high-enough cross-section for activation, or with high-enough half-lifetimes (if profiles are to be measured, several hours are necessary). In Table 5.9, for the example of GaAs, all significant nuclear reactions of Ga and As, as well as several impurity atoms, are listed [785a]. It can be seen that many different reactions must always be expected; most frequently, γ-decays are used for measurement. A compilation of corresponding spectra is available, for example, in the work by Vogg [748a]. In Table 5.10, the half-lifetime, the energy of the main line, and the maximum activity for γ-decays are given for several elements which are important for silicon. Boron, nitrogen and carbon are not suitable for activation analysis since they have a low lifetime as well as a low cross-section; ^{32}P decays with a half-lifetime of 14.3 days while emitting β-particles in sulphur.

Due to the considerable number of lines which appear, it is generally necessary to resort to spectroscopy. Ge(Li) or NaI detectors, combined with multichannel analysers, are used for measurement of γ-spectra. An example for the measurement of contaminations in SiO_2 is given in Figure 5.41. In special cases, for

Table 5.9 Nuclear reactions due to thermal neutrons in GaAs and several doping substances [785a]

Nuclear reaction	Half-life time	Type of decay*	Main energy (keV)	Decay product (stable)	Activation cross-section ($\times 10^{-24}$ cm^{-2})	Activity† (decays/s mg)
Matrix						
^{69}Ga(n, γ) ^{70}Ga	20 m	β$^-$	1650	Ge	1.4	1.85×10^8
		γ	170			
^{71}Ga(N, γ) ^{72}Ga	14.1 h	β$^-$	960	Ge	5.0	1.52×10^8
		γ	630, 840			
^{69}Ga(n, 2n) ^{68}Ga	68 m	β$^+$	1900	Zn	0.55	2.8×10^4
		γ	810–1880			
^{69}Ga(n, p) ^{69}Zn	14 h	IT	440	Ga	0.024	1.1×10^2
^{71}Ga(n, p) ^{71}Zn	3.9 h	β$^-$	1500	Ga	0.0005	2.6×10^1
		γ	390, 490, 610			
^{69}Ga(n, α) ^{66}Cu	5.1 m	β$^-$	2630	Zn	0.105	4.0×10^3
		γ	830, 1040			
^{71}Ga(n, α) ^{68}Cu	30 s	β$^-$	3500	Zn	—	—
		γ	810–1880			
^{75}As(n, γ) ^{76}As	26.5 h	β$^-$	2970	Se	4.5	2.0×10^8
		γ	560–2700			
^{75}As(n, γ) ^{74}As	18 d	β$^+$	900	Ge	0.55	2.8×10^4
		β$^-$	1360	Se		
		γ	600, 640–2530			
^{75}As(n, p) ^{75}Ge	82 m	β$^-$	1200	As	0.118	5.0×10^1
		γ	70–630			
^{75}As(n, α) ^{72}Ga	14.1 h	β$^-$	960	Ge	0.123	5.2×10^1
		γ	630, 840			
		γ	560–2700		—	—
Impurity atoms						
^{197}Au(n, γ) ^{198}Au	2.7 d	β$^-$	1370	Hg	98.8	5.7×10^{12}
		γ	410			
^{50}Cr(n, γ) ^{51}Cr	27.8 d	γ	320	Mn	17	2.8×10^7
^{70}Ge(n, γ) ^{71}Ge	11.4 d	EC		As	3.2	4.6×10^7
^{113}In(n, γ) ^{114}In	50.1 d	IT		Sn	8	2.5×10^7
		γ	190			
^{31}P(n, γ) ^{32}P	14.3 d	β$^-$	1710	S	0.19	2.3×10^7
^{30}Si(n, γ) ^{31}Si	2.7 h	β$^-$	1470	P	0.4	1.85×10^6
^{64}Zn(n, γ) ^{65}Zn	245 d	EC		Ga	0.46	8.1×10^5
		γ	1110			

* β$^-$ = electron; β$^+$ = positron; γ = γ – particles; IT = internal conversion electron; EC = electron capture.
† After 12 hours of irradiation with a thermal neutron flux of 2.5×10^{13} cm^{-2} s^{-1} and a fast flux of 2×10^{10} cm^{-2} s^{-1}.

example, arsenic in silicon (and also, of course, when radioactive isotopes have been implanted), this is not necessary, since the silicon which has also been activated decays with a half-lifetime of $T_{1/2} = 2.7$ h.

Profiles of the entire distribution of doping atoms can be obtained by combining activation analysis with a layer-removal technique, e.g., anodic oxidation in the case of silicon (see Section 5.5.3). Generally, the activity of the removed layer is measured in order to obtain a profile directly. The profile of an arsenic distribution in silicon measured in this way (implantation energy 200 keV, dose 10^{16} cm^{-2}, subsequent diffusion by proton irradiation with approximately 10^{18} cm^{-2} protons at 900 °C) is shown in Figure 5.42. A β-decay of the activated arsenic was used

Table 5.10 γ-decays for doping elements in silicon; according to [748a] and equation (5.54)

Isotope	Act. Isotope	Cross-section (barns)	Half-life time	Energy of the main line (keV)	Activity* (decays/s)
^{27}Al	^{28}Al	0.230	2.27 m	1779	5.92×10^3
^{75}As	^{76}As	4.3	1.1 d	559	314
^{69}Ga	^{70}Ga	1.68	21 m	174	2×10^4
^{71}Ga	^{72}Ga	4.7	14 h	834	177.6
^{113}In	^{114m}In	7.5	50 d	190	555
^{115}In	^{116m}In	42	54 m	417	2×10^6
^{121}Sb	^{122}Sb	6.2	2.75 d	564	1.15×10^3
^{123}Sb	^{124}Sb	4.28	60 d	603	21.1

* Directly after irradiation for 24 hours of 10^{15} atoms with 2.8×10^{13} cm^{-2} s^{-1} thermal neutrons.

for this measurement. A simultaneous β-decay of phosphorus, in this case, limits the measurement sensitivity for arsenic to 5×10^{15} cm^{-3}.

Some of the activated nuclei have an extremely low half-lifetime (a few ms); it is therefore possible to measure during the neutron irradiation. The neutrons are produced by a reactor, and are conveyed into the sample chamber by means of a neutron guide. As an example, the $^{10}B(n, \alpha)^7Li$ reaction will be considered here [162a], [496a], [515a], [611a]; this has such a high cross-section (3.837×10^{-21} cm^2) that it can be used for investigations of boron distributions in silicon. Only He, Be, and Li have similarly high cross-sections. All other cross-sections are approximately 10^7 lower, so that very little background is present.

As a result of the neutron irradiation, the following reactions occur:

$$^{10}B + n \rightarrow {}^7Li\,(839\text{ keV}) + \alpha(1471\text{ keV}) + \gamma(479\text{ keV}) \quad 94\%$$
$$^{10}B + n \rightarrow {}^7Li\,(1014\text{ keV}) + \alpha(1775\text{ keV}) \quad 6\%$$

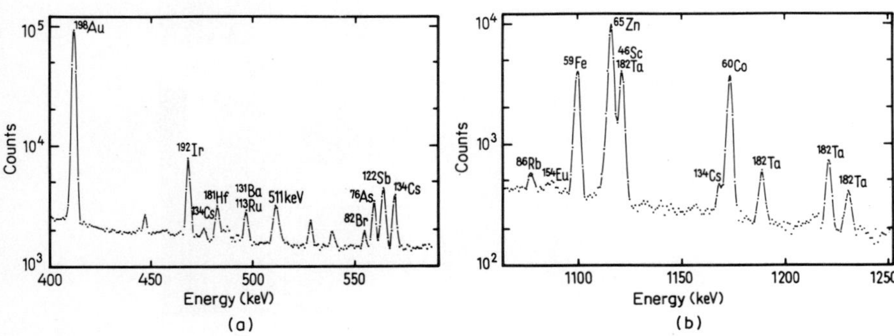

Figure 5.41 Two sections of the γ-spectrum of SiO_2. The spectrum was measured using a very pure SiO_2 sample with trace contaminations between 0.01 and 100 p.p.b. Each point indicates the number of pulses in an energy interval of 1 keV. At 511 keV, a peak occurs because of the annihilation of electrons and positrons [745b]. Reproduced by permission of Philips Technical Review

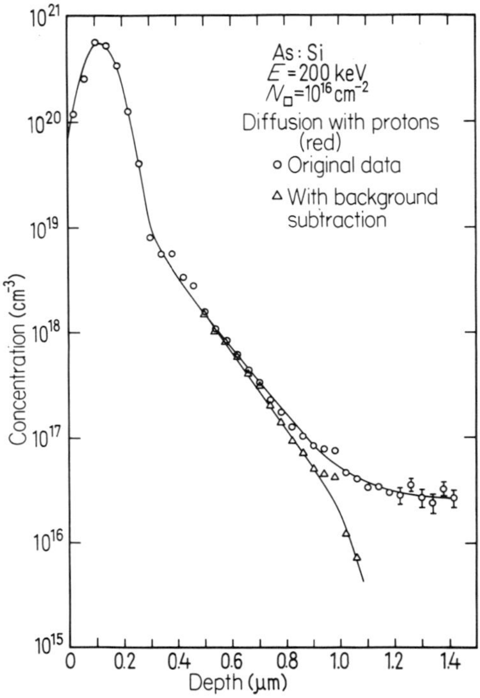

Figure 5.42 Arsenic profile in silicon measured by neutron activation. Implantation energy 200 keV, implantation dose 10^{16} cm^{-2}, subsequent diffusion with protons at 800 °C with a dose of 10^{18} cm^{-2} (radiation-enhanced diffusion)

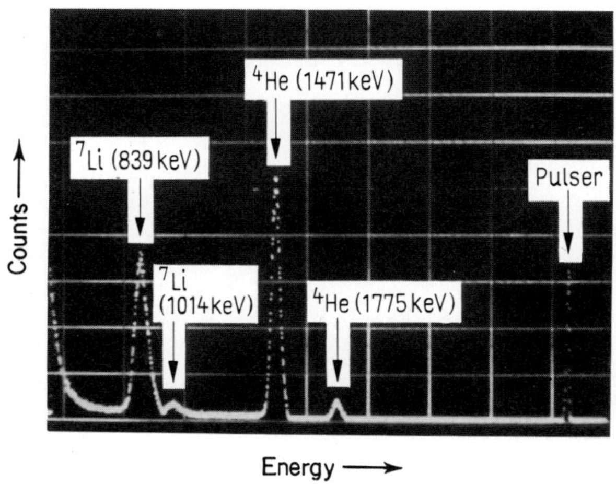

Figure 5.43 Spectrum of a ^{10}B profile determined using the ^{10}B (n, α) ^7Li reaction. The sample was implanted with a dose of 5×10^{14} cm^{-2} at 60 keV, and subsequently annealed in an oxidizing atmosphere for 5 h at 900 °C

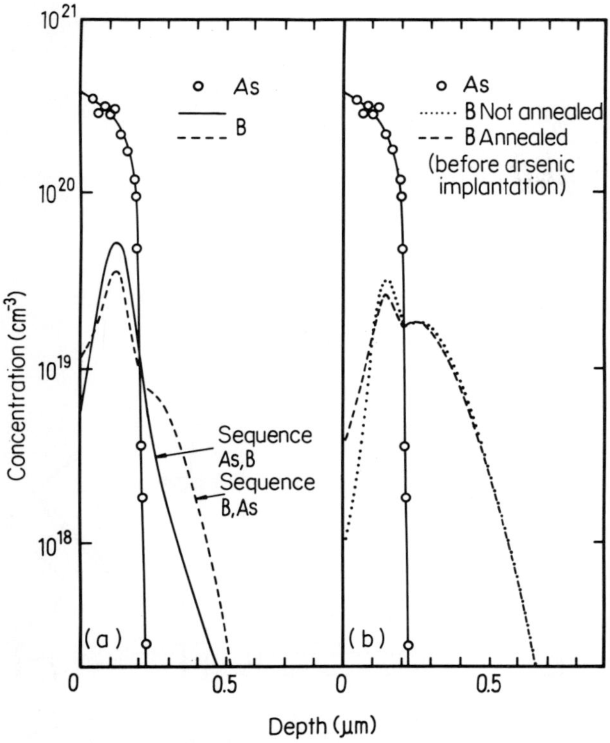

Figure 5.44 Double-diffusion profiles of As and B in silicon. The B-profiles were measured using the $^{10}B\ (n, \alpha)\ ^{7}Li$ reaction. (a) Comparison of different orders of As and B implantation with annealing after each implantation; (b) here, the As was implanted after the B, and the influence of annealing after the B implantation was investigated. B anneal 900 °C, 30 min; As anneal 950 °C, 30 min

The emitted particles are monoenergetic. The energy distribution outside the sample is therefore a measure of the doping distribution within the sample. The energy-depth scale is calculated as in Section 5.8.1. α-particles with 1471 keV are used for measurement, since their intensity is highest. A typical spectrum with all four emitted lines is shown in Figure 5.43.

Evaluated boron profiles are shown in Figure 5.44. This study was made to examine the diffusion behaviour of two-element systems (in this case, arsenic and boron in silicon); this is only possible with great difficulty using electrical measurement methods, especially near the *p–n* junction.

5.10 SECONDARY ION MASS SPECTROSCOPY

Secondary ion mass spectroscopy (SIMS) is a relatively new technique for investigating impurity-atom distributions in solids. The principle of this method is shown in Figure 5.45. The sample is sputtered by irradiation with ions of a few keV energy. Some of the sputtered atoms are electrically charged, and can be

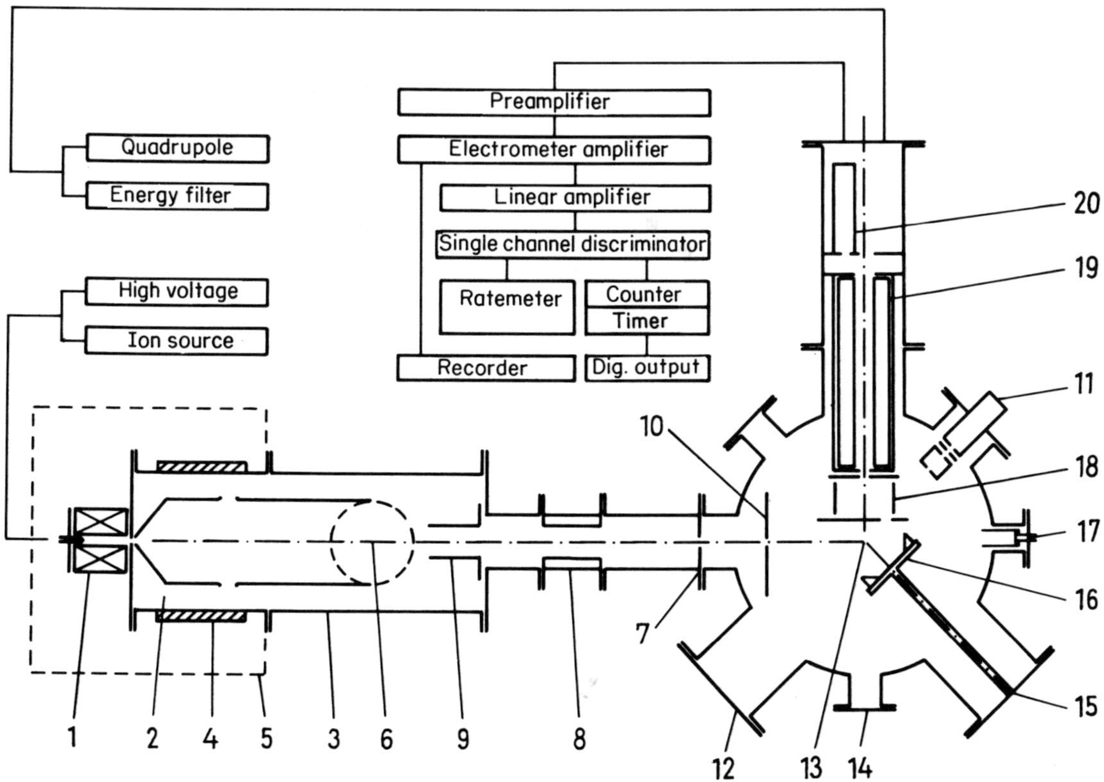

Figure 5.45 Schematic diagram of a SIMS machine.
1. ion source
2. extraction electrode
3. acceleration electrode
4. ceramic insulator
5. protective hood
6. turbo-molecular pump
7. pressure stage
8. valve
9. deflection plates
10. aperture
11. residual-gas analyser
12. viewing port
13. ion-gettering pump
14. current feed through
15. rotation feed through
16. target holder
17. Faraday cup
18. energy filter
19. quadrupole mass filter
20. detector

analysed using a mass spectrometer. Problems with this method are:

(a) Constancy of the sputtering yield,
(b) Planar sputtering,
(c) Constant ionization rate,
(d) Background interference due to primary particles,
(e) Knock-on effects.

The great advantages of this method are that all elements can be measured and that the measurement sensitivity is relatively high, i.e., between 10^{15} cm^{-3} and 10^{18} cm^{-3} depending on the element in question, and even higher in special cases.

In order to achieve a constant sputtering rate, a stable ion source must be used. The sputtering can be accomplished either using a focused beam which is scanned across the area to be analysed, or using a beam having a constant current density across its radius. In the first case, there is the additional possibility of taking 'ion pictures'; in the second, problems regarding a planar surface at great sputtering depths are less serious.

The ionization rate depends greatly upon the gas coverage of the sample [481]. In particular, oxygen can increase the ionization rate up to a factor of 100. In some machines, therefore, sputtering is done using oxygen. If this is not the case, one must always expect a surface effect, i.e., an apparently higher impurity concentration at the surface, caused by the unavoidable oxygen layer. Also very critical for the constancy of the ionization rate is the pressure in the vacuum system, which should be lower than 10^{-5} Pa. Knock-on effects due to the sputtering particles occur at the high acceleration voltages which are needed to achieve high sputtering rates. A compromise can be reached at a maximum energy of the sputtering particles amounting to 3 keV.

If all the negative influences are brought under control, doping profiles are easily obtainable through this method by counting the number of sputtered impurity atoms as a function of the removed depth. Several impurity atom distributions can be measured simultaneously by switching the analyser to various masses. Detailed descriptions of this measurement technique are contained in References [69a], [434], and [435]; and additionally, with an emphasis on implantation, in References [366], [481], [765], and [782]. The construction of a SIMS machine with vacuum chamber, ion source, and analyser is shown in Figure 5.45. Examples of profiles measured by this method are given in Chapter 6.

5.11 OTHER MEASUREMENT METHODS

Many other measurement methods have been developed for examination of ion-implanted layers. For the sake of completeness, several of these methods will be mentioned below, without a detailed treatment.

Radiation damage is not only indicated by changes in electrical or crystallographic parameters, but also by volume changes. This is measurable via the volume change of an implanted area or via the bending of thin layers. The results of such measurements [489] agree well with back-scattering measurements. These effects are important in semiconductor applications where capping layers are necessary for passivation purposes (*cf.* Chapter 3). Due to this volume change (which is eliminated by annealing in most cases, i.e., in elemental semiconductors as well as in most III–V semiconductors), adhesion problems involving passivation layers can occur.

Further methods of determining radiation damage or recrystallization of implanted layers include the Laue technique, examination of pseudo-Kikuchi lines, and especially, electron-microscopical examination [73], [718], [719]. The optical reflection of semiconductors has also been used to determine the damage concentrations [455], [790].

Long established methods of determining crystal defects include electron-spin resonance (ESR) and paramagnetic resonance (PMR); these, of course, can be used for ion-implanted layers as well [162], [323], [672]. They will not be discussed in more detail here due to their limited relevance to electrical parameters and the fact that they are difficult to interpret.

One can recognize amorphous layers by optical inspection in most semiconductors (e.g., silicon, germanium, GaAs, InSb, and probably all others as well), since these layers have a milky appearance (*cf.* Section 3.6) if no secondary effect due to ion etching or ion sputtering has occurred. A change in appearance, in the case of non-amorphous layers, can also sometimes be observed in the form of colours similar to the interference colours.

Measurement methods originally developed for surface analysis have recently come into wide use, in combination with sputtering techniques, for depth analysis as well. The most important of these is the Auger method. This procedure could offer great promise for the future, especially when combined with secondary ion mass spectroscopy (SIMS).

At present, all of these methods have a relatively low measurement sensitivity in common, so that they only appear suitable for examination of high-dose implantations. However, especially for multilayer structures where electrical measurement methods are generally inadequate, these relatively new techniques could be of promise for the future in semiconductor technology and, of course, most of all in application areas outside the semiconductor field, since the use of relatively insensitive non-electrical measurement methods is unavoidable here in any case.

5.12 COMPARISON OF THE VARIOUS MEASUREMENT METHODS

A comparison of the various measurement methods already discussed will be given in the following section. The criteria, for the most part, are related to sensitivity in radiation-damage measurement as well as measurement of the electrical activation of implanted ions. The changes in other physical and chemical properties can be so manifold that it is not possible to discuss them here. A few of these additional ion-implantation effects, however, will be discussed in Chapter 8, which deals with implantation in non-semiconductors.

Thus, the measurement methods will only be compared here in regard to their abilities to determine:

(a) Radiation-damage distributions,

(b) Electrical activation,

(c) Doping profiles.

Depending on the intended application of the implantation, one or another of these aspects will be of more importance. In Table 5.11, the measurement methods discussed in Sections 5.1 to 5.10 are listed, together with a description of their individual characteristics in as much detail as possible.

Based on the data given in this table, the most advantageous methods for measuring ion-implanted semiconductors (aside from unusual problems) are the following:

(a) Sheet resistivity and Hall effect measurement (these determine the sheet resistivity, the carrier concentration, and the mobility);

(b) The same method, but combined with a layer-removal technique, to determine doping profiles and mobility profiles;

(c) Spreading resistance measurement, a very fast method, but relatively inexact (due to the many corrections involved);

(d) The capacitance-voltage method (provides no information on the mobility);

(e) The back-scattering technique, the best method to determine the total concentration of implanted ions (only for $M_{ion} \geqslant M_{target}$) as well as the concentration of radiation damage;

(f) Activation analysis, a versatile method (unfortunately not universal, due to lack of isotopes) for measurement of the total implanted-ion concentration with highest sensitivity;

(g) Secondary ion mass spectroscopy (SIMS), with the disadvantage that its measurement sensitivity depends greatly upon the ionization conditions, making an absolute calibration difficult.

Depending on the specific problem, one or another of these methods will be better suited, or will be the only one usable under the given conditions. In practice, however, it is almost never possible to measure all important parameters using only one of these methods. If ion implantation is used for semiconductor doping

Table 5.11 A comparison of various measurements methods

Measurement method	Measured parameters	Measurement depth (μm)	Diameter of measured area or structure (μm)	Measurement sensitivity		Lateral resolution (μm)	Depth profile	
				Optimal (cm^{-3})	Measurement range		Resolution (μm)	Depth range (μm)
Etching of p–n junctions (5.1)	Steps of conductivity	Space charge	—	?	?	~1	~0.1	10^{-1}–10^5
Thermal probe (5.2)	Conductivity type	1–100	~10^4	~10^{13}	10^7	~10^4	—	—
Capacitance-voltage-measurement (5.3)	Doping	10^{-2}–10 (1)	100–10^4	10^{13}	10^4	—	10^{-2}–10 (1)	10^{-1}–10^3 (2)
Sheet resistivity (5.4)	Resistance	(3)	20–2 × 10^4 (4)	10^{12}	10^8	—	~0.01 (5)	0.1–10^5 (5)
Spreading resistance (5.6)	Resistance	A few microns	~10	10^{12}	10^8	—	—	0.1–10^5
Hall effect (5.5)	Doping; mobility	(6)	20–2 × 10^4 (7)	10^{12}	10^8	—	~0.01 (8)	0.1–10^5 (8)
Current-voltage-measurement (5.7)	Lifetime	0.1–10^2	~10^2	(9)	(9)	100	—	—
Back-scattering technique (5.8)	Impurity-atom distribution radiation damage lattice-site localization	4	10–10^4	10^{18}–10^{20} (10)	10^4	10–10^4	0.03	~4
Nuclear reaction (5.8)	Impurity-atom distribution lattice-site localization	~1–10	10–10^4	10^{15} (11)	(11)	10–10^4	0.01–3 (12)	1–10
Microprobe electrons	Impurity atoms	0.03–1	1	10^{18}–10^{19}	10	1	—	—
Microprobe ions (5.8)	Impurity atoms	1–10	1–10^3	10^{16}	10^3	1–10^3	—	—
SIMS (5.10)	Elements	10^{-2}	1–10^4	10^{15}–10^{18}	10^5	A few microns	—	A few microns
Auger spectroscopy	Elements	10^{-2}	10–100	10^{19}	20	A few microns	—	A few microns
Activation analysis (5.9)	Impurity atoms	(12)	10^4–10^5	10^{12}–10^{17} (13)	(13)	—	0.01	(14)

Table 5.11—*contd*

Detection of elements	Typical analysis time	Reproducibility (%)	Remarks
—	A few hours, including preparation	±5	What is etched or decorated is not certain
—	1 min	—	
—	1–10 min	±2	(1) Depending on Debye length; (2) depending on substrate doping; also combined with layer removal technique
—	1 min; profile: a few hours	±2	(3) Wafer or layer thickness; (4) with special structures 20 μm, otherwise about 2 mm; (5) depth range combined with layer removal techniques depending on wafer thickness
—	Homogeneous material: 10 min; depth profile: 20 min (without preparation of an oblique bevel)	±5	Depth profile when combined with oblique bevels; profile obtainable only through extensive calculations, error in this case approximately ±20 per cent
—	10 min, profile: a few hours	±5	(6) Wafer or layer thickness; (7) special measurement structures are necessary; (8) combined with layer removal technique depth range depends on wafer thickness
—	1 min	±10	Lifetime of approximately 10^{-8} to 10^{-6} seconds
Good (10)	10 min; radiation damage and lattice-site localization: 1 h	±10	Non-destructive; (10) measurement sensitivity depends on the mass of impurity atoms
Good (11)	1 h	—	(11) Depends on the effective cross-section; (12) high resolution when combined with layer removal, otherwise depends on the nature of reaction
$Z \geqslant 4$	1 h	±2	Non-destructive
$Z \geqslant 4$	1 h	±2	Non-destructive
Practically all	Surface: 30 min: a few hours for depth profile	±5	Long measurement time limits range depth; lateral resolution only when small beam diameters are used; matrix effects can occur
All elements except H, He	Surface: 30 min; a few hours for depth profile	±20	Depth profile only when combined with sputtering of the sample; lateral resolution depends on the diameter of the electron beam used
Good	A few hours	±2	(12) Wafer thickness; (13) according to the element; (14) layer removal necessary, otherwise only element analysis

only, the most important methods to examine implanted semiconductor layers are Hall effect and sheet resistivity measurement based on van der Pauw, spreading resistance measurement, and (with certain limitations) the back-scattering technique. Lifetime measurements can best be made indirectly on special *p–n* structures or on finished devices.

6
Properties of Ion-implanted Semiconductor Layers

A great many practical investigations have dealt with the behaviour of implanted ions in semiconductors. The most important aspects in this regard are the number and distribution (profile) of the implanted ions, their electrical activation, and the radiation damage of the crystal lattice. Accordingly, electrical measurements are made for the most part; i.e., Hall effect, sheet resistivity, and back-scattering measurements. A number of indirect methods are used to examine radiation damage in implanted crystals. These include measurements of volume change, electron-spin resonance, paramagnetic resonance, and reflectivity changes, among several others.

This chapter presents a detailed discussion of basic results which are important for the use of ion implantation, especially in the manufacture of electronic devices. Main emphasis will be given here to the most important semiconductor at present, silicon. Next in importance are III–V compound semiconductors such as GaAs and InSb. Studies concerning II–IV and IV–VI compound semiconductors (PbTe, HgCdTe) or germanium will only be considered in passing. Up to now, the compound semiconductors have only found use in some special applications, in spite of their excellent properties in many respects. Their main areas of application are luminescence diodes, semiconductor lasers, Gunn diodes, and integrated circuits using III–V semiconductors, as well as infrared detectors using III–V, IV–VI and II–VI semiconductors. Up to now, GaAs has clearly been the most important of these compound semiconductors. At present, however, interest is shifting (depending on the intended use) in the directions of ternary and quaternary crystals, in which gallium and arsenic are replaced in part by aluminium, phosphorus, indium and antimony.

In the early days of ion implantation, it was widely assumed that its main area of application would be the doping of compound semiconductors, since diffusion in silicon was already well developed. In the meantime, however, the opposite has proven to be the case. Implantation is now being used routinely in the fabrication of a great many silicon devices, thanks to its exact process control and its doping homogeneity as well as its reproducibility. In the meantime, activities in the fields of III–V, II–VI, and IV–VI compound semiconductors have remained at a relatively low level, due mainly to problems with the electrical activation of implanted ions, the annealing of radiation damage, the low range of suitable ions, and the difficult contacting of the implanted layers. Recently, however, implantation into semi-insulating GaAs to form medium and large-scale integrated circuits has gained much interest.

6.1 IMPLANTATION IN SILICON

As mentioned above, most implantation experiments have been carried out using silicon. This is not only due to the importance of silicon in semiconductor technology and the correspondingly great interest in this material, but also (and mainly) to the fact that silicon, as an elemental semiconductor, only consists of one component, so that the recrystallization of radiation damage is relatively simple; therefore, the earliest experiments were already successful.

An important parameter in the use of implantation for the manufacture of devices is the sheet resistivity. If it is assumed that all implanted ions become electrically active and that the radiation damage is completely eliminated by annealing, it is possible to obtain, using the well known relationship between mobility and doping concentration, curves of the theoretical sheet resistivity, depending on the standard deviation of the implanted distribution. In many cases, a knowledge of the sheet resistivity is then sufficient to characterize the implanted material. For the sheet resistivity, ρ_S, the following applies:

$$\rho_S = \frac{1}{\sigma_S} = \frac{1}{q}\left[\int_0^t N(x)\mu_C(x)\,dx\right]^{-1} \tag{6.1}$$

where t is either the wafer thickness for implantation into material of the same conductivity type or otherwise the depth of the p–n junction, σ_S the conductivity of the layer, μ_C the drift mobility, and N the doping concentration. Assuming a Gaussian doping profile, one can obtain curves for various implantation energies, most easily by use of the expression for the mobility by Caughey and Thomas [124]. Curves calculated for p-doped and n-doped elements according to equation (6.1) are shown in Figures 6.1 and 6.2. The broken lines indicate a dose range in which a buried layer is formed; i.e., due to the Gaussian implantation profile, there are two p–n junctions. In Figure 6.3, the substrate doping has been varied for a boron implantation at 100 keV. In the case of a low-doped substrate, hardly any influence can be seen; a pronounced compensation, however, is observable at higher concentrations. In order to achieve these sheet resistivity values for implanted layers in silicon, an annealing step after implantation is generally needed. As a result, due to thermal or (in some cases) enhanced diffusion (*cf.* Section 3.8), a broadening of the profile can occur. At doses greater than 10^{14} cm^{-2}, the maximum solubility can be exceeded and the obtainable sheet resistivity remains higher. Maximum solubilities for the usual doping elements, depending on the temperature, are given in the Appendix. Curves which are

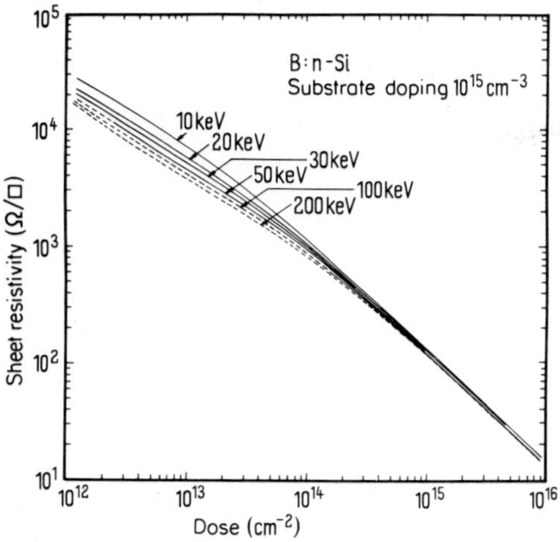

Figure 6.1 Sheet resistivity as a function of implanted dose and energy for boron in silicon, assuming complete electrical activation, substrate doping 10^{15} cm^{-3}; in the area of the broken lines, a buried layer is present

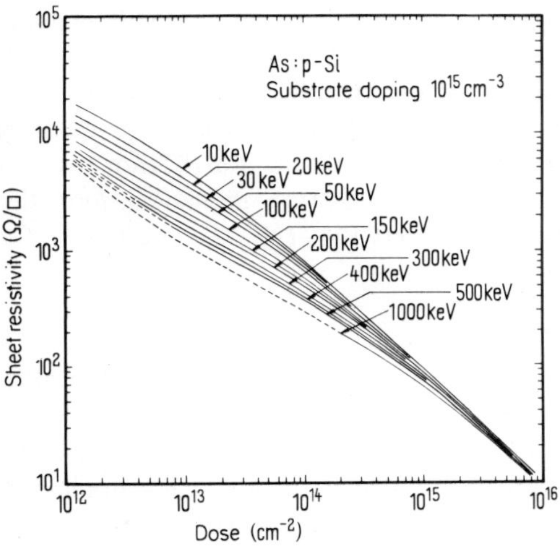

Figure 6.2 Sheet resistivity as a function of implanted dose and energy for arsenic in silicon, assuming complete electrical activation, substrate doping 10^{15} cm^{-3}; in the area of the broken lines, a buried layer is present

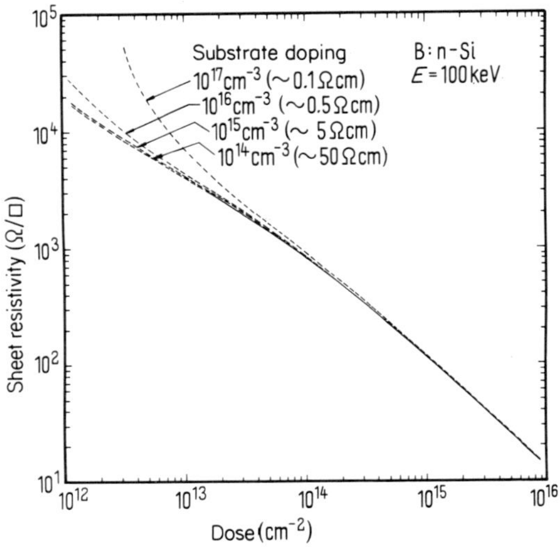

Figure 6.3 Sheet resistivity for a boron implantation into silicon at 100 keV as a function of implantation dose and substrate doping; a buried layer is present in the area of the broken lines

independent of a profile are obtained for standard deviations above approximately 0.15 μm.

In the annealing process with silicon and all its dopants, a distinction is to be made as to whether or not the semiconductor has been made amorphous at the surface by the implantation. In the first case, there occurs an epitaxial recrystallization at about 650 °C with simultaneous electrical activation; in the second, annealing temperatures around 900 °C are necessary. The process of amorphization can be caused by the implanted ions themselves or by an additional implantation of inert ions (Ne, Ar, etc.).

6.1.1 Aluminium

Aluminium is often used in silicon technology as a material for ohmic contacts. It is not very well suited for doping via thermal diffusion, due to its low solubility and the formation of a eutectic with silicon at 577 °C. A doping of silicon with aluminium can be done more easily by using ion implantation. Because of the relatively minor importance of this dopant (with the exception of power devices), relatively few experiments with it have been carried out [59], [114], [385], [470], [552], [704].

Especially at high doses, the electrical activation of aluminium is low, i.e., between 50 per cent (10^{14} cm^{-2} at 60 keV after annealing at 950 °C for 10 min [385]) and approximately 6 per cent (10^{15} cm^{-2} at 30 keV after annealing at 900 °C [59]). The maximum obtainable doping concentration is about 2 ×

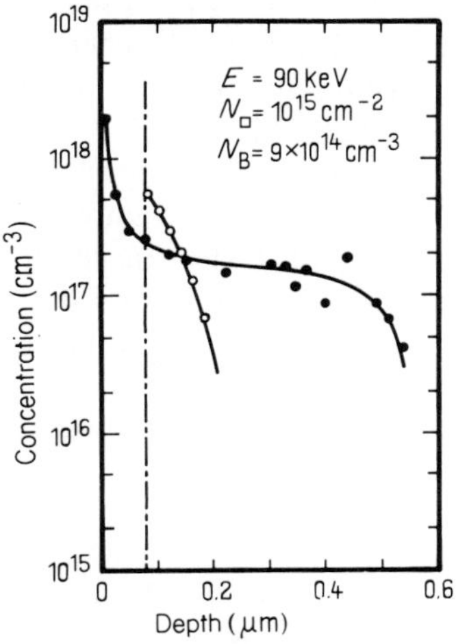

Figure 6.4 Doping profile of an Al-implanted silicon layer, with (○) and without (●) etching before annealing, according to Reference [376]. Reproduced by permission of the American Institute of Physics

10^{18} cm^{-3}, and is thus an order of magnitude below the maximum solubility [731]. Itoh *et al.* [376] have observed an enhanced diffusion effect caused by radiation damage. The doping profile (implanted with 90 keV and a dose of 10^{15} cm^{-2}), after annealing at 800 °C for 20 min, extends up to 0.58 μm (Figure 6.4). If the layer close to the surface is etched off, the depth of the profile is

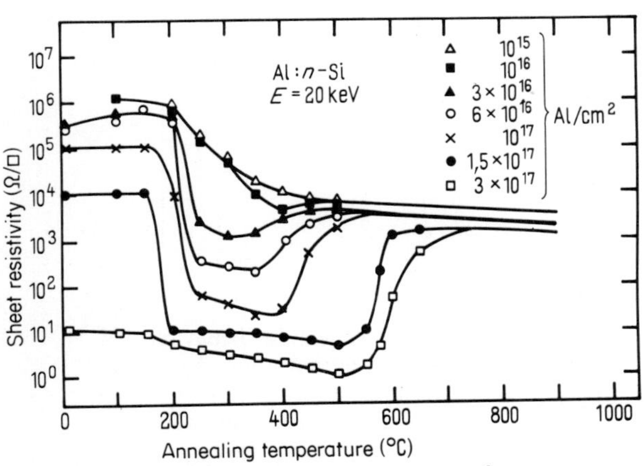

Figure 6.5 Sheet resistivity of Al-implanted silicon layers depending on ion dose and annealing temperature for extremely high doses [418]. Reproduced by permission of Springer-Verlag, Heidelberg

0.26 μm. Itoh explains this in terms of an increase in the diffusion coefficient due to vacancies which are released during annealing of the radiation-damaged surface layer. A similarly deep penetration of aluminium ions has also been measured by Bader and Kalbitzer [51].

Kräutle *et al.* [418] have examined the properties of silicon layers implanted with extremely high doses of aluminium. In Figure 6.5, sheet resistivity measurements depending on dose are given. The transition to 'metallic' conductivity occurs at a dose of approximately 5×10^{16} cm^{-2} (implantation energy 20 keV).

Aluminium seems poorly suited for device applications, though Runge [601] has shown that aluminium-implanted resistors can offer advantages in comparison to boron-implanted ones, due to the lower mobility of the carriers.

6.1.2 Antimony

Antimony is an often used dopant for subcollector diffusions, since it has a very low diffusion coefficient. Recently, however, increasing use has been made of arsenic, because of its higher solubility and the lower sheet resistivities which are therefore possible. Antimony is not used for emitter diffusion or base diffusion.

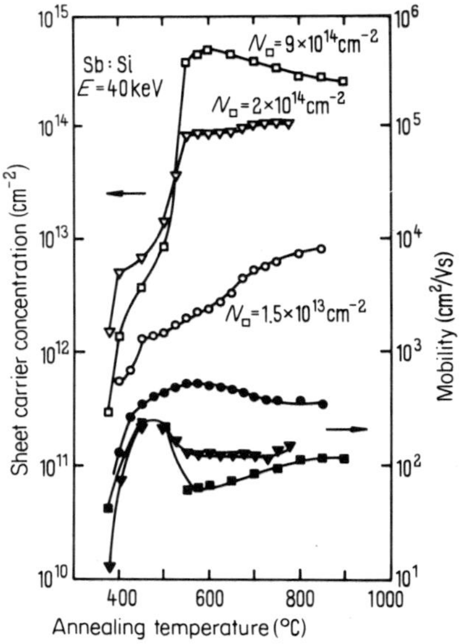

Figure 6.6 Annealing behaviour of sheet carrier concentration $N_{S,eff}$ and sheet Hall mobility $\mu_{S,eff}$, for 100 Ωcm *p*-type silicon samples implanted at room temperature with 40 keV antimony ions. Annealing time 15 min in all cases [385]. Reproduced by permission of Pergamon Press Ltd

Most of the studies made on antimony implantation in silicon are rather old. Their results are somewhat inconsistent. The maximum electrical activation obtained for implantations above the critical dose for amorphization amounts to 100 per cent; the minimum amounts to 3 per cent [59], [160]. Implantations into heated substrates yield poorer results with antimony [159]. Implantations below the critical dose require relatively high annealing temperatures [385].

The percentage of antimony atoms incorporated on to lattice sites depends not only on the ion dose, but also on the annealing temperature. Thus, the percentage of atoms on lattice sites for a 2.6×10^{15} cm^{-2} implantation at 40 keV is about 0 per cent after annealing at 550 °C; this increases to 50 per cent at 650 °C and decreases to 25 per cent at 750 °C [385]. At doses $\leqslant 1.3 \times 10^{15}$ cm^{-2}, above about 600 °C, 90 per cent of the antimony is on lattice sites; no decrease in this value occurs up to 900 °C. In Figure 6.6, the dependence of the sheet carrier concentration and the sheet mobility upon the ion dose and the annealing temperature are shown. After implantation with doses leading to the formation of an amorphous layer ($\geqslant 10^{14}$ cm^2 at room temperature [9]), there is a sudden activation around 550 °C. In Figure 6.7, a measurement of the sheet carrier concentration is shown for an implantation at 350 °C. Since no amorphous layer is

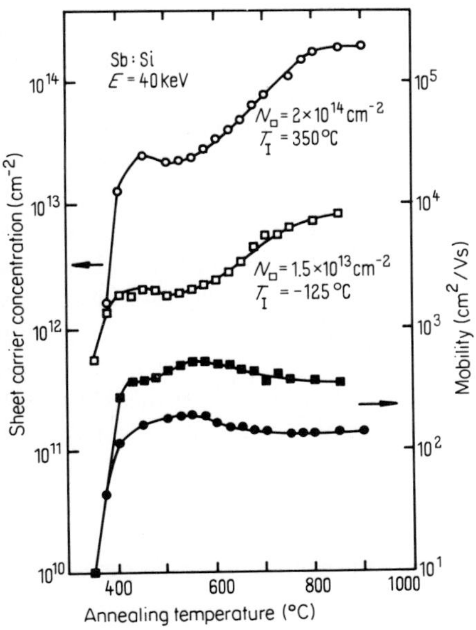

Figure 6.7 Annealing behaviour of sheet carrier concentration $N_{S,eff}$ and sheet Hall mobility $\mu_{S,eff}$ for 100 Ωcm *p*-type silicon samples implanted at 125 °C (□) and 350 °C (○) with 40 keV antimony ions. Annealing time 15 min in all cases [385]. Reproduced by permission of Pergamon Press Ltd

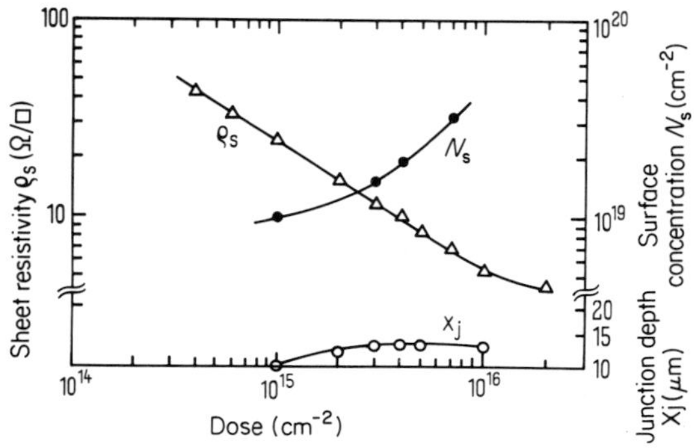

Figure 6.8 Sheet resistivity, surface concentration, and depth of *p–n* junction depending on ion dose for 150 keV antimony in ⟨111⟩-oriented, 10 Ωcm, *p*-type silicon after diffusion at 1250 °C, 2 h in oxidizing atmosphere and 8 h in inert atmosphere [205], reprinted by permission of the publisher, The Electrochemical Society, Inc.

formed, significantly higher temperatures are necessary for electrical activation. Also shown is a low temperature implantation with a dose likewise not causing any amorphous layer formation; the annealing behaviour is therefore very similar.

Profile measurements show that the range agrees with the LSS theory [159], [534]. An example for this can be found in Chapter 5 (Figure 5.34). Based on the many studies on antimony implantation which have been made (in addition to the works already mentioned; see References [157], [174], and [470]), it can be said that antimony seems not very well suited as a dopant in silicon for formation of *p–n* junctions, nor is it well suited for diffusion. Under some circumstances, it could be suitable as a dopant for subcollectors in competition with arsenic. In diffusion technology, as already mentioned, the use of antimony subcollectors is standard. Initial results in this regard are given in a work by Drum [205]. He diffused implanted antimony distributions in oxidizing and inert atmospheres at 1250 °C. Figure 6.8 shows values for the sheet resistivity, the surface concentration, and the *p–n* depth. Sheet resistivities down to 5 Ω/□ could be achieved, a result which is comparable to arsenic-implanted layers. It was possible to grow epitaxial layers of good quality on these layers. In order to avoid out-diffusion, annealing in an oxidizing atmosphere must be done after implanting antimony [519].

6.1.3 Arsenic

Arsenic has only recently—later than boron, phosphorus, gallium, and indium—gained the interest it deserves in connection with ion implantation. Of all dopants, arsenic has the highest solubility in silicon [731], approximately 10^{21} cm^{-3} at 1000 °C. Since arsenic can only be diffused in closed systems (ampoules) or from doped oxide layers, and not (as is usual in silicon planar technology) in open systems, ion implantation offers considerable advantages. Its main applications in integrated circuits are for buried layers, emitters for fast ECL circuits and microwave transistors, as well as for source-drain doping.

Electrical Activation

Arsenic in silicon becomes about 100 per cent electrically active at doses up to 3×10^{15} cm^{-2} after annealing at 800 to 900 °C. Attempts to achieve a high electrical activation without annealing, by implantation at increased temperatures [16], have not been successful. On the contrary, such activation was made more difficult due to the formation of stable defects. Above the critical dose of 2×10^{14} cm^{-2} [9] (at RT), a complete electrical activation is obtained after annealing at 600 °C, due to epitaxial recrystallization of the amorphized crystal lattice. Very high concentrations of implanted arsenic only become electrically active when the concentration at the maximum of the distribution has been decreased so much through a drive-in diffusion step that the solubility of arsenic in silicon is no longer exceeded. As an example, Figure 6.9 shows the time dependence of the electrical activation for an arsenic implantation with a dose of 10^{16} cm^{-2}. At low temperatures, the annealing behaviour of arsenic varies, depending on the dose. It is possible to measure the activation energy of the annealing process by isothermal annealing measurement at low temperatures (see Figure 3.16). If one assumes that no additional energy is needed during epitaxial recrystallization for incorporation of arsenic into the lattice (this assumption is justified by the suitable size of the arsenic atom), an activation energy of 2.75 eV is obtained. At doses below the dose for amorphous layer formation, considerably higher annealing temperatures are needed to reach complete activation. All

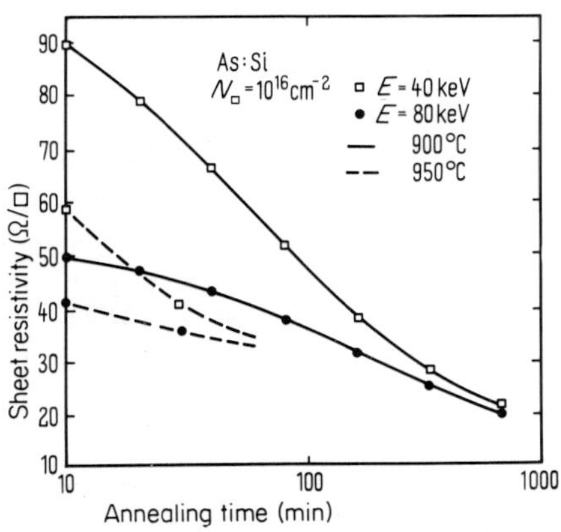

Figure 6.9 Electrical activation of arsenic depending on annealing time and temperature. Implantation energies: 80 and 40 keV; annealing temperatures: 900 and 950 °C [521]

attempts to avoid the necessity of a subsequent annealing step by implantation at higher temperatures (500 to 600°C), with retention of high electrical activation [59], [160], [388], [487], have been unsuccessful; on the contrary, the results were significantly higher sheet resistivities. These findings can be explained by back-scattering measurements [229], which show that only 65 per cent of the arsenic ions are on lattice sites, independent of the dose, if implantation is done at 450°C. Thus, the arsenic ions are incorporated preferentially on to lattice sites during the epitaxial recrystallization of the amorphous layer. These results confirm the assumption made above.

Range and Profiles

Figure 6.10 shows various projected-range and range-straggling measurements, compared with calculations made by Gibbons *et al.* [5]. In spite of the scatter in the values shown, the agreement in regard to R_p is good, while ΔR_p is somewhat higher than predicted by theory. This seems to be a general problem in cases where the implanted ion species is greater in mass than the target atoms.

Doping profiles of arsenic at various doses are shown in Figure 6.11, according to Schwettmann [639]. Also shown is the theoretical distribution according to the LSS theory, for a dose of 10^{15} cm^{-2}. A flat top of the profile at concentrations above 10^{20} cm^{-3}, in cases of higher doses, can be seen clearly; an activation limit exists [243] at approximately 3×10^{20} cm^{-3}, which is below the maximum solubility of about 10^{21} cm^{-3} [616]. At low concentrations, all profiles have an exponential tail, which we attribute (in contrast to Reddi) to the channelling effect prior to the formation of an amorphous layer (compare the results obtained by Blood *et al.* for phosphorus [85], [86]). The reduction to $1/e$ takes place at about 100 nm.

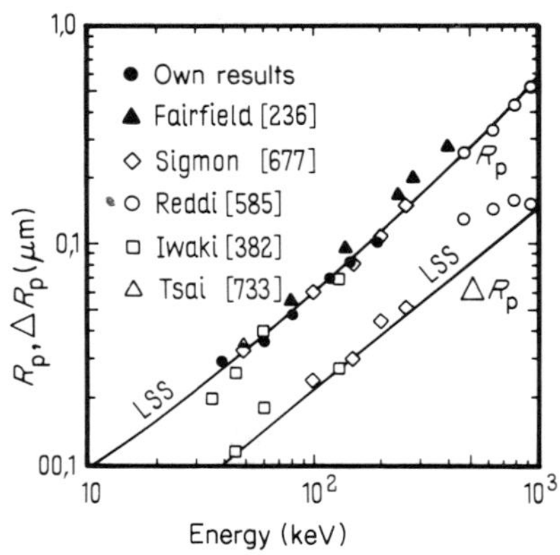

Figure 6.10 Experimental data on R_p and ΔR_p for arsenic in silicon compared to theoretical data according to LSS theory

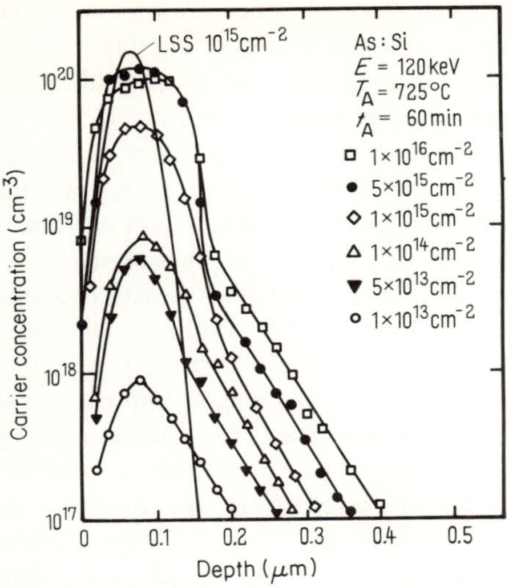

Figure 6.11 Doping profiles of arsenic in silicon for different implantation doses. $E = 120$ keV; $T_A = 725$ °C [639]. Reproduced by permission of the American Institute of Physics

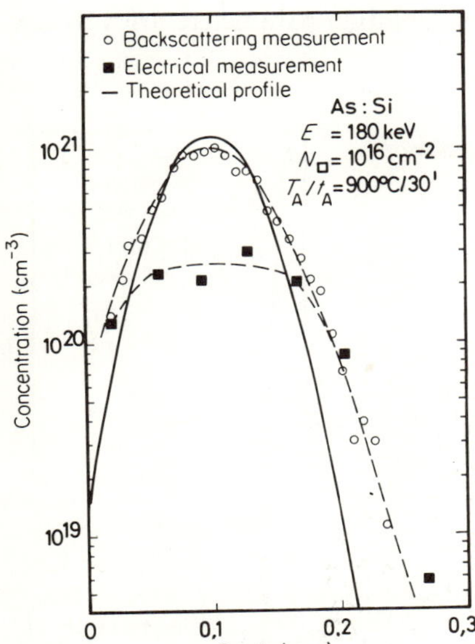

Figure 6.12 Comparison between doping concentration and total concentration. $E = 180$ keV; $N_\square = 10^{16}$ cm^{-2}; $T_A = 900$ °C; $t_A = 30$ min

A comparison between doping concentration and total concentration, as measured using the back-scattering technique, is shown in Figure 6.12. The agreement with the LSS theory is good in regard to the total distribution, as can be seen from the theoretical distribution which is also given. The maximum doping concentration, once again, is only about 3×10^{20} cm^{-3}. Thermal diffusion, resulting in a broadening of the profile, already occurs at the low annealing temperature of 900 °C. The measurement sensitivity of the back-scattering technique is about 10^{19} cm^{-3}, so that profile tails cannot be measured.

If arsenic is implanted in the channelling direction, significantly larger ranges are obtained, though control of the profile is difficult due to its extreme dependence on the angle of incidence. This method is therefore unsuitable for device applications. Figure 6.13 shows a quite impressive example for a profile of a 600 keV arsenic implantation in the ⟨110⟩ direction, compared to the profile of an implantation into a sample tilted by 10°, based on a study by Reddi and Yu [585]. The profiles were measured using the capacitance-voltage method. The channelling profile has a probable range of approximately 4.5 μm, while an average range of only 0.35 μm is the result after tilting the sample. This corresponds very well with the theoretical value of 0.3435 μm given by Gibbons *et al.* [5].

In order to obtain good electrical properties of arsenic-implanted *p–n* junctions, a post-diffusion of the arsenic distribution is necessary to displace the *p–n* junction into the undamaged area (since, apparently, not all of the radiation damage is annealed). Typical diffusion profiles based on implanted distributions are shown in Figure 6.14; the implantation parameters are 120 keV and 10^{16} cm^{-2}. Due to

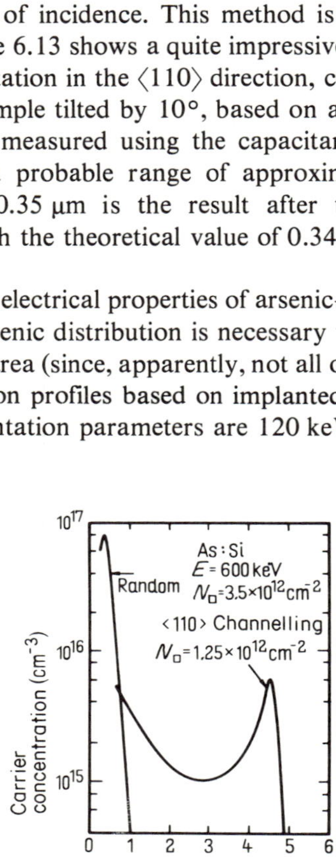

Figure 6.13 Carrier profiles after 600 keV arsenic implantation into ⟨110⟩-oriented or misoriented silicon [585]. Reprinted with permission from *Solid State Technology*, Technical Publishing, a company of Dun & Bradstreet

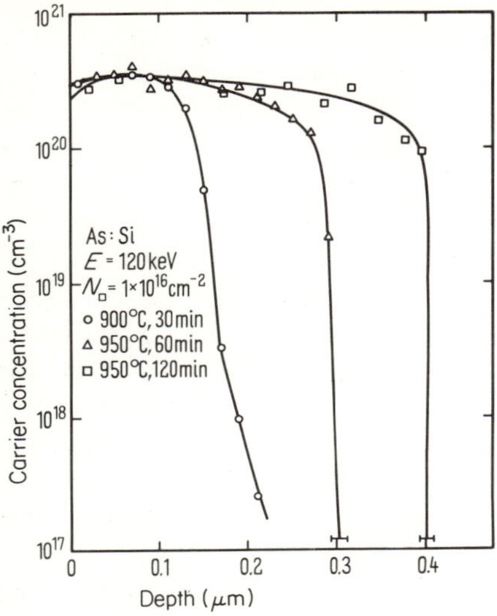

Figure 6.14 Diffusion profiles of an arsenic-implanted silicon sample. $E = 120$ keV; $N_{\square} = 10^{16}$ cm^{-2} [610]

the high vapour pressure of arsenic, annealing in an oxidizing atmosphere or following a coverage with SiO_2 is advisable to prevent out-diffusion. Two characteristic features in the profiles shown in Figure 6.14 are the steep profile slopes into the semiconductor bulk which are caused by the concentration dependent diffusion coefficient (*cf.* Section 3.4), and the almost constant maximum concentration which is due to the so-called activation limit, as well as to continuously occurring activation of the diffusing arsenic. The cause of the activation limit is probably related to the formation of arsenic complexes [235]. A lower maximum concentration only occurs when the arsenic concentration has decreased below this limit. Not so steep profiles are obtained at lower implantation doses, since the concentration dependence of the diffusion coefficient is then smaller.

An example is shown in Figure 6.15, for an implantation energy of 40 keV and a dose of 10^{15} cm^{-2}. The profiles in Figures 6.14 and 6.15 represent annealing in an inert atmosphere, resulting in an arsenic loss of a few per cent. This percentage is smaller than expected, which can be attributed to the natural oxide which forms a diffusion barrier. Further profile measurements on arsenic-implanted layers have been made by Baldo *et al.* [54] and Iwaki *et al.* [382], [382a]. In both of these investigations, the profiles were obtained through neutron activation as well as Hall effect and sheet resistivity measurements. At low implantation doses an exponential tail can be seen, even more pronounced than the one in Figure 6.15. The article by Iwaki *et al.* [382] explains this in terms of interstitial diffusion, but we attribute it, once again, to the scattering of arsenic ions in channels.

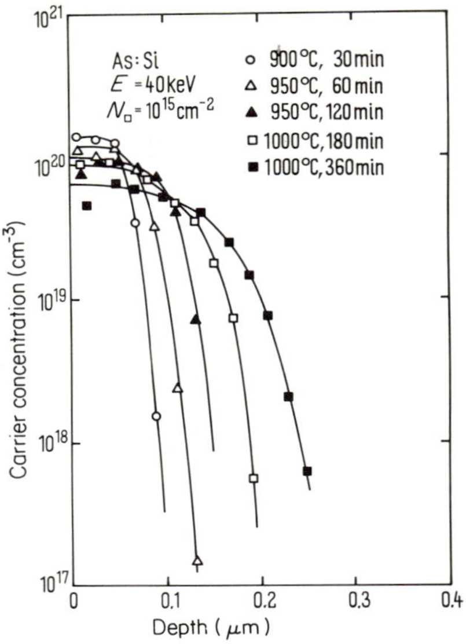

Figure 6.15 Diffusion profiles of an arsenic-implanted silicon sample. $E = 40$ keV; $N_\square = 10^{15}$ cm^{-2} [612]

p–n Junctions

As mentioned above, the *p–n* junction can be improved by displacing it with a drive-in diffusion into an undamaged crystal area. Measurements on ion-implanted diodes have been made by Bogardus and Poponiak [89] as well as Michel *et al.* [497]. The former found that dislocations caused by implantation extend far beyond the *p–n* junction, but have no influence on leakage currents if an additional gettering step is used. It is basically evident that, in order to obtain good *p–n* characteristics, higher temperatures are needed than those required for maximum electrical activation of the implanted ions or good carrier mobility. Figure 6.16 shows measurements of the reverse current density depending on the annealing temperature for arsenic-implanted diodes, compared with the characteristics of a diffused diode [497]. Annealing temperatures above 900 °C are required to obtain minimum reverse currents. This is understandable, since the minority carrier lifetime is much more sensitive to defects than the carrier concentration is.

Epitaxy

Due to its high solubility, arsenic is an especially interesting dopant for sub-collector structures. For this reason, detailed studies on the epitaxial deposition of

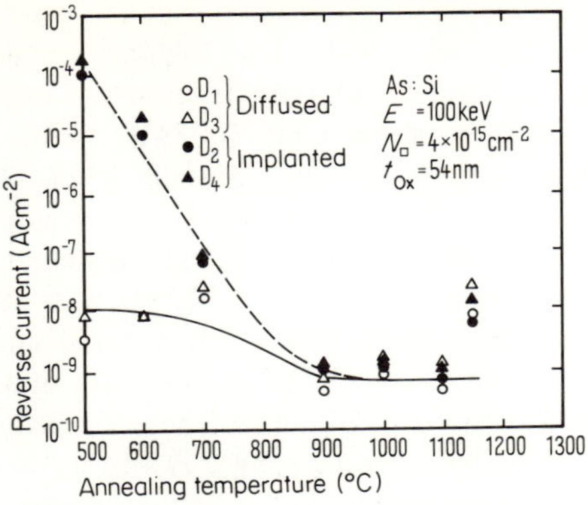

Figure 6.16 Reverse current depending on annealing temperature for an arsenic-implanted *p–n* junction in comparison to a diffused one. $E = 100$ keV; $N_\square = 4 \times 10^{15}$ cm^{-2} [497]. Reproduced by permission of the American Institute of Physics

silicon on arsenic-implanted layers have been made [205], [206], [508]. The conclusion of these studies is that good epitaxial layers can only be produced if the damaged surface is removed by oxidation prior to epitaxy. An etching step with HCl ($\geqslant$0.2 μm) before epitaxial growth is also important. Moline *et al.* [508] used implantation energies of 300 and 150 keV. Before epitaxy, the removal of 0.35 μm SiO_2 (i.e., approximately 0.15 μm Si) was found necessary to obtain good epitaxial layers. It is not very important whether annealing is done in an oxidizing atmosphere or whether the oxidation takes place after annealing in an inert atmosphere. In the first case, a low degree of arsenic loss due to out-diffusion results but, under some circumstances, dislocations grow into the semiconductor (*cf.* Section 3.1). At implantation doses below the critical dose for forming amorphous layers, this oxidation is not necessary. Annealing in an inert atmosphere (1000 °C, 20 min) is then adequate for the deposition of excellent epitaxial layers. Annealing in an oxidizing atmosphere can also be dispensed with if the implantation is made at temperatures between 500 and 600 °C [508] in order to avoid the formation of an amorphous layer. This method, however, does not seem applicable under practical conditions.

Defects in Annealed Layers

As already discussed in Chapter 3, many defects still exist in layers which have been annealed. In the preceding section, it was shown that in order to obtain defect free epitaxy, these layers must be removed by oxidation and etching with

HCl. The problem of secondary implantation (knock-on or recoil implantation) during implantation through SiO_2 layers, with the resulting defects, has also been mentioned in Chapter 3. Most investigations of such damage have been made using transmission electron microscopy (TEM) [456], [456a], [505] or Rutherford back-scattering measurements (Sigmon *et al.* [678]).

Mader and Michel [456], [456a], using TEM, have made a detailed study of defects after arsenic implantation into bare silicon, as well as into silicon covered with SiO_2. They used energies of 40 to 150 keV and doses of 5×10^{15} to 2×10^{16} cm^{-2}; their starting material was $\langle 100 \rangle$-oriented silicon. After annealing at 800 °C, there is a high density of small interstitial dislocation loops (diameter $\approx$ 20 nm, density 300 to 650 μm^{-2}). They are located somewhat deeper than would correspond to the average range. A further annealing at 1000 °C produces larger dislocation structures, whose nature depends on the implantation conditions. If bare silicon is implanted at 40 keV, only a few small dislocation loops (density $\leqslant$0.1 μm^{-2}) are present after annealing at 1000 °C; on the other hand, if implantation is done at 80 keV, large loops are produced, sometimes with planes vertical to the surface. A few structures are described in Table 6.1; Figure 6.17 shows an example for a dose of 5×10^{15} cm^{-2}, with annealing at 1000 °C for 60 min. Only large half-loops (see Figure 6.18) are present after implantation with very high doses. These dislocations are of the misfit type, similar to those produced by phosphorus diffusion. If implantation is done through a SiO_2 layer, climbing dislocation loops of the shear type are produced. Table 6.2 presents some conclusions on defects after arsenic implantation. In addition to dislocations, oxide particles (secondary implantation) are also visible (see Figure 6.19) after implantation at 40 keV through 10 nm of SiO_2. A high density of prismatic interstitial loops, almost parallel to the silicon surface, can be seen after implantation with 80 keV at low doses. If a dose of 10^{16} cm^{-2} is used, irregular meandering dislocation lines are produced (see Figure 6.20). A 3-dimensional tangle of interconnected dislocation lines results at the highest dose used, 2×10^{16} cm^{-2}.

Table 6.1 Dislocation structures in bare Si implanted with 80 keV As^+ ions after drive-in heat treatment [456]

As^+ dose ($\times 10^{16}$ cm^{-2})	Annealing (°C, min)	Dislocations, loop diameters	Depth (nm)
0.5	970, 30	6 loops/μm^2, 80 nm	200
0.5	1000, 60	4 loops/μm^2, 50 nm 1.5 loops/μm^2, 150 nm	
2	970, 30	Irregular half-loops terminating at surface	
2	1000, 60	Regular half-loops edge dislocations	220

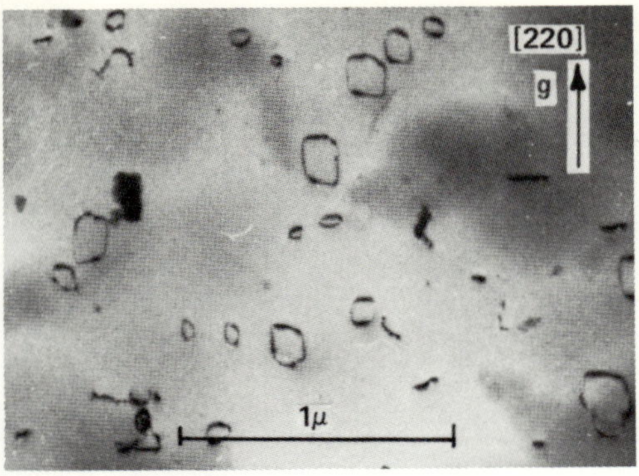

Figure 6.17 Prismatic dislocation loops in silicon implanted with 5×10^{15} cm^{-2} arsenic at 80 keV and annealed at 1000 °C for 60 min [456]. Reproduced by permission of the American Institute of Physics

A further example for the influence of secondary-implanted oxygen is shown in Figure 6.21 (Moline [505]). In this case, a sample only partly covered with SiO_2 was implanted. The upper half of the wafer was not masked with oxide, and shows only a few dislocation lines; in the lower part, however, the dislocation density is extremely high. These dislocations are probably caused by the combined effects of radiation damage, secondary-implanted oxygen, deviation from chemical equilibrium, and (perhaps) a slight change in the lattice constants due to the arsenic.

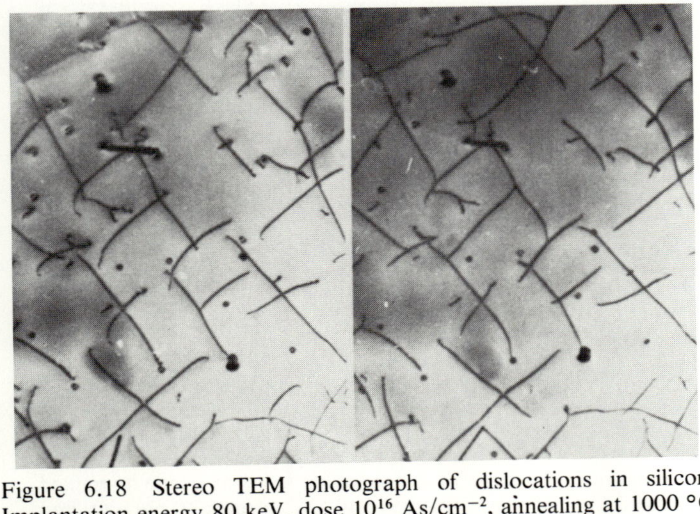

Figure 6.18 Stereo TEM photograph of dislocations in silicon. Implantation energy 80 keV, dose 10^{16} As/cm^{-2}, annealing at 1000 °C for 60 min. Magnification: 15 000× [456]. Reproduced by permission of the American Institute of Physics

Table 6.2 Dislocation structures in Si implanted with As through SiO_2 after heat treatment at 970 °C, 30 min

Dose ($\times 10^{16}$ cm^{-2})	Energy (keV)	Dislocations, diameter of loops	Depth (nm)
1	40	<0.1 loops/μm^2	
2	40	20 loops/μm^2, 50 nm	65
0.5	80	80 loops/μm^2, 100 nm	85
1	80	Meandering shear loops	90
2	80	3-dimensional tangle	

6.1.4 Boron

Boron is the most important dopant for *p*-type layers. Accordingly, in the early days of ion implantation [401], [411], [473], [559], [599], interest in this element was already great. Up to now, it has also become the most widely applied element in semiconductor technology, especially in the areas of integrated MOS circuits and bipolar transistors.

Electrical Activation

The critical dose for boron is between 2×10^{16} and 8×10^{16} cm^{-2} [9]. One can therefore expect relatively low concentrations of radiation damage (whose annealing behaviour, however, is very complex) if the crystal has not been made

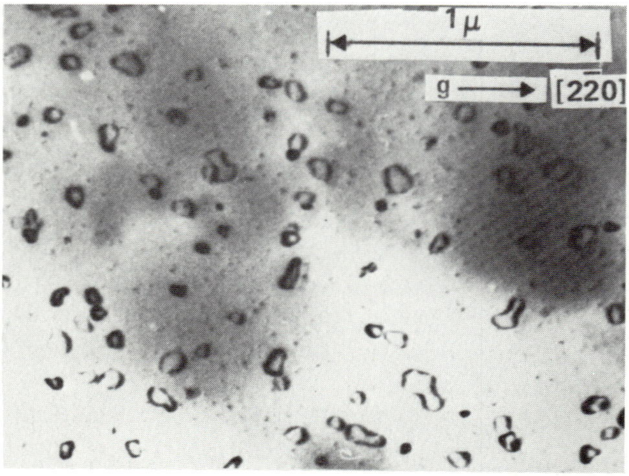

Figure 6.19 Interstitial dislocation loops in silicon implanted with 2×10^{16} cm^{-2} arsenic at 40 keV through 10 nm SiO_2 and annealed at 970 °C for 30 min [456]. Reproduced by permission of the American Institute of Physics

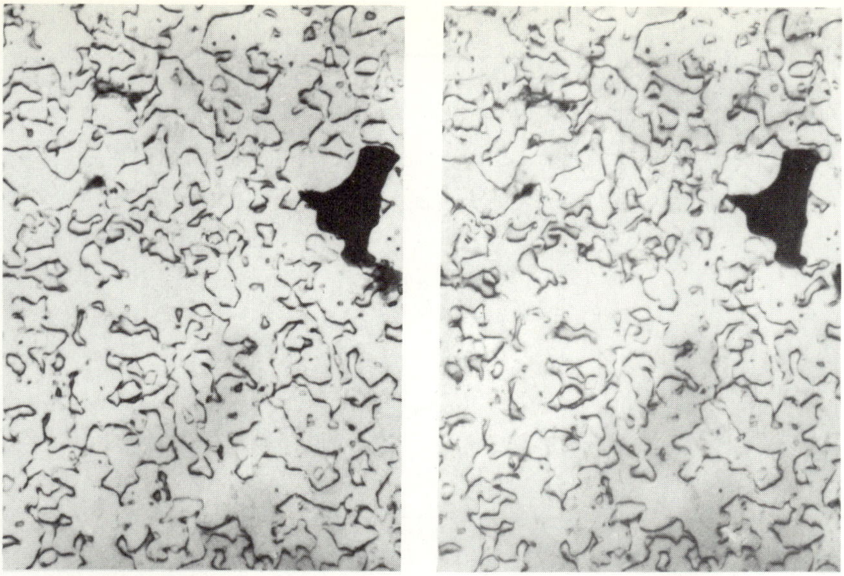

Figure 6.20 Stereo TEM photograph of meandering dislocation loops with shear character in silicon. The sample was implanted with 80 keV As ions with a dose of 10^{16} cm^{-2} through 20 nm SiO_2 and annealed at 970 °C for 30 min. Magnification: 3500× [456]. Reproduced by permission of the American Institute of Physics

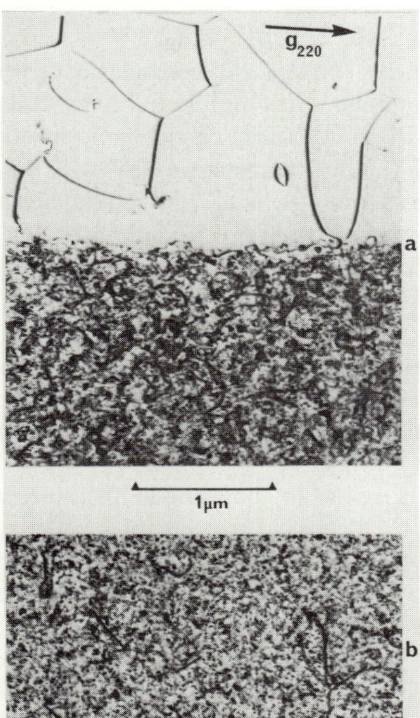

Figure 6.21 (a) Arsenic-implanted and annealed silicon sample. The lower half of the sample was covered with 43 nm SiO_2 [505] and shows many crystalline defects. (b) Silicon sample implanted with As and O and annealed (not oxide covered during implantation). Reproduced by permission of the American Institute of Physics

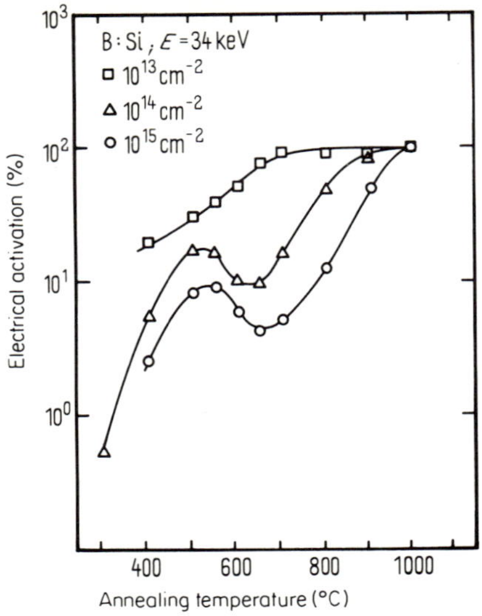

Figure 6.22 Relative increase in electrical activation as a function of annealing temperature for 10^{13}, 10^{14} and 10^{15} cm^{-2} boron in silicon [10]

amorphous by an additional implantation [81], [609], if no molecules have been implanted [522], [609], and if the implantation has not been done at low temperatures [81], [170]. Figure 6.22 [10] shows sheet carrier concentration depending on annealing temperature for an implantation at 34 keV with doses between 10^{13} cm^{-2} and 10^{15} cm^{-2}. At doses above 10^{13} cm^{-2}, a reverse annealing behaviour of the carrier concentration is obtained between 550 and 650 °C. The cause of this behaviour is not yet entirely clear. Blamires [81] assumes a formation of boron pairs in this temperature range. North and Gibson [542] have demonstrated, using channelling measurements, that 30 per cent of the boron ions at room temperature, but only about 5 per cent of these ions at 100 °C, are located on lattice sites; however, after annealing at 1000 °C, the result is more than 90 per cent. This too could be a plausible explanation for the reverse annealing. In sheet resistivity measurements, the effect of reverse annealing cannot be seen so clearly, since changes in mobility as well as in annealing temperature are partly superimposed upon it.

Temperatures of 900 °C are needed to obtain nearly complete activation. If the implantation dose is so low that only very little radiation damage is produced, reverse annealing does not occur. If an amorphous layer is produced by an additional irradiation with other ions, for example, Si, Ne, or O, or by boron implantation at 77 K, reverse annealing does not result, and all implanted ions become electrically active after annealing at about 650 °C. In order to achieve a mobility corresponding to that in homogeneously doped material, however, temperatures of 900 to 950 °C are again necessary. An example of this annealing behaviour, which is typical for amorphous layers, is shown in Figure 6.23. The amorphous layers were produced by implantation of neon, silicon, or BF_2 molecules, or by boron implantation at 77 K. When implanting an additional ion species producing an amorphous layer, special attention must be given to influences upon mobility and lifetime. In the case of the sample made amorphous by neon implantation, for example, a considerably lower mobility can be observed up to annealing temperatures of 1000 °C.

Detailed studies on the annealing behaviour of layers implanted with boron or boron molecules (BF, BF_2, BCl, BCl_2) have been made by Beanland [66]. A comparison for various combinations is shown in Figure 6.24. Reverse annealing is seen in cases where boron alone has been implanted. Since amorphous layers are formed in the molecule implantations, as well as in those with boron and chlorine or boron and fluorine, no reverse annealing occurs in these cases. All in all, it can be seen that the fluorine compounds of boron are better suited for implantation. Molecule implantations will be considered in more detail in the following section.

For the formation of buried layers, implantations at very high energies are necessary. The annealing behaviour of the sheet resistivity for a high energy implantation at 8 MeV is shown in Figure 6.25. An example of an E–dE/dx detector employing such a layer is given in Section 7.3.5.

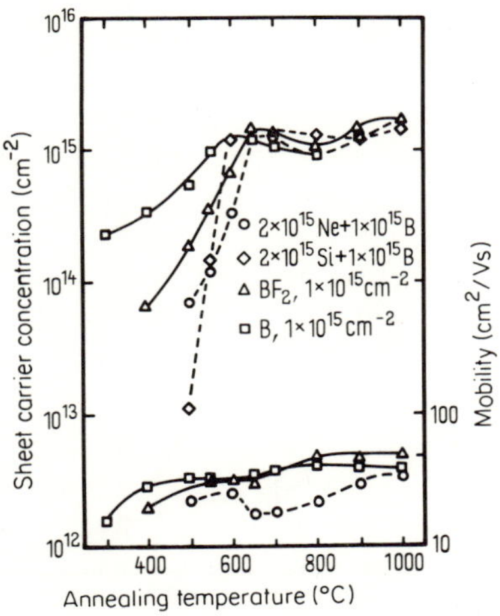

Figure 6.23 Annealing behaviour of sheet carrier concentration ($N_{S,\text{eff}}$) and sheet Hall mobility ($\mu_{S,\text{eff}}$) for boron-implanted silicon samples made amorphous through implantations of neon, silicon, BF_2 or boron (the latter at 77 K) [10]

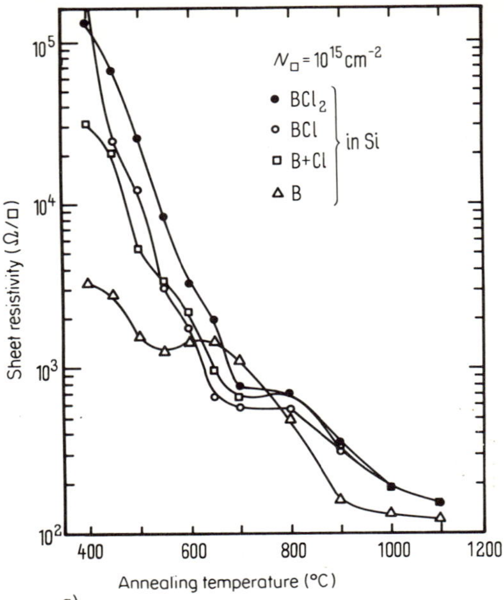

a)

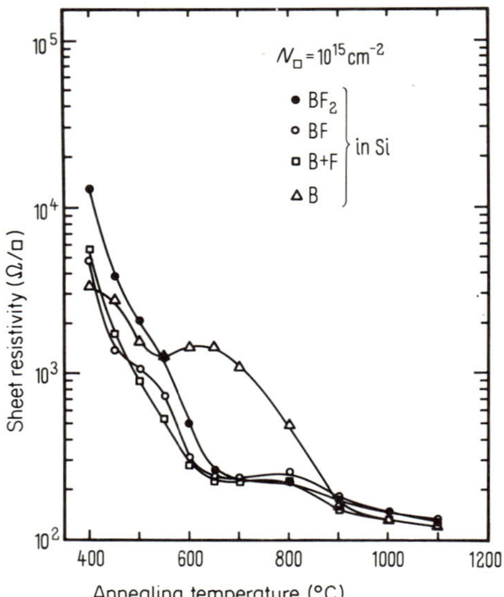

b)

Figure 6.24 Isochronal annealing behaviour of sheet resistivity for implanted silicon: B^+ 25 keV; Cl^+ 80 keV + B^+ 25 keV; BCl^+ 105 keV; BCl_2^+ 184 keV; F^+ 43 keV + B^+ 25 keV; BF^+ 68 keV; BF_2^+ 111 keV; according to Reference [66]

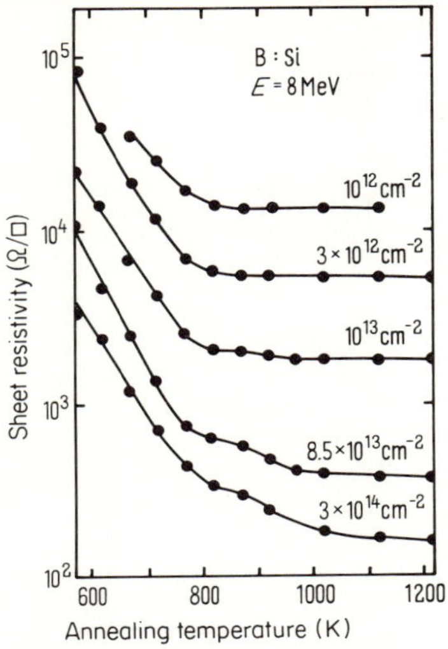

Figure 6.25 Annealing behaviour of sheet resistivity for silicon implanted with 8 MeV boron ions and doses of 10^{12} to 3×10^{14} cm^{-2} [417]

Range and Profiles

Measurements of range and range straggling are compared with theoretical values according to LSS [442] in Figure 6.26. The correlation is good, in spite of the relatively great spread in the measured values. The recent calculations by Gibbons *et al.* have been used for the LSS values [5]; these calculations eliminate a previous discrepancy between experiment and theory by using experimental values for S_e published by Eisen [221] instead of theoretical values based on LSS. Brice has also used these values for S_e in his calculations [102], [103].

Hofker *et al.* [337], [338] have measured profiles in amorphous and polycrystalline silicon. The layers were produced by either neon implantation or chemical deposition; the boron profiles were measured using the SIMS technique. The ranges are identical within 2 per cent and, due to the lower density, are somewhat greater than in monocrystalline silicon. Values for the probable range (which, for the asymmetrical profiles being measured, is not identical to the mean range) are given in Table 6.3. Figure 6.27 shows several profiles measured in amorphous silicon, which have been fitted by Pearson functions [32]. It is possible in this way to determine moments of higher order. The profiles clearly show a slow decrease towards the semiconductor surface, which is inconsistent with the results based on LSS, but agrees with the more exact calculations of higher moments by Winterbon [779].

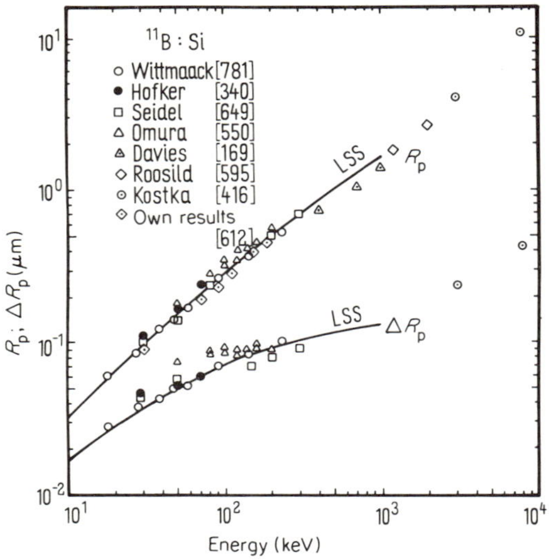

Figure 6.26 Experimental range and range straggling data of boron in silicon

In single crystalline silicon, the profiles which are obtained always have tails into the bulk of the semiconductor [609], [649]. These tails can be attributed to ions which are scattered into channels and therefore have a greater range. They can therefore not be eliminated by sample tilting. Figure 6.28 shows profiles by Seidel [649], measured by the capacitance-voltage method. The exponential profile tails show a decrease to $1/e$ after about 60 to 100 nm. The influence of the Debye length (*cf.* Section 5.3.1) on the measurement is not adequate to explain this effect. If the implantation temperature is lowered and BF_2 molecules are

Table 6.3 Comparison of ranges for boron in amorphous, polycrystalline, and monocrystalline silicon [337]

Energy (keV)	Range (nm)		
	Amorphous Si	Polycrystalline Si	Monocrystalline Si
30	117	117	112
50	187	190	188
70	261	251	244
100	351	339	336
150	487		
200	583	597	572
300		787	768
400		960	970
600		1295	1260
800		1573	1538

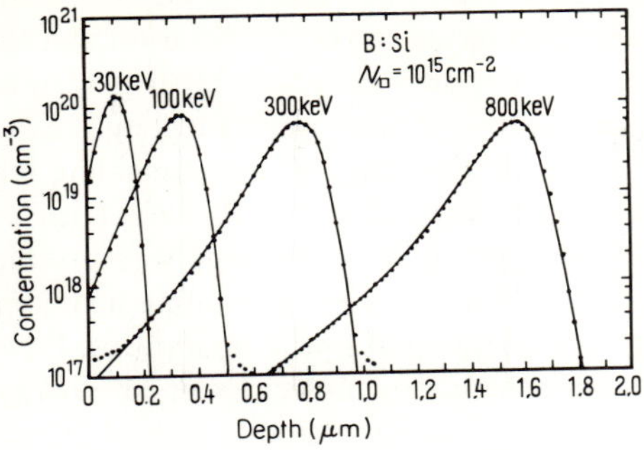

Figure 6.27 Boron profiles in amorphous silicon measured by SIMS [337]. The solid lines are fitted Pearson distributions. Reproduced by permission of Gordon and Breach

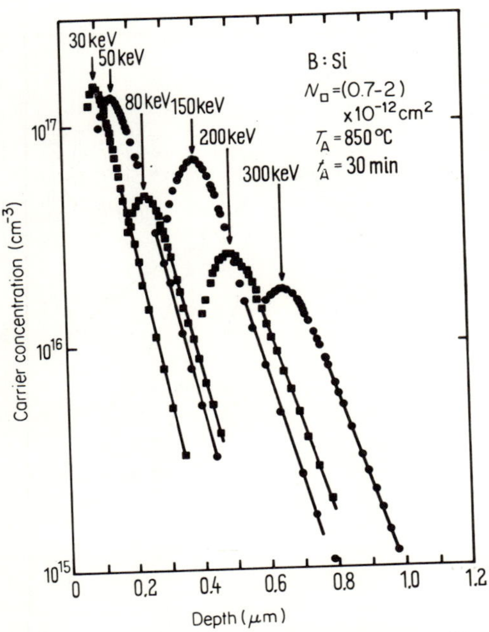

Figure 6.28 Boron profiles in single-crystalline silicon measured by the capacitance-voltage method. The exponential tails decrease at 60 to 100 nm to $1/e$ [649]. Reproduced by permission of Springer-Verlag, Heidelberg

implanted [522], profiles of the same gradients are also obtained at doses below 10^{14} cm^{-2}. Interstitial diffusion or enhanced diffusion can therefore be ruled out. If the ion dose is so high that an amorphous layer is formed, this channelling is largely suppressed, due to the greater ion straggling.

The dependence of the profile shape at low annealing temperatures upon the radiation damage has been determined by Ryssel *et al.* under various experimental conditions [609]. The radiation damage was produced by an additional implantation of neon or silicon, either by implanting heavy molecular compounds of the doping ion or by implanting at low temperatures. Figure 6.29 shows, as an example, an implantation of boron molecules at an energy of 34 keV. At low annealing temperatures, electrical activation is only obtained in the layer close to the surface, which has been made amorphous by the implantation and already recrystallizes at low temperatures; however, in the region where the radiation damage was too low to produce amorphousness, a temperature of approximately 900 °C is necessary to obtain complete electrical activation of the implanted ions.

Hofker *et al.* [339] have demonstrated, by comparisons between SIMS and electrical profile measurements, that at very high boron concentrations (implantation dose 10^{16} cm^{-2}), some of the boron is not electrically active, nor does it diffuse during annealing (Figure 6.30). An inactive portion of the ions in the profile tail towards the semiconductor bulk, which they found at doses between

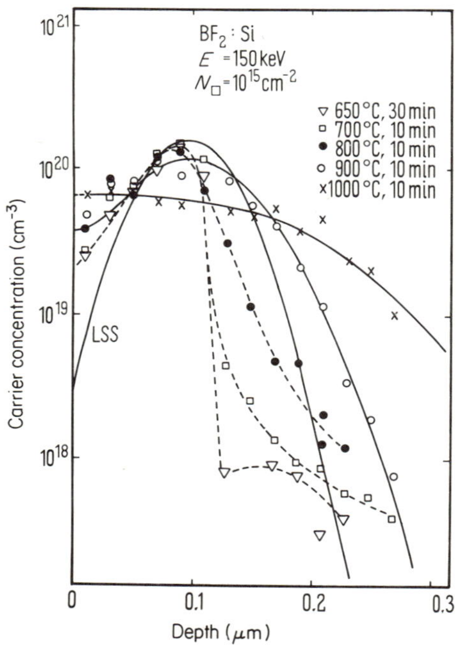

Figure 6.29 Radiation-damage dependent electrical activation of boron in silicon after a BF_2 implantation at 150 keV energy and 10^{15} cm^{-2} dose [609]

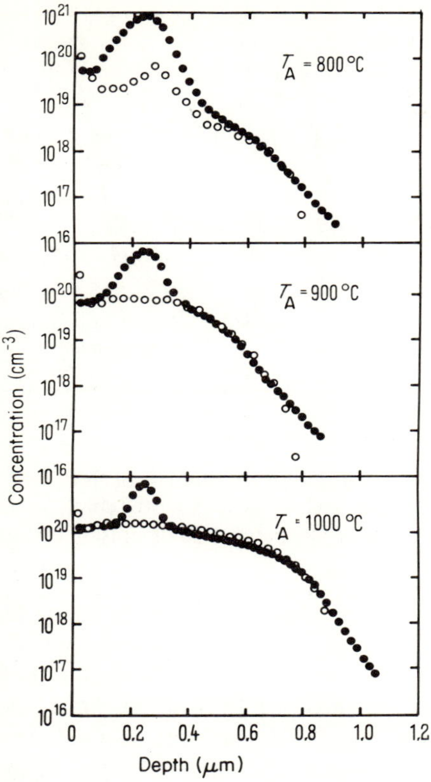

Figure 6.30 Comparison between doping concentration (○) and total impurity concentration (●) after a 10^{16} cm^{-2} boron implantation at 70 keV into silicon: (a) $T_A = 800$ °C; (b) $T_A = 900$ °C; (c) $T_A = 1000$ °C, according to [339]. Reproduced by permission of the American Institute of Physics

10^{14} cm^{-2} and 10^{16} cm^{-2}, is explained in terms of rapidly diffusing boron on interstitial sites. At implantation doses of 2×10^{15} cm^{-2} to 6×10^{15} cm^{-2} (energy 82 keV, annealing temperature 800 °C), Schwettmann [640] found an approximately constant maximum concentration of about 10^{20} cm^{-3}, which corresponds roughly to the solubility.

Figure 6.31 shows two profiles as examples of high energy implantations; these were made with a boron dose of 3×10^{13} cm^{-2} at 3 MeV as well as 8 MeV [416]. The profiles were measured by etching and staining of the *p–n* junctions; a capacitance-voltage measurement is also given for comparison, showing a qualitative agreement at higher concentrations only. The continuous lines are Gaussian curves fitted to the measurement points. The ranges were found to be 4 μm and 10.4 μm; the values for range straggling were 0.25 μm and 0.43 μm.

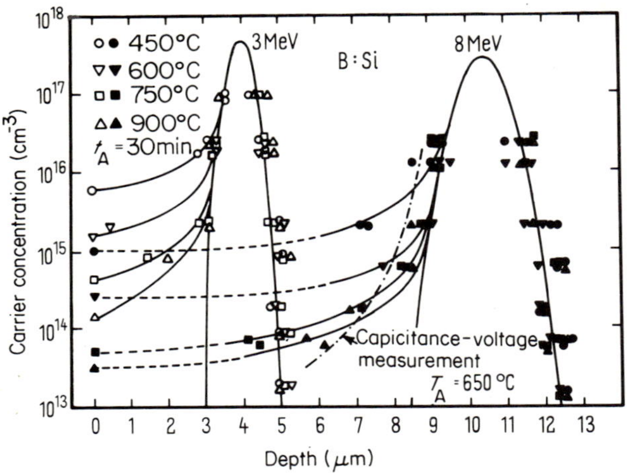

Figure 6.31 High-energy profiles of boron in silicon, energies 3 and 8 MeV, dose 3×10^{13} cm^{-2}, according to [416]. Reproduced by permission of Gordon and Breach

For the application of ion implantation in silicon planar technology, a matter of interest is not only the range of boron in silicon, but also its range in SiO_2. Wittmaack *et al.* [781] have investigated boron profiles in SiO_2 by means of the SIMS technique. They obtained profiles with the same shape as those found by Hofker in amorphous silicon [337], and were also able to fit them by means of Pearson functions. Measured profiles are shown in Figure 6.32, implanted at

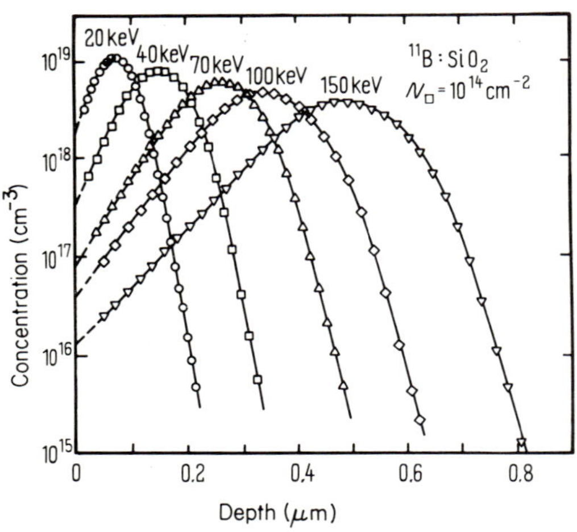

Figure 6.32 Boron profiles in SiO_2 measured by SIMS, according to [781]. The solid lines are fitted Pearson distributions

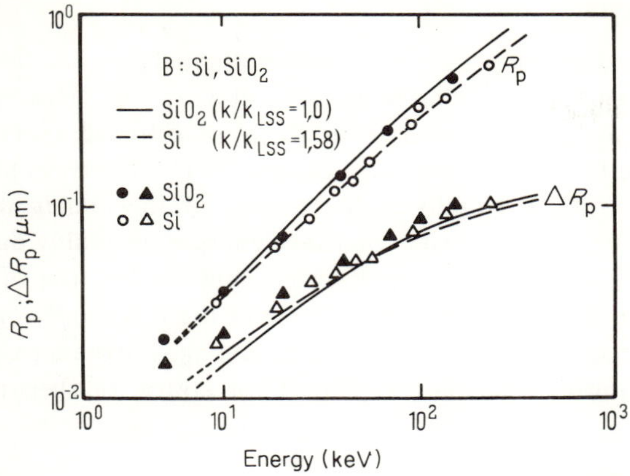

Figure 6.33 Range and range straggling for boron in SiO_2 and amorphous silicon. Theoretical curves according to References [387] (SiO_2), [103], [269] and [527] (silicon), experimental data from [781]. k is the proportionality factor of the electronic stopping, k_{LSS} the theoretical value according to equation (2.16)

energies between 20 and 150 keV and doses of 10^{14} cm^{-2}. Their values for range and range straggling in SiO_2 and silicon, together with theoretical values [103], [269], [387], [527], are given in Figure 6.33. The agreement is quite good, although the experimental range in SiO_2 appears lower than according to theory, and the straggling of boron in SiO_2 as well as in silicon is somewhat higher than theory.

For implantation, boron is often obtained from BF_3 molecules, in which case the BF_2 current is always greater than the boron ion current. Due to the greater mass of BF_2 in comparison to boron, this offers a simple means of achieving not only a higher current, but also shallower boron distributions as well as amorphous layers even at room temperature [522]. The following applies for the effective boron energy during a molecule implantation of BF_2 [522]:

$$E_B = \frac{M_B}{M_{BF_2}} E_{BF_2} \tag{6.2}$$

One therefore obtains, for example, in the case of an implantation with BF_2 ions at 150 keV, the same profile as that resulting from an implantation with boron ions at 34 keV. If the amorphizing dose for BF_2 (approximately 5×10^{14} cm^{-2} [517] were to be exceeded, however, the exponential profile tails would be suppressed.

p–n Characteristics

Even at low annealing temperatures, boron-implanted *p–n* junctions provide good reverse as well as forward characteristics [408], [704]. However, in order to obtain reverse currents comparable to those of diffused diodes, annealing at 900 °C or above for 30 minutes is necessary [497]. Diodes implanted with BF_2 (Figure 6.34) also show equally good results; for low annealing temperatures, however, they have a high density of surface charges, probably due to the fluorine component [497]. If boron implantation is made in the area of a *p–n* junction (source or drain), as for instance in the case of the threshold voltage adjustment for MOS transistors (see Chapter 7), higher leakage currents result due to radiation damage, requiring annealing at 900 °C or above for their reduction [497].

Defects in Annealed Layers

Many defects after annealing can also be seen in boron-implanted layers. Problems connected with annealing in an oxidizing atmosphere have already been mentioned in Section 3.1. Figures 6.35 and 6.36 show typical dislocation lines obtained after annealing in an oxidizing atmosphere at 1100 °C [579]. The doses were 10^{14} and 10^{15} cm^{-2} at an energy of 30 keV. A detailed investigation of defects following implantations of B, BF, BF_2 and F has been made by Prussin [579a], with doses between 10^{14} and 10^{15} cm^{-2}. The energies were selected according to equation (6.2) so that they were equivalent to a 30 keV boron implantation. Annealing in an inert atmosphere at 1100 °C for 1 hour with subsequent oxidation prevents formation of dislocations in almost all cases. A few dislocations are present after annealing in an inert atmosphere at 1000 °C; however, after immediate annealing in an oxidizing atmosphere at 1100 °C, a great variety of

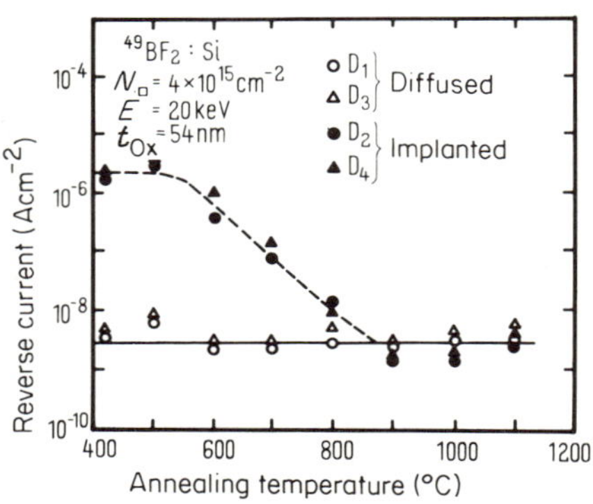

Figure 6.34 Reverse-current density of silicon diodes produced through BF_2 implantation. By annealing at 900 °C or above, results as good as those for diffusion [497] are obtained. Reproduced by permission of the American Institute of Physics

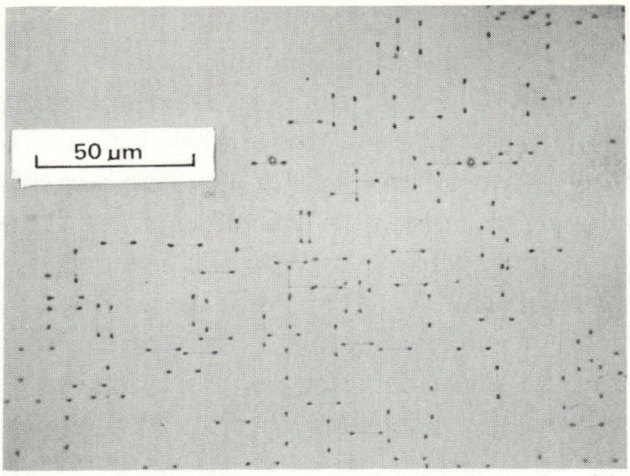

Figure 6.35 Silicon sample implanted with boron ($E = 30$ keV, $N_{\square} = 10^{14}$ cm^{-2}, $\langle 100 \rangle$ orientation) after wet oxidation at 1100 °C. The dislocations were made visible by Sirtl etching for 30 s [579]. From Prussin and Fern, *J. Electrochem. Soc.*, **122** (1975), 830, reprinted by permission of the publisher, the Electrochemical Society, Inc.

dislocations and stacking faults can be seen, depending on the dose. No significant difference was found between implantations of B, BF and BF_2.

Further studies on this topic have been made by Comer and Roosild [149], Madden and Davidson [455], Hasegawa *et al.* [322], [323], Tamura *et al.* [718], Lecroisnier *et al.* [425] and Prussin [580]. Precipitations of boron silicide [149]

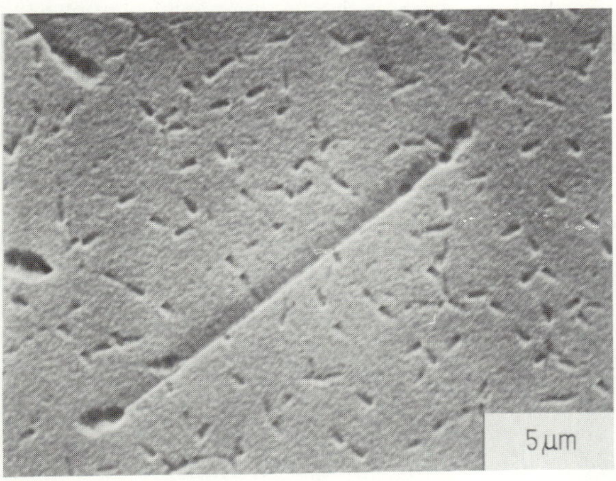

Figure 6.36 SEM photograph of dislocations (boron in silicon, $E = 30$ keV, $N_{\square} = 10^{15}$ cm^{-2}, $\langle 100 \rangle$ orientation) after wet oxidation at 1100 °C for 1 h and Şirtl etching for 15 s [579]. From Prussin and Fern, *J. Electrochem. Soc.*, **122** (1975), 830, reprinted by permission of the publisher, the Electrochemical Society, Inc.

were found; these are formed as a result of annealing above 1000 °C, during which the dislocations are annealed. Dislocations resulting from annealing in an inert atmosphere at 1000 °C for 30 minutes, at various depths, can be seen in Figure 6.37. The implantation was made at 1 MeV with a dose of 10^{15} cm^{-2}. No dislocations are visible up to a depth of 1.5 μm; however, dislocation loops and rod-shaped dislocations can be seen at 1.6 and 1.7 μm. The number of defects decreases again at greater depths. In an extensive study, Prussin [580] has investigated defect generation during annealing and diffusion of boron-implanted layers. Small dislocation loops were found to develop during implantation. These reach their maximum size at the beginning of annealing, due to the incorporation of point defects. During further annealing, several effects occur:

(a) Recombination of vacancies and interstitial atoms,
(b) Shrinkage of intrinsic or extrinsic dislocation loops, with simultaneous generation of vacancies or interstitial atoms,
(c) Accumulation of vacancies or interstitial atoms at dislocations,
(d) Influence of surface effects, leading to a reduction of dislocations and interstitial atoms.

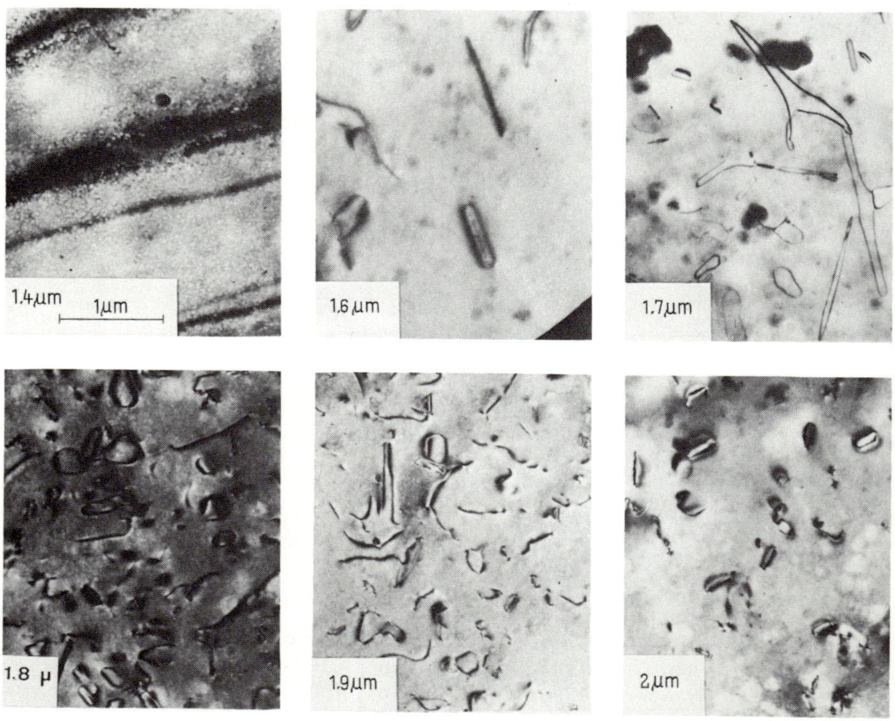

Figure 6.37 Microphotographs of defects at various depths after a boron implantation at 1 MeV with a dose of 10^{15} cm^{-2}; annealing at 1000 °C for 30 min [425a]

Effects (a), (c), and (d) would tend to restore the equilibrium of the vacancy-interstitial atom concentrations. During annealing in an oxidizing atmosphere, extrinsic dislocations cannot anneal, due to the inadequate number of vacancies.

6.1.5 Gallium

Gallium has relatively little importance in semiconductor technology, since it is not masked by SiO_2 and has a low solubility in silicon (1.4×10^{19} cm^{-3} at 1000 °C [731]). It is used for deep, large-area diffusions, for example, for thyristors. However, quite a few implantation experiments have been carried out with gallium [59], [385], [470], [485], partly because this element is very easily measurable using the back-scattering technique. The electrical activation of gallium is low, i.e., maximally 30 per cent [59], due to its low degree of incorporation on to lattice sites. A back-scattering spectrum for a gallium implantation in ⟨110⟩ direction and in random direction, following an annealing step at 900 °C, is shown in Figure 6.38. It appears that all atoms are on random sites. By etching off thin layers and making repeated measurements, however, it is found (Figure 6.39) that only the gallium atoms in the layer near the surface are on random sites; at greater depths, they are partly on regular interstitial sites and regular lattice sites. The authors explain this phenomenon in terms of vacancy-enhanced diffusion.

Gallium profiles have been measured by Crowder [157], Gamo *et al.* [275] and Dearnaley *et al.* [184]. The range and the standard deviation correspond roughly with the LSS theory; however, considerable profile tails occur. In Figure 6.40, a profile according to Dearnaley *et al.* [184] is shown, obtained by implantation of radioactive ^{72}Ga ions. The implantation energy was 45 keV, the dose was 10^{13} cm^{-2}, and the sample was tilted by 10° to avoid channelling. Dearnaley *et al.* explain the profile tails in terms of ions scattered into channels (axial or

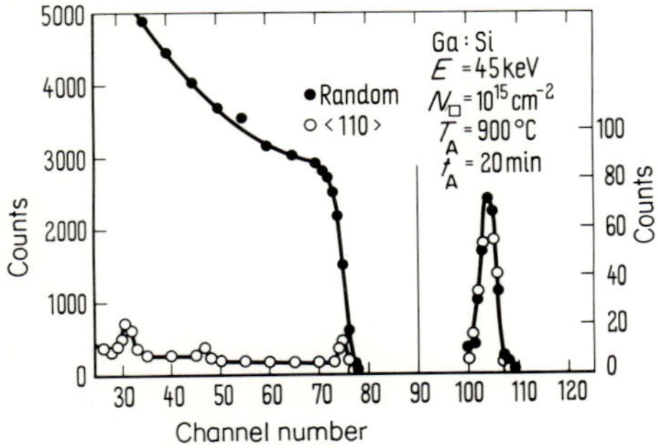

Figure 6.38 Back-scattering spectrum of a silicon sample implanted with gallium in random and ⟨110⟩ direction. $E = 45$ keV, $N_{\square} = 10^{15}$ cm^{-2}, $T_A = 900$ °C [476]. Reproduced by permission of Springer-Verlag, Heidelberg

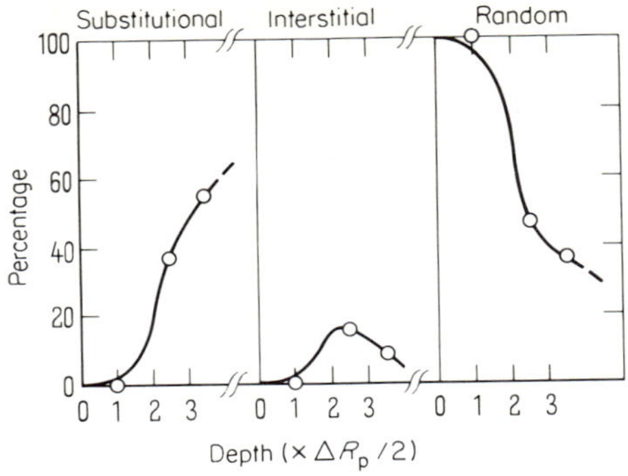

Figure 6.39 (a) Number of atoms on lattice sites, (b) interstitial sites and (c) random sites, depending on depth, for silicon layer implanted with gallium. $E = 45$ keV, $N_{\square} = 10^{15}$ cm^{-2}, $T_A = 900$ °C [476]. Reproduced by permission of Springer-Verlag, Heidelberg

planar), and suggest that this is also the cause of the profile tails found by Gamo [275] and Crowder [157], rather than an enhanced interstitial diffusion. In agreement with us, they consider the great majority of similar profile tails to be due to the scattering of ions into channels, and not to any specific diffusion effect. The temperature dependence observed in some cases (e.g., Gamo *et al.* [275]) is due

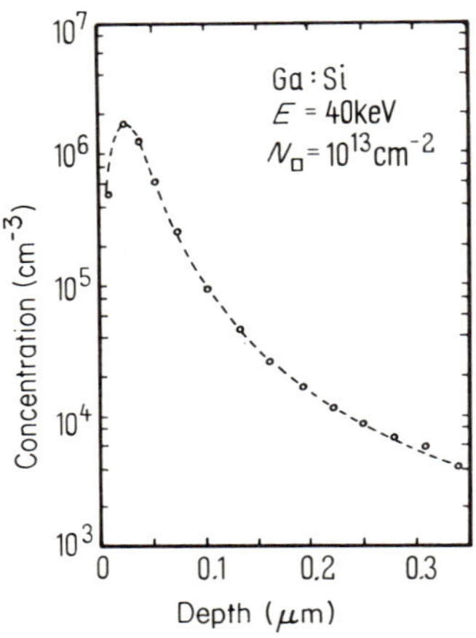

Figure 6.40 Depth distribution of ^{72}Ga ions implanted into a misoriented silicon crystal at 40 keV with a dose of 10^{13} cm^{-2} at room temperature, according to [184]. Reproduced by permission of the American Institute of Physics

to the lower radiation-damage concentration which results at higher implantation temperatures. Stephen and Grimshaw [704] made implanted diodes using gallium, but poor junction properties were found. In addition, research has been carried out on the adjustment of the threshold voltage [602], the implantation of resistors [552], and on nuclear-radiation detectors [349]. Attempts have also been successful [758] to produce solar cells having an efficiency of 6 per cent without an anti-reflection coating.

All in all, it can be said in regard to gallium that this element seems better suited than aluminium for the *p*-doping of silicon. However, the data available to date indicate that an adequate electrical activation can only be expected at low concentrations.

6.1.6 Indium

Most studies on the implantation of indium into silicon have not dealt with the aspect of electrical activation, but rather, with lattice location or diffusion effects

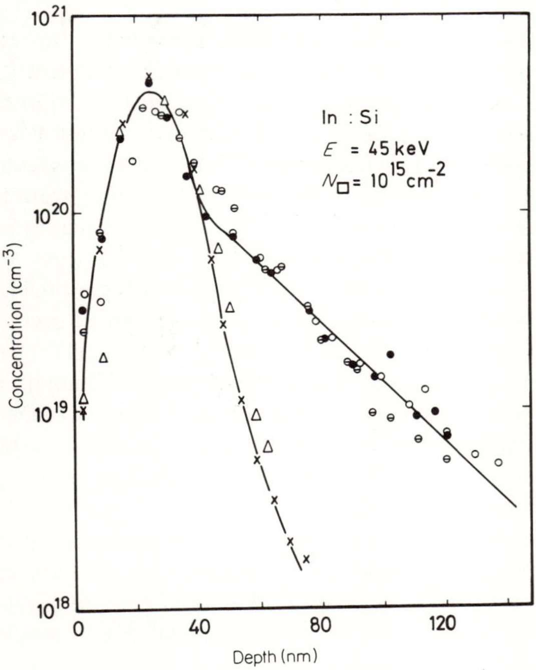

Figure 6.41 Concentration profiles of 45 keV In ions implanted with 10^{15} cm^{-2} at 100 °C (×), 200 °C (△), 300 °C (⊖), 400 °C (●), and 500 °C (○). An enhanced diffusion independent of temperature is obtained at temperatures above 300 °C [273]. Reproduced by permission of Springer-Verlag, Heidelberg

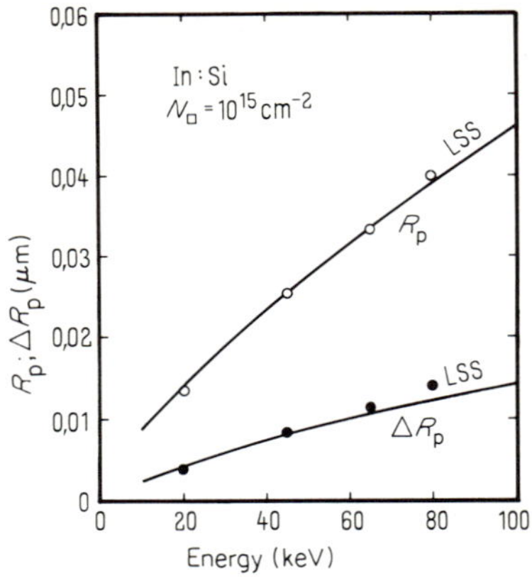

Figure 6.42 Range and range straggling of indium in silicon [274]. Solid lines according to Gibbons *et al.* [5]

[114], [229], [476], [598]. Indium is only incorporated on to regular lattice sites to a very minor extent (10 per cent at a dose of 2×10^{15} cm^{-2}, energy 45 keV, implantation at room temperature; see Table 6.4). Depth-dependent back-scattering measurements [476] have indicated an increase in the portion on regular interstitial sites up to a maximum of 50 per cent, at greater depths, due to interstitial diffusion. An abnormal diffusion effect resulting from high-temperature implantation has been determined by neutron-activation measurements [273]. An example for indium implantation at 45 keV with a dose of 10^{15} cm^{-2}, at temperatures between 100 and 500 °C, is shown in Figure 6.41. Above 300 °C, an enhanced diffusion independent of temperature can be seen, probably due to an interstitial mechanism. The likelihood of the latter model is strengthened by the possibility of suppressing the diffusion effect by an additional implantation with nitrogen ions; however, the model by Dearnaley *et al.* [184] referred to in the previous section could also serve as an explanation. Measured values for the range [274] are in very good agreement with the LSS values according to Gibbons *et al.* [5]; these are shown in Figure 6.42.

Not very many electrical measurements have been carried out [384], [556]. The electrical activation is low ($\leqslant$1 per cent); the junction characteristics of implanted diodes are extremely poor [2]. Research data up to the present indicate that indium is only suitable as a dopant for special silicon devices (e.g., extrinsic photoconductors).

6.1.7 Phosphorus

Besides arsenic, phosphorus is the most important group V dopant for implantation into silicon. For this reason, many experiments with this element were carried out quite early in the history of ion implantation [289], [292], [309], [412], [462], [463].

Electrical Activation

If the implantation dose is greater than the amorphizing dose (3×10^{14} cm^{-2} [9]), phosphorus too results in the usual steep annealing behaviour, with nearly complete electrical activation at 600 °C. Lower doses lead to a gradual increase in the activation up to 300 °C. Figure 6.43 shows sheet resistivity as a function of ion dose and annealing temperature [662]. The annealing stage at 600 °C can be seen clearly; still higher annealing temperatures lead to a slight additional increase in the electrical activation. High-dose implantations ($\geqslant 10^{16}$ cm^{-2}) only attain a complete electrical activation after a diffusion step, since the solubility is exceeded in the range of the maximum concentration immediately following the implantation.

Implanting at higher temperatures avoids formation of an amorphous layer, but produces many defects which are difficult to anneal, so that all in all, higher annealing temperatures are required than those needed after room temperature implantation [662], [770]. In this case, furthermore (similarly to boron), there occurs a reverse annealing between 400 and 600 °C [791]. An example of a

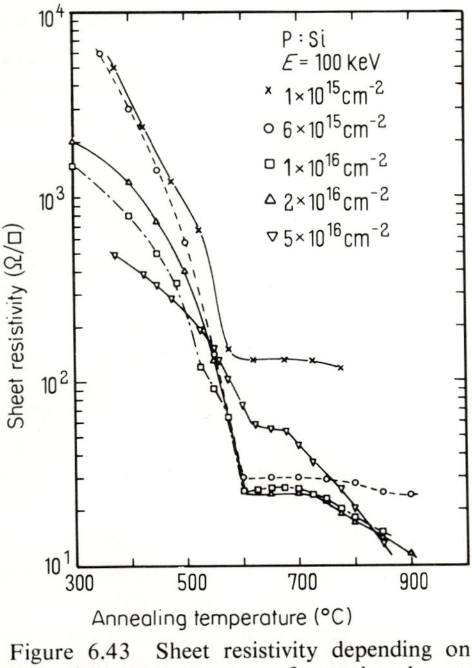

Figure 6.43 Sheet resistivity depending on annealing temperature for phosphorus-implanted layers with various doses at 100 keV [662]. Reproduced by permission of Gordon and Breach

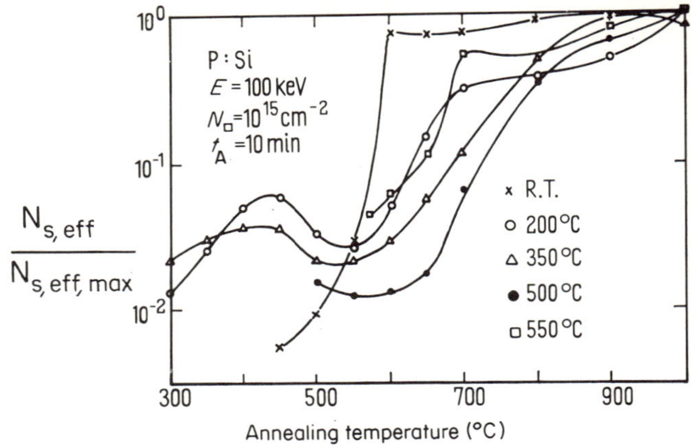

Figure 6.44 Isochronal annealing of phosphorus-implanted silicon. $E = 100$ keV; $N_\square = 10^{16}$ cm^{-2}; implantation temperature between room temperature and 550 °C [791]. Reproduced with permission

100 keV implantation with a dose of 10^{16} cm^{-2}, at various implantation temperatures, is shown in Figure 6.44. Correlated with the differences in annealing behaviour, various defects occur, as verified by TEM investigations [791]. Similarly to arsenic and boron, phosphorus reaches an approximate 100 per cent electrical activation following annealing between 600 and 900 °C, if the solubility limit (see Appendix) is not exceeded.

Range and Profiles

Phosphorus profiles always show a channelling tail, or a second maximum in the concentration distribution, if the samples have not been tilted out of the channelling direction during implantation. Figure 6.45 shows a compilation of published range measurements. This figure only gives values for the 'random' portion of the profile. The agreement with theoretical calculations is good, as in the other cases discussed. The alteration of phosphorus profiles by rotation and tilting of the samples has already been illustrated in Figure 4.21. These experiments show clearly that a penetration of ions into channels cannot be suppressed completely, even when the crystal is optimally misoriented.

Very pronounced channelling profiles have been measured by Reddi and Sansbury [584]. They implanted in the channelling direction, using a focused beam deflected up to $\pm 1.5°$ during implantation of the silicon wafers. In this way, a perfect channelling in certain regions of the wafer could be assured. The profiles were measured by the capacitance-voltage method using small diodes, and the region of the wafer where the ions had penetrated optimally in the channelling direction was determined. Several of these profiles [584] are given as examples in Figure 6.46. Implantation was made at energies between 30 keV and 600 keV,

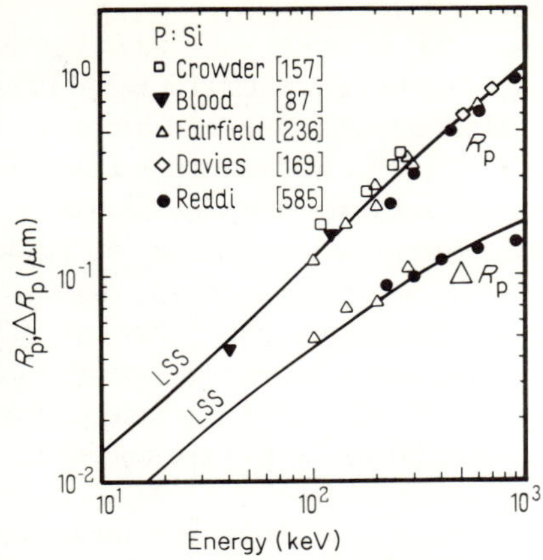

Figure 6.45 Range and range straggling of phosphorus in silicon

into ⟨100⟩-oriented silicon. The *p–n* junction used for the capacitance-voltage measurements was located at a depth of 0.15 (±0.02) μm. The probable channelling range at the lowest energy measured (30 keV) is higher by a factor of approximately 20 than the projected range according to the LSS theory. These

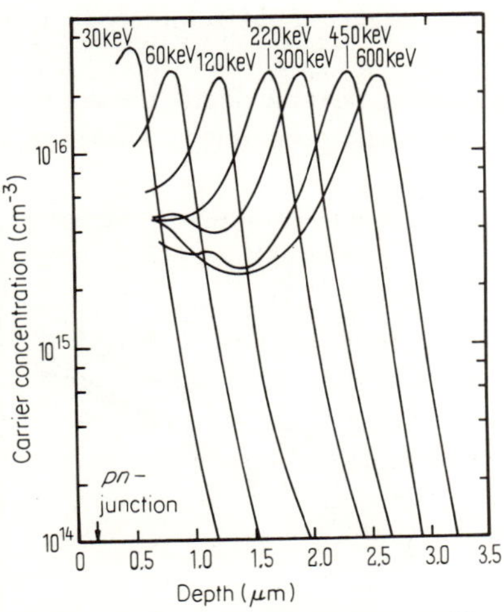

Figure 6.46 Channelling profiles measured after phosphorus implantation between 30 keV and 600 keV into ⟨110⟩-oriented silicon [584]. The profiles were obtained by capacitance-voltage measurements. Reproduced by permission of the American Institute of Physics

experiments were carried out for ⟨110⟩, ⟨111⟩, and ⟨100⟩ orientation. As expected, the range in the ⟨110⟩ direction was higher than that in the ⟨100⟩ and ⟨111⟩ directions. Figure 6.47 shows the results for the ⟨110⟩ and ⟨111⟩ directions, together with results by Moline [506], Goode *et al.* [301], Reddi and Sansbury [584] and Galaktionova *et al.* [272]. Also shown is a straight line proportional to $E^{1/2}$; in the case of electronic stopping alone (*cf.* Chapter 2), the range should be directly proportional to the energy in this manner. As Figure 6.47 shows, this is indeed the case for the ⟨100⟩ direction at energies above 100 keV, and for the ⟨100⟩ and ⟨111⟩ directions at energies above 200 keV. The deviations at low energies indicate that in these cases the mechanism of stopping due to nuclear collisions also plays a significant role. In order to reduce measuring errors, R_{max} in Figure 6.47 were measured at 50 per cent of the maximum value of the distribution.

Many additional channelling profiles have been measured by Dearnaley *et al.* after tracer implantations of ^{32}P [183]; a few of these measurements have already been presented in Chapter 2 in explaining the channelling effect. Cembali *et al.* [125], [126] have investigated the annealing behaviour of channelling profiles as well as radiation damage. Blood *et al.* [85], [86], [87] have demonstrated that profile tails which occur in the case of misoriented silicon are caused by ions scattered into channels. They used thin silicon foils (about 0.2 μm) placed on thick silicon substrates, and measured phosphorus profiles in the substrates after implantation, i.e., the ions implanted into the substrates through the thin silicon wafers, so that all diffusion effects could definitely be ruled out as a cause of the profile tails.

For reasons of the greater achievable depth, the use of channelling distributions for devices would be an interesting possibility. However, due to the precise wafer

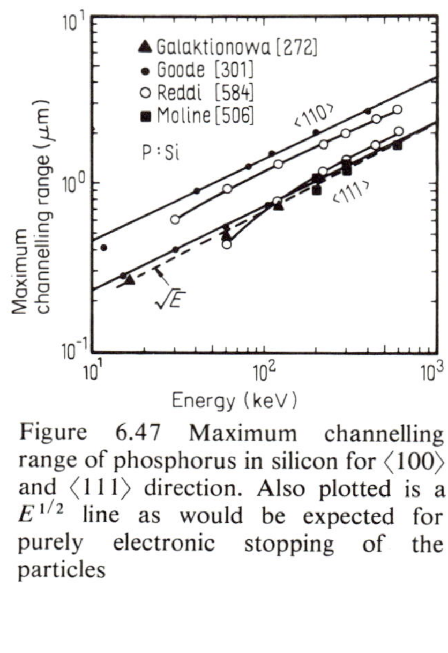

Figure 6.47 Maximum channelling range of phosphorus in silicon for ⟨100⟩ and ⟨111⟩ direction. Also plotted is a $E^{1/2}$ line as would be expected for purely electronic stopping of the particles

orientation that is necessary, to the low degree of tolerable beam divergence, and to the critical dependence upon ion dose and implantation temperature, this is not yet practicable at present.

Junction Characteristics

A great many measurements have been carried out upon phosphorus-implanted *p–n* junctions [82], [290]. In this case too, an annealing step at 900 °C is necessary to obtain minimum reverse currents [497]. The lowest values reached are 3×10^{-10} A/cm^2 at a reverse voltage of 5 V. These values are better than those for corresponding diodes produced by boron implantation [497]. Values of the leakage current for implanted diodes, depending on annealing temperature, are shown in Figure 6.48.

Defects in Annealed Layers

In the case of phosphorus-implanted layers too, a great many defects occur during the annealing process. Tamura [717] has made an investigation of such defects, using a transmission-electron microscope. His results are shown in Figure 6.49, based on implantations at 100 keV with a dose of 5×10^{14} cm^{-2} into ⟨111⟩-oriented material for various implantation temperatures. In addition, photographs of defects at three different temperatures can be seen in Figure 6.50. Point defects are visible following implantation at room temperature; these form dislocation loops at higher temperatures, which disappear again above 1000 °C. Implantations between 200 and 600 °C result in a great variety of defects, which can only be annealed with difficulty. The occurrence of rod-shaped defects is correlated with reverse annealing. In the case of implantation above 600 °C, dislocation

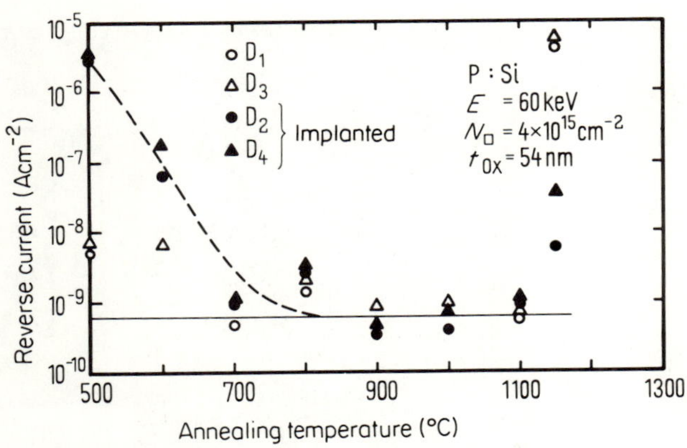

Figure 6.48 Leakage of phosphorus-implanted *p–n* junctions depending on annealing temperature, in comparison with diffused diodes [497]. Reproduced by permission of the American Institute of Physics

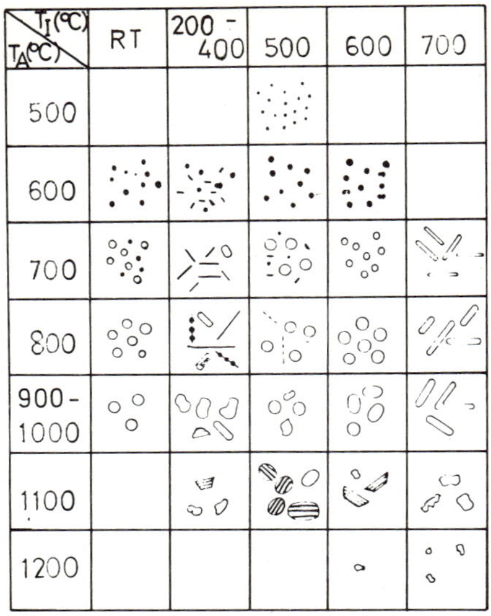

Figure 6.49 Defects in phosphorus-implanted silicon depending on implantation and annealing temperature. Implantation energy: 100 keV; dose: 5×10^{14} cm^{-2} [717]. Reproduced by permission of the American Institute of Physics

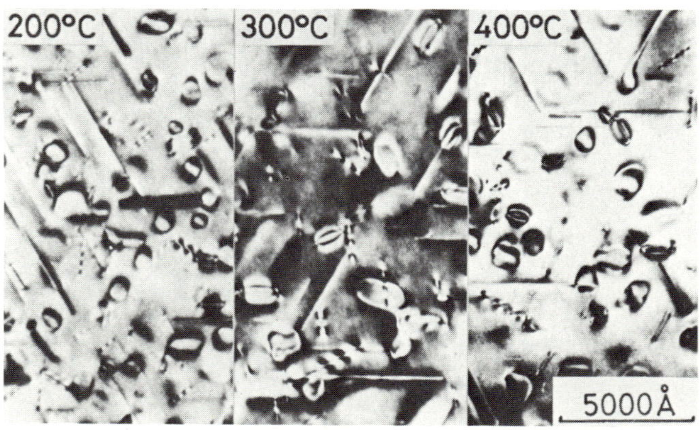

Figure 6.50 Photographs of defects after annealing at 800 °C. The samples were implanted at 200 to 400 °C [717] (*cf.* Figure 6.49). Reproduced by permission of the American Institute of Physics

loops are already present after implantation; these remain even after annealing at 1200 °C. Tamura *et al.*, in a further study [719], have also investigated the influence of the annealing atmosphere and have found that, for the case of annealing in an oxidizing atmosphere, dislocations grow into the semiconductor bulk. Further studies on defects in phosphorus-implanted layers can be found in References [322], [659], and [718].

Along with boron and arsenic, phosphorus (in comparison to all other elements of groups III and V) is best suited for the doping of silicon by ion implantation. However, due to the more pronounced channelling than for boron or arsenic, special care has to be taken in the fabrication of devices (e.g., compensation of the base through profile tails of the emitter). Moreover, phosphorus shows a highly non-linear diffusion which is not yet well understood, and which makes an exact profile control difficult.

6.1.8 Implantation of Other Elements

In addition to the elements already discussed, a great many others have also been implanted into silicon. Many of these have had no practical importance up to the present, for example, bismuth [157], [468] (although electrical activity up to 60 per cent has been achieved [157]). Other elements have been investigated in regard to their electrical properties, for example, hydrogen [449], [548], which produces a shallow donor level at 26 meV in the case of low doses, or oxygen and carbon (in connection with studies on the effects of radiation damage [379], [691]). In the latter case, the radiation damage produces *p*-type layers in silicon [691]. By irradiating silicon (*p*-type, 1 Ω cm) with high doses of hydrogen (4×10^{16} cm^{-2}, 1 MeV), it can be made high resistive [642]. The spreading resistance reaches 10^7 Ω; after annealing at 600 °C, it can be reduced to 5×10^3 Ω. A low-dose irradiation with nitrogen ions produces *n*-type conductivity; high doses cause formation of a Si_3N_4 layer [258], and show good promise for device application, e.g., for buried insulating layers for fully isolated integrated circuits. The implantation of gold [635] has been the subject of much interest in recent years, for the purpose of minority carrier lifetime adjustment.

6.1.9 Gettering

The most important gettering mechanisms are the accumulation of impurities at crystalline defects and formation of chemical compounds. The best-known methods for the gettering of undesired impurities [295], [528] utilize phosphorus glass or boron glass [369], [491], or radiation damage [350], [478]. Hsieh *et al.* [350] have investigated the gettering properties of implanted phosphorus, argon and arsenic, by measuring the reverse currents of silicon multidiode targets. Doses of 10^{16} P/cm^2, 10^{15} As/cm^2, or 3×10^{15} Ar/cm^2 were adequate to achieve the same effect as a conventional phosphorus gettering. Buck *et al.* [110], [111], using the back-scattering technique, have investigated the gettering effect of a

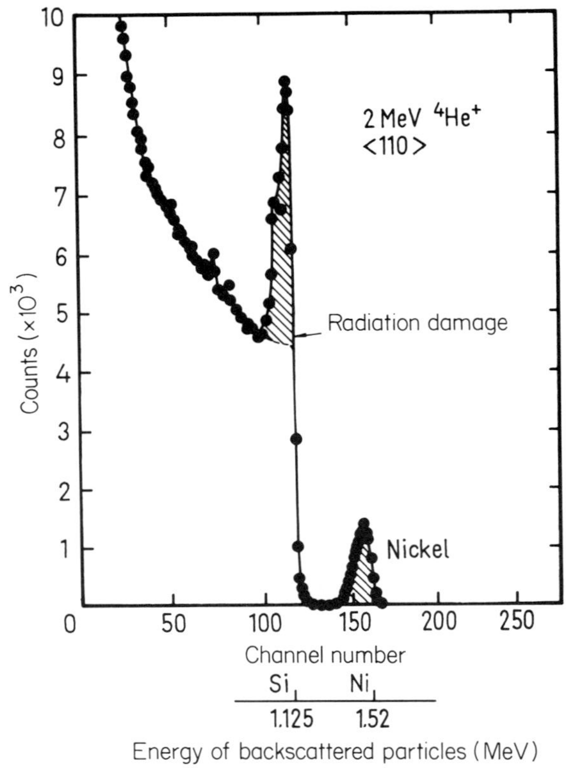

Figure 6.51 Gettering of nickel. Back-scattering spectrum (2 MeV He ions) of an Si layer damaged by implantation of 100 keV Si ions into the front side. Before annealing (900 °C, 30 min), the nickel was deposited by evaporation on to the reverse side [111]. Reproduced by permission of North-Holland Physics Publishing

damaged silicon layer. The damage was produced by irradiation with 10^{16} cm^{-2} Si ions at 100 keV. Iron, cobalt and gold were gettered relatively slowly, copper and nickel much more quickly. The concentrations in the damaged layer were 10^{13} cm^{-2} to 10^{14} cm^{-2} for iron, cobalt and gold; in the cases of copper and nickel, they were between 6×10^{14} cm^{-2} and 5×10^{16} cm^{-2}. The fact that the various elements behave differently can be explained in terms of their solubilities and their diffusion coefficients via interstitial lattice sites.

Figure 6.51 shows a typical example for the gettering of nickel. In the channelling-measurement diagram, a nickel distribution can be seen on the right; this has been gettered in the damage distribution which is seen to the left of it. Seidel *et al.* [653], [654] have investigated the gettering effects of various ions upon gold. They used ion doses of 10^{16} cm^{-2}, at an implantation energy of

200 keV. The samples were investigated using the back-scattering technique, and also in some cases with a transmission electron microscope. Argon damage gettered considerably better than a phosphorus diffusion at temperatures below 1000 °C, and just as well as one at 1150 °C. The other ions gettered at 1000 °C more poorly than the phosphorus diffusion, the order of their effectiveness being Ar ⩾ O ⩾ P ⩾ Si ⩾ As ⩾ B. All in all, one can say that the gettering effect depends on the quantity and the nature of the radiation damage which remains after annealing. At temperatures below 1000 °C, therefore, argon irradiation is very well suited for the effective gettering of gold, while the *p*- or *n*-doping elements, especially arsenic, are less suitable.

6.1.10 Oxidation of Implanted Silicon

The oxidation rate of implanted silicon can be increased either through high concentrations of radiation damage or through the chemical effects of doping; using the latter, the oxidation rate can also be decreased. This case has been the subject of some interest, since in this way, selective oxidation could be possible [263], [752], [582a]. Experiments with oxygen, nitrogen and carbon have been carried out. In Figure 6.52, the time dependence of the oxide growth is shown for a nitrogen implantation at 30 keV, with a dose of 2.4×10^{17} cm^{-2}. Oxidation was carried out at 950 °C. A dose of 2.4×10^{17} cm^{-2} is sufficient to form a Si_3N_4 layer with a width of approximately 64 nm, as measured by ellipsometry. After an oxidation at 950 °C for 15 h the implantation synthesized Si_3N_4 still masks the silicon against oxidation. The oxide thickness was 1.3 μm in case of ⟨100⟩ and more than 1.4 μm in case of ⟨111⟩ silicon, whereas the oxidized nitride had a thickness of 600 to 700 Å depending on the preoxidation heat treatment. It was

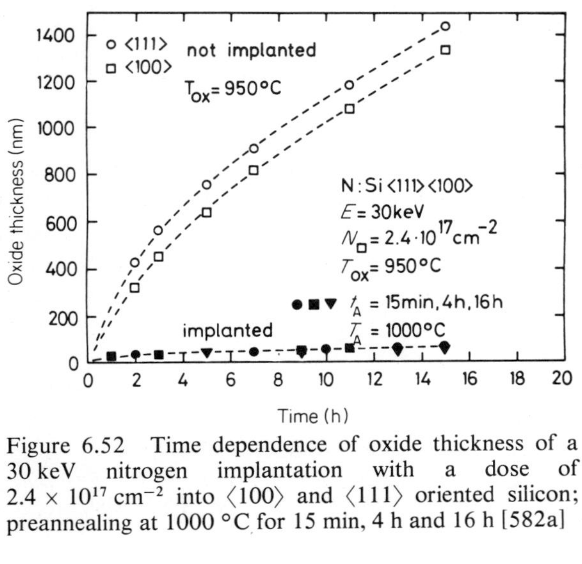

Figure 6.52 Time dependence of oxide thickness of a 30 keV nitrogen implantation with a dose of 2.4×10^{17} cm^{-2} into ⟨100⟩ and ⟨111⟩ oriented silicon; preannealing at 1000 °C for 15 min, 4 h and 16 h [582a]

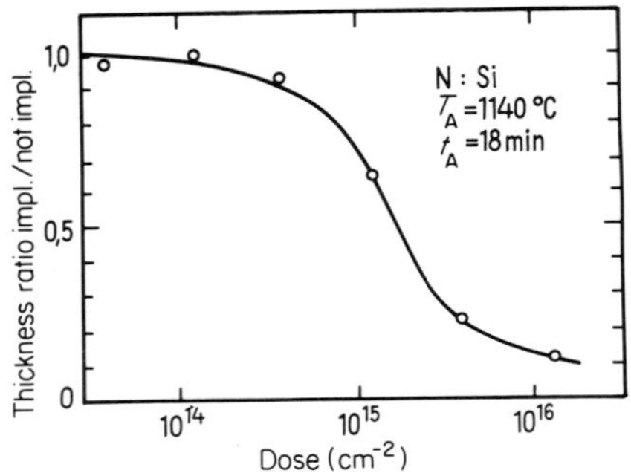

Figure 6.53 Dose dependence of oxidation after nitrogen implantation of silicon |263|. From Fritsche and Rothemund, *J. Electrochem. Soc.*, **122** (1975), 1234, reprinted by permission of the publisher, The Electrochemical Society, Inc.

found by IR spectroscopy that no heat treatment is necessary to form the Si_3N_4 before the thermal oxidation starts; however, a treatment for 16 h slightly improves the oxidation resistance. The relationship in oxide thickness for implanted and non-implanted samples following dry oxidation at 1140 °C (18 min in all cases), depending on implantation dose, is shown in Figure 6.53. At relatively low doses, a clear reduction in the oxidation rate is already apparent.

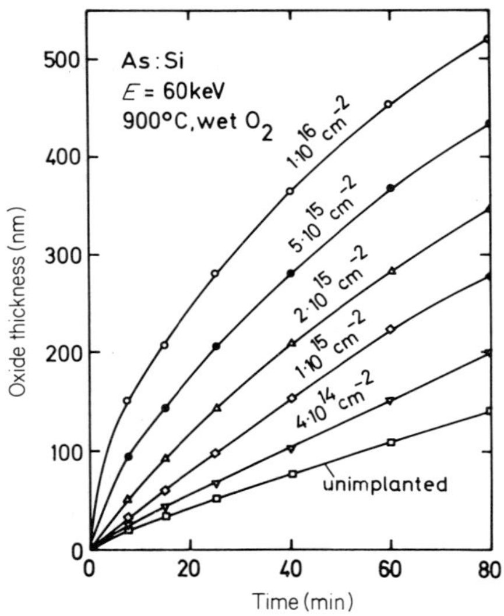

Figure 6.54 Oxide thickness versus oxidation time for 900 °C wet oxidation with implantation dose as parameter |302b|

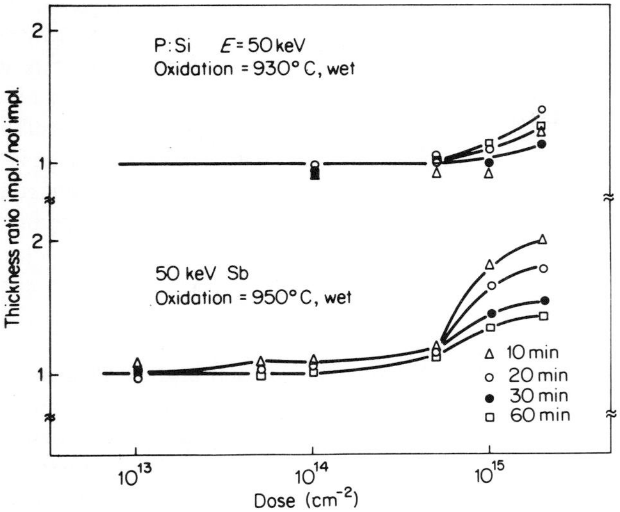

Figure 6.55 Oxide thickness ratio for implanted and non-implanted areas with Sb and P, depending on ion dose for wet oxidation [541]

The implantation of carbon only reduces the oxidation rate to a small extent. In the case of oxygen, the situation is more complex. The oxidation rate increases at high concentrations, but decreases at low doses.

The oxidation rate increase as a result of irradiation with antimony, phosphorus, tin, and argon ions has been studied by Nomura *et al.* [541], and for arsenic and antimony by Götzlich *et al.* [302b]. Figure 6.54 shows measurements of the oxide thickness following wet oxidation of arsenic-implanted samples at 900 °C depending on the dose. Figure 6.55 shows the thickness relationship for antimony-implanted and phosphorus-implanted layers, depending on oxidation time and implantation dose. The increase in the oxidation rate, regardless of the ion species, begins at a dose of 5×10^{14} cm^{-2}. By preannealing between 400 °C and 1100 °C in dry nitrogen, it could be shown that the increase in the oxidation rate is not a damage effect, but rather, a chemical effect due to the implanted ions. Apparently, the oxidation is only influenced in the region of the linear growth rate; in the parabolic growth-rate region, a constant oxidation rate is again seen (see Section 3.8.3). The etch rate of the oxide layers is not influenced by the enhanced oxidation.

6.1.11 Laser Annealing

All results presented up to now have been obtained using thermal annealing. The results obtainable by means of laser annealing will be discussed in the following section. Laser annealing is a relatively new technique and it is uncertain at present whether it will become a standard annealing method. Nevertheless, a presentation of some information on this topic seems desirable within the context of this book.

With laser annealing, a distinction is made between two different modes. Using pulsed lasers (e.g., Nd:YAG or ruby), a melting of the top layer takes place; this layer subsequently recrystallizes epitaxially. Such a process results in a perfect crystal lattice under optimum conditions [529a]. If a CW laser is used, on the other hand, a solid-phase epitaxy similar to thermal annealing occurs [279a], [776a]. The basic differences between these two methods have been outlined in Section 3.3.

Pulsed-laser Annealing

In Figure 6.56, profiles of boron, phosphorus and arsenic, measured after pulsed-laser annealing with an energy of 1.7 J/cm^2, are shown in comparison to 'as-implanted' profiles. The profiles were measured using SIMS. A ruby laser with a pulse length of 60 ns was employed in all three cases. One can see very clearly the diffusion during the melting phase. In the case of boron and arsenic, theoretical diffusion profiles are also included; these have been calculated assuming diffusion lengths of 655 Å and 944 Å, respectively. For boron, this means a diffusion coefficient of 2.4×10^{-4} cm^2/s, with a melting phase duration of 180 ns [755a]. This diffusion coefficient for boron in liquid silicon is approximately 7 orders of magnitude higher than that in the solid at the melting point. In contrast to the diffusion in the solid, no concentration dependence of the diffusion coefficient is seen in the liquid case.

Figure 6.57 shows a TEM measurement of boron-implanted sample annealed with a Q-switched ruby laser (pulse duration ~50 ns), operating in multimode with 1.5 J/cm^2. There is no damage visible down to the resolution of the microscope (~10 Å). In contrast, dislocations and dislocation loops can be seen in the thermally annealed sample [791b]. To obtain such a good epitaxial regrowth, it is necessary to melt through the damaged layer. In this case, the melting depth was down to approximately 0.9 μm. Many calculations have been performed to determine the melting depth [51b], [67a], [633a]. In Figure 6.58, time-dependent temperature distributions at different depths are shown for a 1.5 J/cm^2 pulse [51b]. The surface layer melts and remains liquid for about 130 ns. Since the evaporation temperature ($T_v = 3513$ K) is not reached in the melt, there is no loss of silicon. The deep layers (4500 Å and 6000 Å) remain solid at all times.

Although the crystal lattice can be recrystallized perfectly, a certain concentration of defect levels in the band gap results [41d], [385a], which could be annealed out if necessary using an additional thermal annealing step at 800 °C. This annealing would not introduce defects, however, as in the case of purely thermal annealing [301a].

At high concentrations of implanted ions, the solubility of the dopant can be greatly exceeded [296a], [340a], [586], [586a]. This can be interpreted as being

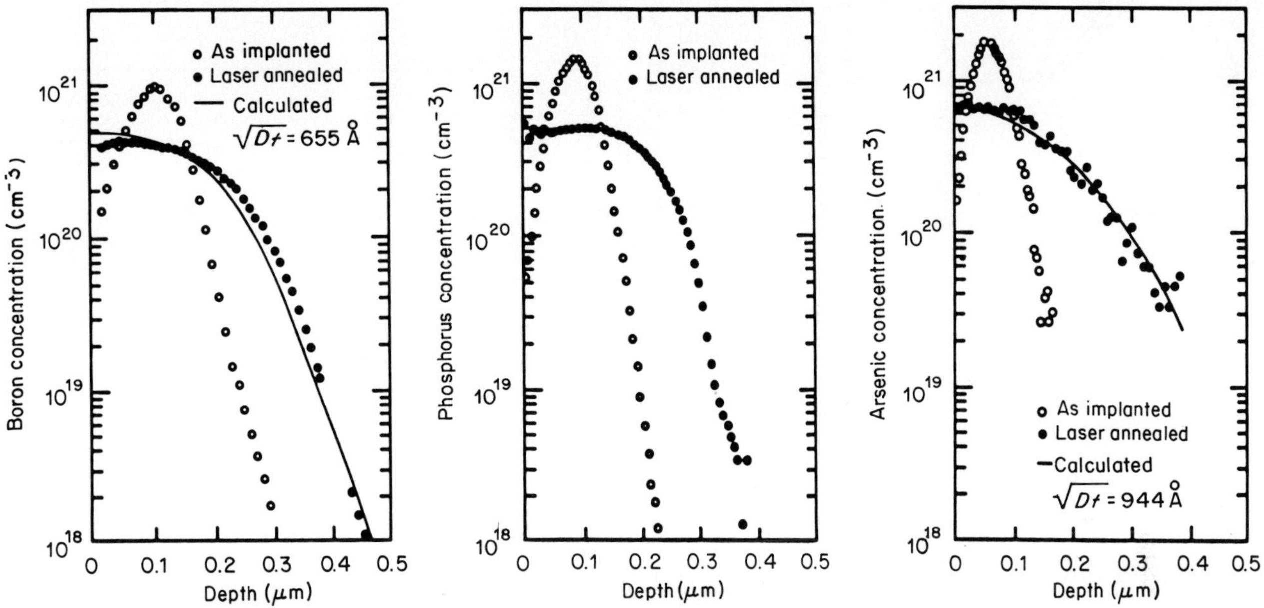

Figure 6.56 Profiles for B (35 keV, 10^{16} cm^{-2}), P (80 keV, 1.16×10^{16} cm^{-2}), and As (100 keV, 1.86×10^{16} cm^{-2}) implanted silicon, before and after laser annealing with 1.7 J/cm^2. In the cases of B and As, theoretical calculations are also given [768a], [768b]. Reproduced by permission of the American Institute of Physics

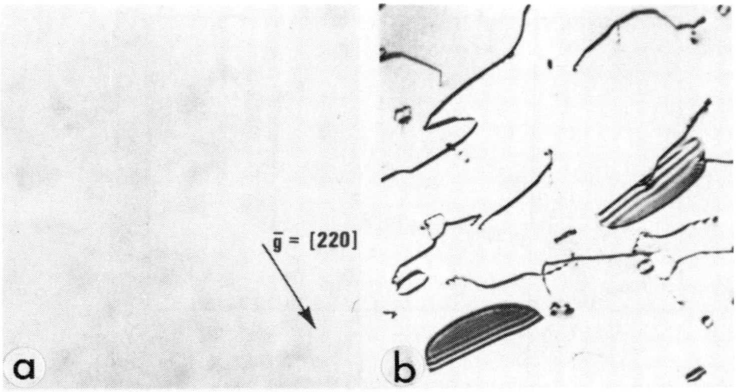

Figure 6.57 Comparison between laser and thermal annealing for boron-implanted ($E = 35$ keV, $N_{\square} = 3 \times 10^{15}$ cm^{-2}) silicon: (a) laser-annealed with 1.5 J/cm^2 (ruby laser); (b) thermally annealed at 900 °C for 30 min [791b, 529a]

due to a supersaturation caused by the fast regrowth, which takes place at a rate of 2.7 m/sec. The regrowth time is about 1 μs [41d]. By a subsequent thermal annealing, however, the solubility equilibrium is established again, and defects can result. A further effect is due to the solubility being exceeded. X-ray measurements show that boron and phosphorus doping lead to a contraction of the lattice,

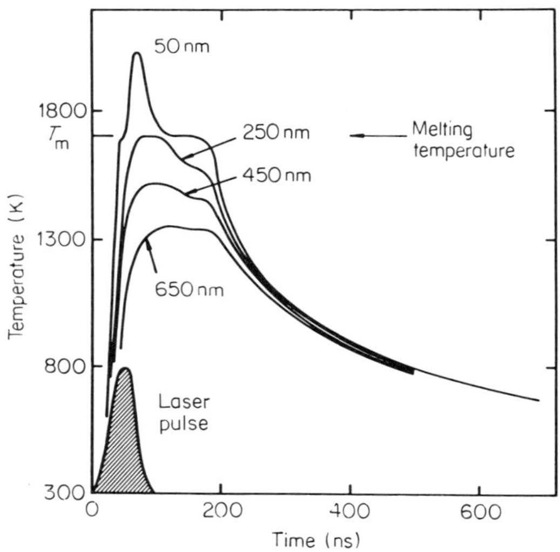

Figure 6.58 Temperature–time profiles at several depths inside the amorphous silicon layer. The energy density of the ruby pulse is 1.5 J/cm^2 [51b]. Reproduced by permission of the American Institute of Physics

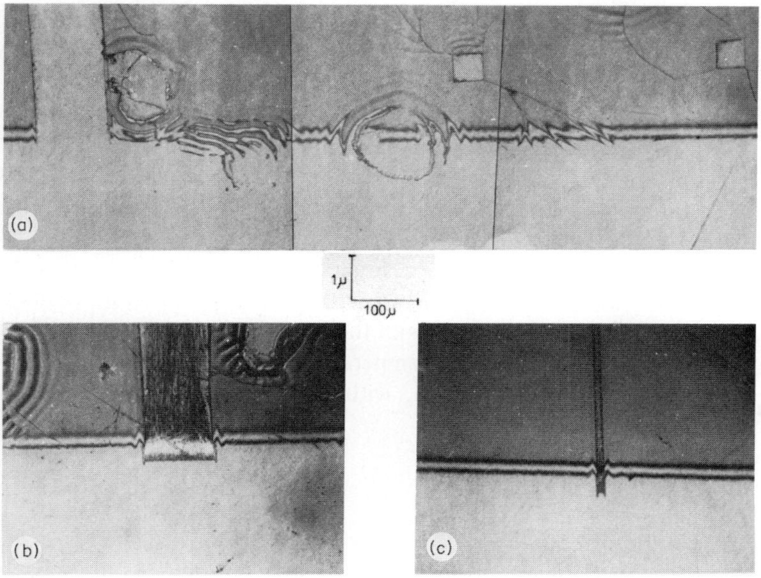

Figure 6.59 Bevel-sectioned laser-annealed silicon. Oxide thickness: 0.35 µm; energy density: 2.2 J/cm^2 (ruby laser). a, unimplanted region, b, implanted, 100 µm window, c, implanted, 10 µm window. Reproduced by permission of the American Institute of Physics [334a]

while antimony doping expands the lattice, and arsenic, which is nearly the same size as silicon, causes no essential change [41d]. The contractions are as high as 1.32 per cent (2.5×10^{16} B/cm^2) and the expansion is 0.36 per cent (1.6×10^{16} Sb/cm^2).

One problem with pulsed-laser annealing is connected with the masking by SiO_2 layers. Either due to acoustic shock waves, or due to the melting of the silicon, the SiO_2 tends to blister off. At lower energies, only a ripple can be seen. Furthermore, interference effects can enhance or suppress the energy density reaching the silicon substrate. Especially at mask windows, this effect causes severe deterioration of the surface and the junction depth. An example for a decorated bevel section of such a wafer is shown in Figure 6.59 [334a]. The ripple is as large as 1 µm. The junction depth is greater at the window corners.

With lasers having an energy higher than the band-gap energy (e.g., ruby-laser $\lambda = 0.63$ µm; frequency-doubled Nd:YAG-laser $\lambda = 0.53$ µm), no influence of the implanted dose is found. The wavelength of the Nd:YAG laser ($\lambda = 1.06$ µm) is very close to the band gap. Therefore, the absorption depends on the carrier concentration and the crystalline structure. Amorphous silicon has a higher absorption coefficient than monocrystalline silicon [215a], [716a]. The same is true for high carrier concentrations. Therefore, a selective annealing would seem to be possible.

CW-laser Annealing

During CW-laser annealing, the surface layer of the sample is heated without melting; here, a solid-state epitaxy takes place, as in the case of thermal annealing. Up to the present, Ar, Kr and CO_2 lasers [167c], [124b], [715a], [279a] ($\lambda = 0.515$ μm, $\lambda = 0.468$ μm and $\lambda = 10.6$ μm, respectively) have been used for this purpose. The dwell time is between a few milliseconds and several seconds, depending on the laser power and the wavelength.

In Figure 6.60, the time necessary to regrow 500 Å of amorphous silicon is shown as a function of the substrate temperature. These calculations were carried out using steady-state conditions, which are reached within approximately 50 μs [776a]; this is short in comparison with the laser dwell time (⩾1 ms). The hatched bands in Figure 6.60 indicate that temperatures between 850 and 950 °C lead to regrowth of the layer in 1–10 ms, with an absorbed power of about 1.7×10^3 W/cm.

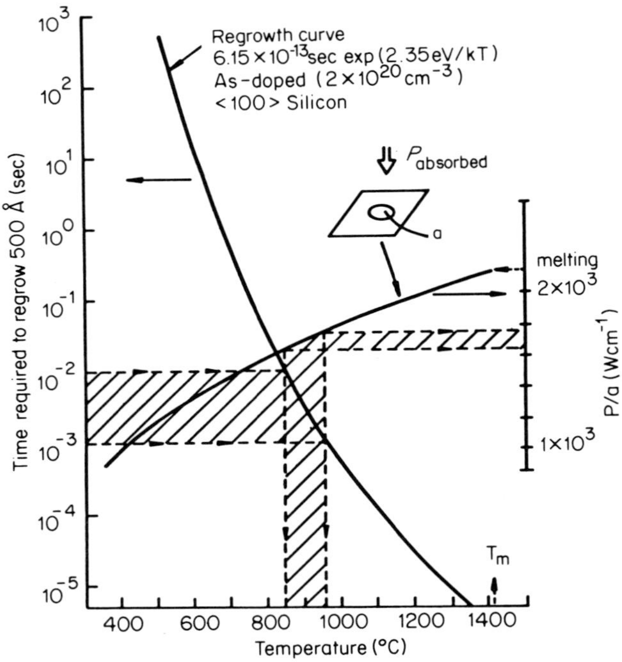

Figure 6.60 Time required for the epitaxial regrowth of 500 Å amorphous silicon as a function of annealing temperature and power density. The shaded areas define the temperature regions and laser powers needed for thermal regrowth of the layer [776a]. Reproduced by permission of the American Institute of Physics

For CO_2 lasers, energy absorption takes place through the mechanism of free carrier absorption, requiring higher power levels depending on the carrier concentrations, which can be adjusted by irradiation at increased temperatures.

Figure 6.61 shows a comparison between arsenic profiles for the as-implanted case, for laser annealing with an argon laser (scan rate 2.76 cm^2/s, beam diameter 22 μm, power level 7 W), and for thermal annealing at 1000 °C for 30 min [279b]. The laser annealed profile is identical to the as-implanted profile, which can be perfectly matched using a Pearson-VI distribution. The thermally annealed profile shows the well known concentration-dependent diffusion. The avoidance

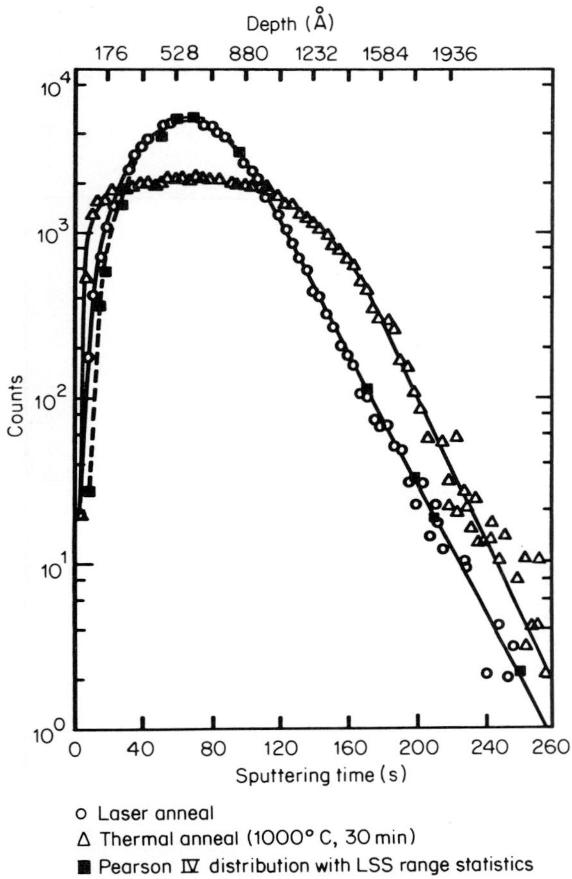

Figure 6.61 Arsenic-concentration profiles in ⟨100⟩ silicon as-implanted, laser-annealed, and thermally-annealed; $E = 100$ keV, $N_{\square} = 5 \times 10^{14}$ cm^{-2} [279b]. Reproduced by permission of the American Institute of Physics

of diffusion, with the type of laser annealing just discussed, makes it a perfect tool for the tailoring of implantation profiles.

In the case of CW annealing, the solubility can also be exceeded. This is shown in Figure 6.62 for an arsenic implant [586]. In this figure, the electrical carrier concentration is given following argon-laser annealing. For comparison, results obtained using e-beam and thermal annealing are also shown. The absorbed power per length P/a was 0.17 W/μm for the argon laser and 0.14 W/μm for the e-beam, at substrate temperatures of 350 and 50 °C, respectively. After laser or e-beam annealing, the solubility of 4×10^{20} cm^{-3} is exceeded by a factor of 3. Obviously, the recrystallization takes place more quickly than the establishment of the equilibrium solubility. No difference is seen between laser and e-beam annealing. The equilibrium concentration of arsenic on lattice sites, however, is established in the case of thermal annealing, indicating that the high electrical activation in the case of CW-laser or e-beam annealing is due to supersaturation.

In the case of CO_2 laser annealing, the radiation is primarily absorbed through the photoexcitation of free carriers. It therefore depends greatly on the carrier

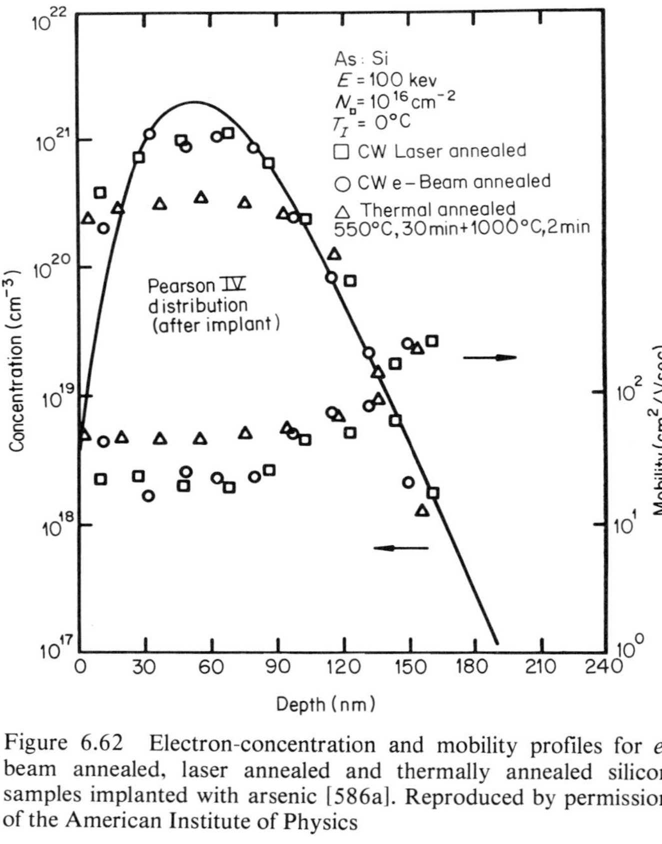

Figure 6.62 Electron-concentration and mobility profiles for e-beam annealed, laser annealed and thermally annealed silicon samples implanted with arsenic [586a]. Reproduced by permission of the American Institute of Physics

concentration of the sample, which in turn depends on the dopant concentration as well as the temperature. With a laser power of 27 W, areas between 1 and 3 mm in diameter can be annealed within several seconds, using substrate temperatures from 200 to 400 °C to enhance the absorption [124c]. During CO_2-laser annealing, the entire thickness of the sample is heated, since the absorption coefficient is so low (~10 cm^{-1} at 10^{16} cm^{-3}). Therefore, this type of annealing is intermediate between furnace annealing on the one hand, and argon- and krypton-laser as well as pulsed-laser annealing on the other, all of which heat only a surface layer.

6.2 IMPLANTATION INTO GERMANIUM

During the last several years, the importance of germanium in semiconductor technology has decreased considerably. Therefore, relatively little research has been done on the suitability of ion implantation for the doping of this semiconductor. Since radiation damage can be annealed here at rather low temperatures—the recrystallization of an amorphous layer is already complete at about 420 °C—the implantation temperature is an important parameter as well as the cause of the partly contradictory experimental results.

All elements of groups III and V have been implanted to study radiation damage and sputtering effects [36], [47], [80], [328], [342], [343], [344], [389], [483], [545], [574], [728], and germanium itself. Several studies have dealt with implantation of rectifying contacts for nuclear-radiation detectors, e.g., Reference [574]; others have treated implanted microwave transistors [625] and implanted germanium junction field-effect transistors [756].

Herzer and Kalbitzer [328] have made a detailed investigation of annealing behaviour and implantation profiles for arsenic, boron, phosphorus and gallium in ⟨111⟩- and ⟨110⟩-oriented germanium. The implantations were made without tilting the samples, i.e., in the channelling direction. Boron reaches its maximum electrical activation immediately after implantation; for the other ion species, though, an annealing step between 400 °C (gallium) and 550 °C (arsenic, phosphorus) is necessary to obtain the maximum possible activation. For implantation doses of 10^{14} cm^{-2}, the following electrical activations were measured after annealing at 400 °C: boron 20 per cent, phosphorus 20 per cent, gallium 80 per cent. Figure 6.63 shows boron profiles as examples for these measurements; the profiles were obtained after implantation into ⟨110⟩-oriented germanium at 10 and 20 keV with a dose of 10^{14} cm^{-2}, with and without annealing. As expected, an especially high penetration depth is reached in the ⟨110⟩ direction; thermal diffusion does not occur. The relatively high deviation from the theoretical range according to LSS [5] is probably due to channelling; under some circumstances, it could also be due to radiation-enhanced diffusion during implantation [328]. The phosphorus-implanted and gallium-implanted profiles clearly show thermal diffusion.

Alton and Love [36] have obtained very similar results on annealing behaviour; their studies also included aluminium, antimony and germanium.

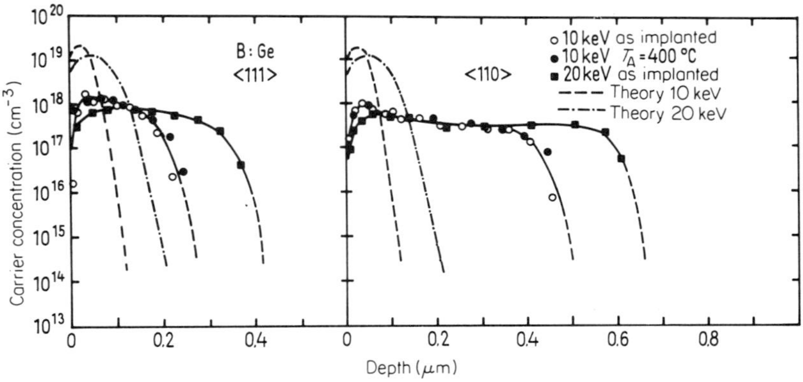

Figure 6.63 Boron profiles in ⟨111⟩- and ⟨110⟩-oriented germanium, for the unannealed case as well as after annealing at 400 °C, in comparison with theory [328]. Reproduced by permission of Springer-Verlag, Heidelberg

Antimony requires a temperature of 625 °C for maximum activation, while aluminium has a steep annealing step at 500 °C. Contrary to the above-cited study, boron too was found by them to have an annealing step at about 450 °C. This discrepancy is probably due to a dose-rate effect. The germanium-implanted samples had unannealed radiation damage up to temperatures of 700 °C.

Björkquist *et al.* [80] have studied the occupation of lattice sites by implanted ions in germanium. The number of ions on lattice sites is considerably higher in all cases than corresponds to the solubility in thermal equilibrium. In contrast to the situation for silicon implantation, no interstitial components at all could be found. The results of this study are summarized in Table 6.4, with comparison of results for silicon. The implantation temperatures were 300 to 350 °C (Ge) and 350 to 450 °C (Si).

Schmid *et al.* [625] have investigated the applicability of implantation for microwave-transistor fabrication in germanium. Their studies concentrated on the characteristics of boron, aluminium and arsenic. The electrical activation of boron, depending on the dose, is shown in Figure 6.64. The implantation energies (68 keV to 78 keV) were chosen in such a way that the *p–n* junction was always located at the same depth (500 nm). These measurements show that the solubility of boron is about 2×10^{19} cm^{-3}.

Lattice localization by means of the $^{11}B(p, \alpha)^{8}Be$ reaction indicated that only about 5 per cent of the ions were on lattice sites, for a dose of 10^{16} cm^{-2} and an annealing temperature of 600 °C; this agrees with the data in Figure 6.63. Ponpon *et al.* [574] also examined the behaviour of implanted boron ions. Using back-scattering measurements, they determined that practically no radiation damage remains after annealing at 170 °C. In contrast, Schmid *et al.* [625] found practically no radiation damage even at a dose ten times higher (10^{16} cm^{-2}, as

Table 6.4 Comparison of lattice-site position for implanted ions in germanium and silicon [80]

Ion	Germanium Per cent in direction ⟨111⟩	Germanium Per cent in direction ⟨110⟩	Silicon Per cent in direction ⟨111⟩	Silicon Per cent in direction ⟨110⟩
Sb	90	85	89	87
Bi	76	80	87	86
In	75	75	52	25
Tl	65	35	84	46
Sn	—	—	92	96
Pb	87	85	—	—

Per cent in ⟨110⟩ direction ≈ per cent on lattice sites.
Per cent ⟨111⟩ – per cent ⟨110⟩ ≈ per cent on interstitial sites.

opposed to 10^{15} cm^{-2}), while Sigurd *et al.* [687] even obtained amorphous layers for corresponding doses. These discrepancies too are probably the result of current-density differences during implantation. According to both studies just mentioned, the level of electrical activation is practically unaffected by annealing; at low doses, though, the mobility decreases at about 200 °C with subsequent recovery.

Considerably better results have been obtained with aluminium. Here, a pronounced annealing step at 500 °C was found. The good electrical activation

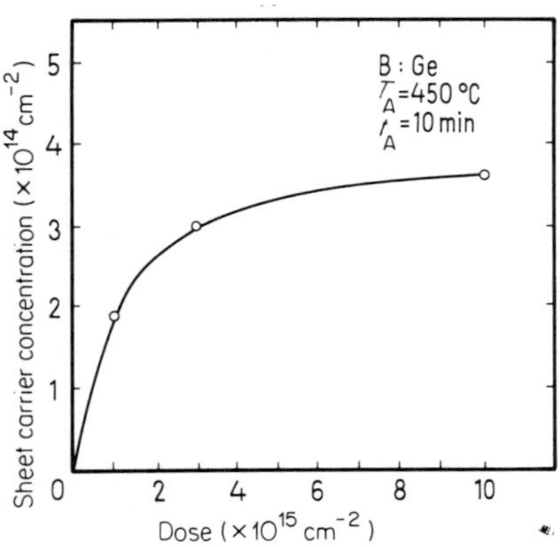

Figure 6.64 Electrical activation of boron in germanium depending on ion dose. Implantation energies: 78 keV (10^{15} cm^{-2}); 72 keV (3×10^{15} cm^{-2}); 60 keV (10^{16} cm^{-2}) [625]

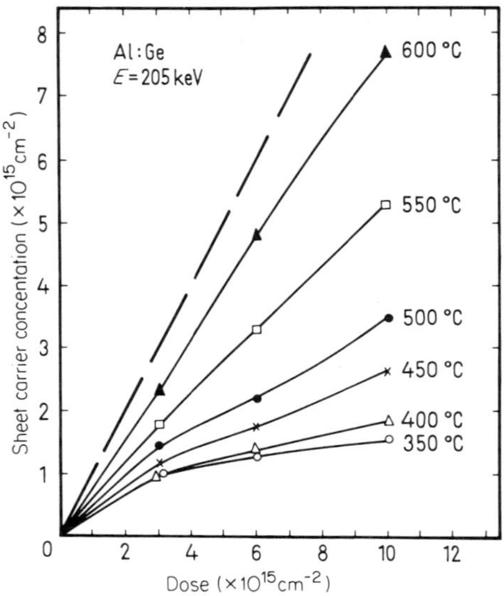

Figure 6.65 Electrical activation of aluminium in germanium depending on dose and annealing temperature [625]. The broken line corresponds to 100 per cent electrical activation

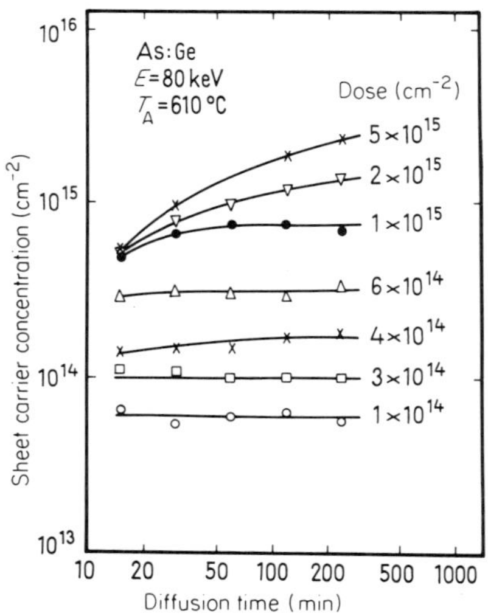

Figure 6.66 Electrical activation of arsenic in germanium depending on ion dose and diffusion time at 610 °C [625]

level, even at high doses, is attributed to the high solubility of 4×10^{20} cm^{-3} [243]. Figure 6.65 shows sheet carrier concentrations depending on ion dose and annealing temperature.

Because of its high mass, arsenic can only be used for predeposition in device applications [625]. Due to its high solubility (approximately 10^{20} cm^{-3}) and high enough diffusion coefficient, this element (the same as aluminium) is very well suited for the implantation of transistors. The electrical activation obtainable for a drive-in diffusion at 610 °C is shown in Figure 6.66. For low doses it is about 80 per cent; at higher doses it only increases to this value after diffusion, due to the solubility limit. Figure 6.67 shows junction depth depending on diffusion time at 610 °C, for doses between 10^{14} cm^{-2} and 5×10^{15} cm^{-2}. A SiO_2 masking layer was used during diffusion to prevent out-diffusion of the arsenic. The concentration dependence of the diffusion coefficient can be seen very clearly.

Similar measurements were made by Tinsley and Jones [728]. They implanted at room temperature; after annealing between 450 and 500 °C (for bismuth and antimony) or between 550 and 600 °C (for indium), they obtained a maximum lattice-site occupation of 75 to 80 per cent with a subsequent sharp drop at 700 °C (implantation dose 5×10^{14} cm^{-2}). They explain this effect through exceeding the solubility limit as well as precipitation of excessive ions at higher temperatures. Electrical activations of 50 per cent (Ga), 80 per cent (Bi) and 15 per cent (In) were measured. The latter figure is contradicted by back-scattering measurements, which indicate 75 per cent of indium ions on lattice sites. An explanation could be low ionization according to Fermi–Dirac statistics.

Mobility measurements of indium-implanted and gallium-implanted layers have shown that the mobility corresponds to that of the 'bulk' material [389], although

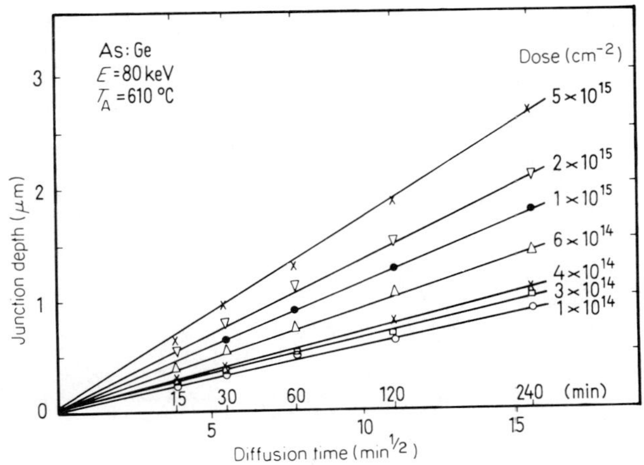

Figure 6.67 Depth of p–n junctions in arsenic-implanted germanium depending on diffusion time at 610 °C [625]

Table 6.5 Diffusion behaviour of implanted ions in germanium [47]

Element	Energy (keV)	D (cm^2/s)	Annealing time (min)	Proportion not out-diffused (%)	SiO_2 mask (nm)
Sb	700	1.8×10^{-13}	20	60	0
	700	1.8×10^{-13}	80	35	0
	200	1.8×10^{-13}	30	20	0
	200	1.8×10^{-13}	60	15	0
	200	1.2×10^{-13}	30	30	300
	200	1.2×10^{-13}	60	23	300
As	700	9×10^{-14}	45	60	0
	200	9×10^{-14}	45	40	0
P	200	4×10^{-14}	20	100	0
	200	4×10^{-14}	80	40	0

The implanted doses were between 10^{12} and 10^{13} cm^{-2}.

the electrical activation after annealing at 500 °C for 45 min is relatively low (50 per cent for gallium, 10 to 15 per cent for indium).

Axmann *et al.* [47] have obtained doping profiles of Sb-implanted, As-implanted and P-implanted layers. Implantation was made with doses between 10^{12} cm^{-2} and 10^{13} cm^{-2}, at 200 to 700 keV. By capacitance-voltage measurement, they determined the out-diffusion of these elements during annealing as well

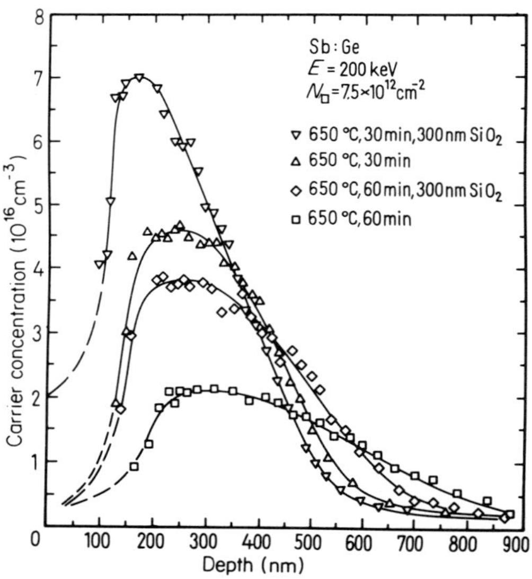

Figure 6.68 Diffusion profiles of antimony in germanium [47]. Reproduced by permission of the American Institute of Physics

as the corresponding diffusion coefficients. The out-diffusion was reduced by using SiO_2 masking layers. Several typical results after annealing at 600 °C are listed in Table 6.5, and a typical doping profile is shown in Figure 6.68.

6.3 III–V SEMICONDUCTORS

The III–V semiconductors, especially GaAs, GaP, $GaAs_{1-x}P_x$ and $Ga_{1-x}Al_xAs$, are currently along with silicon the most important semiconducting materials. The reasons for this include the direct band-to-band transition of many of these compounds which enables radiative recombination with high efficiency, the high mobility of the electrons which allows higher cut-off frequencies in devices and the Gunn effect [308] through which simple high-frequency oscillators are possible.

Surveys on the state of the art in implantation of III–V semiconductors have been compiled by Eisen [219], Degen [186c] and Donnelly [202b]; a great many studies on this topic were also presented at the 1974 Implantation Conference in Osaka [24].

6.3.1 Gallium Arsenide

The most important compound semiconductor by far for device applications is GaAs. The greatest number of implantation experiments has therefore been made using this material. As *p*-type doping elements beryllium [364], cadmium [355], [361], [363], magnesium [364], and especially zinc [361], [363], [444], [462], [487], [589], have been studied. Implanted *p–n* junctions could be applied for luminescence diodes. *n*-type doping is important for microwave devices such as Gunn diodes and Schottky field-effect transistors, due to the high electron mobilities. Sulphur [208], [361], [430], [618], selenium [248], silicon [617], [618], tellurium [217], [315], [486] and tin [355] have been investigated for these purposes.

Masking Layers

Due to the ease with which GaAs decomposes at higher temperatures, the use of masking layers to prevent evaporation of a semiconductor component is absolutely essential. This applies for practically all III–V semiconductors. Above about 600 °C, arsenic evaporation becomes significant, thus altering the stoichiometry at the surface. Instead of annealing under a masking layer, the latter could be done under high arsenic pressure. From the viewpoint of implantation technology, however, this would be a step backwards.

SiO_2, either pyrolytically deposited at 350 to 450 °C or sputtered, represents the simplest type of masking layer; a great many experiments with it have been carried out [67], [312], [315]. SiO_2 masks excellently against out-diffusion of arsenic; however, it has a high diffusion coefficient for gallium, so that gallium vacancies can form through out-diffusion. In the case of *p*-type doping, this could ease incorporation of doping elements on to gallium sites, but the existence of gallium vacancies is clearly disadvantageous for the incorporation of impurity atoms on to arsenic sites, especially considering the possibility of a reaction between gallium vacancies and doping atoms. Another possibility is offered by Si_3N_4; this too is widely used in silicon technology [67], [312], [315], [787]. Ion sputtering is best suited as the deposition method, since pyrolytic layers must be deposited at temperatures above 700 °C at which GaAs can already decompose. Additional masking layers which are either possible or have already been experimented with are AlN [764], Al_2O_3 [363] and SiO_2–Al_2O_3 mixed layers, silicon [502], and gallium oxide [327], [645], as well as metals [646], [647] which do not alloy with GaAs at the annealing temperatures used. In summary, of importance for the successful doping of GaAs and other III–V semiconductors are the investigation and selection of suitable masking layers. A nearly ideal method for examining the passivating effect of masking layers is the back-scattering technique.

The partly inadequate passivation effect of SiO_2 upon GaAs was first determined by Gyulai *et al.* [312] through back-scattering measurements. A back-scattering spectrum showing out-diffusion of gallium in SiO_2, following an annealing step at 850 °C, has already been given in Figure 3.39. None the less, adequate electrical characteristics of the implanted layers can be obtained in the case of *p*-type doping. The main reason for this (aside from the sometimes eased incorporation of doping atoms on to gallium vacancies) is that the atoms, during the necessary annealing, always diffuse deeply into the crystal (compared to the relatively low implantation depth), while the surface properties of the crystal play no important role.

Woodcock *et al.* [787] have investigated the masking effect (e.g., against out-diffusion of the implanted components) and the passivating effect of sputtered SiO_2 and Si_3N_4, as well as of pyrolytically (at 750 °C) deposited Si_3N_4 layers, by means of photoluminescence measurements. After annealing in a vacuum at 600 °C, the SiO_2-masked layers show a complete degradation of the band-edge emission, while no change occurs below the nitride layers. For the pyrolytic layers, it was important to bring the samples up to these temperatures very rapidly, due to the high deposition temperature.

Sato [619] has observed formation of a *n*-type layer during annealing (750 °C) of semi-insulating GaAs masked with pyrolytic SiO_2. Formation of this layer could be prevented by the use of Si_3N_4. However, this effect can also occur in the case of Si_3N_4 masked layers; this is possibly due to material changes during annealing.

Bell *et al.* [67] have determined that gallium oxide forms after implantation of 10^{14} cm^{-2} tellurium under a SiO_2 masking layer. During electrical measurements of *p*-type layers, no significant differences could be found between SiO_2 masking and Si_3N_4 masking [796]. However, Eisen *et al.* [217], [315] measured a higher activation for tellurium-implanted layers beneath Si_3N_4 masking layers. For other ion species, the results in this connection allow no clear conclusions. For sulphur-implanted selenium-implanted layers, practically the same activity could be measured under SiO_2 and Si_3N_4 masking layers [796] after annealing at 850 °C for 30 min.

A reason for the partly contradictory results could be the low annealing temperatures of 700 to 750 °C which were apparently used to obtain masking-layer adhesion; these temperatures are not adequate to obtain complete electrical activation. In the following sections, based on several special cases, additional reference will be made to masking-layer problems.

A very interesting possibility for masking has been investigated by Davies *et al.* [172], [180]. Using gallium-saturated and arsenic-saturated SiO_2 layers, it is possible to influence the out-diffusion of gallium from the substrate. In a comparison with pure SiO_2 and Si_3N_4 layers, they determined various achievable activation levels depending on the dopant-atom incorporation into the lattice. In the case of sulphur, for example, Ga–SiO_2 is the best suited masking layer; for silicon, though, pure SiO_2 as the masking layer is best. This is because in the first case an incorporation on to As sites is promoted by the avoidance of Ga vacancies, while in the second case an incorporation on to Ga sites is promoted by the occurrence of Ga vacancies.

n-Type Doping

n-type doping in the melt is problem free and represents the current state of the art; doping by diffusion, however, is quite difficult due to the low diffusion coefficient. For several years, therefore, attempts have been made to achieve *n*-type doping of GaAs through the use of ion implantation. The main goal is the production of contacts for Gunn elements and Schottky field-effect transistors, i.e., a high electrical activation of the implanted layers is necessary. Hunsperger [357] has studied the influence of the substrate material on the maximum electrical activation obtainable in semi-insulating GaAs. For sulphur, he determined values between about 7 per cent and 46 per cent using constant implantation parameters (energy: 30 keV or 150 keV; doses: 5×10^{12} cm^{-2} or 10^{13} cm^{-2};

Table 6.6 Influence of semi-insulating GaAs on the efficiency of sulphur [357]

Sample	$N_{S,eff}$ (cm^{-2})	$\mu_{S,eff}$ (cm^2/Vs)	Doping efficiency (%)
A_1 surface	3.0×10^{12}	2910	20
A_1 back surface	2.1×10^{12}	2660	14
A_2 surface	5.4×10^{12}	2620	36
A_2 back surface	3.6×10^{12}	2920	24
B_1	4.1×10^{12}	2840	27
C_1	4.4×10^{12}	3170	29
C_2 surface	5.0×10^{12}	3620	33
C_2 back surface	4.4×10^{12}	3410	29
C_3 surface	4.4×10^{12}	3360	29
C_3 back surface	3.8×10^{12}	3350	25
D_1 surface	1.0×10^{12}	2540	6.7
D_1 back surface	1.2×10^{12}	2410	8

Dose and energy: 5×10^{12} cm^{-2} (30 keV) and 10^{13} cm^{-2} (150 keV); annealed for 20 min at 800 °C.

annealing at 800 °C for 20 min). His results appear in Table 6.6, and show clearly that at the present development stage in crystal growing, measurements in GaAs can only be compared to one another if they have been made using the same material.

In order to obtain a high electrical activation, it seems advantageous in the case of *n*-type doping to implant at relatively high substrate temperatures. Harris *et al.* [316] have found that the electrical activation is lower if an amorphous layer is produced during implantation. Good results for tellurium were obtained after implantation at 150 °C [315]; for sulphur implantation, the temperature was between 150 and 450 °C [172]. Back-scattering measurements by Gamo *et al.* [277] for tellurium-implanted and cadmium-implanted layers indicate an optimum incorporation on to lattice sites after implantation between 200 and 300 °C. These measurements, depending on crystal orientation, are shown in Figure 6.69. Further studies on the lattice-site localization of tellurium have been carried out, among others, by Takai *et al.* [714].

Very detailed studies on the *n*-doping of GaAs have been made by Woodcock *et al.* [787] and Davies *et al.* [172], [180] for selenium, tin, silicon and sulphur, as well as by Eisen *et al.* [217], [315] for tellurium. For passivation, Davies *et al.* used undoped, gallium doped, and arsenic doped SiO_2 layers, as well as AlN and Si_3N_4; they implanted between room temperature and 650 °C. An example of their results, concerning the dependence of the sheet carrier concentration on the implantation temperature in the case of sulphur, can be seen in Figure 6.70 [172]. At temperatures between 200 and 500 °C, $N_{S,eff}$ is 2.5 times higher than with unheated samples. The decrease at higher temperatures is probably due to gallium out-diffusion.

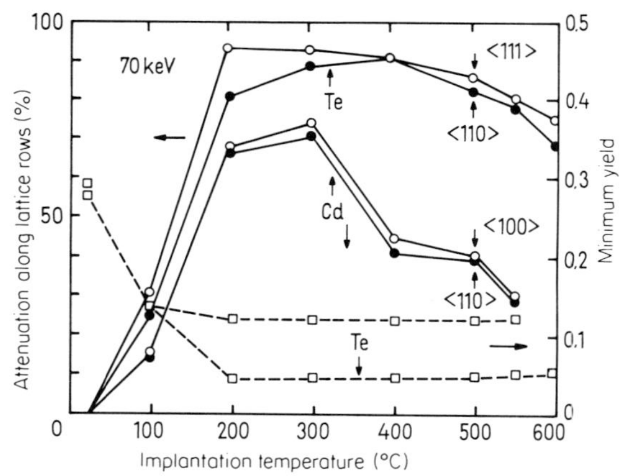

Figure 6.69 Lattice incorporation of implanted cadmium and tellurium in GaAs depending on implantation temperature and crystal orientation [277], [715]

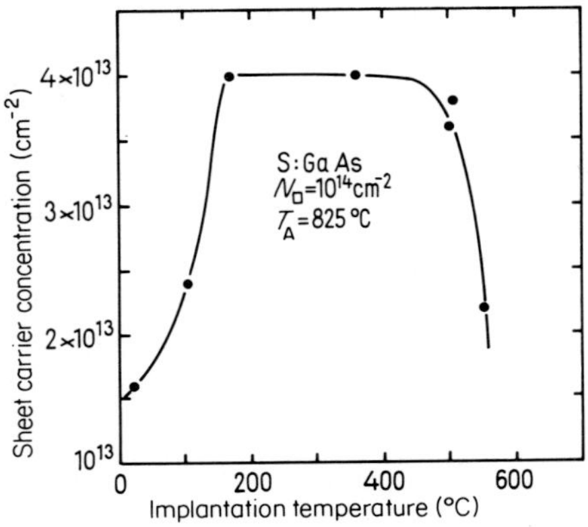

Figure 6.70 Influence of implantation temperature on electrical activation of implanted sulphur in GaAs [172]. Reproduced by permission of Pergamon Press

The dependence of $N_{S,eff}$ on the annealing temperature under a Ga–SiO_2 masking layer, for a sample implanted with sulphur at 500 °C (the temperature is somewhat higher than optimal), is shown in Figure 6.71. The highest activation, 40 per cent, was reached under a Ga–SiO_2 masking layer for a dose of 10^{14} cm^{-2} at 825 °C. For Si_3N_4, the activation is 25 per cent; for pure SiO_2, it is slightly

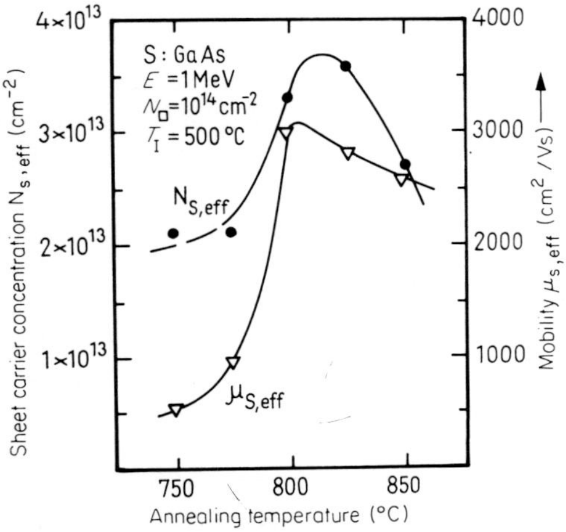

Figure 6.71 Dependence of sheet carrier concentration and mobility on annealing temperature for sulphur-implanted GaAs masked during annealing with Ga-doped SiO_2 [180], reprinted by permission of the publisher, The Electrochemical Society, Inc.

Table 6.7 Effects of implantation temperature and masking layer on the activity of S, Se and Si in GaAs [180]

Ion	T_I (°C)	T_A (°C)	Masking layer	$\mu_{S,eff}$ (cm^2/Vs)	$N_{S,eff}$ (cm^{-2})	$N_{S,eff}$ (hot impl.) / $N_{S,eff}$ (RT impl.)
S	360	825	Ga–SiO_2	3100	4.0×10^{12}	2.5
	20	825	Ga–SiO_2	3000	1.6×10^{12}	
Se	280	825	Ga–SiO_2	2050	2.9×10^{12}	3.2
	20	825	Ga–SiO_2	1650	0.9×10^{12}	
Si	235	825	As–SiO_2	2750	2.2×10^{12}	1.2
	20	825	As–SiO_2	2700	1.9×10^{12}	
Si	295	810	SiO_2	3000	2.3×10^{12}	1.3
	20	810	SiO_2	3350	1.8×10^{12}	

lower; for AlN, it is under 10 per cent. Apparently, gallium in SiO_2 effectively prevents gallium out-diffusion from the substrate. Similar results were found for selenium-implanted samples, but in this case the mobility is lower, and the optimum annealing temperature is 850 °C for a dose of 10^{14} cm^{-2}. In the case of silicon-implanted samples (dose: 10^{14} cm^{-2}; implantation temperatures between 250 and 300 °C), the results are exactly the opposite. Ga–SiO_2 masking layers only result in a maximum activation of 6 to 8 per cent after annealing at 800 °C; for pure SiO_2, though, the maximum activation is up to 24 per cent, while AlN and Si_3N_4 are in between. In this case, a certain amount of gallium out-diffusion apparently enhances incorporation of silicon into the crystal lattice. The sheet carrier concentration increases proportionally to the annealing temperature, up to the point of masking-layer destruction (about 900 °C). Optimum results for implantation at room temperature as well as higher temperatures, under optimum annealing conditions, are compared in Table 6.7. For sulphur and selenium, a clear increase in $N_{S,eff}$ is seen for high temperature implantation, while for silicon this effect hardly occurs at all.

Woodcock *et al.* [787] have also compared implantations at room temperature and higher temperatures. Annealing was done at 700 °C under a Si_3N_4 masking layer; at this temperature the layers are not yet activated optimally. In all cases (selenium, tin, sulphur, silicon), profile measurements showed considerable deviations from the LSS theory. Examples can be seen in Figures 6.72 to 6.74. At this relatively low annealing temperature, better results by implanting into heated samples were only obtained with selenium and tin. Implantations at increased temperatures and high doses lead to deeper distributions. For tin, a maximum concentration of 10^{18} cm^{-3} was obtained after implantation of 2.5×10^{15} cm^{-2} at 200 °C.

In the case of tellurium, for doses of 10^{14} to 10^{15} cm^{-2}, the obtainable values of $N_{S,eff}$ amount to approximately 3×10^{13} cm^{-2}. For this purpose, an annealing temperature of 900 °C under Si_3N_4 masking is required. The adhesion of the layers is a great problem in this case. At lower doses, higher activation values can

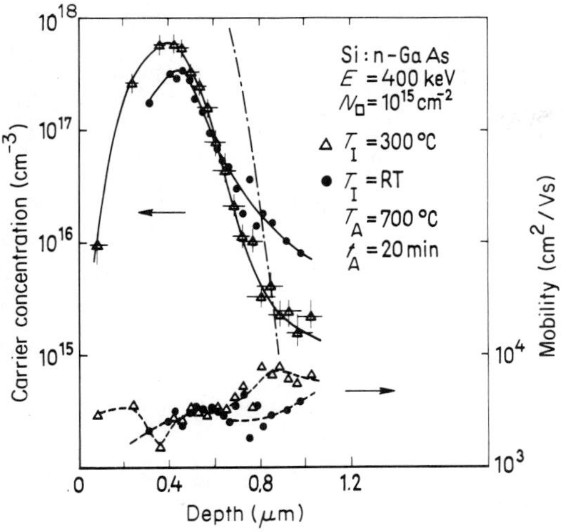

Figure 6.72 Doping and mobility profiles for silicon-implanted GaAs layers using 400 keV and doses of 10^{13}, 10^{14} and 10^{15} cm^{-2}. The theoretical distribution is also shown [787]. Annealing at 700 °C for 20 min. Reproduced by permission of Pergamon Press

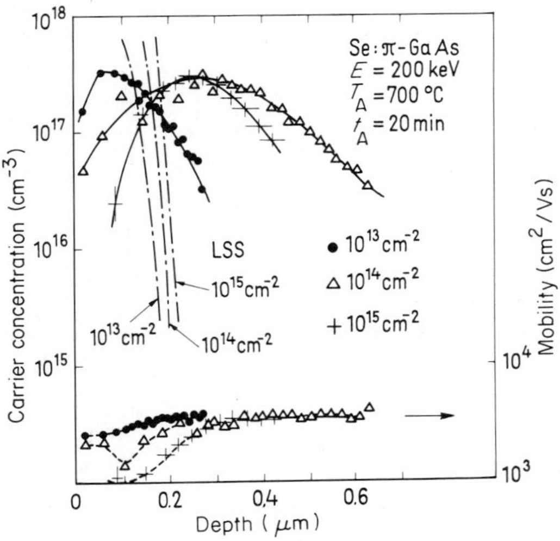

Figure 6.73 Doping and mobility profiles of selenium-implanted GaAs layers using 200 keV and doses between 10^{13} and 10^{15} cm^{-2}. The theoretical LSS-distributions are also shown. Annealing at 700 °C for 20 min. [787]. Reproduced by permission of Pergamon Press

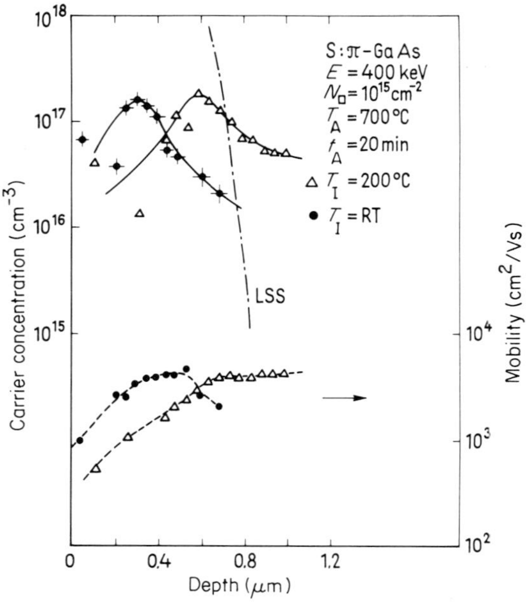

Figure 6.74 Doping and mobility profiles of sulphur-implanted GaAs layers using 400 keV (10^{15} cm^{-2}) at room temperature and 200 °C [787]. Annealing at 700 °C for 20 min. Reproduced by permission of Pergamon Press

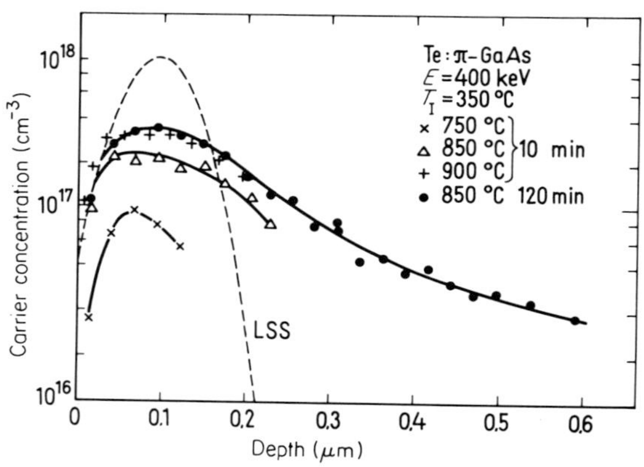

Figure 6.75 Tellurium profiles in GaAs for various annealing times and temperatures. $E = 400$ keV; $N_{\square} = 10^{13}$ cm^{-2} [520]

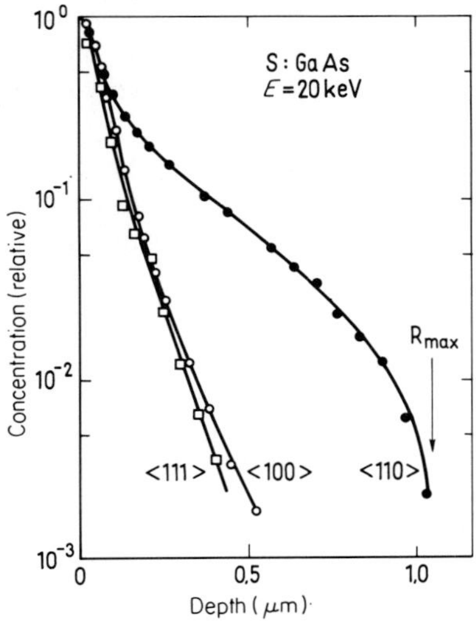

Figure 6.76 Influence of crystal orientation on channelling profiles of 20 keV sulphur in GaAs [769]. Reproduced by permission of Gordon and Breach

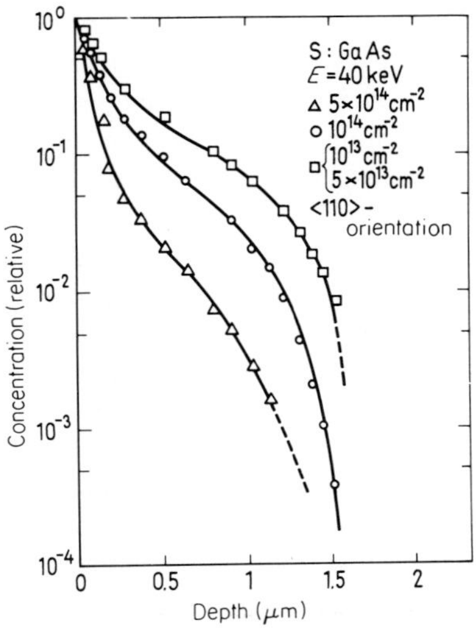

Figure 6.77 Influence of dose on channelling of 40 keV sulphur in ⟨110⟩-oriented GaAs [769]. Reproduced by permission of Gordon and Breach

be obtained. For a dose of 10^{13} cm^{-2} with annealing at 850 °C for 120 min (masking layer: Si_3N_4), Müller *et al.* [520] measured an activation of approximately 75 per cent. Several profiles with very deep profile tails can be seen in Figure 6.75.

Several studies have also been made on the channelling effect in GaAs. Whitton and Bellavance [769] have measured profiles in GaAs using radiotracer implantations of the isotope ^{35}S. Figure 6.76 shows the dependence of the range on the crystal orientation. As in the case of silicon, the probability of channelling is highest in the ⟨110⟩ direction. The dependence of the channelling effect on the ion dose is shown in Figure 6.77. No dependence can be seen in the dose range up to 5×10^{13} cm^{-2}, but the channelling effect decreases greatly at higher doses, since the scattering of the ions is more pronounced due to the increase in radiation damage.

p-Type Doping

Using thermal diffusion, GaAs is always doped with zinc in an ampoule under high arsenic pressure. The first implantation experiments for *p*-type doping were made with zinc and cadmium [359], [360], [363], [377], [486], [632]. All experiments were done using SiO_2 masking layers, unless explicitly stated otherwise. In order to prevent formation of amorphous layers, most early experiments were made at 400 °C [359], [363], [486]. *p*-type layers were formed, but insulating layers were simultaneously formed too, extending up to 160 μm into the semiconductor bulk depending on substrate doping and annealing temperature, with junction depths of a few micrometres [377]. Itoh *et al.* [377] have demonstrated that this effect is due to the influence of arsenic vacancies which form during implantation and diffuse into the semiconductor during annealing. The formation of these semi-insulating layers can be prevented by an additional implantation of arsenic. Hunsperger and Marsh [360] have investigated the dependence of this effect on substrate concentration and implantation temperature. They found that the width of these semi-insulating layers depends on the reciprocal value of the substrate doping. At low substrate doping levels the width is increased by annealing; at higher levels it is reduced. After room temperature implantation, it is significantly smaller (<0.2 μm, substrate doping $\approx 2 \times 10^{18}$ cm^{-3}) and sometimes non-existent; after annealing at 800 °C it disappears entirely. More recent measurements indicate that this effect probably depends greatly on the substrate material used and that it does not occur in the case of good material, at least not after room temperature implantation. In any case, the conclusion can be drawn from these measurements that it is best to implant *p*-doping elements at room temperature.

Figure 6.78 shows a comparison of annealing behaviour for various zinc implantations, in all cases masked with SiO_2 during annealing [792], [796]. These measurements show clearly how much caution is called for when comparing implantation experiments. In the study by Yuba *et al.* [792] for a dose of 10^{16} cm^{-2}, a maximum surface carrier concentration of approximately 10^{15} cm^{-2}

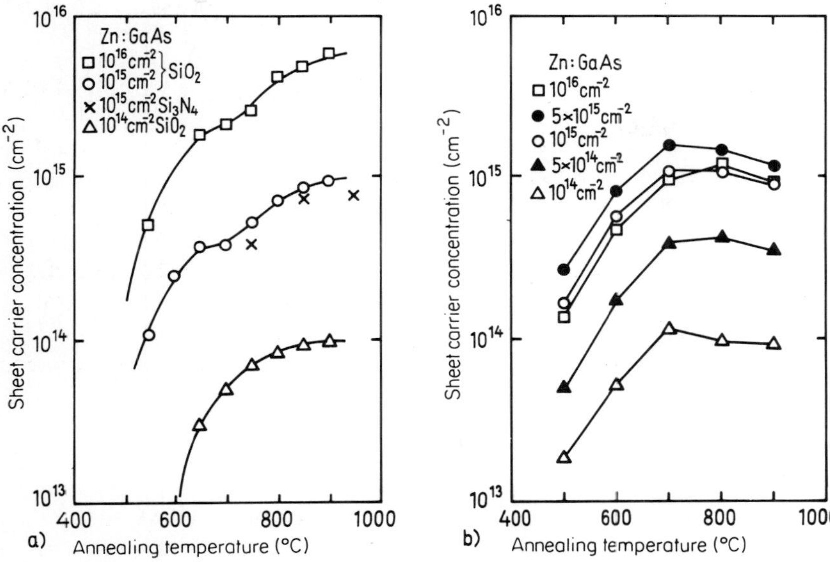

Figure 6.78 Comparison of isochronal annealing behaviour for implanted zinc in GaAs at various doses under an SiO_2 masking layer, according to Zölch [796] ($t_A = 30$ min) (a), as well as Yuba *et al.* [792] ($t_A = 30$ min) (b). Also shown is a measurement under an Si_3N_4 masking layer [796]

was found, but for the same dose, Zölch [796] measured 2×10^{15} cm^{-2}. The differences in the curve shapes, with a steady increase on the one hand [796] and a carrier-concentration maximum between 700 and 800 °C on the other [792], are also quite surprising and cannot be explained at present. Zinc has an electrical activation of approximately 100 per cent up to implantation doses of 10^{15} cm^{-2}, and maximum doping concentrations are reached which correspond to the chemical solubility (about 1.7×10^{20} cm^{-3} at 800 °C [130]). Figure 6.79 shows several doping profiles, depending on the annealing temperature, for an implantation at 150 keV with a dose of 10^{15} cm^{-3}. At low temperatures, activation is still incomplete; diffusion starts from 750 °C. This diffusion, however, is significantly lower than in the case of an ampoule diffusion at a corresponding temperature. This is probably due to different diffusion mechanisms (substitutional diffusion [796] in the case of implanted zinc). The mobility only reaches values which correspond to the doping after annealing above 800 °C.

Cadmium only shows a complete electrical activation up to doses of 10^{14} cm^{-2}, and requires very high annealing temperatures. Measurements by Zölch for doses between 10^{14} cm^{-2} and 10^{16} cm^{-2} are compared in Figure 6.80 [796]. Especially interesting is the low mobility, which indicates an inadequate annealing of defects. This finding has also been corroborated by back-scattering measurements. In further agreement, studies on the electroluminescence of cadmium implanted diodes have shown a very low luminescence efficiency [612].

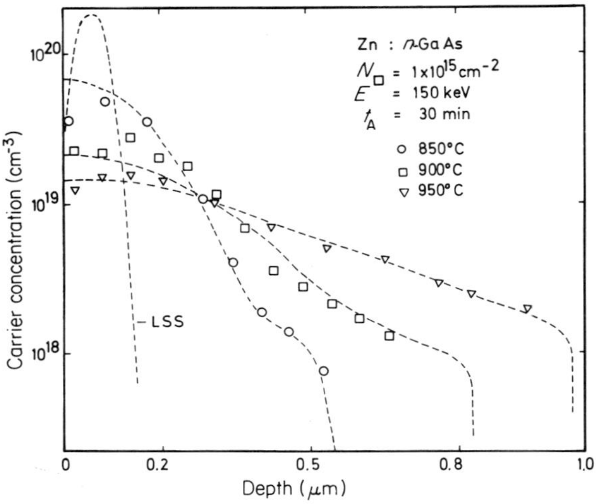

Figure 6.79 Doping profiles of zinc-implanted GaAs after annealing at 850 to 950 °C; energy: 150 keV; dose: 10^{15} cm^{-2} [796]

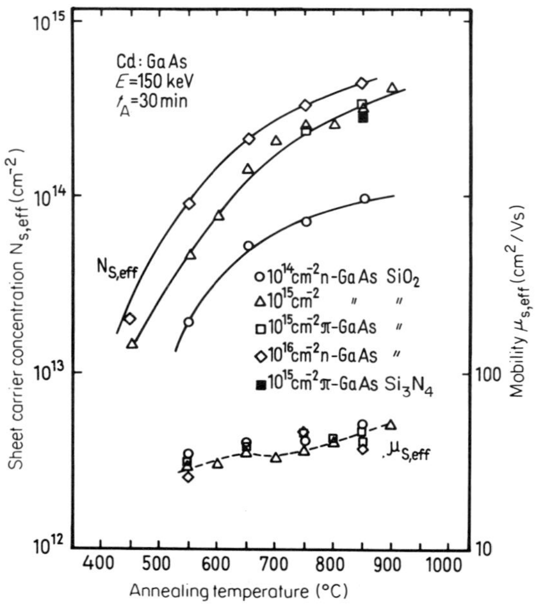

Figure 6.80 Electrical activation of cadmium-implanted GaAs [796]. Energy: 150 keV; dose: 10^{14} to 10^{16} cm^{-2}; annealed with Si_3N_4 or SiO_2 masking layer

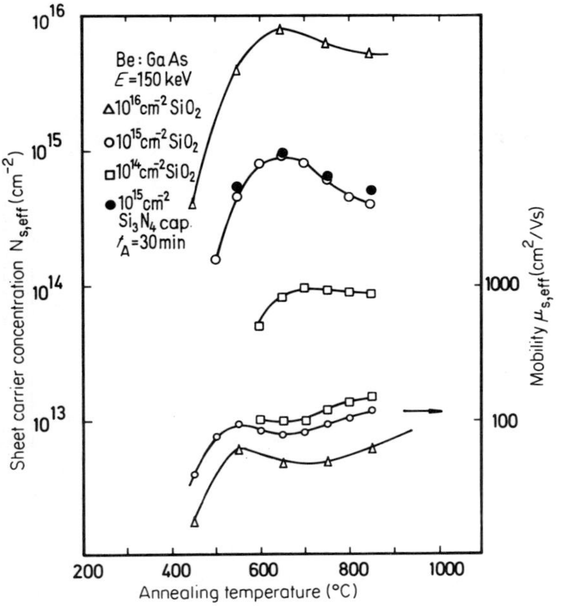

Figure 6.81 Electrical activation of beryllium-implanted GaAs [796]. Energy: 150 keV; dose: 10^{14} to 10^{16} cm^{-2}; annealed with Si_3N_4 or SiO_2 masking layer

Beryllium and magnesium have been studied as additional *p*-type doping elements [796]. Beryllium has about 100 per cent electrical activation for doses between 10^{14} cm^{-2} and 5×10^{15} cm^{-2} after annealing at 650 °C for 15 to 30 min (Figure 6.81). At higher annealing temperatures, this value decreases again, regardless of whether SiO_2 or Si_3N_4 is used as the masking layer. For low doses, temperatures of 700 °C are required to obtain 100 per cent electrical activation of the implanted ions. A maximum activation after annealing at temperatures as low as 650 °C is obtained for beryllium at high doses as well. This is perhaps explainable by the low damage concentration due to the low mass of beryllium in comparison to GaAs. Unexplained, however, are the characteristics for low doses, as well as the decrease in the carrier concentration for high annealing temperatures and high doses. The results under SiO_2 versus Si_3N_4 masking layers are about the same, with a tendency to slightly higher activation for Si_3N_4. Hunsperger *et al.* [364] have also obtained comparable results. Doping profile measurements too do not allow any clear interpretation. Doping and mobility profiles are shown in Figure 6.82. Implantation was done at 150 keV with a dose of 10^{15} cm^{-2}; the annealing temperature was varied between 600 and 850 °C with a constant annealing time of 30 min. The profile has an approximately Gaussian shape after annealing at 700 °C; it then becomes wider with further annealing, during which

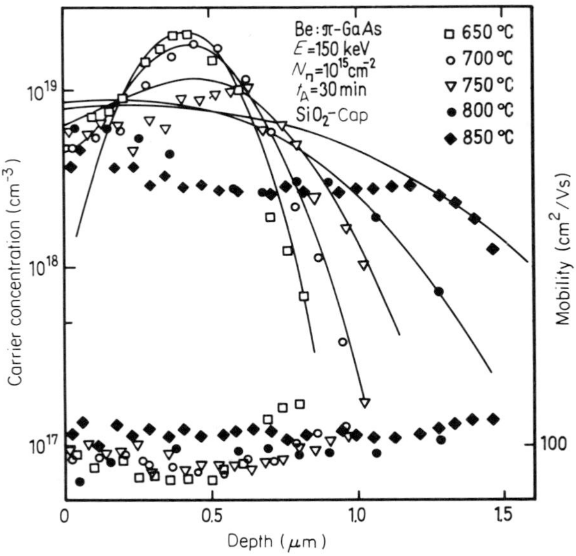

Figure 6.82 Doping profiles and mobility of beryllium-implanted GaAs after annealing at 650 to 850 °C; energy: 150 keV; dose: 10^{15} cm^{-2} [796]

beryllium is lost either due to out-diffusion or to complex formation. A mobility corresponding to the doping is already obtained after annealing at 600 °C. Chatterjee *et al.* [132] have investigated beryllium implanted layers by photoluminescence measurement under various annealing conditions. The intensity was better for Si_3N_4 masking than for SiO_2 or for annealing at high arsenic pressures.

The annealing characteristics of magnesium are different from those of beryllium [364], [796]. After annealing at 600 °C, only between 0.4 per cent (dose: 10^{15} cm^{-2}) and 5 per cent (dose: 5×10^{13} cm^{-2}) of the implanted ions are electrically active. At doses $<5 \times 10^{13}$ cm^{-2}, 100 per cent activation is only obtained after annealing at 900 °C or above [364]; at higher doses, a maximum occurs after annealing between 650 and 750 °C, with a subsequent decrease in the activation. The maximum obtainable surface carrier concentration is approximately 2×10^{14} cm^{-2}. The luminescence efficiency of magnesium implanted diodes is considerably worse than that of zinc or beryllium implanted diodes [612]. A comparison of the maximum achievable sheet carrier concentrations for various doping elements as a function of the dose, at an implantation energy of 150 keV is shown in Figure 6.83.

In an interesting study, Kachurin *et al.* [280], [392] have investigated the influence of vacancies on the diffusion of doping atoms. For cadmium and zinc, they were able to reduce diffusion during annealing very greatly through additional implantation of arsenic. They explain this in terms of a trapping of the

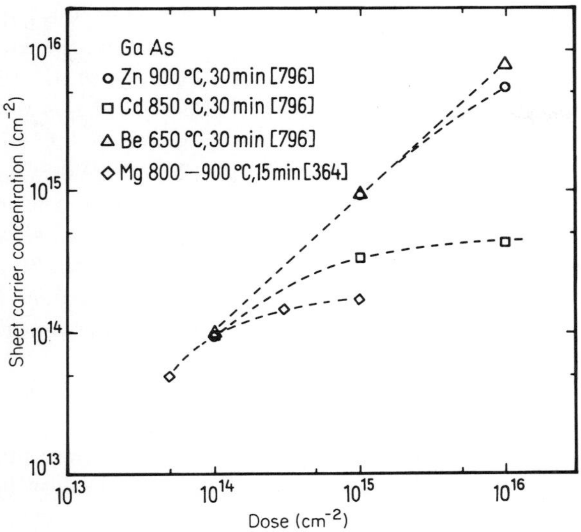

Figure 6.83 Comparison of maximum obtainable sheet carrier concentrations for Zn, Cd, Be and Mg in GaAs [364], [796]

zinc or cadmium atoms at the excessive Ga vacancies produced by the arsenic implantation.

Mixed Crystals

By implanting elements belonging to group III or group V of the periodic system (e.g., Al, In, P, Sb) into GaAs, mixed crystals can be produced. Because of the very high doses necessary, relatively few experiments have been carried out [68], [358], [512], [690], [696], [762]. Using implantations of aluminium and zinc, Hunsperger and Marsh [358] have fabricated diodes emitting light in the forward direction with a maximum at 1.48 eV. The dose used was approximately 2×10^{16} cm^{-2} at 30 keV, at an implantation temperature of 400 °C. The intensity of the visible radiation was relatively low, in spite of the annealing temperature of 900 °C. Corresponding experiments with phosphorus [762] have also shown a low intensity of the short-wave radiation. Skolnik *et al.* [690] as well as Spitzer [696], using infrared absorption measurements, have shown for aluminium and phosphorus implantations with doses between 2×10^{16} cm^{-2} and 3.7×10^{16} cm^{-2} at 10 keV that, after annealing at 900 °C for one hour, practically all implanted phosphorus and aluminium ions are incorporated on to gallium and arsenic sites, respectively. However, a high luminescence efficiency was not attained, apparently due to unannealed radiation damage. Monteith *et al.* [512] have demonstrated by photosensitivity measurements that, for GaAs diodes implanted with indium and zinc (indium dose: 10^{17} cm^{-2}, energy: 60 keV; zinc dose: 10^{14} cm^{-2}, energy: 60 keV), up to 40 per cent of the indium ions are actively incorporated into the GaAs. An article by Belyi *et al.* [68], which also cites some additional Soviet conclusions, reports on experiments in the production of $GaAs_{1-x}P_x$ and $Al_xGa_{1-x}As$ through implantation of aluminium and phosphorus ions (at room temperature or 420 to 500 °C), respectively. Incorporation of the ions was verified by photoluminescence measurements. After annealing at 400 °C following room temperature implantation, as well as directly after high temperature implantation, an incorporation of about 30 per cent of the ions on to active lattice sites was measured. The discrepancy between these results and those of Hunsperger and Marsh [358] has not yet been explained.

6.3.2 Other III–V Semiconductors

GaP, as well as GaAs–GaP mixed crystals, are very important materials for the fabrication of luminescence diodes emitting red, green, and yellow light. However, relatively few studies have been made on the doping of these semiconductors by ion implantation [131], [774], while considerably more research has dealt with back-scattering measurements in GaP [122], [569], [769]. The reason for this low interest is probably the fact that zinc diffusion is relatively well understood. In addition to the implantation of *p–n* junctions [368], [669], [774], the implantation of nitrogen into GaP and $GaAs_{1-x}P_x$ to increase the luminescence efficiency has also been investigated [298], [490]. The exact process control and precise doping provided by ion implantation would offer advantages in both of the above applications, for example, for optimization of the doped layer in regard to emission and absorption.

Wiemer *et al.* [774] were the first to demonstrate that zinc implanted $GaAs_{1-x}P_x$ diodes ($x = 0.4$) can attain luminescence properties equal to those of diffused diodes. They used very high doses (10^{16} cm^{-2}) and high annealing temperatures (920 °C for 30 min). A gettering step before implantation, as well as an avoidance of excessive annealing times, were found to be important. Itoh and Oana [378], [381] have measured the annealing characteristics and doping profiles of zinc implanted $GaAs_{0.62}P_{0.38}$ layers. Implantation was made at 20 keV between room temperature and 450 °C. The temperature dependence and dose dependence of the sheet resistivity can be seen in Figure 6.84. A saturation effect can only be seen above doses of 5×10^{15} cm^{-2}. The effective carrier concentration, $N_{S,eff}$, of the zinc implanted layers is abnormal in behaviour for room temperature implantation. At low doses, the activity of the zinc was approximately 60 per cent (annealing at 800 °C for 40 min), as opposed to approximately 150 per cent for a dose of 2×10^{15} cm^{-2} and 300 per cent for 5×10^{15} cm^{-2}. At higher implantation temperatures, this effect was reduced [378]. Correlated with this were a decrease in the junction depth and an increase in the luminescence efficiency. The cause of these effects could not be determined. Ono *et al.* [551], [669] have studied the behaviour of zinc implanted layers under defined conditions. All experiments were made with junction depths of about

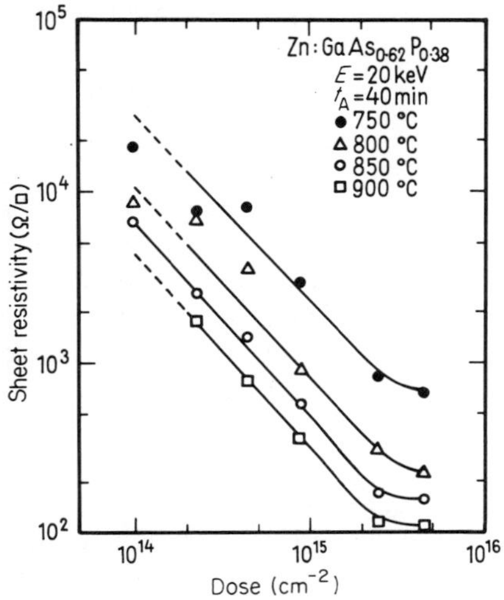

Figure 6.84 Sheet resistivity of zinc-implanted $GaAs_{0.62}P_{0.38}$ layers depending on dose, at various annealing temperatures [381]. Reproduced by permission of the American Institute of Physics

2 μm, which results in an optimum external quantum efficiency. Table 6.8 shows the exact junction depths, depending on dose and diffusion time. Implantation was done at 650 keV with doses between 2×10^{15} and 16×10^{15} cm^{-2}; annealing and diffusion were carried out in a quartz ampoule under a SiO_2 masking layer at 750 °C. In contrast to the diffused sample used for comparison, the implanted layers have a p^- layer at the *p–n* junction. The relative luminescence intensity as a function of dose and implantation temperature can be seen in Figure 6.85. The maximum occurs between about 4×10^{15} and 8×10^{15} cm^{-2}. The cause of the high efficiency is the relatively low zinc concentration at the *p–n* junction. The maximum light output of the diodes was 500 cd/m^2 at 8.9 A/cm^2.

Table 6.8 Depth of the *p–n* junction, depending on implantation dose and annealing temperature [551]

Dose ($\times 10^{15}$ cm^{-2})	Annealing time (h)	*p–n* junction (μm)	Width of p^--layer (μm)
2	28	1.6	1.0
4	18	2.0	1.1
8	15	1.3	1.1
16	8	2.4	0.8
Diffusion	0.5	2.4	0

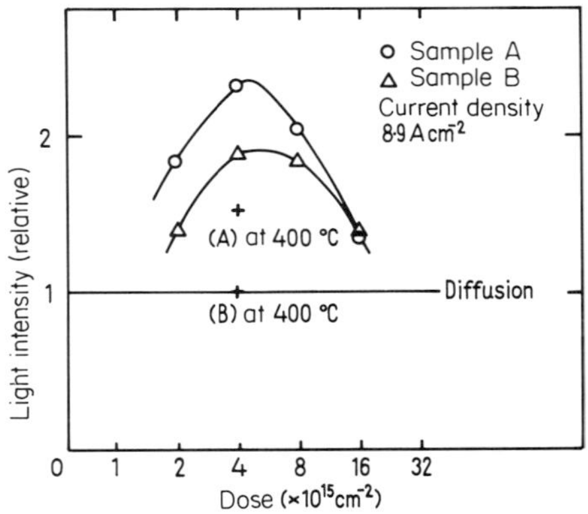

Figure 6.85 Relative luminescence intensity depending on ion dose, compared with diffused diodes. Implantation energy: 150 keV. Diodes implanted at 400 °C have a lower efficiency [551]

Chatterjee *et al.* [131] have obtained similarly good results with beryllium implanted diodes. Beryllium has a lower diffusion coefficient than zinc, so that higher annealing temperatures are possible without the *p–n* depth being raised above the optimum value of about 2 μm.

Inada and Ohnuki [368] have implanted magnesium and zinc into *n*-type GaP at 15 keV. Magnesium is an especially attractive element for *p*-type doping, due to its high solubility and its high green-to-red luminescence relationship [298]. Prior to annealing, the samples were passivated with SiO_2 in a nitrogen atmosphere. To obtain *p–n* junctions, annealing at 900 °C for 30 min was necessary. The *p–n* depth was then 12.2 μm for zinc and about 0.1 μm for magnesium. The luminescence characteristics of such diodes are comparable to those of diffused diodes.

Nitrogen implantation to increase the luminescence efficiency of the indirect semiconductors with the composition $GaAs_{1-x}P_x$ ($0.45 \leqslant x \leqslant 1$), through formation of isoelectronic traps, is an alternative to doping with NH_3 during crystal growth. It has been found that high temperature implantation causes an increase in luminescence efficiency [298], [490], [709]. A detailed study of this has been made by Gonda *et al.* [298]. They determined that for $x = 0.52$, the integral intensity is higher by a factor of up to 10^5 than in the unimplanted case (implantation temperature: 350 °C, dose: 10^{17} cm^{-2}, annealing temperature: 800 °C); i.e., implanted nitrogen atoms are incorporated on to phosphorus sites and form isoelectronic traps. Similar results have been found for $Al_xGa_{1-x}As$ [297], [460].

Using laser annealing, even better results than those for thermal annealing have been obtained. Takai and Ryssel [715a], [715b] have found that surface deterioration effects can be avoided by the use of CO_2-laser annealing. In Figure

6.86, a comparison between laser-annealed and thermally-annealed luminescence spectra shows the improvement achievable in the former case.

Not very many experiments have been carried out with the other III–V semiconductors. The greatest interest has been given to InSb [250], [448] and InAs [171], [448], [449]. Other studies have been made with GaN [474]. InSb and InAs are materials for infrared detectors and infrared emitters at wavelengths up to 5.3 μm and 3.1 μm, respectively. By implanting sulphur into InSb and InAs, as well as zinc into InSb, McNally [448] has produced *p–n* junctions with good reverse as well as forward characteristics. The range of zinc ions in InSb corresponds approximately to the LSS theory [449]. By proton bombardment of *p*-type InSb, it is possible to produce *n*-type layers [250] and obtain good *p–n* characteristics. This *n*-type doping is due to radiation-damage effects. Hurwitz and Donnelly [367] have produced excellent infrared detectors by implanting beryllium into InSb (details in Section 7.3.5). Beryllium seems to be very well suited for doping of III–V semiconductors through ion implantation, as already shown by the results with GaAs and $GaAs_{1-x}P_x$.

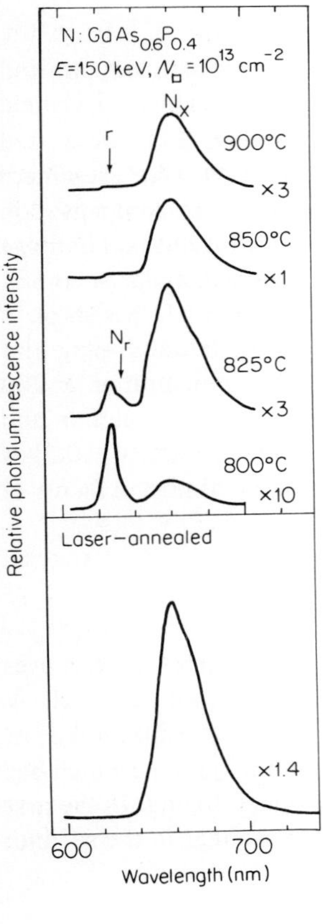

Figure 6.86 Comparison of 77 K photoluminescence spectra for samples annealed thermally at 800, 825, 850 and 900 °C for 1 h, and for samples annealed by a laser beam for 3 s with a power density of 133 Wcm^{-2}. The samples were implanted with 150 keV nitrogen to a dose of 10^{13} cm^{-2}. Multiplication factors indicate the values of the expanded scale

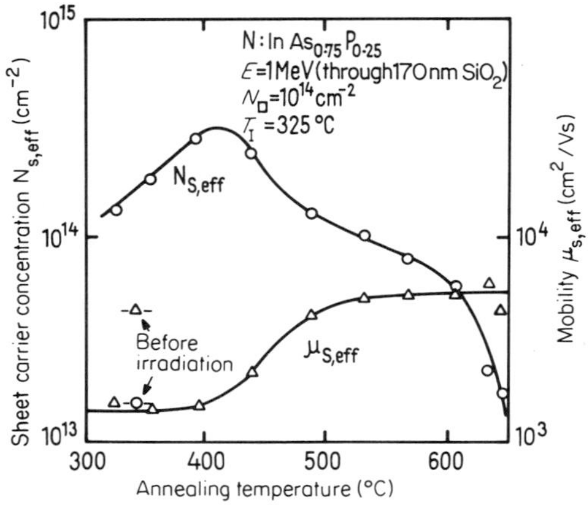

Figure 6.87 Annealing of radiation-damage-induced changes in carrier concentration and mobility, for nitrogen-implanted $InAs_{0.75}P_{0.25}$ [171]. Reproduced by permission of the Institute of Electrical Engineers

Davies *et al.* [171] have examined the doping of $InAs_xP_{1-x}$ with sulphur and silicon. It was found that the masking-layer problem is considerably less critical than in the case of GaAs. SiO_2, Si_3N_4, and gallium saturated SiO_2 were used with good results. Depending on the *x*-value (0.2, 0.5 and 0.75 were examined) and thus the band gap, the electrical properties are changed from compensation characteristics for a composition approaching InP ($x = 0.2$) to radiation-damage produced *n*-type conductivity for a narrow-band composition (similar to InSb). An example showing these changes can be seen in Figure 6.87, based on a nitrogen implantation (10^{14} cm^{-2} at 325 °C). A return to the initial doping and mobility values only occurs after annealing at 650 °C. After implantation at 200 to 250 °C and annealing temperatures between 675 and 725 °C, sulphur and silicon have an electrical activation of up to 80 per cent ($x = 0.75$ and $x = 0.5$) or 50 per cent ($x = 0.2$), respectively; these values are higher by almost a factor of 10 than the corresponding values for GaAs.

6.3.3 Insulation Through Ion Implantation

An entirely different application of ion implantation, in contrast to the uses already discussed, is the production of insulating layers in GaAs and other III–V semiconductors by irradiation with hydrogen [210], [238], [251], [698], [783] or oxygen [88], [239]. Hydrogen implantation produces radiation damage which has a compensating effect; oxygen implantation causes a chemical doping. In the first case, therefore, the compensation can be eliminated by annealing; in the second, the annealing is necessary to obtain the desired effect.

These methods can be used, for example, to fabricate guard-ring diodes [251], field-effect transistors [577] or lasers [88]. Using the capacitance-voltage method, Foyt *et al.* [251] have measured doping profiles in GaAs after implantation of 10^{13} cm^{-2} protons at various energies. An example for an energy of 100 keV is given in Figure 6.88. Without annealing, the sample has a carrier concentration below 10^{11} cm^{-3} up to the approximate range of the ions. The substrate had an *n*-type doping of 2×10^{16} cm^{-3}. In the vicinity of the average proton range, the carrier concentration increases within a small area above the substrate concentration. The results for *p*-type layers were similar. Dyment *et al.* [210] have made a detailed study of the electrical and optical properties of proton insulated *p*-type GaAs layers, in connection with stripe-structure lasers, using doses between 10^{13} cm^{-2} and 10^{17} cm^{-2} at a fixed energy of 300 keV. The substrates had doping levels between 2×10^{18} cm^{-3} and 1.4×10^{19} cm^{-3}. The absorption for light having an energy below the band gap is considerably higher. At 0.936 μm, for example, ≈89 per cent of the otherwise transmitted light is absorbed (dose: 10^{16} cm^{-2}). Depending on the dose, this increased absorption is reduced after annealing between 300 and 600 °C, without the occurrence of pronounced annealing steps. The average specific resistivity after proton irradiation of *p*-type GaAs at various doses is shown in Figure 6.89. The specific resistivity reaches a maximum of 2.5×10^5 Ωcm at a dose of 3×10^{15} cm^{-2}. The reason for this is not clear, although similar findings have been made for *n*-type GaAs [577] as well as *n*-type and *p*-type GaP [698]. An estimate of the carrier concentration for the same dose indicates a value below 10^{10} cm^{-3}. A comparison between optical and electrical measurements shows that the increased optical absorption is reduced at considerably lower temperatures than those related to carrier-compensation

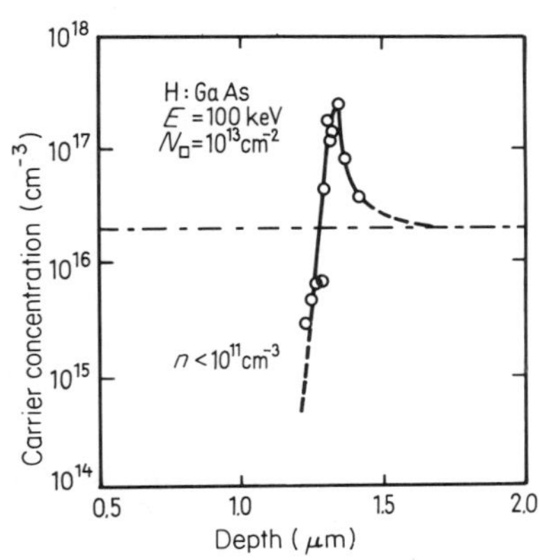

Figure 6.88 Carrier concentration in *n*-type GaAs ($n \approx 2 \times 10^{16}$ cm^{-3}) after proton irradiation with 100 keV and 10^{13} cm^{-2} [251]. Reproduced by permission of Pergamon Press

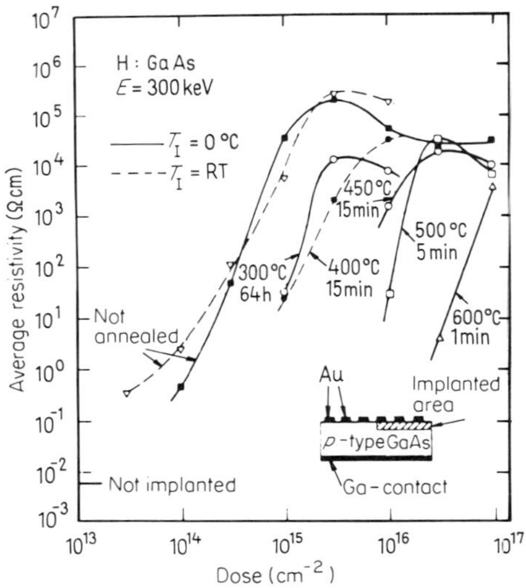

Figure 6.89 Average specific resistivity of proton-implanted GaAs layers depending on dose and annealing temperature after implantations at 0 °C and room temperature [210]. Reproduced by permission of the American Institute of Physics

reduction. It is thus possible to produce highly resistive material with low absorption if an annealing step at an appropriate temperature is applied.

Proton insulation is not only applicable with GaAs, but also with several other III–V semiconductors. Research on GaP [698] as well as $Ga_{1-x}Al_xAs$ [238] has been published. With InSb, in contrast to those compounds, *n*-type layers are produced through proton irradiation [250]. The specific resistivity for *p*-type and *n*-type GaP, after proton irradiation with 10^{11} to 2×10^{17} cm^{-2} at 300 keV, is shown in Figure 6.90 [698]. At a dose of about 4×10^{14} cm^{-2}, a maximum value of about 10^{14} Ωcm occurs, decreasing to about 10^{10} Ωcm at higher doses. The results for *p*-type and *n*-type substrates are practically identical. At annealing temperatures up to 400 °C, a sheet resistivity of at least 10^{10} Ωcm is retained. As in the case of GaAs, the thickness of the compensated layer is approximately 1 μm per 100 keV of implantation energy.

Especially important is the use of proton insulation for hetero-laser structures made from $Ga_{1-x}Al_xAs$. Favennec and Diguet [638] have shown for layers with $x = 0.35$ that the resistivity is saturated at a dose of 10^{13} cm^{-2}, and that it is up to 3500 times higher than the initial resistivity. In annealing experiments, the layers were stable up to 180 °C. In the case of compensation through oxygen implantation, a genuine doping process is involved; therefore, the semi-insulating layers are stable at higher temperatures than those produced by proton insulation. The

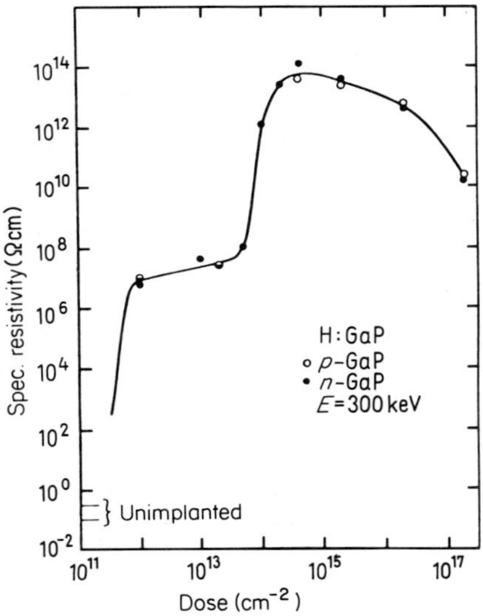

Figure 6.90 Specific resistivity as a function of proton dose for implantation into GaP at 300 keV [698]. Reproduced by permission of the American Institute of Physics

optical characteristics are greatly improved by annealing [88]. The main disadvantage of this method is the high acceleration energy needed to obtain an adequate depth of the compensated layer. Using protons in GaAs and GaP, approximately 1 μm per 100 keV can be made semi-insulating, while in the case of oxygen, an acceleration energy of 2.5 MeV is needed to obtain a range (R_p) of 2 μm in GaAs.

6.4 IMPLANTATION INTO II–VI AND IV–VI SEMICONDUCTORS

The most important semiconductors in this group are cadmium sulphide, cadmium telluride, zinc sulphide and zinc selenide. Various ternary compounds have also been the subject of great interest in recent years, since they offer the possibility of adjusting the band gap almost as desired by variation of the composition. Some examples include cadmium mercury telluride, lead tin telluride, and lead tin selenide, just to name a few. Depending on their band gaps, these semiconductors are suitable for a great variety of applications, for example, for light emitters (especially ZnS), solar cells (CdS) or infrared detectors (PbS, PbSnTe, HgCdTe, etc.).

Initial studies in this area were made using zinc oxide [666], [667] and zinc telluride [347]. ZnTe becomes semi-insulating, similarly to GaAs, through proton irradiation [196]. The depth of the insulated area is again about 1 μm per 100 keV. Through implantation of fluorine [347] and chlorine [467], *n*-type layers can be produced. Gettings and Stephens [283] have investigated semi-insulating CdTe layers implanted with argon, indium, tellurium and bismuth. The highest electrical activation (30 per cent) was obtained through an indium implantation with a dose of 10^{14} cm^{-2} at 200 °C and annealing at 400 °C. Bismuth resulted in a very low electrical activation; for tellurium and argon, there was none at all. The annealing characteristics of radiation damage were investigated through back-scattering measurements; a partial correlation with the electrical measurements was obtained. SiO_2 layers were found to be unsuitable for passivation during annealing. *p*-type conductivity was determined following implantation of phosphorus [393] or arsenic [194]. Due to its band gap of 2.4 eV, CdS is interesting as a material for luminescence diodes. It is normally *n*-type and its doping cannot be reversed by diffusion. Using implantation of phosphorus [41], [348], bismuth [135], [223] or nitrogen [668], however, a reversal of the conductivity type has been obtained. Shiraki *et al.* [668] have fabricated various types of diodes emitting green, yellow, orange, and red light when operated in the forward direction, using implantations between 50 keV and 200 keV with doses between 10^{15} cm^{-2} and 5×10^{16} cm^{-2}. All diodes were implanted at room temperature; annealing was done at 400 °C in atmospheres of helium or argon. An oxide deposited at 200 °C in a glow discharge was used as the masking layer.

During the last few years, small band-gap semiconductors have become important in the fabrication of infrared detectors. Since it is difficult or impossible to produce *p–n* junctions by other methods, ion implantation offers a promising alternative. Initial studies have dealt with *n*-type doping through proton irradiation of $Hg_{1-x}Cd_xTe$ ($x = 0.5$; 0.31; 0.25) [249] for the spectral range between 1.6 and 6 μm, as well as PbTe and $Pb_{1-x}Sn_xTe$ ($x = 0.12$) [193], [199] for ranges up to 5 or 8 μm, respectively. These studies were related to the fabrication of photo detectors. Accordingly, only optical measurements (detectivity) and in some cases *p–n* characteristics were reported on, and not the obtained doping concentrations, the junction depths or the mobilities. The doping effect through protons is due to radiation-damage effects.

For this group of semiconductors, there are two further doping methods: alteration of the stoichiometry and impurity atom doping. For both methods, ion implantation can be used with good results. Through implantation of antimony into *p*-type PbTe [197], [198], good-quality infrared detectors have been made. Up to now, $Hg_xCd_{1-x}Te$ ($x = 0.18$ to 0.30) is the only material for the 10-μm range whose doping could be changed by impurity atom doping with copper, silver, aluminium, zinc, phosphorus, arsenic and fluorine [466], [610a]. The doses used were between 10^{13} cm^{-2} and 10^{16} cm^{-2}.

A doping effect based on alteration of the stoichiometry through implantation of a semiconductor component has been obtained for $Hg_{1-x}Cd_xTe$ ($0.19 \leqslant x \leqslant 0.5$, corresponding to 2–14 μm at 77 K) using mercury implantation (dose: 10^{12} to 10^{13} cm^{-2}; energy: 30 keV) [241]. No annealing was needed to obtain *n*-type conductivity. $PbS_{1-x}Se_x$ (selenium implantation) and $Pb_{1-x}Sn_xTe$ ($x = 0.1$; tellurium implantation) have also been doped to obtain *p*-type conductivity [200], [399]. Annealing after implantation was necessary in both cases. In the case of $Pb_{1-x}Sn_xTe$, the junction depth was measured as approximately 1 μm after annealing at 450 °C for 15 min in a hydrogen atmosphere. At an implantation energy of 120 keV and a dose of 10^{15} cm^{-2}, the projected range is about 0.04 μm; i.e., the tellurium already diffuses considerably at this annealing temperature. Back-scattering measurements showed that after annealing at 450 °C, practically all radiation damage was annealed. For comparison, nitrogen was implanted into *p*-type material. As expected, *n*-type conductivity was first obtained due to radiation-damage effects, but after annealing at 400 °C this conductivity was no longer present.

6.5 SILICON CARBIDE

SiC is a very promising material for devices operating at high temperatures. In this case, though, diffusion requires temperatures above 2000 °C. The usual masking technologies (SiO_2, Si_3N_4) are therefore no longer usable for device structuring, which means that ion implantation would be a very well suited doping method here. Several experiments in this direction have already been made [33], [95], [117], [469], [480], [597]. The studies by Marsh and Dunlap [469] were the most comprehensive. They implanted arsenic, antimony, beryllium, boron, gallium, helium, indium, phosphorus, nitrogen, hydrogen, tellurium and bismuth; by the use of Hall effect or thermoprobe measurements, they detected formation of *n*-type layers in the cases of nitrogen, phosphorus, antimony and bismuth. In one case, implantation of aluminium led to *p*-type doping. For the *n*-type dopants the activation amounted to 30–50 per cent after annealing between 1100 and 1500 °C, but good mobility values (at a constant doping concentration) could only be reached beginning at 1600 to 1700 °C.

Addamiano *et al.* [33] have made a study on doping profiles and optical properties of boron-implanted, aluminium-implanted and nitrogen-implanted layers in hexagonal 6 H SiC. The profile measurements were made using SIMS. Annealing at 1400 °C eliminated radiation damage and re-established the original luminescence at 77 K. The crystalline structure remained intact, as the authors concluded from the occurrence of the typical boron luminescence after implantation and annealing at 1400 °C. Nitrogen implanted samples yielded highly *n*-type layers after annealing at only 1400 °C. Through aluminium implantation, it was possible to change the doping of *n*-type material after annealing at 1400 °C (dose: 5×10^{15} m^{-2}). At lower doses, only a reduction of the original *n*-type conductivity was obtained. As a measurement example, two boron profiles (implantation at 30 to 60 keV) are compared with the theoretical LSS distribution in Figure 6.91. The range is about 20 per cent higher than the theoretical LSS range. Profile tails of the distribution, probably due to channelling, can be seen clearly. After

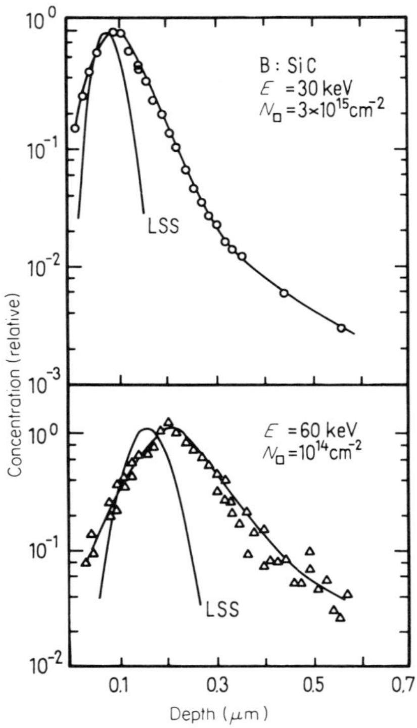

Figure 6.91 Boron profiles in silicon carbide. Energy: 30 keV or 60 keV; dose: 3×10^{15} cm^{-2} or 10^{14} cm^{-2}; no annealing after implantation [339]

annealing at 1400 °C the boron is no longer measurable, due to the considerable out-diffusion. The aluminium profiles showed a high concentration of aluminium at the surface after annealing at 1400 °C.

Campbell *et al.* [117] have investigated the radiation-damage distribution and lattice localization of nitrogen-implanted α-SiC at implantation temperatures between 20 and 450 °C, using the $^{15}N(p, \alpha)$ reaction. For higher implantation temperatures, a reverse annealing of the radiation damage occurred at 800 °C, i.e., a decrease in ions on lattice sites and an increase in the radiation-damage concentration. Further annealing up to 1485 °C reduced the radiation-damage concentration, but the amount of nitrogen on lattice sites was hardly increased. All in all, the radiation-damage concentration is considerably lower after implantation at higher temperatures, and the annealing presents fewer problems as well since no amorphous layers are formed. The incorporation on to lattice sites after annealing at 1485 °C amounted to about 39 per cent for room temperature implantation and about 47–51 per cent for implantation into heated substrates

In addition, several attempts have been made to produce SiC through carbon implantation into silicon [95], [597], [643]. After annealing at 800 °C, the presence of some individual SiC regions was found by infrared absorption measurement. By implanting at various energies chosen so that a region with a relatively constant carbon concentration was produced, Rothemund and Fritzsche [597] have shown that a layer having a different composition is formed. Measurements of electron-diffraction patterns have led to results similar to those obtained for mixtures of silicon and cubic graphite. It has not yet been possible to determine whether or not a special modification of SiC is the reason for this.

7
Devices

In this chapter, no attempt will be made to provide a complete survey of all previously-implanted devices; rather, with the use of several typical examples, the advantages offered by implantation will be demonstrated and new applications suggested. From case to case, the various individual characteristics of implantation play more or less important roles. The most important advantages, from the viewpoint of device characteristics, are the following:

1. Doping uniformity, i.e., homogeneity of an implanted sample, and reproducibility from sample to sample (see Sections 7.1, 7.2, 7.3, 7.4).
2. Controlled introduction of small doping concentrations, by exact measurement of the ion current (see Sections 7.1 and 7.2).
3. Minimal lateral spread of the implanted profile (see Section 7.1).
4. Low processing temperatures (see Section 7.1).
5. Introduction of elements which are not soluble in the solid, or which can only be introduced with difficulty (e.g., under high pressure; see Section 7.4).
6. Variation of the doping profile through multiple implantation, or the attainment of profiles with the concentration maximum inside the solid (see Sections 7.3 and 7.4).

In many cases, several of these points are of importance. Especially the factor of implantation uniformity is becoming more and more important as device characteristics are improved and the packing density is increased. The uniformity can also be decisive for the replacement of a diffusion step through ion implantation, even without any further implantation related advantages.

Further technical processing advantages, which are not necessarily related to device characteristics but which can enable a more efficient production, are as follows:

(a) Simple masking by the use of photoresist or metal layers (see Section 7.1).
(b) Implantation through thin surface layers (oxides, nitrides).
(c) Flexibility in regard to the choice of the dopant.
(d) Speed of the doping process.
(e) Good applicability for self-aligned process steps.

The first industrial use of ion implantation for semiconductor devices was in the field of MOS transistors; this is still its main area of application. The reasons lie in the great advantages offered here by implantation, especially for adjustment of the threshold voltage and for well doping in CMOS technology.

7.1 MOS DEVICES

For MOS devices, almost all the important advantages of ion implantation can be put to good use. The most important of these are the exact control of doping concentration by integration of the beam current, the implantation homogeneity, the low lateral spread of the ions, and the low processing temperature. Practically all MOS devices are now made using one or more implantation steps. In addition to conventional MOS transistors, this section will also deal with CMOS and charge-coupled devices (CCD).

7.1.1 Self-aligned Gate

The switching speed of MOS transistors is limited, among other factors, by the drain-gate overlap capacitance (Miller capacitance). Due to overlay inaccuracies during lithography, as well as to lateral diffusion, such an overlap is unavoidable for diffused transistors.

Figure 7.1(a) shows a schematic cross-section of a transistor in which this overlap is quite pronounced. Ion implantation offers a simple possibility for the reduction of this problem. The gate electrode can be used as the implantation mask, while the diffused source and drain areas are located further away from the gate (Figure 7.1(b)). In this way, a perfect adjustment between gate, source and drain can be achieved [97], and inaccuracies in adjustment or differences in the lateral diffusion are no longer of any importance. Using an improved technique, implantation can be done through the thin gate oxide (100 nm), which extends up to the source and drain (see Figure 7.1(c)) [98], [663]. In this way, the semiconductor surface is already passivated before doping. In Figure 7.2, the reduction of the gate-drain overlap capacitance is shown for implanted transistors, conventionally diffused transistors and transistors with a polysilicon gate without implantation [191]. Complete implantations of source and drain have also been carried out [446], [686]. Because of the high ion dose and high annealing temperatures which are necessary to obtain low sheet resistivities, this method has not yet come into general use; for VLSI circuits, however, it is the main technology which is to be employed.

Since the gate metallization must be applied before implantation, only temperatures below 577 °C (the eutectic temperature for aluminium-silicon) can be used for the activation of the implanted ions—boron for *p*-channel transistors, phosphorus for *n*-channel transistors—unless another gate material (polysilicon, molybdenum) can be applied. In general, 500 °C is considered adequate. Replacing aluminium by polysilicon results not only in a lower threshold voltage (see Section 7.1.2) due to the higher annealing temperatures which can be used [190],

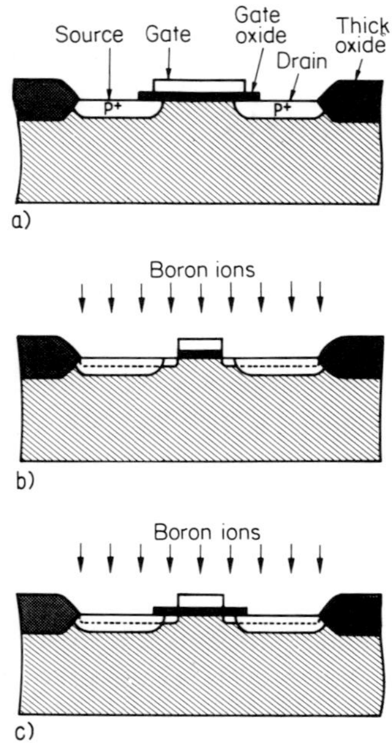

Figure 7.1 Schematic illustration of a MOS transistor: (a) conventional, considerable gate-drain overlap; (b) self-adjustment of gate and drain by means of ion implantation; (c) same as (b), but with implantation through a passivating oxide

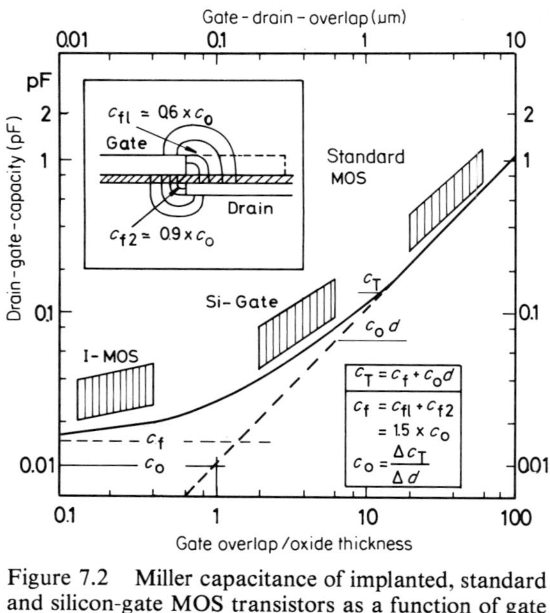

Figure 7.2 Miller capacitance of implanted, standard and silicon-gate MOS transistors as a function of gate overlap [191]. Reproduced by permission of Gordon and Breach

[540], but also in a reduction of the $1/f$ noise [540] as well as lower sheet resistivities. A combination of Schottky contacts (as source and drain) with the self-aligned gate technique has also been tried with good results [663].

The dose range for these applications is between 10^{14} cm^{-2} and 5×10^{15} cm^{-2}. The ion energy is chosen so that the gate metallization is adequate to stop all ions, while the gate oxide is penetrated by most of the ions. For a gate-oxide thickness of 100 nm, energies between 30 and 60 kV are generally used for boron and between 80 and 160 kV for phosphorus.

7.1.2 Threshold-voltage Adjustment

The most important use of ion implantation is for adjustment of the threshold voltage through implantation into the channel [45]. The threshold voltage is obtained as follows:

$$V_T = \Phi_{MS} + 2\Phi_F - \frac{Q_{SS} + Q_B}{\varepsilon_0 \varepsilon_{r,ox} C_{ox}} \tag{7.1}$$

where Φ_{MS} is the difference between the work functions of gate and silicon, Φ_F the Fermi potential, ε_0 the permittivity of free space, $\varepsilon_{r,ox}$ the dielectric constant of the oxide, Q_{SS} the sheet charge at the interface ($Q_{SS} = qN_{SS}$; N_{SS} is the surface state density), Q_B the space-charge and C_{ox} the gate capacitance per unit area ($C_{ox} = \varepsilon_0 \varepsilon_{r,ox}/t_{ox}$; t_{ox} oxide thickness).

Φ_F and Q_B are obtained as follows:

$$\begin{aligned} \Phi_F &= \frac{kT}{q} \ln \frac{N_B}{n_i} \\ Q_B &= \sqrt{[2q\varepsilon_0 \varepsilon_{r,Si} N_B (2\Phi_F + V_R)]} \end{aligned} \tag{7.2}$$

$\varepsilon_{r,Si}$ is the dielectric constant of silicon, N_B the doping concentration, and V_R the substrate voltage relative to the source.

The reduction of the threshold voltage in integrated circuits is especially necessary to obtain compatibility with bipolar integrated circuits (e.g., TTL), as well as to reduce the power consumption. It can be seen from equation (7.1) that there are several methods to reduce V_T. The simplest way is to implant boron (in the case of p-channel transistors) into the channel through the thin gate oxide, prior to the gate formation. By means of this 'quasi-sheet charge', Q_{SS} is compensated, and V_T is thus reduced. The change in the threshold voltage is given approximately by:

$$\Delta V_T = Kq \frac{N_\square}{C_{ox}} = K \frac{t_{ox} N_\square}{\varepsilon_0 \varepsilon_{r,ox}} \tag{7.3}$$

where C_{ox} is the specific oxide capacitance and $N_\square$ the implanted ion dose. The constant K indicates which portion of the implanted ions influences the threshold voltage, i.e., which portion is in the channel and is electrically active. The various implantation possibilities are shown in Figure 7.3. Profile (a) is very close to the model for compensation of Q_{SS}; with (b), a lower dose is required to obtain the same change in V_T; with (c), the influence of oxide thickness variations is very small; therefore, this method seems to be the best one. The main advantage of implantation in this process is the fact that the implanted dose can be adjusted precisely. By means of capacitance-voltage measurements on a test wafer before implantation, it is possible to determine the exact value of the dose which is needed to obtain a given value of V_T.

The various possibilities for adjusting V_T can also be seen from equation (7.1). In contrast to the standard *p*-channel MOS process using ⟨111⟩-oriented silicon, the ⟨100⟩ process utilizes ⟨100⟩-oriented material, which has a lower Q_{SS} value. However, the carrier mobility in the channels is lower in the case of this process, and lower cut-off frequencies are obtained. By using Si_3N_4/SiO_2 sandwich structures, the dielectric constant can be increased, and V_T can thus be reduced by a factor of approximately 1.5. The silicon-gate process mentioned above arrives at a reduced V_T value by means of the smaller difference between the work functions Φ_{MS}. The reduction amounts to approximately 2 V. With this process, the parasitic Miller capacitance is also reduced, since the polycrystalline gate can be used as a 'self-adjusted' diffusion mask for both source and drain diffusion. Especially together with ion implantation, this technique is of great importance in MOS technology, since it simultaneously offers the possibility of a third 'wiring plane'.

During the reduction of the threshold voltage of *p*-channel transistors, a doping change takes place; therefore equation (7.3) is only valid approximately. More

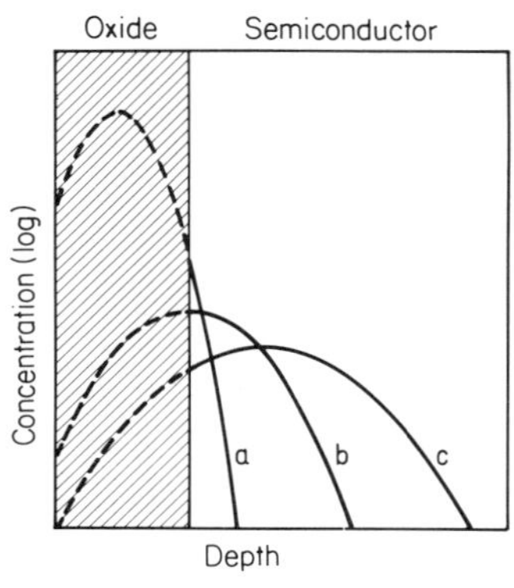

Figure 7.3 Comparison of possibilities for threshold-voltage adjustment. (a) Maximum of the distribution in the oxide; (b) maximum at the SiO_2–Si interface; (c) maximum within the semiconductor

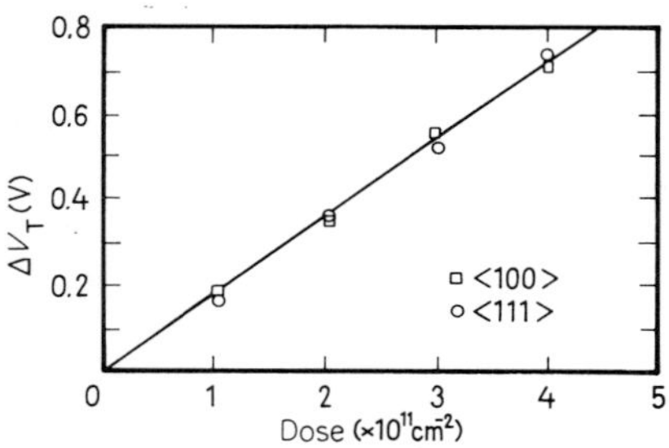

Figure 7.4 Reduction of threshold voltage depending on boron dose for ⟨100⟩- and ⟨111⟩-oriented silicon. Implantation at 30 keV through 100 nm SiO_2 [450]. Reproduced by permission of the American Institute of Physics

precise calculations have been made by Sigmon [680], Swanson and Meindl [712], as well as Höfflinger and Gabler [336]. Assuming a rectangular doping profile with the concentration $N_A = KN_\square/x_j$, the following is valid for ΔV_T (modified according to Reference [680]):

$$\Delta V_T = Kq\frac{N_\square}{C_{ox}} + \frac{kT}{q}\ln\frac{N_B}{N_A} + q\frac{KN_\square}{2\varepsilon_0\varepsilon_{r,Si}}\left(1 - \frac{Q_B}{KqN_\square}\right)^2 \tag{7.4}$$

where x_j is the depth of the p–n junction and N_B is the background doping. The last expression in equation (7.4) represents the band bending at the surface. If this is too high, the channel can no longer be switched off.

It is also possible to increase the threshold voltage through an implantation into the channel (e.g., boron for n-channel transistors and phosphorus for p-channel transistors). In this case, equation (7.3) is a good approximation if the range of the

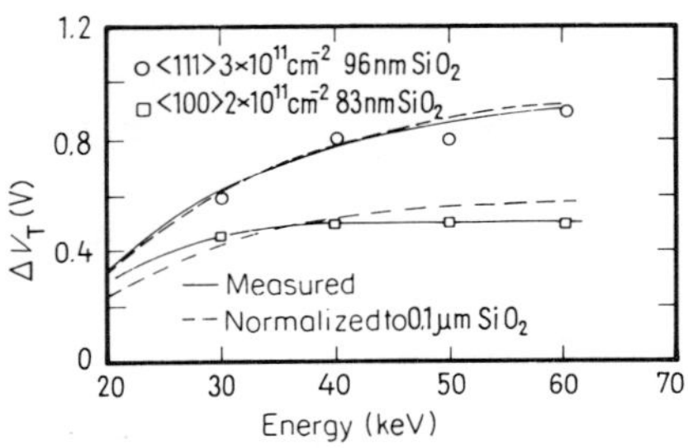

Figure 7.5 Saturation of ΔV_T with increasing implantation energy. The broken lines correspond to an oxide thickness of 100 nm [450]. Reproduced by permission of the American Institute of Physics

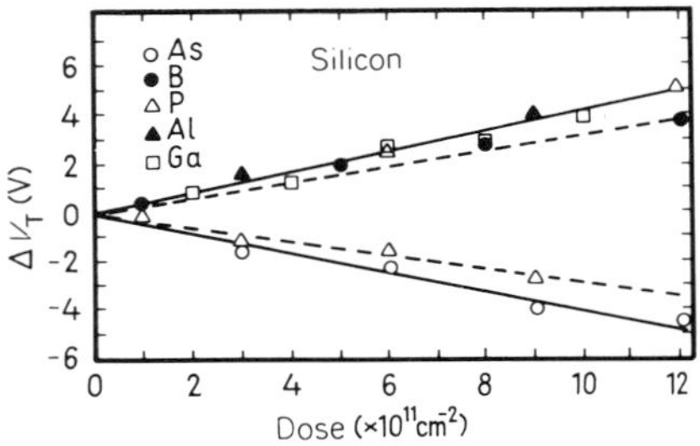

Figure 7.6 Alteration of the threshold voltage through implantations of B (35 keV), Al (100 keV), P (207 keV) and As (270 keV) [602]. The broken lines correspond to a 100 per cent ion activation

ions is lower than the width of the space-charge layer [591], [712]. However, the reversal of the sign in equation (7.3) should be noted.

The reduction of the threshold voltage, depending on the boron dose for ⟨100⟩- and ⟨111⟩-oriented silicon, is shown in Figure 7.4 [450]. The implantation energy is about 30 kV; the oxide thickness is 0.1 μm, i.e., the maximum of the distribution lies roughly at the SiO_2 interface. The influence of the implantation energy is shown in Figure 7.5. A saturation of ΔV_T is reached if almost all ions come to rest within the silicon.

Many studies have dealt with special problems of threshold-voltage adjustment through ion implantation, as well as with improved models, for example, References [336], [353], [414], [477], [591], [680], and [757]. Of special interest is a result by Sigmon [680], who found that the ion drift in the gate oxide is reduced considerably by implantation of boron through the gate oxide. He explains this in terms of an ion gettering at implantation produced defects in the oxide. Runge [602] has investigated the adjustment of the threshold voltage through implantations of boron, aluminium, gallium, phosphorus and arsenic in silicon. His results are shown in Figure 7.6. All curves can be explained by the simple model of equation (7.3). Surprisingly, however, values of K greater than 1 were obtained in some cases for the heavy ions. Runge explains this in terms of trap levels caused by incomplete annealing of radiation damage.

7.1.3 Depletion-load Transistors

It is only a small step from threshold-voltage reduction to the fabrication of depletion-load transistors. A high-enough dose is implanted into the channel so that, even at a gate voltage of 0 V, a conducting channel is produced. Depletion-load transistors are used, for example, as load transistors in inverters; in comparison to conventional enhancement transistors, they increase the switching speed by a factor of 2 to 3. In addition, power dissipation as well as dimensions are reduced.

Such transistors can be fabricated together with enhancement transistors on the same wafer, if the gate area of the enhancement transistors is masked (e.g., with photoresist) during implantation. The typical fabrication sequence for depletion-load transistors is shown in Figure 7.7. After opening contact windows in the oxide (mask I), the p^+ doped areas are diffused in an oxidizing atmosphere. Gate and contacts are opened by means of mask II. A further oxidation step is taken to produce the gate oxide (50 to 120 nm). The threshold voltage is reduced to the desired value by a boron implantation. Here, the field oxide (about 1500 nm) masks the areas which should remain undoped; therefore, no additional masking step is needed. The photoresist of the third masking step serves to shield enhancement transistors against the second boron implantation. By means of this implantation, depletion transistors are formed which can conduct at a gate voltage of zero volts. The further processing steps are standard. Figure 7.8 shows alteration of threshold voltage for *p*-channel transistors from enhancement to depletion operation, depending on ion dose. The same procedure, of course, can be applied to *n*-channel transistors [237], [246].

Figure 7.9 shows a cross-section through two transistors [237], where the enhancement transistor as well as the depletion transistor has a self-aligned gate. By use of high-resistivity substrate material (200 Ωcm), very fast circuits (inverter switching times of 115 ps) can be fabricated. The boron implantation for the depletion-load transistor serves to reduce the substrate-voltage effect. Typical threshold-voltage values, depending on the ion dose, are shown in Figure 7.10. The relationship between dose and V_T is non-linear for the implantation at 50 keV, since most of the ions come to rest outside of the depletion layer. The

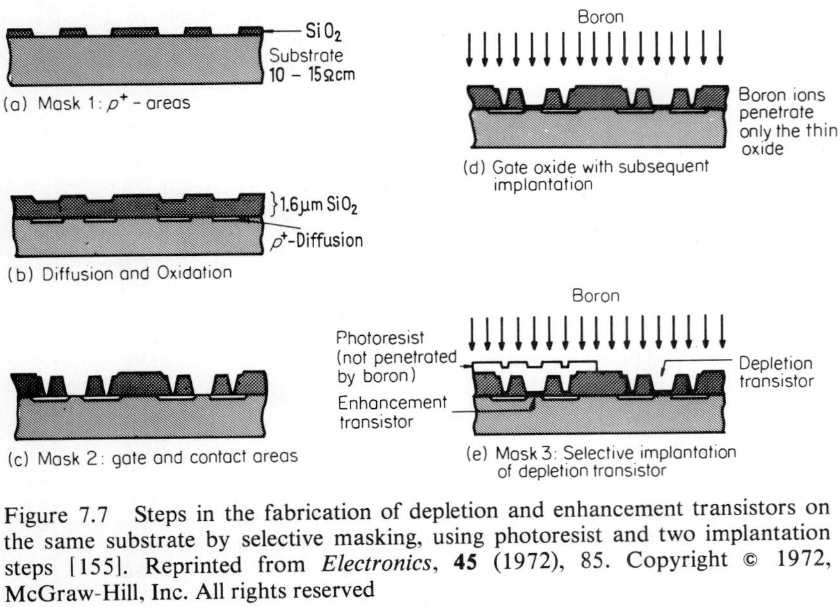

Figure 7.7 Steps in the fabrication of depletion and enhancement transistors on the same substrate by selective masking, using photoresist and two implantation steps [155]. Reprinted from *Electronics*, **45** (1972), 85.

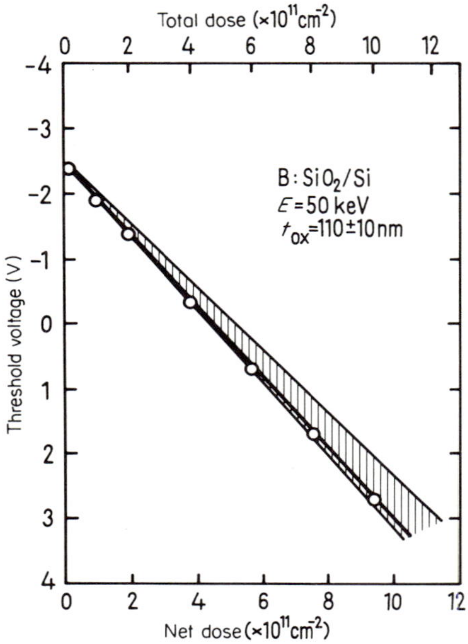

Figure 7.8 Dose dependence of the threshold voltage for *p*-channel enhancement and depletion transistors after implantation of 50 keV boron ions through 110 ± 10 nm SiO_2. The shaded area is determined by the oxide thickness error of ± 10 nm [414]. Reproduced by permission of the American Institute of Physics

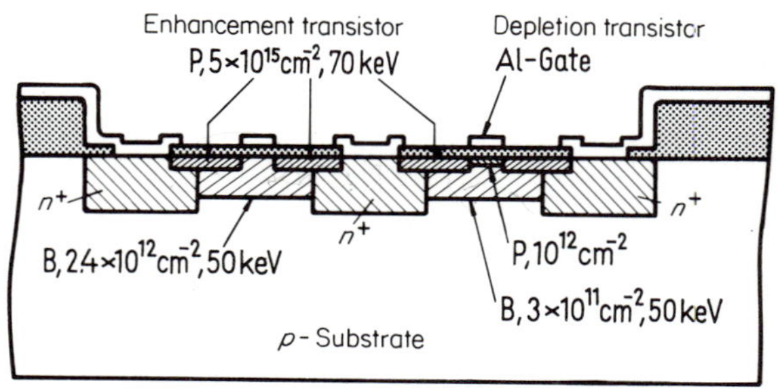

Figure 7.9 Cross-section through enhancement and depletion transistors with *n*-channel technology: according to Reference [237]. Reproduced with permission from Fang and Rupprecht, *IEEE J. Solid State Circuits*, **Sc-10** (1975), 205. Copyright © 1975 IEEE

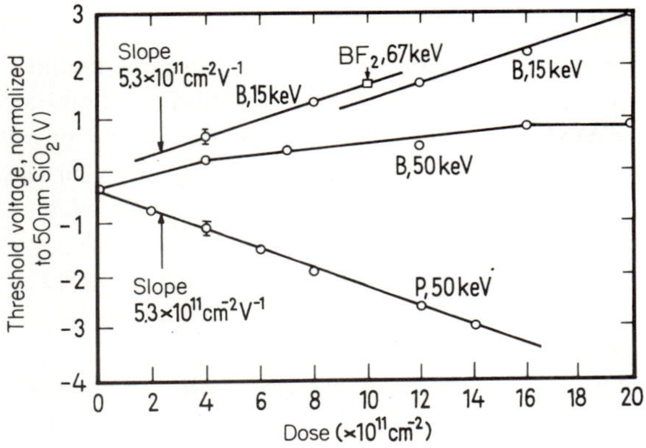

Figure 7.10 Threshold voltages as a function of the ion dose, for enhancement (^{11}B or $^{49}BF_2$) and depletion (^{31}P) transistors with *n*-channel technology [237]. Reproduced with permission from Fang and Rupprecht, *IEEE J. Solid State Circuits*, **Sc-10** (1975), 205. Copyright © 1975 IEEE

discontinuity of the curve in the 15 keV case is due to a change in the technological process. The same results could be achieved with the use of 15 keV boron ions or 67 keV BF_2 ions having an equivalent range distribution (*cf.* equation (6.2)). The typical fabrication sequence for these *n*-channel MOS structures is shown in Table 7.1. Of special interest in this process are the use of only 50 nm of gate oxide and that of high-resistivity substrate material, as well as the double implantation of the depletion transistor.

Table 7.1 Steps in the fabrication of *n*-channel enhancement and depletion transistors [237]

(1)	Initial oxidation 5000 Å.
(2)	Define and open the source and drain contact regions.
(3)	Phosphorus diffusion to 1 μm and reoxidation.
(4)	Open all gate regions and grow 500 Å gate oxide.
(5)*	Apply photoresist (∼1 μm), open the depletion type channel region, implant B^{11} at 70 keV to 5×10^{11} cm^{-2}, and P^{31} at 50 keV to 1.6×10^{12} cm^{-2}.
(6)*	Remove resist and reapply 1 μm photoresist, open the enhancement type channel region, implant B^{11} at 15 keV to 5×10^{11} cm^{-2}.
(7)	Remove resist and deposit about 50–70 Å of phosphorus-silica glass (PSG) for stabilization.
(8)	Post implant anneal at 970 °C for 30 min in N_2.
(9)	Open contact holes.
(10)	Aluminium metallization.
(11)	Shallow source and drain extension implant: P^{31} at 50 keV to 4×10^{15} cm^{-2}.
(12)	Post implant anneal at 515 °C for 15 min in N_2.

* These two steps may be interchanged.

7.1.4 Complementary MOS Transistors

A further application of ion implantation are complementary MOS devices (CMOS), where *p*- and *n*-channel transistors are fabricated on the same substrate. The advantage of CMOS technology is its extremely low power consumption, which makes such integrated circuits especially well-suited for electronic watches and memories. Especially in connection with new methods such as the silicon gate technique, which makes an additional 'wiring plane' possible, or the oxide insulation technique (Locos, Planox, Isoplanar), a high packing density can be achieved. With only a few exceptions, all integrated circuits for electronic watches are currently being made using CMOS technology.

The starting material for CMOS devices is usually *n*-doped silicon. In order to fabricate *n*-channel transistors, a *p*-well must first be produced in which the complementary transistor can then be made [204], [712]. This well doping can be done using various methods: selective epitaxy, standard diffusion, diffusion from oxide or ion implantation. The advantage of implantation, once again, is the exact process control which is possible. The range of the ions, of course, is not sufficient to dope a volume which is deep enough. Therefore, the implantation only serves the purpose of predeposition. Following this, the desired distribution is obtained through a 'drive-in' diffusion. The main fabrication steps are shown in Figure 7.11. After implantation of a *p*-type layer (typical dose: 10^{13} cm^{-2} boron), the *p*-well is made by drive-in diffusion at about 1200 °C for several hours. The source and drain areas for *p*- and *n*-channel transistors are then diffused selectively or

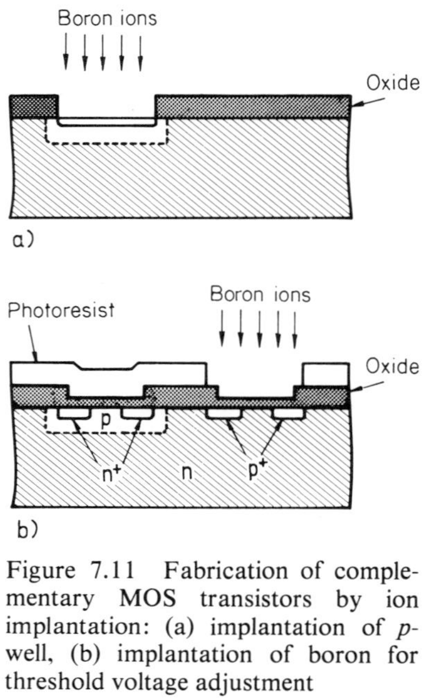

Figure 7.11 Fabrication of complementary MOS transistors by ion implantation: (a) implantation of *p*-well, (b) implantation of boron for threshold voltage adjustment

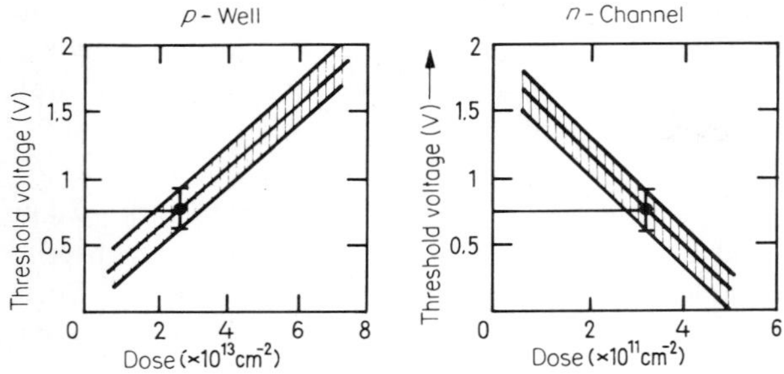

Figure 7.12 Dependence of threshold voltage on dose of the well implantation (*n*-channel transistor) or dose of the channel (*p*-channel transistor) [189a] for CMOS transistors; well: 60 keV boron through 100 nm SiO_2 in 3 Ωcm ⟨100⟩ Si with drive-in diffusion at 1200 °C for 16 h; *p*-channel: 50 keV boron through 100 nm SiO_2–Si_3N_4 in 3 Ωcm ⟨100⟩ Si with annealing at 525 °C for 10 min. Reprinted with permission from *Solid State Technology*, Technical Publishing, a company of Dun & Broadstreet

implanted. The threshold voltages for the two transistor types can be adjusted independently. For the *n*-channel transistor, it is adjusted by variation of the dose for the well, or by means of an implantation; the *p*-channel transistor always requires a separate implantation for this purpose (*cf.* the previous section). In Figure 7.12, for example, the relationship between dose and threshold voltage is

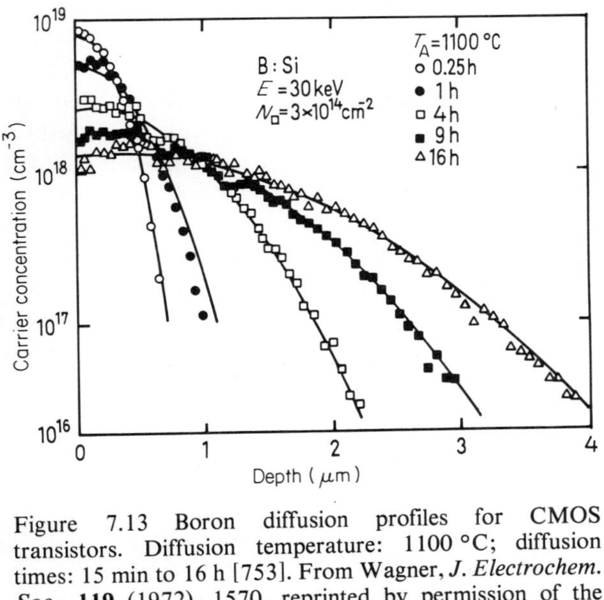

Figure 7.13 Boron diffusion profiles for CMOS transistors. Diffusion temperature: 1100 °C; diffusion times: 15 min to 16 h [753]. From Wagner, *J. Electrochem. Soc.*, **119** (1972), 1570, reprinted by permission of the publisher, The Electrochemical Society, Inc.

shown for the n-channel and p-channel transistors of a CMOS integrated circuit intended for use in electronic watches [189a]. In this case, only a very low supply voltage is needed, and there is a low power consumption as well. Implantation doses of boron for this purpose are between 5×10^{12} cm^{-2} and 5×10^{13} cm^{-2}, at energies between 20 and 100 keV. The diffusion time is 10 to 20 h at 1000 °C. The reproducibility is generally ± 5 per cent, compared with ± 20 per cent for conventional thermal diffusion. Examples of boron diffusion profiles are given in Figure 7.13 for times between 15 min and 16 h at 1100 °C. Because of the high diffusion depth, the profiles have practically no correlation with the original implantation profiles.

In the case of n-channel transistors, problems result from leakage currents, which are due to the inversion of the low doped p-type substrates at the surface. By means of a so-called field implantation of boron, this inversion can be prevented [246]. This implantation can be made into the entire area through the thick oxide. Typical doses are between 10^{12} and 10^{13} cm^{-2}. A very detailed study on this topic has been published by Douglas and Dingwall [203], including a treatment of p-channel transistors. Alternatively, this implantation can be performed before the field oxide is grown. In this case, the loss of boron during oxidation has to be considered.

7.1.5 Charge-coupled Devices

A significant development in recent years are charge-coupled devices (CCD), a concept based on MOS structures which collect minority carriers and can store them in localized potential wells at the SiO_2–Si interface [56], [99]. The same as with the other MOS devices discussed in these pages, ion implantation has several important applications in this new field too.

A diagram showing the design and the operation of a simple three-phase CCD structure is given in Figure 7.14 [56]. The charge can be injected for such a device either in series via a p–n junction, or via the absorption of photons in the vicinity of the potential wells. By applying suitable voltages at the electrodes (*cf.* Figure 7.14), these charges can be shifted and then read via a diode. CCDs can be used as digital as well as analogue memories, and are especially promising for optical applications, since they are simultaneously usable for detection, storage and reading of the signal. The dependence of the potential at the surface (i.e., the depth of the potential well) upon the gate voltage, the doping concentration, the oxide thickness and the flat-band voltage is shown in equation (7.5) [415a]:

$$\Phi_S = V_G + \frac{Q_{SS}}{C_{ox}} - B\left\{\sqrt{\left[1 + \frac{2}{B}\left(V_G + \frac{Q_{SS}}{C_{ox}}\right)\right]} - 1\right\}$$

$$B = \frac{qN_B\,\varepsilon_0\,\varepsilon_{r,Si}\,t_{ox}^2}{\varepsilon_{r,ox}^2} \qquad (7.5)$$

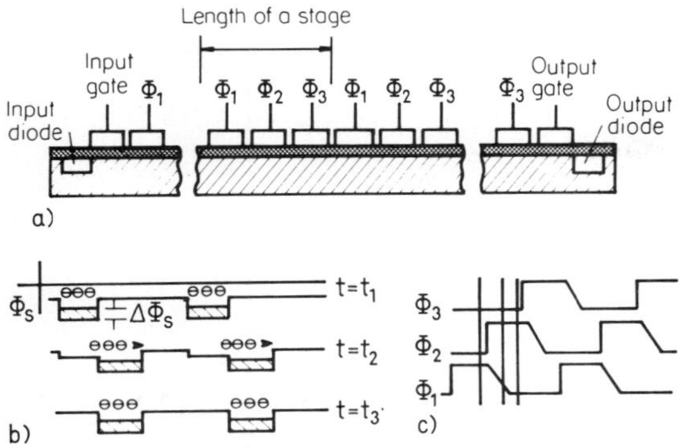

Figure 7.14 Three-phase charge-coupled device with *n*-channel technology: (a) cross-section; (b) surface potential at three different times; (c) voltages at the electrodes [56]. Reproduced with permission from Barbe, *Proc. IEEE*, **63** (1975), 38. Copyright © 1975 IEEE

where $\varepsilon_{r,Si}$ and $\varepsilon_{r,ox}$ are the relative dielectric constants of the semiconductor and the oxide, respectively, ε_0 is the absolute dielectric constant (permittivity), V_G the gate voltage, Q_{SS} the oxide charge per unit area, N_B the substrate doping and t_{ox} the oxide thickness.

When the charge (Q) is stored in the potential well, its depth is reduced by $\Delta\Phi_S$:

$$\Delta\Phi_S = \frac{Q}{C_{ox} + C_S} \tag{7.6}$$

where C_S is the space-charge capacitance.

In order to allow a charge transfer with only two electrodes (two-phase CCD), it is possible to use different oxide thicknesses under the electrodes, thus producing a potential barrier. A simpler solution is the use of an implanted potential barrier at the edge of the electrodes, as shown in Figure 7.15. Typical phosphorus doses for *p*-channel elements are between 10^{12} and 2×10^{12} cm^{-2}.

For a suppression of conducting channels at the surface, implantation can again be used (channel stop). A problem in optical applications occurs when strong signals cause collected charges to drain from a filled potential well into surrounding ones (blooming). This blooming effect can be reduced by the use of implanted 'drain electrodes' [658]. A possible way to reduce surface effects (trapping losses during charge transfer, etc.) is the use of buried channels which can be produced by ion implantation. A typical boron dose, for *p*-channel devices, is 5×10^{12} cm^{-2} [671]. A schematic cross-section through a three-phase structure is shown as an example in Figure 7.16. In this case, a channel stop with a

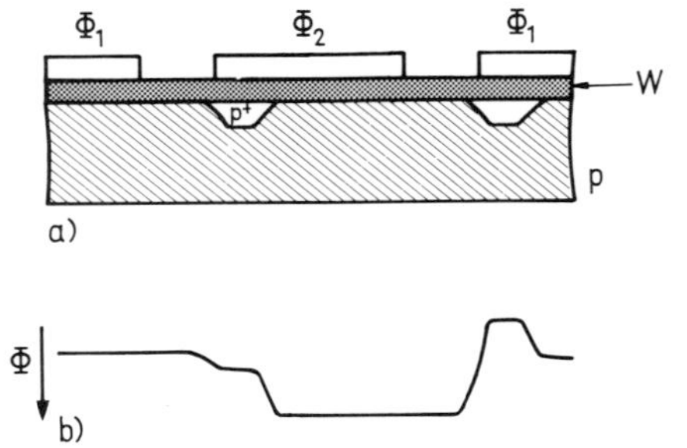

Figure 7.15 Implanted two-phase CCD structure: (a) cross-section; (b) surface-potential profile with inhomogeneous barriers, produced by a p^+ implantation [56]. Reproduced with permission from Barbe, *Proc. IEEE*, **63** (1975), 38. Copyright © 1975 IEEE

phosphorus dose of 10^{12} cm^{-2} was also implanted. Further advantages of buried channels are their shorter transfer times, due to a higher potential gradient in the channel, as well as their lower capacitances [56]; however, the charge which can be stored is also lower.

7.2 RESISTORS

In addition to transistors and diodes, integrated circuits also require a relatively large number of resistors. The precise implantation of very low doping concentrations makes it possible to fabricate resistors with tight tolerances and high values. Since integrated circuits are usually fabricated on *n*-type layers, the resistors are *p*-type; boron is generally used for this purpose.

7.2.1 Resistivity Range

Diffusion fabricated resistors can have values of up to about 1 kΩ/□; those made by ion implantation, though, can be up to 20 kΩ/□ with tolerances between 1 and

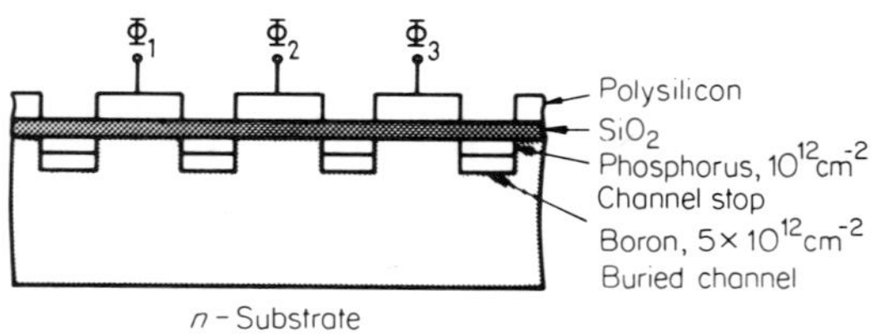

Figure 7.16 Schematic cross-section through a three-phase charge-coupled device with buried channel according to Reference [671]. Reproduced by permission of the American Institute of Physics

Table 7.2 Typical resistor characteristics [538]

Resistor	Sheet resistivity (kΩ/□)	Maximum annealing temperature	Linearity	Temperature coefficient (p.p.m./°C)	Remarks
Standard-diffused	0.01–0.8	1 200 °C	Standard	⩾400, min. at 50 Ω/□	
Boron-implanted					
(a) 'Drive-in' diffused	10–20	1 200 °C	Standard	⩾400, min. at 50 Ω/□	
(b) High temperature annealing	50–50	1 000 °C	Standard	⩾400, min. at 400 Ω/□	Predictable value
(c) Low temperature annealing	50–500	500 °C	Standard	0 at 3 kΩ/□	Sensitive to dose rate during implantation
Special implantation					
(a) Masked implantation	10–100	1 000 °C	Poor	⩾400	Two masking steps
(b) Radiation damage	100–500	500 °C	Very good	−15.000	Combined with boron
(c) Deep level (In)	1–200	900 °C	Good	−10.000	implantation for TK adjustment

3 per cent, while for less stringent requirements, up to 500 kΩ/□ are possible [404], [446], [596]. Boron implanted resistors are annealed at different temperatures depending on fabrication procedure: (a) low-temperature annealing (400 to 500 °C); (b) high-temperature annealing (800 to 1000 °); (c) predeposition (1100 to 1200 °C). The choice of temperature is determined by the technology being employed, for example, low-temperature annealing is used for MOS devices if metallization has been applied before the annealing, thus ruling out higher temperatures for reasons of alloy formation. In this case, implantation is usually made through a thin oxide passivation layer. The annealing temperature is chosen in such a way that the resistivity is altered as little as possible, in order to obtain best possible reproducibility and homogeneity. For these purposes, the implantation temperature too must be kept constant, since various annealing effects during implantation could otherwise occur. This is especially important in high current implantation.

Typical values obtainable for all three temperature ranges are compared with diffused resistor values in Table 7.2 [538]. The sheet resistivity of resistors annealed at high temperatures (⩾900 °C) is well predictable from theory (*cf.* Figures 6.1 to 6.3). The upper limit of the sheet resistivity is determined by the influence of charges in the oxide, as well as by fluctuations in the substrate doping. For doses between 10^{13} and 10^{14} cm^{-2}, the influence of surface charges

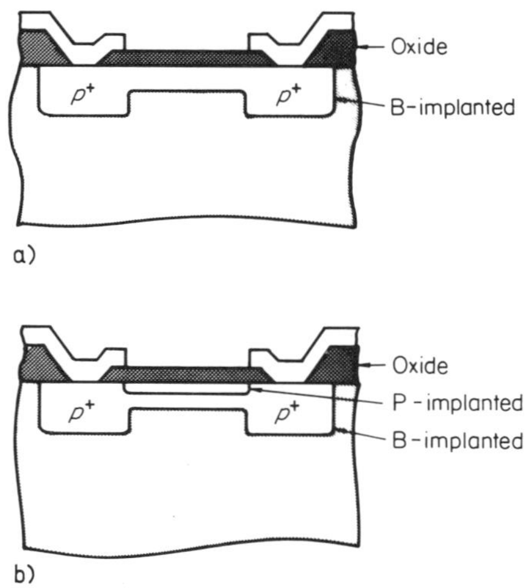

Figure 7.17 Cross-section through boron-implanted resistors: (a) oxide passivation, danger of surface influences; (b) oxide passivation and phosphorus-implanted protective layer (from [427a]). Reproduced with permission from Lee *et al.*, *Proc. IEEE*, **62** (1974), 1241. Copyright © 1974 IEEE

can be neglected; at lower doses, however, Q_{SS} plays an important role. By an additional implantation into layers near the surface with ions of the opposite conductivity type, this effect can be eliminated. The resistor is thereby buried, and surface influences are eliminated. Seidel and Gibson [650] have fabricated resistors using this method with 20 MΩ and sheet resistivities between 1 kΩ/□ and 120 kΩ/□. A schematic cross-section through resistors with and without buried structure is shown in Figure 7.17. As a result of this additional doping, however, the linearity is influenced negatively, so that buried structure resistors can only be used for low voltages. High resistivities, along with reduced sensitivity to oxide charges, can also be obtained through radiation damage produced by an additional implantation.

7.2.2 Temperature Coefficient

The temperature coefficient of implanted resistors depends greatly on the annealing conditions, and can be selected to a certain degree by appropriate choice of annealing parameters [404], [552]. Boron-implanted resistors annealed at low temperatures change the sign of their temperature coefficient from negative to positive at about 3 kΩ/□. Values for different fabrication procedures are compared in Figure 7.18. The higher the annealing temperature, the more similar are the characteristics of implanted and diffused resistors. By means of a double implantation of boron and phosphorus [446], or by radiation damage produced by a further implantation [447], the temperature coefficient can be determined within broad limits (positive, negative, or 0). It has recently been found that the temperature coefficient can be influenced greatly by traps in the band gap [538],

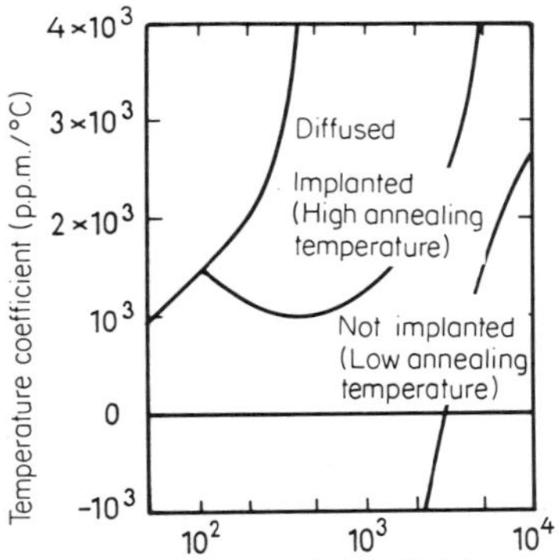

Figure 7.18 Temperature coefficients of diffused and implanted resistors [538]. Reproduced by permission of *Acta Electronica*

and that negative values up to 10^4 p.p.m./°C are possible. Applicable ion species include indium, thallium, sulphur, tellurium and molybdenum.

7.2.3 Linearity

With high resistivity resistors, the linearity poses a problem. At higher voltages the resistivity increases, due to the larger space charge between the substrate (*n*-type) and the resistor (*p*-type). With an additional implantation of neon [539], damage can be produced, thus lowering the mobility and linearizing the resistor. Very accurate ($\leqslant 1$ per cent) as well as linear resistors are required, for example, for digital-analog converters. Such double-implanted layers are usually annealed at only 500 °C in order to activate enough of the boron on the one hand, and to retain enough radiation damage on the other.

7.3 DIODES

Regardless of their application, this section deals mainly with devices consisting of only one *p–n* junction. In order to obtain good electrical properties in this case, the doping profile or the doping homogeneity are usually the important factors. For precise control over both of these, ion implantation can be used.

7.3.1 Varicap Diodes

Varicap diodes are *p–n* or Schottky diodes utilizing the alteration of the capacitance as a function of the reverse bias, for example, for tuning purposes in radio and television receivers, for microwave mixers, and for parametric amplifiers. The sensitivity m of varicap diodes depends on the profile at the *p–n* junction. It is defined as:

$$m = d(\lg C)/d(\lg V) \tag{7.7}$$

where C is the capacitance of the *p–n* junction for the voltage V. Abrupt junctions have a sensitivity of 0.5. Hyperabrupt junctions have a higher value for m; this is desirable to obtain a high capacitance variation for a low voltage change. For this purpose, it is necessary to have a doping profile which decreases from the *p–n* or Schottky junction. This can only be obtained with a complicated multiple diffusion step, or more simply, by the use of ion implantation. If the doping has the following depth dependence:

$$N_D - N_A = Kx^n$$

where x is the distance from the *p–n* junction and n is a constant, then the following is valid for m:

$$m = \frac{1}{2-n} \tag{7.8}$$

and for the relationship between the differential capacitance and the voltage:

$$C \sim (V_D - V)^{-m} \tag{7.9}$$

(V_D is the diffusion voltage.)

A linear relationship between frequency and bias would be desirable for tuning applications, i.e.:

$$N_D - N_A = Kx^{-3/2}$$

or (7.10)

$$C \sim (V_d - V)^{-2}$$

Values of m between 2.1 and 2.5 have been obtained by Brook and Whitehead [105] as well as by MacRae [451].

The implanted ion species is generally phosphorus. It is either implanted at a low energy for predeposition, with subsequent diffusion (only the exact dose control offered by implantation is taken advantage of in this case), or it is implanted at various energies up to about 400 keV to obtain the profile needed for the desired capacitance-voltage characteristic directly. For varicap diodes, the matching of the various diodes is extremely important. Implantation is therefore better than diffusion even if it is only used for predeposition. For example, MacRae [451] has obtained a spread in the m value of less than 3 per cent. Similarly good results have recently been obtained by means of molecular beam epitaxy [138].

7.3.2 IMPATT Diodes

The use of ion implantation for IMPATT diodes is only of interest at frequencies above 10 GHz, since its better process control only brings advantages in this range. Conventional diodes have a p^+nn^+ structure. To obtain higher efficiencies, a structure with drift regions for holes and electrons is used at high frequencies. The term 'double drift region (DDR) diodes' is then used. Figure 7.19(a) shows a schematic comparison between DDR and conventional IMPATT diodes. For the fabrication of such structures, epitaxy is in competition with ion implantation. Studies in this area have been made mostly at the Bell [501], [651] and Hughes [426], [427], [428] laboratories. A doping profile of a DDR diode, according to Seidel *et al.* [651], is shown in Figure 7.19(b). This diode is fabricated in a complementary manner (n^+npp^+ structure) to take advantage of the higher range of boron. The p-type drift region was made by a boron implantation at 200 keV (dose: 4.5×10^{12} cm^{-2}); the n-type drift region was made by a double phosphorus implantation at 150 keV (dose: 2.4×10^{12} cm^{-2}) and 240 keV (dose: 3.5×10^{12} cm^{-2}). The contact implantation was made with arsenic at 50 keV (dose: 5×10^{14} cm^{-2}), in order to obtain a shallow doping. Annealing was done at 850 °C for 30 min. This diode had an efficiency of 7.4 per cent at 92 GHz and an output of 0.18 W. Through a combination of implantation technology and double

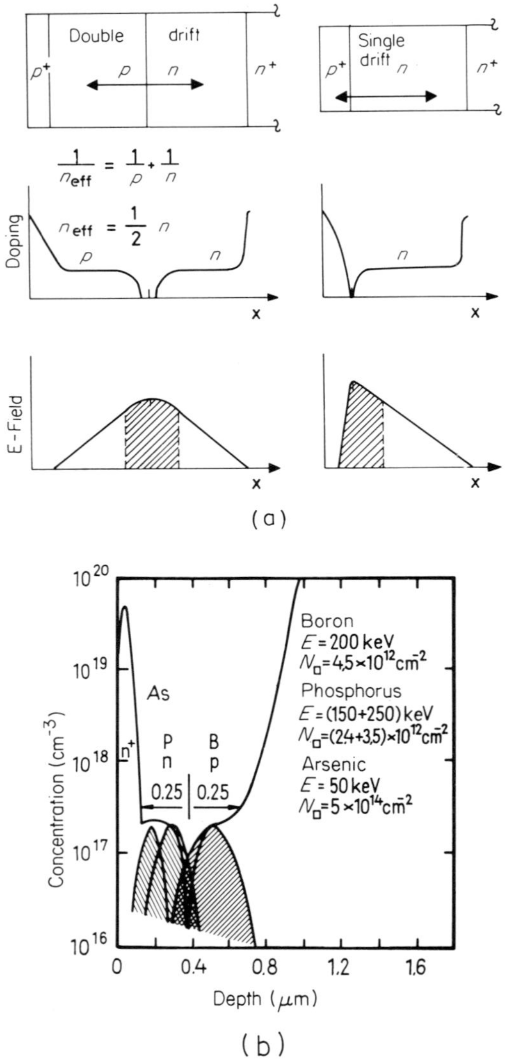

Figure 7.19 (a) Schematic comparison between a DDR diode and a conventional IMPATT diode; (b) doping profile of a symmetrical 100-GHz DDR–IMPATT diode [651]. Reproduced with permission from Seidel *et al.*, *Trans. Electron. Devices*, **ED-21** (1975), 523. Copyright © 1975 IEEE

epitaxy, it has been possible to fabricate DDR diodes with Read structure, output up to 6.8 W [501], and 12 per cent efficiency for the X-band.

7.3.3 Silicon Multidiode Targets

With this device, ion implantation makes a significant yield increase possible by a simplification of the technology as well as a better homogeneity. The silicon multidiode target (vidicon) consists of about 10^6 diodes on a single silicon wafer. The failure of one single diode would make the target unusable, due to the clear visibility of this defect ('white video defect'). Only a single photolithographic step is necessary with the use of ion implantation [568]. Following oxidation, the windows for the individual diodes are etched; these are then covered again by a thin oxide by further short oxidation (Figure 7.20). The implantation boron ions penetrate only the thin oxide. These ions are then activated by annealing and the boron atoms are diffused deeper into the silicon (drive-in diffusion). Etching of the wafer in buffered fluoric acid, as the final step, exposes the diodes.

White video defects in diodes are caused by high leakage currents. A gettering effect by radiation damage produced by ion implantation can reduce these leakage currents, thus leading to a significant reduction of such defects [350]. For this purpose, implantation can be made with high doses of phosphorus (10^{16} cm^{-2}), arsenic (10^{15} cm^{-2}) and argon (10^{15} cm^{-2}). Such a gettering, of course, can also be used for other devices. Detailed studies on this topic, using the back-scattering technique, have been made by Buck *et al.* [110].

Blooming effects cause problems in the case of this device too, but not as much as with charge-coupled devices (CCD). A method to reduce this effect is offered by the control of surface recombination at the side of the light incidence through ion implantation [688]. The effective surface-recombination velocity, as an approximation, is inversely proportional to the implanted ion dose.

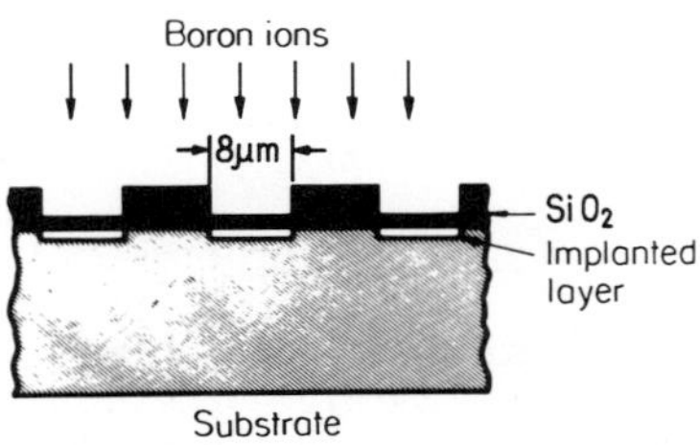

Figure 7.20 Cross-section through an implanted silicon multidiode target. Following boron implantation, the thin protective oxide is removed in buffered fluoric acid without a masking step [568]. Reproduced by permission of the American Institute of Physics

7.3.4 Solar Cells

The fabrication of solar cells by ion implantation was one of the earliest applications of the latter technique. The advantage here (Gusev *et al.* [308a], King *et al.* [411]) is due to the possibility of producing shallow *p–n* junctions of 0.25 μm, with good sensitivity in the blue region of the solar spectrum as well as efficiencies above 10 per cent. Recently, cells with efficiencies up to 15 per cent have been verified [204a]. At present, the most essential use of solar cells is for space technology; if, though, they come into wide use in the future for purposes of pollution free energy production, the use of implantation as their doping process will no doubt increase, since it is better suited as a large-scale, automated method for high throughput. The main advantages of implantation are good doping homogeneity, extremely shallow *p–n* junctions (important for high efficiency blue-sensitive cells), and simplified process technology through elimination of the need for reverse-side etching.

7.3.5 Nuclear Detectors

Detectors for nuclear radiation require shallow *p–n* junctions ('thin windows') in order to avoid a 'dead layer' which does not contribute to the signal. Therefore, the implantation energy chosen is generally very low (2 to 10 keV). A great advantage of implantation, once again, is its low process temperature. The implanted detectors are annealed at only 300 to 400 °C to avoid diffusion as well as deterioration of material characteristics (specific resistivity, minority carrier lifetime) [37], [394], [472], [494]. The first experiments were made by Alvaeger and Hansen [37] as well as by Martin *et al.* [472]. Very good detectors have been fabricated by Meyer [497] and by Kalbitzer *et al.* [394] through reduction of radiation damage by implanting in a channelling direction, which is very important at the low annealing temperatures used.

Position sensitive detectors can be made easily by using an implanted layer as a resistive layer for measurement of the impact location of a particle [395], [423]. A

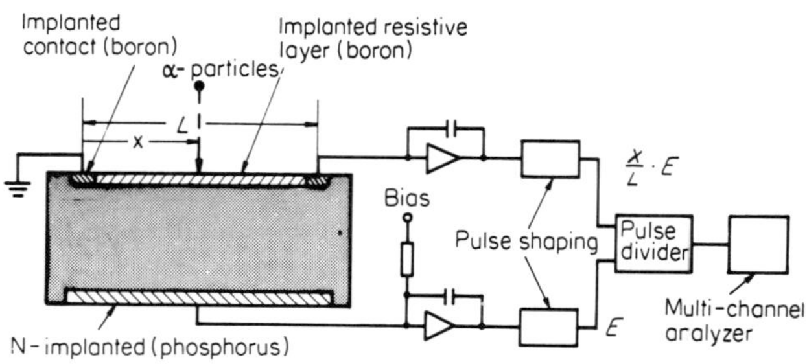

Figure 7.21 Cross-section through a position-sensitive nuclear detector, with block diagram of the electronics [423]. Reproduced by permission of North-Holland Physics Publishing

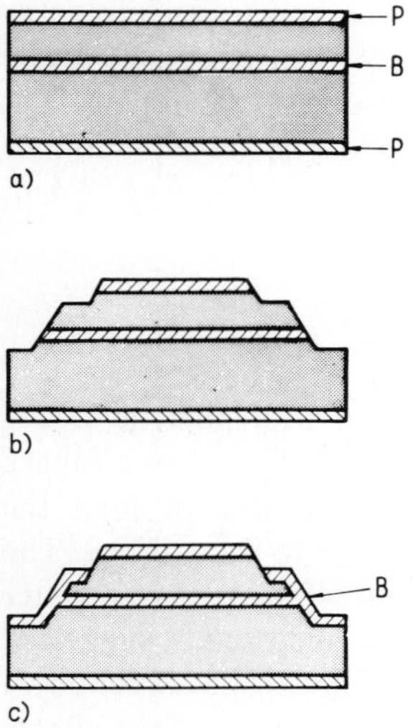

Figure 7.22 Fabrication process of an integrated E–dE/dx detector according to Kostka and Kalbitzer [417]: (a) phosphorus implantation into front as well as reverse sides (60 keV, 10^{14} cm^{-2}); high-energy boron implantation (4–8 MeV, 3×10^{13} cm^{-2}); (b) mesa etching; (c) contact boron implantation (20 keV, 3×10^{14} cm^{-2})

detector according to Laegsgaard [423] is shown in Figure 7.21. The boron-implanted layer (4 mm × 14 mm) simultaneously serves to form the p–n junction, and also works as a resistor. Contacts have been fabricated at both ends and on the reverse side by implantations of boron and phosphorus, respectively, at doses of 10^{16} cm^{-2}. All implantations were made at 60 keV. The resistivity of the contact is typically 20 to 50 kΩ, and is determined by the implanted dose. In this case, once again, very low annealing temperatures (400 °C, 5 min) have been used.

Another interesting use for ion implantation is the production of E–dE/dx detectors through high energy implantation [417], [471]. The fabrication method

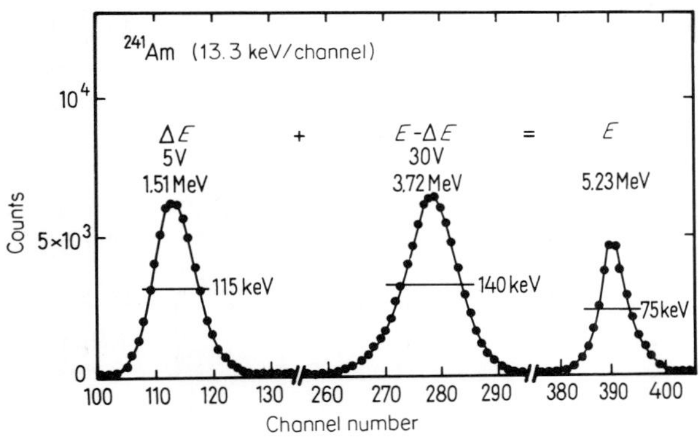

Figure 7.23 α-Spectrum from both sections of the integrated E–dE/dx detector. The total spectrum is also given for comparison [417]

for such a detector is shown in Figure 7.22 [417]. A high-energy boron implantation (4–8 MeV) forms an isolated *p*-type layer, which is then contacted by a second boron implantation (20 keV, 3×10^{14} cm^{-2}), following mesa etching. The front-side and reverse-side contacts are formed through phosphorus implantation (6 keV, 10^{14} cm^{-2}). Annealing is done at 900 K for 30 min. A spectrum of ^{241}Am, measured with this detector, is shown in Figure 7.23.

7.3.6 Photodiodes

An interesting application of implantation for the production of avalanche diodes has been studied by Gibbons [286]. By implanting boron at 400 keV (see Figure 7.24) through a pyrolytic SiO_2 masking layer, it was possible to limit the breakdown to the area where the boron distribution is close to the surface. The *p–n* junction was made by phosphorus implantation at 40 keV. Even better characteristics have been obtained by Berchtold [70], using a structure suggested by Ruegg [600]. Data on the structure used are given in Figure 7.25. Breakdown only occurs between the high-doped areas (n^+ and *p*). Edge and surface effects are thus eliminated. The *p*-type multiplication area was produced by ion implantation, with a subsequent drive-in diffusion. Precise data have not been published. The n^+ layer was made by a shallow phosphorus diffusion ($x_j = 0.3$ μm). A typical breakdown voltage is 300 V, for a π-layer doping of 4.5×10^{14} cm^{-3}. Gain-bandwidth products up to 300 GHz have been measured. For low light levels, multiplication factors above 10 000 have been achieved.

In recent years, interest in photodetectors for the near and medium infrared regions has increased. Suitable semiconductor materials include InSb (up to 5.5 μm), $Hg_{1-x}Cd_xTe$, and various binary and ternary chalcogenides. The doping effect of radiation damage has been utilized for the fabrication of detectors in $Hg_{1-x}Cd_xTe$ [249], $Pb_{1-x}Sn_xTe$ [201], PbTe [193], and InSb [250]. In other implantation experiments, doping has been carried out either chemically or by altering the stoichiometry. Examples for this are PbTe [193], PbS [202], $PbS_{1-x}Se_x$ [200], $Pb_{1-x}Sn_xTe$ [399], and $Hg_{1-x}Cd_xTe$ [241], [466], [610a]. The

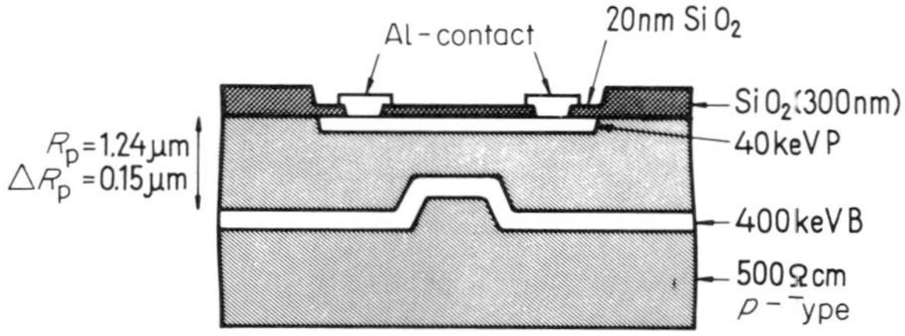

Figure 7.24 Avalanche photodetector, consisting of germanium with a deep boron implantation for definition of the multiplication region; dose 5×10^{12} cm^{-2} to 3×10^{13} cm^{-2} [286]. Reproduced by permission of Springer–Verlag, Heidelberg

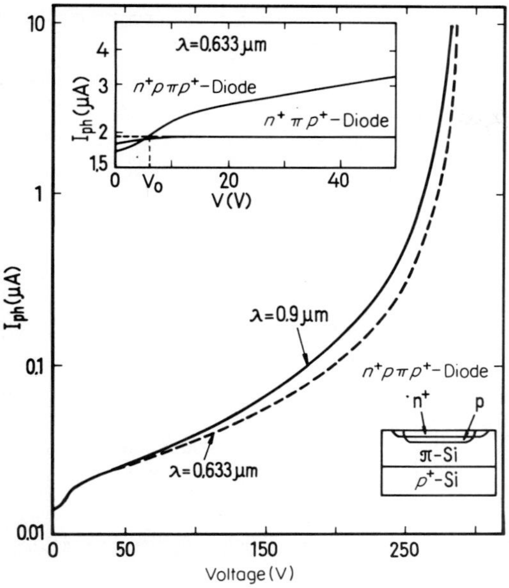

Figure 7.25 Reverse current of a boron-implanted avalanche silicon photodiode [70]. Reproduced by permission of the American Institute of Physics

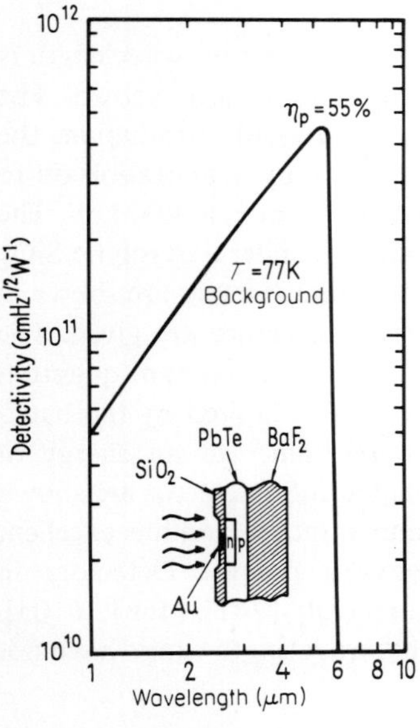

Figure 7.26 Detectivity as a function of wavelength for a 380 μm × 380 μm epitaxial PbTe diode formed by Sb implantation [202a]. Ion dose: 10^{14} to 2×10^{14} cm^{-2}; implantation energy: 400 keV; annealing at 340 to 400 °C. Reproduced by permission of the American Institute of Physics

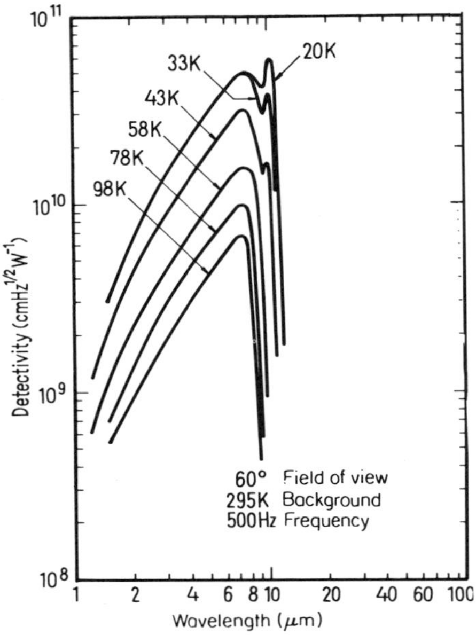

Figure 7.27 Detectivity as a function of wavelength and temperature for a $Pb_{0.88}Sn_{0.12}Te$ diode with a diameter of approximately 125 μm [199] formed by proton implantation ($N_{\square} = 5 \times 10^{14}$ cm^{-2}, $E = 200$ keV). No annealing was done after implantation

detectivity of an antimony implanted PbTe detector depending on wavelength is given in Figure 7.26, in which the structure of the diode is also shown. The maximum quantum efficiency is 55 per cent. For reverse-side irradiation, the result is 51 per cent. The PbTe was deposited epitaxially by evaporation on to BaF_2. The ion doses used were between 10^{14} and 2×10^{10} cm^{-2}, at 400 keV. The layers were annealed at 340 to 400 °C for several minutes under a pyrolytic SiO_2 layer. Values measured for a detector made by proton bombardment are shown in Figure 7.27, where detectivity curves depending on temperature are given. The maximum detectivity is higher than 10^{12} cm$^{1/2}$ W^{-1} and the maximum quantum efficiency is 37 per cent. At large angles, the detectivity is limited by the background radiation. The implantation dose was 5×10^{14} cm^{-2} at an energy of 200 keV. Current-voltage characteristics of an implanted InSb detector are shown in Figure 7.28. The diode was made by beryllium implantation, and has excellent characteristics in both the reverse and the forward ranges. Detectors in $Hg_{1-x}Cd_xTe$ up to 14 μm have also been fabricated recently [241], [466], [610a]. In the case of lead chalcogenides, experiments in this wavelength range have not yet been successful.

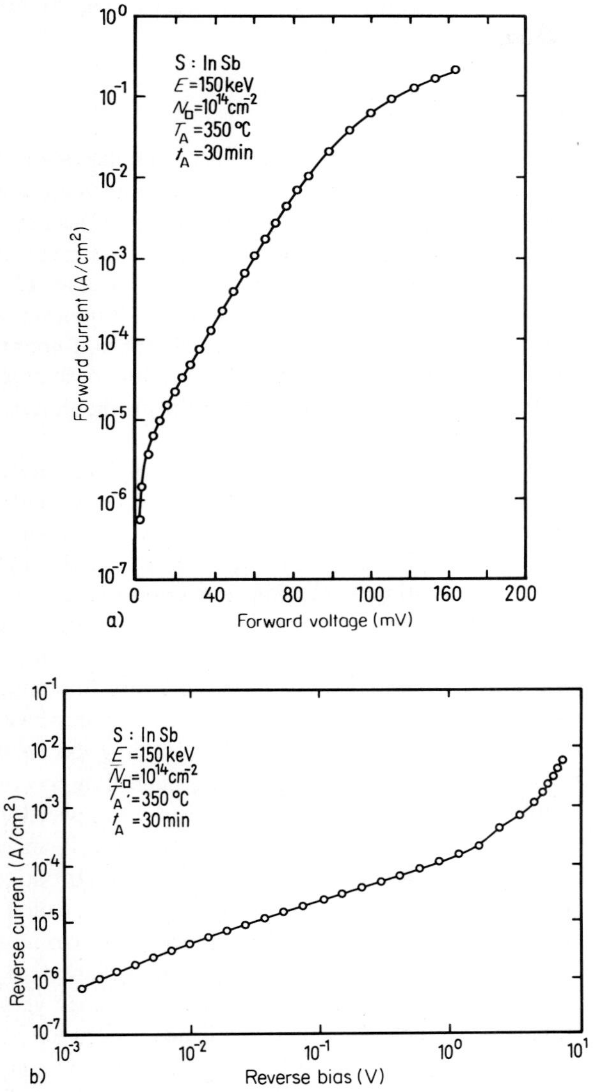

Figure 7.28 Current-voltage characteristics of a sulphur-implanted InSb detector [77]: (a) forward characteristic; (b) reverse characteristic

With the use of proton insulation, it has been possible to fabricate avalanche Schottky photodiodes in GaAs with gain-bandwidth products above 50 GHz [437]. The semi-insulating GaAs served as a guard ring to prevent surface breakdown.

7.3.7 Luminescence Diodes and Laser Diodes

In this field, there are three possibilities for the use of implantation: fabrication of *p–n* junctions, definition of the recombination area through implantation of protons or oxygen, and increase of luminescence efficiency in GaP or $GaAs_{1-x}P_x$ by nitrogen implantation. Recently, there has been an increase in published studies on the implantation of *p–n* junctions. Wiemer *et al.* [774] have demonstrated for the first time that Zn-implanted luminescence diodes in $GaAs_{1-x}P_x$ ($x \approx 0.4$) are equal in quality to diffused diodes. Fundamental experiments have been discussed in Section 6.3.2. Several Japanese studies have also shown the high suitability of zinc implantation for the production of luminescence diodes [551], [774].

In Figure 6.85, the relative luminescence intensity was shown for diodes implanted with 150 keV zinc ions, depending on dose and implantation temperature, with a comparison to diffused diodes. It is seen that an optimum dose is between 4×10^{15} and 8×10^{15} cm^{-2}. The reason for the high efficiency is a relatively high lifetime, resulting from a low zinc concentration at the *p–n* junction. The maximum brightness of the diodes was 5140 cdm^{-2} at 8.9 A/cm^2. Following implantation, the diodes were diffused under a SiO_2 masking layer in a vacuum at 750 °C until the *p–n* depth amounted to 2 μm. Implantation at higher temperatures resulted in a lower luminescence efficiency. In contrast to this study, Itoh *et al.* had to anneal at 900 °C [378], [381] to obtain good results, and found an improvement with implantation at 400 °C. Although they too obtained better results through implantation, their luminescence yield was lower by a factor of 2 to 3. The relatively high sheet resistivity of the implanted layer seems to be a problem which can lead to an inhomogeneous current distribution even under optimum conditions, as shown in Figure 7.29. An additional shallow implantation or diffusion could help alleviate this problem under some circumstances.

A number of studies have been made on beryllium implantation in GaAs and $GaAs_{1-x}P_x$ [131], [132]. Chatterjee *et al.* [131] have obtained very good results with beryllium-implanted diodes. They implanted doses between 1.5×10^{14} and 6×10^{14} cm^{-2} at room temperature into $GaAs_{1-x}P_x$ ($x \approx 0.38$), at an energy of 130 keV. Annealing was done between 800 and 900 °C under a Si_3N_4 masking layer. Best results were obtained after annealing at 900 °C for 1 hour. The maximum light intensity was 83 μcd at 10 mA.

Regarding the increase of the luminescence efficiency by formation of isoelectrical traps in GaP and $GaAs_{1-x}P_x$ through ion implantation, only fundamental photoluminescence-measurement results are available; these have already been discussed in Section 6.3.2.

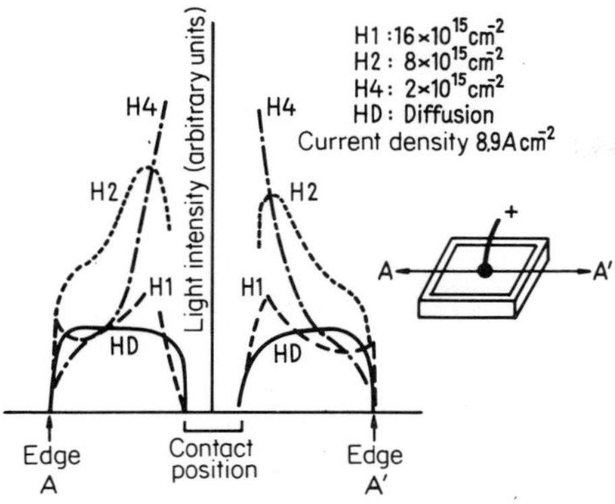

Figure 7.29 Spatial distribution of luminescence in implanted $GaAs_{1-x}P_x$ diodes depending on the ion dose [551]. Reproduced by permission of the Japanese Journal of Applied Physics

Barnoski *et al.* [57] have fabricated laser diodes through zinc implantation. At room temperature and 20 keV, they implanted a dose of 10^{16} cm^{-2} into *n*-type GaAs doped with 10^{18} cm^{-3} of tellurium. After annealing at 900 °C for 3 h under a SiO_2 masking layer, the junction depth was 1 to 2 μm. At 77 K, the threshold current density was 2000 Acm^{-2}.

The formation of semi-insulating layers to define the recombination area of $Ga_{1-x}Al_xAs$ heterolasers is an important use of ion implantation with III–V semiconductors. The laser must be made with a stripe structure for attainment of stable light emission [168], [553]. The four main possibilities for fabrication of such stripe structures are compared in Figure 7.30. Least favourable is structure (a), since in this case the current flow is only concentrated in the centre by the oxide layer [592]. A zinc diffusion, as shown in (b), yields better results; a simpler solution, though, is the insulation of the areas to the left and the right of the stripe through proton irradiation (c) [209]. This can be accomplished, for example, by a simple masking with tungsten wires, thus making possible (as with structures (a) and (b)) a system fabrication with the active layer below, for better heat dissipation (a so-called upside-down mounting). The best results in regard to the threshold current, however, can be obtained with the technologically difficult to produce mesa structure [736] shown in (d).

The great advantage of proton insulation is the simplicity of the process: no separate diffusion, as in case (b), is necessary; one photoresist process is eliminated (a), (b), (d); and the heat dissipation is better than for structures (a) and (d). A disadvantage is posed by the higher optical losses in semi-insulating material, although these can be greatly reduced by an appropriate annealing step. After a proton implantation with an energy of 300 keV and a dose of

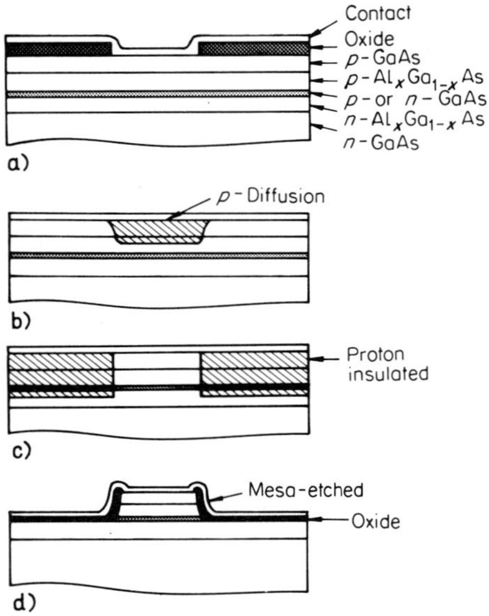

Figure 7.30 Fabrication of lasers by: (a) contact structure; (b) contact diffusion; (c) proton isolation; (d) mesa etching

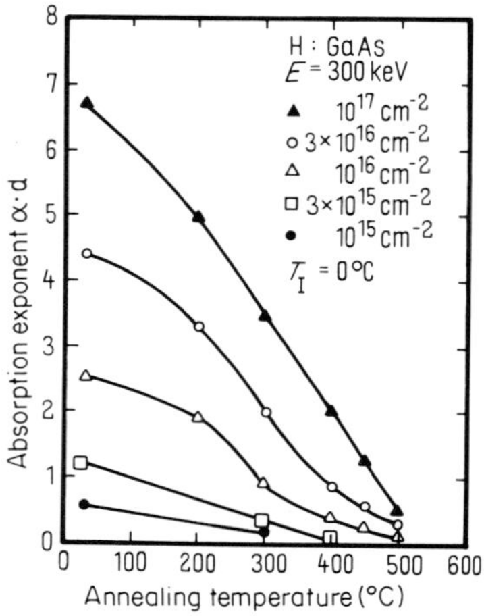

Figure 7.31 Isochronal annealing behaviour of optical absorption depending on the ion dose of proton-implanted GaAs; annealing time 15 min [210]

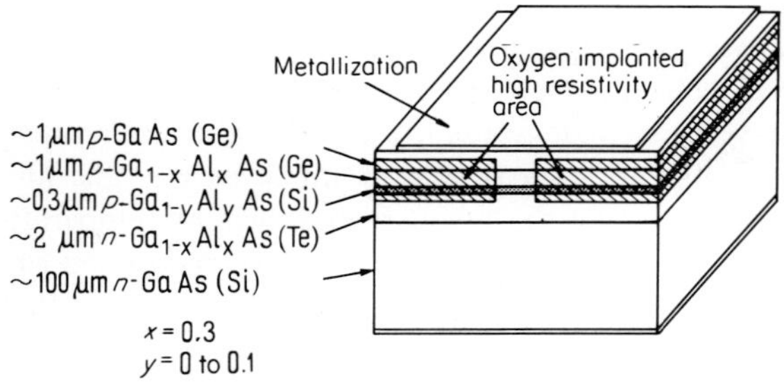

Figure 7.32 Schematic illustration of an oxygen-implanted heterolaser [88]

3×10^{15} cm^{-2}, the optical losses can be reduced to the pre-implantation value by annealing at 450 °C for 15 min, while the resistivity is still 200 times the original value (1 Ωcm *p*-type after annealing; 10^{-2} Ωcm *p*-type before implantation). Figure 7.31 shows absorption values for various ion doses at 300 keV. In Figure 6.81, the sheet resistivity behaviour of *p*-type GaAs has already been shown, depending on proton dose and annealing, according to the same article by Dyment *et al.* [210]. The width of the semi-insulating layer, depending on the energy, is about 1 µm/100 keV. The resistivity is approximately constant over the entire insulating layer. Due to the high proton range, practically all important insulation depths can be reached.

A further possibility for the fabrication of stripe lasers is offered by oxygen implantation, which also makes GaAs semi-insulating [329]. Since this is a case of chemical doping, the annealing is not especially critical [88]. A typical structure is shown in Figure 7.32, where (as opposed to proton insulation) the insulating region does not extend up to the surface. The most favourable dose was found to be 10^{14} cm^{-2}. A disadvantage of this method is posed by the extremely high acceleration voltages. The projected range for 2.5 MeV oxygen ions is only about 2 µm [88]; this means that standard ion accelerators are not usable for this purpose.

7.4 BIPOLAR TRANSISTORS

The first attempts to fabricate bipolar transistors by ion implantation into silicon [408], [740] were not very successful. The reasons for this were inadequate knowledge of the doping profiles, especially regarding the tails of emitter distributions in the base region. A breakthrough was achieved with the use of arsenic instead of phosphorus as the emitter dopant [58], [561], [562], since arsenic has less pronounced profile tails. With few exceptions, only *n–p–n* transistors in silicon have been implanted up to the present. Following initial experiments with combinations of diffusion and implantation, it is now usual to implant the base (boron) as well as the emitter (arsenic). It has also been possible to fabricate radio-frequency transistors with excellent characteristics in germanium through implantation of the base (arsenic) and the emitter (aluminium) [625]; as usual for germanium, however, these are *p–n–p* transistors.

Several different concepts have been investigated experimentally. On the one hand, attempts have been made to use low annealing temperatures [561], [679] in order to retain the implantation profile as much as possible and thus achieve improved process control; on the other, high annealing temperatures have been used to diffuse the emitter distribution beyond its profile tail and displace the emitter-base junction into undamaged material having a high lifetime [562], [585]. The latter concept is now the generally accepted one, due to the better emitter characteristics made possible. At the high concentrations necessary for emitters, arsenic has a highly concentration-dependent diffusion coefficient (*cf.* the profiles in Section 6.1.3). It is therefore possible to eliminate the profile tails which are also present for arsenic, probably due to channelling, by means of a diffusion step. At the same time, profiles are thus obtained which are steeper than those possible with ion implantation alone.

The sequence of base and emitter fabrication can be chosen freely with ion implantation. If the base is implanted before the emitter, one must accept the fact that the base is diffused too during the necessary emitter diffusion. Therefore, it is clearly better to implant the emitter first, to diffuse at relatively high temperatures (950–1000 °C), and then to implant the base and activate it for only a short time by a low-temperature annealing (850–900 °C) to ensure good process control over the base width. In addition, by using this method, a coupled diffusion of the boron and the arsenic (which could have an adverse effect on the transistor parameters) is also avoided [794]. Regardless of these facts, quite a few experiments have been made with implantation of the base prior to the emitter. The reason for this is compatibility with transistor fabrication using conventional diffusion methods.

A schematic cross-section through a typical high-frequency transistor with oxide isolation technology, together with the related doping profiles for the emitter and base, are shown in Figure 7.33 [303]. Two implantations are generally used for the base: a shallow contact implantation, which simultaneously serves to prevent lateral injection (dose: 5×10^{13} to 5×10^{14} cm^{-2}; energy: 10–50 keV), and a deeper implantation to form the active base (dose: 3×10^{12} to 5×10^{14} cm^{-2}; energy: 20–300 keV). The emitter is implanted in a single step with energies of 40–150 keV and doses of 4×10^{15} to 2×10^{16} cm^{-2}. The wide range of doses for the active base implantation is due to differences in the desired transistor data, as well as to the extent of base profile and emitter profile overlapping through implantation and annealing.

Through the use of ion implantation, it is possible within broad limits to vary the transistor parameters with relative independence of one another. The collector current I_C depends on Q_B, the integral of the base doping over the base width, and is given by:

$$I_C = \frac{q n_i^2 A_E D_n}{Q_B} \exp\left(\frac{qV_{BE}}{kT}\right) \tag{7.11}$$

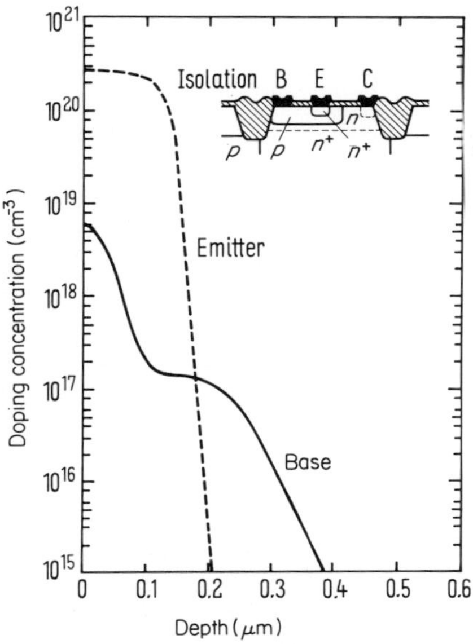

Figure 7.33 Cross-section through an implanted radio-frequency transistor, with its doping profile [303]

where A_E is the emitter area, V_{BE} the base emitter voltage, n_i the intrinsic carrier concentration, D_n the electron diffusion coefficient in the base, and Q_B/D_n the base Gummel number. The base current is given by:

$$I_B = \frac{qn_i^2 A_E}{(Q_E/D_p)_{eff}} \exp\left(\frac{qV_{BE}}{kT}\right) \tag{7.12}$$

where $(Q_E/D_p)_{eff}$ is the effective emitter Gummel number. For the current gain, one obtains via equations (7.11) and (7.12):

$$h_{FE} = \frac{I_C}{I_B} = \frac{(Q_E/D_p)_{eff}}{Q_B/D_n} \tag{7.13}$$

Since the base doping is proportional to the implantation dose of the deep base implantation, the following is valid:

$$h_{FE} \sim \frac{1}{N_{\square}} \tag{7.14}$$

Thus, the current gain can be adjusted by varying the dose of the active base implantation, with a constant emitter efficiency. Figure 7.34 shows an experimental example for this relationship, based on equation (7.14) [562]. A more exact treatment of these relationships is given in a work by Archer [42].

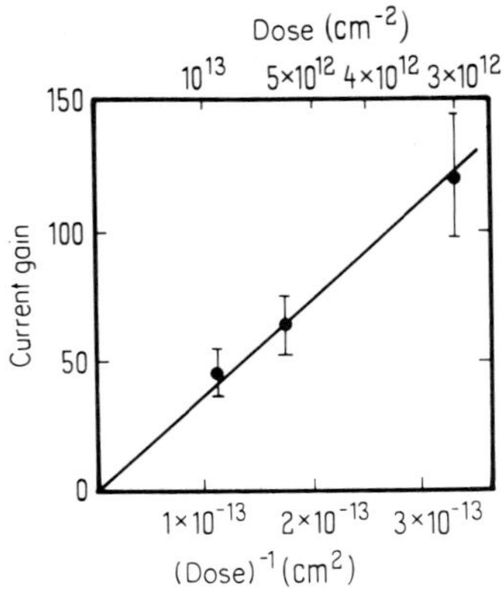

Figure 7.34 Current gain h_{FE} as a function of the active-base implantation dose for a 5 GHz transistor [562]. Reproduced with permission from Payne *et al.*, *IEEE Trans. Electron Devices*, **ED-21** (1974), 273. Copyright © 1974 IEEE

The cut-off frequency can be influenced by the energy of the active base implantation, since the base transit time (τ_B) is a function of the base width and thus a function of the implantation energy. An example for this is shown in Figure 7.35. Here, the current gain is changed somewhat too, since during variation of the energy, more or fewer ions are implanted into the active base. A complete separation of cut-off frequency and current gain is therefore not possible. The

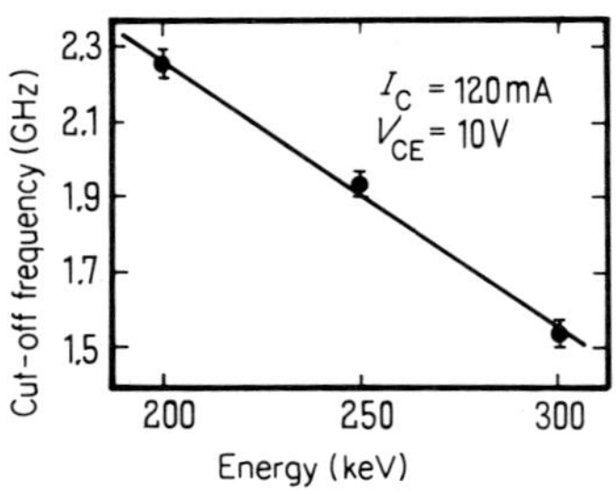

Figure 7.35 Cut-off frequency f_T depending on energy of the active-base implantation for a 2 GHz transistor [562]. Reproduced with permission from Payne *et al.*, *IEEE Trans. Electron Devices*, **ED-21** (1974), 273. Copyright © 1974 IEEE

emitter-base breakdown voltage is determined by the dose of the shallow base implantation, and can therefore be adjusted too within certain limits.

The main advantages of implanted transistors, due to the high degree of flexibility in determining their characteristics as well as the good reproducibility obtainable, are a significant reduction in parameter variations and, as a result, a very considerable increase in the yield. A further improvement of high-frequency transistors might be possible through the fabrication of 'pedestal' collectors, using high-energy implantation [795] or radiation-enhanced diffusion [610].

7.5 FIELD EFFECT TRANSISTORS

Field effect transistors have been fabricated through ion implantation into silicon [425], germanium [756] and especially GaAs [218], [333], [356], [357], [402], [707], [741], [748]. Lecroisnier and Pelous [425] have produced a power field effect transistor in silicon through implantation of a buried finger-shaped structure for the gate. A diagram of this structure is shown in Figure 7.36. Through the use of ion implantation, fabrication of this device is quite simple, though it is theoretically possible to produce it by conventional means too. Boron ions at 600 and 900 keV were used for implantation of the gate; the total dose was 6×10^{14} cm^{-2}. A microphotograph of a bevelled cross-section through this finger-shaped structure is shown in Figure 7.37; also given here are the dimensions of the structure. An output power of 200 mW at 2.7 GHz, with a power gain of 6 dB, were obtained with this transistor. Similar research has been carried out by Dolan [192].

Because of the higher electron mobility in GaAs as compared to silicon, field effect transistors in n-type GaAs are important devices for microwave frequencies; they are generally fabricated as Schottky field effect transistors. Ion

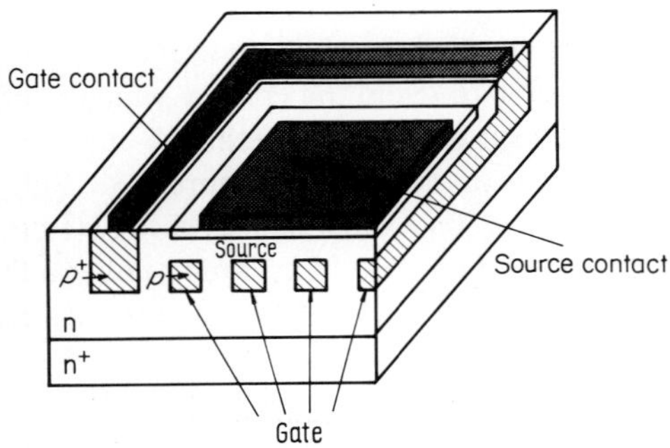

Figure 7.36 Schematic cross-section through an implanted power field effect transistor with a finger-shaped gate structure [425]. Reproduced with permission from Lecroisnier and Pelous, *IEEE Trans. Electron Devices*, **ED-21** (1974), 113. Copyright © 1974 IEEE

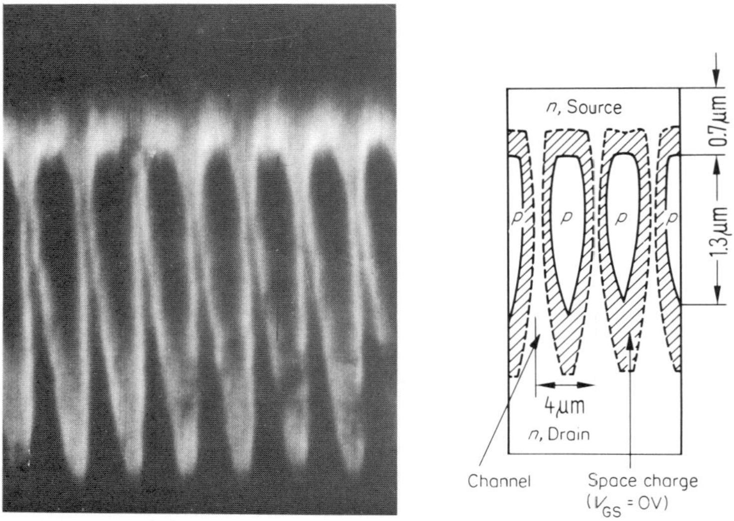

Figure 7.37 SEM photograph of a 2° bevel cross-section through a power field effect transistor. The space-charge appears white (potential contrast) [425]. Reproduced with permission from Lecroisnier and Pelous, *IEEE Trans. Electron Devices*, **ED-21** (1974). Copyright © 1974 IEEE

implantation can be used in three cases: for the doping of the source and drain areas, for channel implantation in semi-insulating or low-doped GaAs, and for definition of structures by proton insulation. Source and drain require high concentrations. For *n*-type doping elements (e.g., selenium and tellurium), no doping above 10^{19} cm^{-3} has been reached as yet (*cf.* Section 3), so that ion implantation for this use is still in the research stage [218]. The results are considerably better for attempts to produce the channel through ion implantation [333], [356], [357], [402], [418]. It is thus possible to avoid the epitaxy which is otherwise needed, as well as to produce selectively conducting areas. A mesa-etched transistor by Hunsperger [357] is shown as an example in Figure 7.38. Sulphur was implanted with a dose of 10^{13} cm^{-2} at 150 keV as well as 5×10^{12} cm^{-2} at 30 keV. The electrical activation after annealing (800 °C, 20 min, under a SiO_2 masking layer) depended very greatly on the substrate used, and was between approximately 7 per cent and 36 per cent.

Stolte [707] has shown that, in semi-insulating GaAs with a low chromium content as well as in low-doped GaAs after implantation of low sulphur doses, an 85 per cent electrical activation is obtainable (for implantation at 500 °C). He also determined that the Hall mobility is higher than the theoretical mobility. In both cases, the use of sulphur seems to yield somewhat better results than tellurium. For such experiments, the use of good semi-insulating or undoped epitaxial substrate material is quite important. Higgins *et al.* [333] have implanted

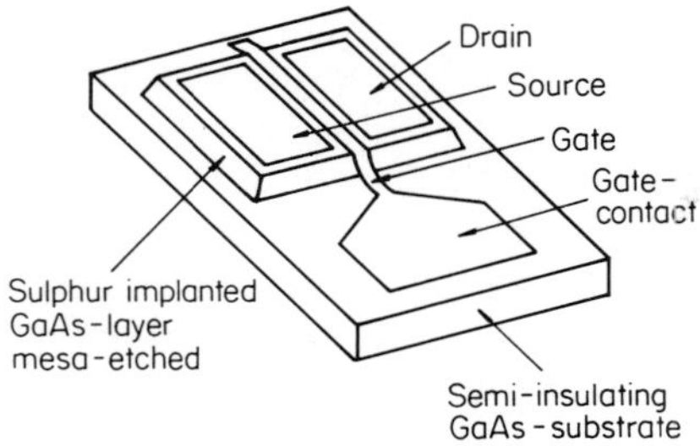

Figure 7.38 Illustration of an implanted mesa-etched Schottky field-effect transistor in GaAs [357]. Reproduced by permission of Pergamon Press

sulphur at 100 keV and selenium at 400 keV, with annealing under Si_3N_4 passivation layers at 850 °C for 30 min. The doses were 9×10^{12} cm^{-2} for sulphur and 2×10^{12} cm^{-2} for selenium. The activation was above 90 per cent for selenium but about 20 per cent for sulphur. For this reason, different doses were used in transistor fabrication to enable comparison of the results. In contrast to epitaxial transistors, the selenium-implanted or sulphur-implanted transistors showed no gain drift and no noise dependence on the drain voltage, although there was a slightly higher noise level. The implanted transistors were just as good, with a gain of 11 dB at 10 GHz. Kellner *et al.* [402] have even achieved a 100 per cent electrical activation for sulphur at a dose of 5×10^{12} cm^{-2}, with an annealing temperature above 860 °C. However, their transistors produced more noise than those fabricated epitaxially. Tung-ho *et al.* [741], using sulphur implantation (energy: 100 keV, dose: 1.5×10^{13} cm^{-2}), obtained activations of 8.8 per cent after annealing at 750 °C, 30.3 per cent for 800 °C, and 68.4 per cent for 900 °C. In contrast to the other experiments, annealing was done in these cases under arsenic pressure for 15 min.

Junction field effect transistors have been fabricated by Vodicka and Zuleeg [748] through implantations of sulphur and tin. Sulphur was implanted at 30 and 150 keV with doses of 10^{13} cm^{-2} at 350 °C, and tin at 30 to 40 keV with doses of 10^{14} cm^{-2}, into Cr-doped semi-insulating material with annealing at 750 °C under a SiO_2 layer for 10 min.

Definition of field effect transistor structures in GaAs by proton insulation has not yet been carried out, although its applicability has been shown for the case of planar Gunn elements [742].

7.6 VARIOUS SEMICONDUCTOR DEVICES

In addition to the above-mentioned uses of implantation for the development and fabrication of devices, there are also a number of other promising applications;

these will be touched on briefly in this section. In the field of III–V semiconductors, there are especially a great many possibilities for the future of implantation.

By the use of proton bombardment, Stoll *et al.* [706] have developed photodetectors which can be integrated into optical GaAs light guides. For this purpose, they increased the absorption coefficient under a Schottky contact by means of proton bombardment. At 1.15 μm and room temperature, the external quantum efficiency was approximately 16 per cent. This integrated detector is shown in Figure 7.39.

Mizutani and Kurumada [503] have implanted sulphur at 200 keV with a dose of 5×10^{12} cm^{-2} at room temperature into Cr-doped semi-insulating GaAs. The Cr content of the substrate material was lower than 10^{16} cm^{-3}. After implantation, annealing was done under a SiO_2 passivation layer at 800 °C for 210 min. The doping efficiency was 90 per cent. With this material, they were able to produce planar Gunn elements for digital applications. The structuring of the elements was done with mesa etching. The nL and nd products were 4×10^{14} cm^{-2} and 4.4×10^{12} cm^{-2}, respectively. The characteristics of these Gunn elements were better than those of correspondingly doped epitaxial layers. The authors have suggested the fabrication of complete planar structures through selective implantation, as well as the monolithic production of field-effect transistors together with Gunn devices on the same substrate, for use in integrated pulse regenerators and other applications.

Upadhyayula *et al.* [742] have fabricated planar Gunn elements with the use of proton insulation. They implanted doses of 10^{13} cm^{-2} at energies of 280 keV, 200 keV and 100 keV into epitaxial material. During implantation, masking was done with phosphorus-silica glass, since thicker masking layers can be deposited in this way than with pure SiO_2. This technique seems very promising for planar structures in GaAs.

Pruniaux *et al.* [578] have also produced MIS transistors in GaAs using proton insulation. For a gate of 5 × 500 μm, a transconductance of 5 mA/V was

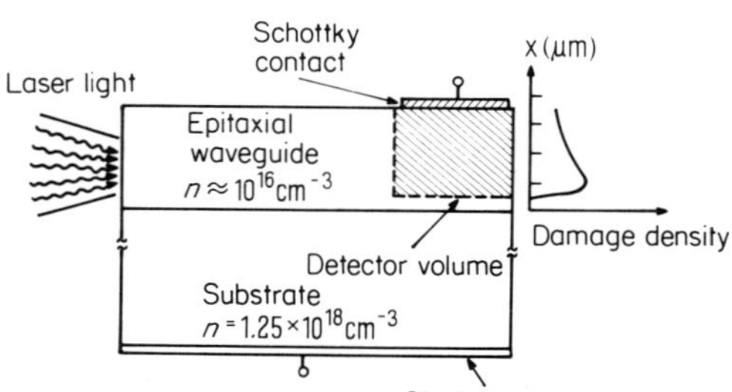

Figure 7.39 Cross-section through a proton-implanted optical wave-guide detector [706]. Reproduced by permission of the American Institute of Physics

obtained. Lepselter *et al.* [432] have combined Schottky barriers with boron-implanted source and drain areas in silicon to produce MOS transistors. Platinum–silicide Schottky contacts were used as the source-drain contacts, reinforced by a titanium–platinum–gold layer which also served as the gate.

Many experiments have also been made on the use of silicon to produce insulating layers through implantation [641], [644], i.e., by irradiation with oxygen and nitrogen as well as protons. The results, though, were not as good as those obtained for several III–V semiconductors.

8

Implantation into Non-semiconductors

In recent years, implantation into non-semiconductors has become increasingly important, aside from the classical applications of ion implantation for semiconductor doping. Here, initial attempts to produce SiO_2 or Si_3N_4 through oxygen or nitrogen implantation into silicon [258], [759] have been of some interest, but much more attention has been given to the following two fields:

(a) implantation into metals, and
(b) implantation into optical materials.

Much research has been done in both of these areas, but only a few of these studies will be mentioned in this chapter to give a general idea of the many possibilities of ion implantation—from surface hardening of razor blades to passivation of uranium—and also, perhaps, to suggest future applications. Detailed reviews on the use of implantation beyond the field of semiconductors have been published by Thomson [726a] and Grant [302a].

8.1 IMPLANTATION INTO METALS

In this section, the main influences of implantation on metals will be discussed. Implantation has only been applied to metals in the relatively recent past, since the ion doses needed are comparatively high. Although doses of 10^{11} to 10^{16} cm^{-2} are adequate in semiconductor technology, doses in the area of 10^{17} cm^{-2} to 10^{18} cm^{-2} are generally needed for metals. In addition, the number of usable ion species is considerably larger than in semiconductor technology.

There have been several international conferences dealing exclusively with ion-beam applications for metals [29], [30a]. In the book entitled *New Uses of Low-Energy Accelerators* [182a], Dearnaley provides a general survey on implantation into metals, with the emphasis on corrosion. All aspects of ion implantation into metals are treated in a volume edited by Hirvonen [13a]. Picraux [572a] has published a comprehensive work on solubility, precipitation, lattice incorporation, and diffusion after implantation of ions into metals. The influence of implantation on tribological properties of metals was discussed in detail in a recent conference [30a].

8.1.1 Corrosion

The main impetus for intensive research in the field of implantation into metals was given by the investigation of Crowder and Tan [159a] on copper passivation through boron implantation, following the initial results of Trillat and Haymann [730b] on uranium passivation through argon irradiation. The main reason for the increased research has been the interest in materials for fusion reactors, especially in respect to their behaviour under extreme radiation levels, as well as in production of corrosion resistant, surface passivated metal layers.

The most obvious method for surface passivation is the implantation of ions which prevent or reduce corrosion as constituents of alloys, for example, nickel or chromium in steel. In addition, implantation is especially well suited for systematic investigations of corrosion mechanisms, since it offers the possibility to implant practically any substance with a known concentration.

Corrosion is an extremely complicated process which depends on the electrochemical properties of the metal constituents and of the oxidation medium, as well as on grain boundaries and grain boundary diffusion, mechanical tension, temperature and atmosphere. Possibilities for investigating these effects are offered especially by the back-scattering technique, electron microscopy and activation analysis. In the following, several practical studies in these areas will be outlined.

Detailed studies of oxidation of titanium and stainless steel have been made by Dearnaley *et al.* [182a], [182c]. Ten different ion species were investigated in regard to their oxidation reducing effects. After implantation, the samples were

Table 8.1 Effects of ion implantation upon the oxidation of stainless steel [182a]

Sample number	Ion species	Ion dose	Oxidation effect
S1	Ca	10^{15} cm^{-2}	Increased 20%
S2	Ca	10^{16} cm^{-2}	Increased 40%
S3	Ca	10^{16} cm^{-2}	Increased 51%
S4	Ca	5×10^{16} cm^{-2}	Increased 100%
S5	Y	10^{16} cm^{-2}	Reduced 30%
S6	Y	5×10^{16} cm^{-2}	Reduced 50%
S7	Eu	5×10^{16} cm^{-2}	Increased 50%
S8	Eu	5×10^{16} cm^{-2}	Increased 62%
S9	Bi	10^{16} cm^{-2}	Reduced 28%
S10	Bi	10^{16} cm^{-2}	Reduced 31%
S11	Bi	5×10^{16} cm^{-2}	Reduced 50%
S12	In	2×10^{16} cm^{-2}	Reduced 32%
S13	Al	10^{17} cm^{-2}	Reduced 52%

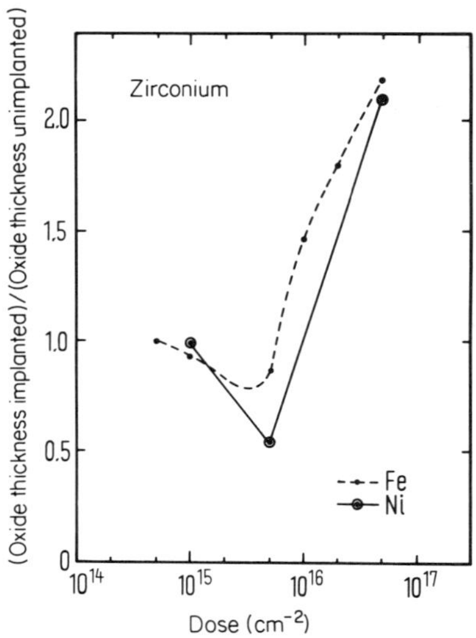

Figure 8.1 Influence of Fe and Ni implantation on oxidation of electropolished zirconium as a function of ion dose. The minimum corresponds to a surface concentration of approximately 3000 p.p.m. [182c]

oxidized in a dry oxygen atmosphere (600 °C for titanium, 800 °C for steel) for 25 to 30 min. Calcium ions were found best suited for preventing titanium oxidation. Higher doses (up to 5×10^{16} cm^{-2}) increased this effect. Europium had a similar influence. All other elements (Zn, In, Ce, Al, Y, Ni and Bi) increased the oxidation rate. For stainless steel, opposite results were obtained. Calcium and europium increased the oxidation rate; the other ion species decreased it. This effect is highly correlated with the electro-negativity of the elements. In Table 8.1, several results for stainless steel are shown.

Similar investigations have been made for aluminium [115a], [730a]. In this case, the oxidation follows the Wagner–Hauffe rule [325a], i.e., the oxide thickness is proportional to the square root of the time (diffusion-limited oxidation). Gold and silver, for example, reduce oxidation, while lead and bismuth increase it. At high doses of bismuth, oxidation already takes place at room temperature. In the case of chromium, it was found that the oxidation behaviour of implanted layers is mainly determined by radiation damage [523a]. With nickel too, radiation damage has a great influence on the oxidation behaviour [298a]; in addition, differences in atomic radius are especially important. Further experiments have been made with copper [590] and zirconium [182a]. With zirconium, all implanted elements increase the oxidation rate, except for the transition elements (Fe, Ni, Cr) and niobium [182a]. The oxidation experiments were made in dry oxygen at temperatures between 380 and 400 °C. The oxygen incorporation

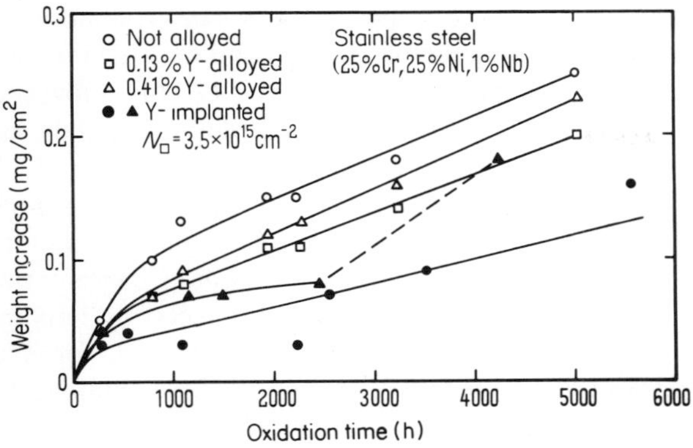

Figure 8.2 Influence of yttrium implantation with a dose of 3.5×10^{15} cm^{-2} on the oxidation behaviour of stainless steel in a CO_2 atmosphere at 700 °C compared to alloyed samples [41b]

was determined using the ^{16}O (p, p) reaction. The dose dependence of the oxide thickness for iron and nickel is shown in Figure 8.1. The optimum dose is obtained at about 3×10^{15} cm^{-2}. In the case of alloys (Zirkaloy 2), similar concentrations are required for optimum oxidation reduction [398].

A number of experiments have dealt with corrosion in aqueous solutions, in order to determine specific boundary conditions. Salt-spray apparatus was used in some cases; in others, oxidation potentials in electrolytes (e.g., 0.1 M NaOH + Na_2SO_4) were determined [43b], [43c].

The formation of surface alloys by ion implantation whose characteristics are already known through conventional production techniques is a more direct method than the previously mentioned experimental one. The implantation of yttrium into stainless steel [41b], for example, can reduce oxidation at high temperatures. Ion implantation is especially well suited here since for homogeneous alloys, hardness and strength are poor, probably due to yttrium deposition at grain boundaries. The passivating effect of implanted yttrium on austenitic steel (25 per cent Cr, 25 per cent Ni, 1 per cent Nb) in a CO_2 atmosphere at 700 °C is compared to yttrium-alloyed steel in Figure 8.2. The dose was 3.5×10^{15} cm^{-2}; multiple implantation was performed at various energies between 80 and 410 keV to dope a 0.2 μm thick layer as homogeneously as possible [41b]. The weight increase through oxidation is highest for unalloyed steel and lowest for the implanted samples. The passivating effect of yttrium was even more pronounced at 800 and 850 °C.

8.1.2 Investigation of Reactor Materials

A problem with nuclear reactors is the high level of particle irradiation within the reactor. Especially for fusion reactors currently being planned, comprehensive material studies are needed. For example, through irradiation with energetic helium atoms generated in the reactor during various (n, α) reactions, blisters or bubbles are produced in the metals, since helium has only a low solubility. These blisters can greatly change the properties of the reactor materials, especially those of steel. Through irradiation with fast neutrons, small 'empty' blisters (10 nm diameter) are formed in steel; these are called 'voids'. Ion implantation makes it possible to examine materials within relatively short times regarding their suitability for reactor use, in the first case by helium implantation, and in the second (in order to distinguish between the two effects) by implantation of ions which are practically inert in the various types of steel, for example, nickel, chromium or iron. In order to obtain relatively deep distributions, implantation is often done at very high energies with multiply-charged ions. The doses used are between about 10^{16} cm^{-2} and 10^{19} cm^{-2}. The most important methods to investigate these effects are transmission and scanning electron microscopy.

Photographs of niobium surfaces, after helium irradiation at 900 °C with an energy of 500 keV and a dose of about 5×10^{17} cm^{-2}, at various dose rates, are shown in Figure 8.3 [167b]. An increased blistering at the surface can be seen at high dose rates. An even clearer situation results after implantation into molybdenum using helium doses of 2.5×10^{18} cm^{-2}, at a current density of 1.2×10^{-4} A/cm^2 (Figure 8.4). In this case, nearly circular blisters are formed, from which the 'lid' has been pressed off in some cases [745a]. More or less similar results are obtained after irradiation of practically all reactor construction materials; only the dose dependence varies.

As concerns the conditions for blister formation in metals, on the one hand the depth at which the blisters form must be greater than the thickness of the sputtered metal layer up to a critical dose N_C, while on the other, the mobility and solubility of the implanted ions must be low [596a]. If the depth at which the blisters form is set equal to the projected range R_p, it follows for the start of blister formation based on the first condition that:

$$S \leqslant R_p \frac{N}{N_C} \tag{8.1}$$

where N is the atomic density of the substrate and S is the sputtering yield.

Equation (8.1) is satisfied for all light ions such as helium and hydrogen; in cases of heavy ions (e.g., argon), blister formation only occurs at high energies at which the penetration depth increases and the sputtering rate decreases. This can be realized experimentally through irradiation at various angles. The greater the deviation between angle of incidence and normal, the fewer blisters will be formed. The second condition explains why very high doses ($\geqslant 5 \times 10^{18}$ cm^{-2}) are needed in the case of hydrogen irradiation for blister formation, and why a high dose-rate dependence exists [596a]: hydrogen has a very high diffusion coefficient in most metals. The 'deckeldicke' (lid thickness) as well as the blister diameter increase with rising acceleration energies. Das *et al.* [167a] have found a very good correlation between average range and the deckeldicke in aluminium,

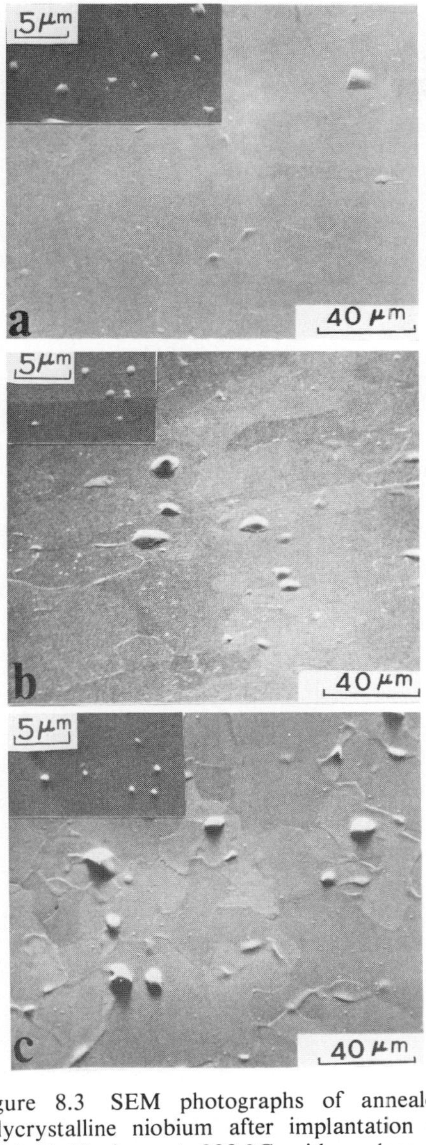

Figure 8.3 SEM photographs of annealed polycrystalline niobium after implantation of 500 keV He ions at 900 °C with a dose of 0.1 Cb cm^{-2}, for fluxes of (a) 10^{13} cm^{-2} s^{-1}, (b) 10^{14} cm^{-2} s^{-1}, (c) 10^{15} cm^{-2} s^{-1} [167b]

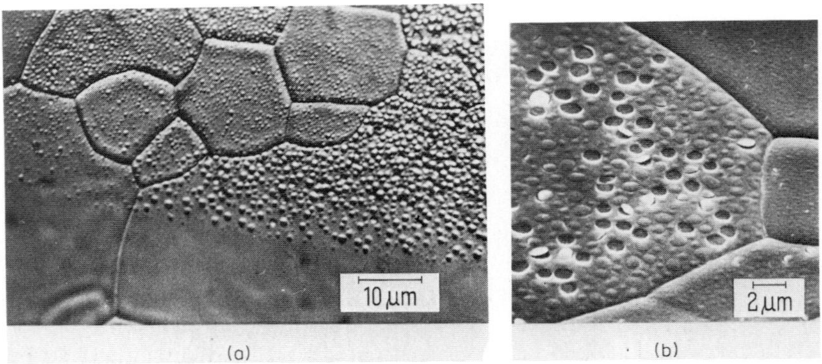

Figure 8.4 Optical photograph (a) and SEM photograph (b) of an annealed Mo sample implanted at 15 keV with $2.5 \times 10^{18}\ cm^{-2}$ He ions; current density: $1.2 \times 10^{-4}\ Acm^{-2}$ [745a]

vanadium and niobium. It is assumed that the explanation for blister formation is the accumulation of gases in the form of bubbles, which finally press away the material above them [62], [396]. At low energies, mechanical tension can also cause greater deckeldicke than those corresponding to the ion range [596a]. Figure 8.5 shows a diagram of this mechanism. It has been assumed here that the

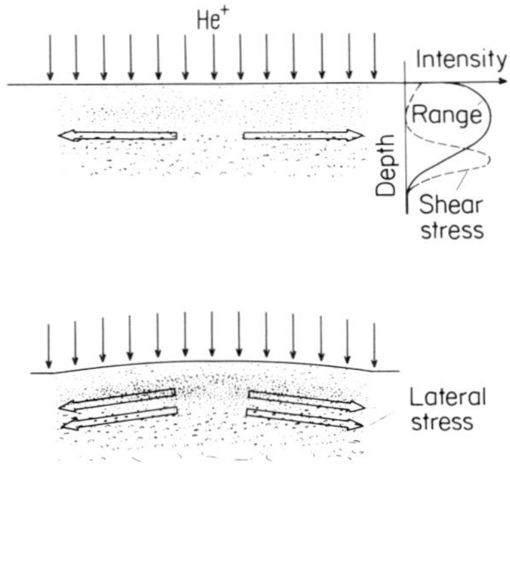

Figure 8.5 Mechanism of blister formation. Hydrostatic gas pressure and implantation-induced stress lead to the formation of lids ('Deckel') [596a]

shear stress is proportional to the concentration gradient. Several recent studies on blister formation have been published in the conference proceedings *Application of Ion Beams to Materials* [30] and *Application of Ion Beams to Metals* [29].

Much research has also dealt with the formation of voids. A good survey has been provided by Johnston and Rosolowski [388a] as well as by Kulcinski [419a]; both also quote many older studies. Through irradiation with high energy neutrons, large numbers of atoms are displaced and vacancies as well as interstitial atoms are produced. These recombine or are trapped at sinks (the sample surface, grain boundaries, dislocations or existing voids). Since interstitial atoms are trapped at least at one of these sinks, most likely at dislocations, the vacancies collect at other positions and finally precipitate as voids. Helium is not necessary to produce voids, but void formation increases greatly if helium is present in the metal. Regarding theories on formation of voids, see References [43a], [154a], [576a].

The investigation of voids by ion implantation was originally suggested by Nelson and Mazey [536a]; since then, a great many experiments have been carried out. A listing of already used ion species is given in Table 8.2 [388a], [419a]. Figure 8.6 shows a stainless steel sample implanted with 10^{17} cm^{-2} nickel atoms at 625 °C. The irradiated area has increased in volume and has a different structure than after blister formation. In crystalline metals, a formation of superlattice structures can occur if the dose is high enough ($\geqslant$1 dpa = displacement per atom) and if the temperature is close to the threshold temperature for void formation [388a]. Volume increase of stainless steel after irradiation with 20-MeV C^{++} ions is shown in Figure 8.7 [2].

Table 8.2 Ion species used for investigation of void formation; according to [388a] and [419a]

Implanted element	Energy (MeV)	Remarks
H	0.1–1.2	
D	12.3	
C	0.1–20	Also double-charged ions
N	2.0	
O	0.1	
Al	0.4–2	
Fe	0.1–36	Up to octuple-charged ions
Ni	0.1–46.5	Also double- and sextuple-charged ions
Cu	0.5–15	Also double- and triple-charged ions
Se	6–11	
Nb	15	
Ta	7.5	

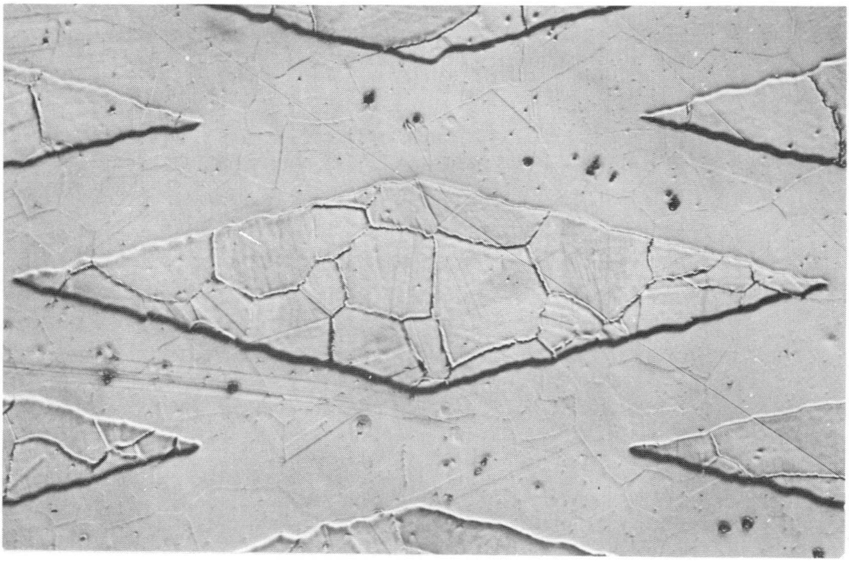

Figure 8.6 Implantation of nickel into stainless steel at 625 °C with a dose of 10^{17} cm^{-2} with a mesh mask [388a]

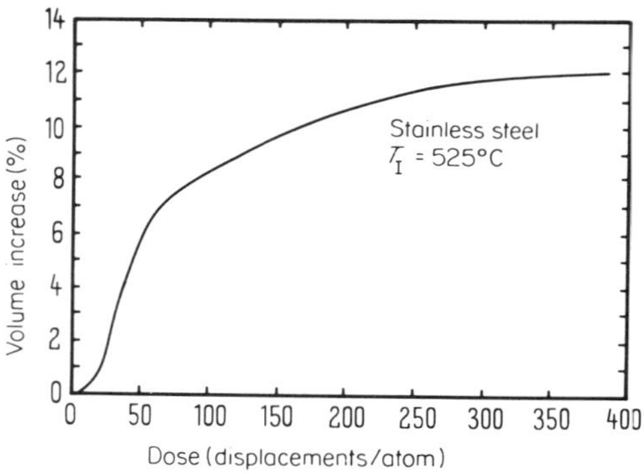

Figure 8.7 Dose dependence of the volume increase for stainless steel after nickel implantation at 525 °C [2]

8.1.3 Alteration of Mechanical Surface Properties

Alteration of the mechanical surface properties of metals has also been the subject of much research. Good reviews of recent results have been compiled by Hartley [318], Dearnaley [186a], [186b], Hirvonen [13a] and Ashworth *et al.* [30a].

An increase in the hardness of steel has been observed after implantation of the following ions: nitrogen, argon, boron and carbon [397], [557], [713a]. This can

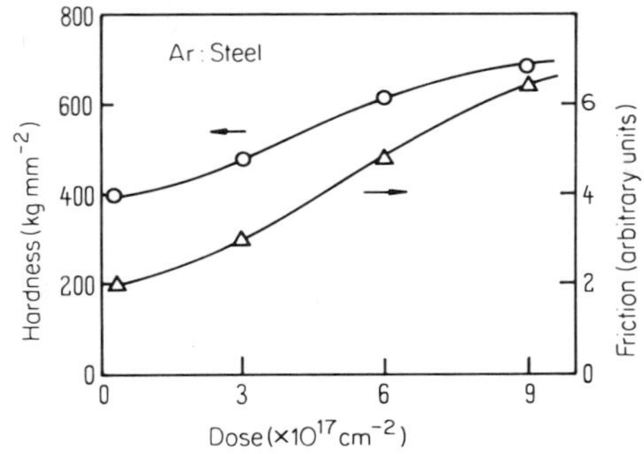

Figure 8.8 Alteration of microhardness (upper curve) and friction, as a function of dose, for argon implantation into steel at 40 keV [318]

be explained in terms of radiation-damage effects extending far beyond the projected range [285], [299] as well as chemical effects produced by the ions. As an example, the increases in microhardness and friction coefficient after irradiation of steel with argon ions are shown in Figure 8.8. The increases amount to more than two or three times the initial values, respectively; the doses, though, are extremely high. The implantation of nitrogen has been the subject of especially intensive research. Nitrogen increases the hardness of steel by radiation damage as well as nitride formation [271], [397], [557]. For example, the Vickers hardness of steel can be increased from 300 kg to 400 kg through irradiation with 10^{17} cm^{-2} of nitrogen. Boron too is quite well suited for increasing the hardness and friction resistance of iron [713a]. The dose needed in this case is surprisingly low: 10^{15} cm^{-2}. Double implantations of nitrogen together with nitride-forming elements, such as Cr, V, Mo and Al, could also be of interest for hardness alteration [318].

Material wear due to friction can also be decreased by implantation, often by up to a factor of 30 [318], [319]. This effect is especially pronounced in the case of heavy ions, but is apparently independent of a chemical effect. An example for aluminium implanted with H, C or Ar at 40 keV [557] is shown in Figure 8.9. The amount of material wear is a function of the hardness as well as the amount of friction, although the relationship is not a simple one. Therefore, the implantation of metallic and non-metallic ions also has a great influence on the friction coefficient. In this case, the different effects obtained with different ion species are partly due to the influence of the implanted ion on the oxidation behaviour [318]. In part, the effects produced by implantation can be attributed to better 'lubrication' by chemical material changes at the surface. For example, Hartley [318], [319] has been able to reduce the friction of steel considerably by implantation of molybdenum with the corresponding double dose of sulphur. The use of either of these elements alone leads to significantly less friction reduction. The formation of MoS_2, however, could not be verified.

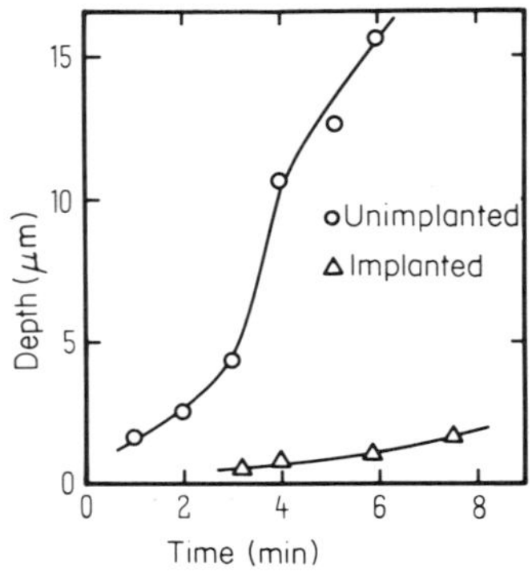

Figure 8.9 Decrease in wear for implanted aluminium. (O): not irradiated; (Δ): irradiated with N, C or Ar at 40 keV; according to Reference [557]. Reproduced with permission

8.1.4 Formation of Superconductors

By implanting ions into superconducting metals, it is possible to alter the transition temperature or the critical current. These changes are attributable either to radiation-damage effects or to the formation of alloys or metastable phases [710]. Increases in the transition temperature for molybdenum after implantation of nitrogen, carbon or sulphur, depending on the impurity concentration, are shown in Figure 8.10. The maximum transition-temperature increase amounts to 6 K. In order to obtain a high transition temperature, alloys must be found with a high electron–phonon coupling constant (i.e., a high electron state density at the Fermi energy), a low average phonon frequency, and a high degree of phonon–electron interaction. The difficulties in manufacturing materials with high transition temperatures are connected with crystal-lattice instabilities occurring with such alloys, which are especially well suited for superconduction because of their phonon spectra. Since many different alloys must be prepared and their properties have to be measured, ion implantation again offers advantages. In the case of implantation, the solubility can also be exceeded. However, in order to alter the transition temperature by room temperature implantation, the diffusion coefficient of the implanted substance has to be low, so that it does not precipitate at grain boundaries. As a solution, low temperature implantation (down to helium temperatures) can be applied [112], [326]. In this way, it is possible to form metastable phases far beyond solubility limits; for example, the solubility of lead in manganese is 1 p.p.m., whereas by implantation, approximately 1200 p.p.m. can be incorporated [112]. The ease with which the concentrations of the implanted ions can be varied makes implantation an especially outstanding method for investigating superconduction.

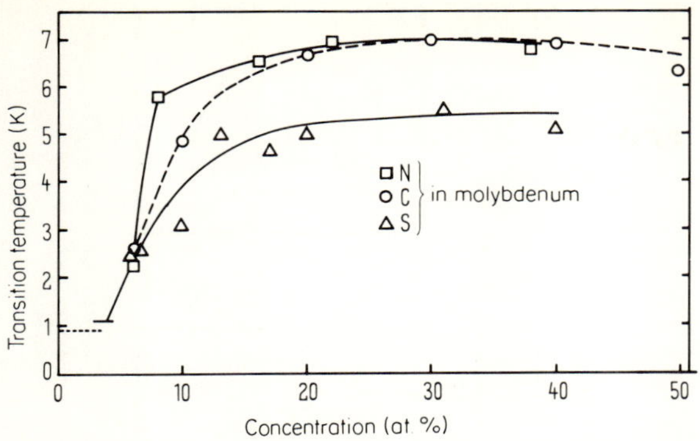

Figure 8.10 Increase in transition temperature of molybdenum after implantation of sulphur, nitrogen or carbon as a function of impurity-atom concentration [495]. Reproduced by permission of Kernforschungszentrum, Karlsruhe

Depending on the alloy, an annealing step may be needed to incorporate the implanted atoms into the metal lattice (A 15 alloys) [710]. Annealing should in no case be carried out for the case of metastable alloys, or for that of an increase in the transition temperature due to radiation damage. In Figure 8.11, an example for $NbC_{0.89}$ after an additional carbon implantation is shown. The dose was so

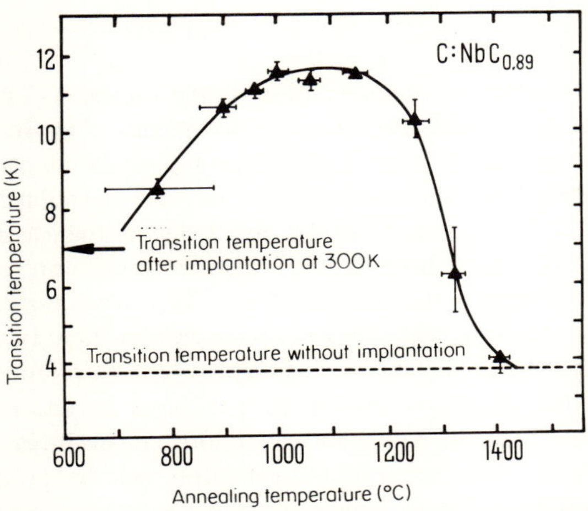

Figure 8.11 Transition temperature of a $NbC_{0.89}$ crystal after implantation of additional carbon depending on annealing temperature. The arrow indicates the transition temperature after implantation at 300 K. Prior to implantation, the transition temperature was about 4 K [281]. Reprinted with permission from Geerk *et al.*, *Solid State Communications*, **23** (1977), 83. Copyright © 1977 Pergamon Press Ltd.

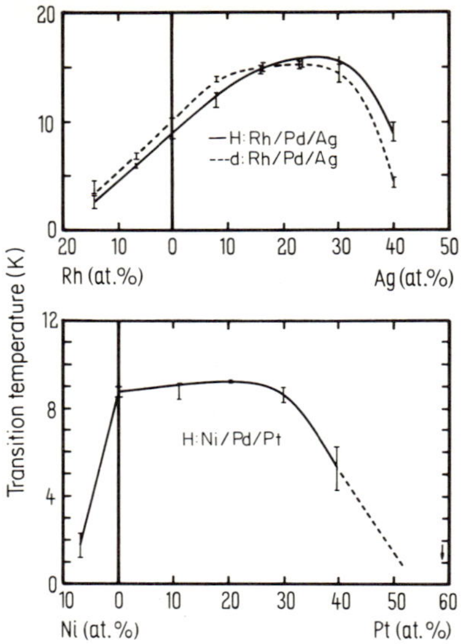

Figure 8.12 Maximum transition temperatures in the Rh–Pd–Ag and Ni–Pd–Pt systems after implantation of optimum doses of hydrogen (solid line) and deuterium (broken line) [710]

high that the stoichiometric relationship of 1 was exceeded. The annealing time was 200 s at all temperatures. The single crystal becomes superconducting at 4 K, while the implanted layer reaches a maximum transition temperature of 11.6 K. At higher temperatures, the out-diffusion of carbon reduces the transition temperature again. This increase was not obtainable by the use of other methods.

Metastable alloys have been formed mainly by proton implantation into palladium alloys at temperatures below 4 K [710]. Transition temperatures for Pd–Rh and Pd–Ag alloys, depending on their composition as well as hydrogen and deuterium irradiation, are shown in Figure 8.12. An optimum increase results for a concentration of approximately 25 per cent. Similar results have been obtained with Pb–Cu alloys. Alloys formed at low temperatures, however, are not necessarily unstable. Quasiamorphous structures can be produced in molybdenum by implantation at 4 K, which are stable even at room temperature and possess high transition temperatures. Examples for various group V elements implanted at 4 K are shown in Figure 8.13.

Within the scope of this chapter, it is not possible to present a detailed survey of this field. However, publications by Heim and Stritzker [326], Stritzker [710], and Meyer *et al.* [496], [496b] discuss the use of ion implantation for superconductors in more detail.

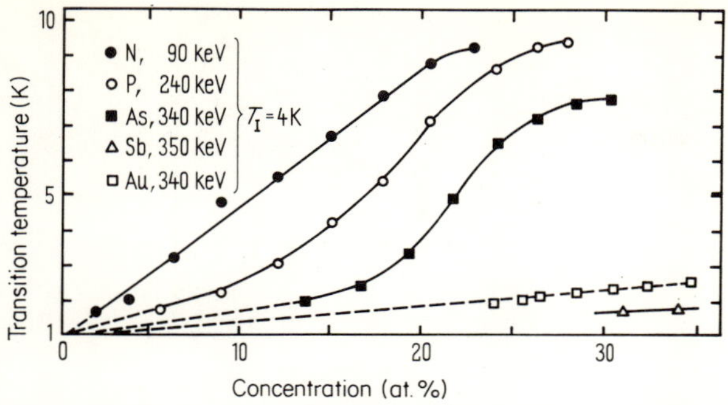

Figure 8.13 Increase in transition temperature depending on implanted concentrations of N, P, As, Sb and Au in molybdenum [496]

8.2 IMPLANTATION INTO OPTICAL MATERIALS

The least problematical use of implantation in this field is the fabrication of light guides by alteration of the refractive index. The light is conducted in the region where the refractive index has been increased. Since its maximum alteration occurs at the average radiation-damage range, X_D (if radiation damage is the cause of the refractive index change), or at the average range, R_p (in the case of chemically active ions), the fabrication of low-loss, buried light guides is thus possible without difficulty. Light having a wavelength λ_0 will be conducted in a light guide of thickness t under the following condition [279]:

$$\Delta\varepsilon/\varepsilon_0 > (\lambda_0/4t)^2 \tag{8.2}$$

where ε_0 is the absolute dielectric constant (permittivity) and $\Delta\varepsilon$ is the change in the relative dielectric constant.

Due to radiation-damage effects, an increase in the refractive index [64], [700] amounting to several per cent is usually found after irradiation of glass or quartz with ions. Through implantation of chemically active ions, for example, nitrogen into SiO_2, considerably greater changes in the refractive index are possible. The latter changes steadily in this case from 1.46 to 1.96, due to oxynitride formation. In some cases, on the other hand, the refractive index can be decreased by radiation-damage effects, for example, up to 10 per cent in $LiNbO_3$ [763]. The change in the refractive index for SiO_2 by implantation of Li atoms is shown in Figure 8.14 [700]. The refractive index increase is linear in this region and even after annealing at 500 °C, a change in the refractive index is retained. At the same time, the optical losses are reduced very significantly by the annealing step. The ion doses used are between about 10^{15} cm^{-2} and several 10^{16} cm^{-2}, depending on the ion species. The implantation energy is chosen according to the desired light-guide dimensions. The various possibilities for fabrication of light guides are shown in Figure 8.15. Structures (a) and (b) in Figure 8.15 can be obtained by ion

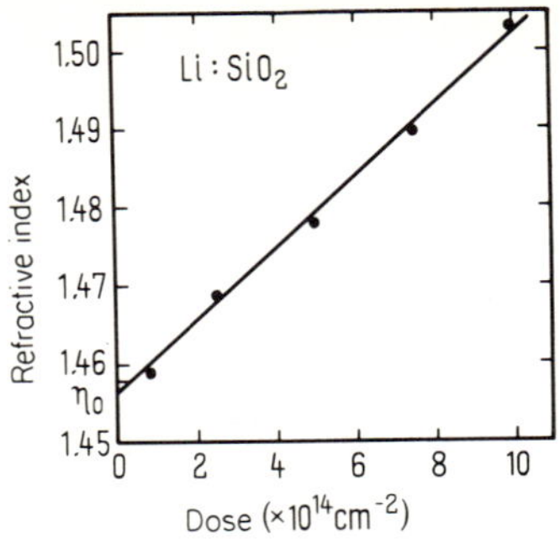

Figure 8.14 Change in the refractive index of quartz as a function of Li ion dose [700]. Reproduced by permission of the Optical Society of America

implantation; structure (c) can be made by etching, following a whole-area implantation, or by means of a deposition process. In addition to quartz and glass, GaAs and other III–V semiconductors are promising materials for light guides, since it is possible under some circumstances through their use to combine elements such as lasers, detectors, and couplers with light guides on the same substrate. The simplest fabrication method, besides epitaxy, is the proton insulation of GaAs (see Section 6.3.3). There is a very low carrier concentration in the implanted layer, and the optical dielectric constant increases as follows [279]:

$$\Delta\varepsilon = q^2 \frac{\Delta N}{\omega^2 m^*} \tag{8.3}$$

where ΔN is the difference in the carrier concentration between the substrate and

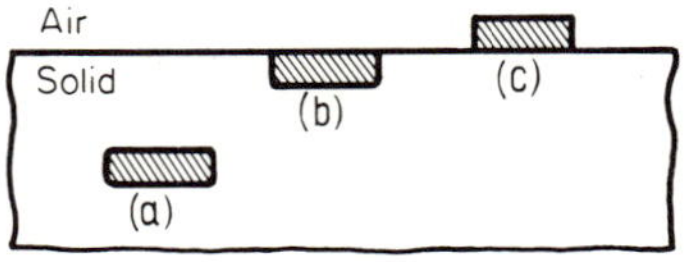

Figure 8.15 Schematic illustration of various types of light guides: (a) buried light guide (produced by implantation); (b) light guide at the surface (produced by implantation), (c) light guide on top of the surface (produced by implantation and etching, or by deposition with an appropriate masking procedure)

the implanted layer, m^* is the effective mass of the carriers, and ω is the optical frequency. Because of the high optical losses, the implanted structures have to be annealed; typical parameters are 500 °C for 30 min [279]. Equation (8.3), of course, is also valid for epitaxially produced light guides. If the doping of the compensating layer is very low, and if only the slowest light modes are to be transmitted [789], the following applies for the substrate:

$$N_B > (2s-1)^2 \frac{\pi^2 c_0^2 m^* \varepsilon_0}{4q^2 t^2} \tag{8.4}$$

Parallel light guides which are coupled by overlapping modes transmit power and form a directional coupler [499]. Somekh *et al.* [693] have produced couplers in GaAs by proton irradiation with doses of 6×10^{14} cm^{-2}. They used 2.4 μm wide light guides at a distance from each other of 6.4 μm. After 2 mm travel of the light, the total energy could be transmitted to the neighbouring light guides. If two light guides are connected by a ring coupler, the result is a frequency filter [464]. A band-blocking device, for example, can be made by using grid structures perpendicular to the light guide. A further application, the production of demodulators by implantation, has already been discussed in connection with electron devices. A few of these structures are shown in Figure 8.16. A disadvantage of implanted waveguides, compared to epitaxial or deposited guides, is their high loss, which still requires a detailed investigation.

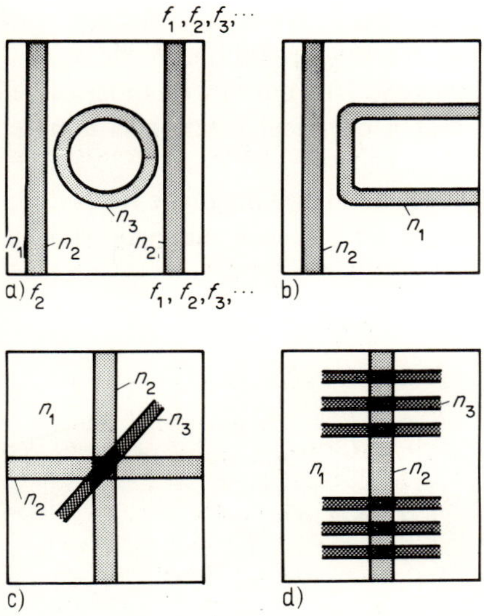

Figure 8.16 Opto-electronic light-guide devices: (a) frequency filter; (b), (c) directional couplers; (d) resonator

The use of colour centres in alkali halogenides for storage of optical information is well known [629]. Such colour centres can be fabricated selectively by ion implantation into NaF [459]. Li^+, B^+, C^+, O^+ and F^+ have been used with energies of 280 keV to 3 MeV and doses of 10^{13} cm^{-2} to 5×10^{15} cm^{-2}. Only the results with lithium were satisfactory. With colour centres made in this way, it was possible to store holograms for several years without losses.

8.3 FURTHER USES OF ION IMPLANTATION FOR NON-SEMICONDUCTORS

In addition to the above-mentioned uses of ion implantation outside of the semiconductor field, a great many experiments have also been carried out in other directions.

A subject of relatively great interest has been implantation into magnetic garnet crystals. Such crystals are suitable for non-volatile storage of digital information. Usually, epitaxial magnetic layers (e.g., $Tb_{2.4}Er_{0.6}Fe_5O_{12}$, or in general $A_{3-x}B_xFe_{5-y}Ga_yO_{12}$, where A and B are rare earths or yttrium) on a non-magnetic garnet substrate (e.g., $Sm_3Ga_5O_{12}$) are used. Without an external magnetic field, the domains in the epitaxial layer are randomly oriented (Figure 8.17(a)). Under the influence of a magnetic field, they form cylindrical regions having the same orientation ('magnetic bubbles'), with a diameter of a few microns (Figure 8.17(b)). By the use of evaporated NiFe contacts, these bubbles can be moved magnetically. In this way, storage of digital information (presence or absence of a bubble) is possible.

Ion implantation can be used in different ways in this connection. Implantation of protons, by reduction of stress between substrate and epitaxial film, leads to the necessary reorientation of the magnetization. Wolfe *et al.* [784] have compared the effects of protons (300 keV, 10^{17} cm^{-2}) with those of helium (also at 300 keV, 10^{17} cm^{-2}), and found that a chemical reaction is apparently the cause of the reduced stress.

'Hard' bubbles, which require higher magnetic fields ($\geqslant$55 Acm^{-1}) for their formation and which are based on a special magnetic spin in the domains, can be

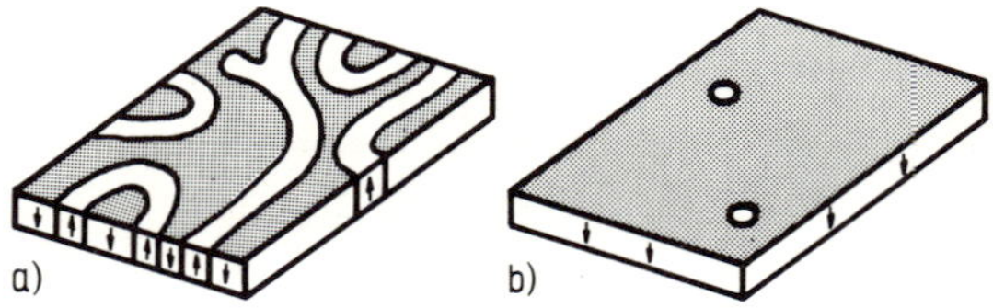

Figure 8.17 Structure of domains in magnetic garnet layers: (a) without magnetic field; (b) with magnetic field

eliminated by hydrogen irradiation [785]. The usable energy range, for $(YGdTm)_3(FeGa)_5O_{12}$, is between 25 and 300 keV at doses of 10^{16} to 10^{17} cm^{-2}. In all these experiments, annealing temperatures up to 1000 °C had no effect on the implantation produced results.

Conducting paths for bubbles can be produced in two ways by implantation (Johnson *et al.* [386]): by alteration of the magnetic properties along the path, or by an etching step, taking advantage of the extremely increased etch rate of garnet films after implantation of H, He or Ne ions. The etch rate in phosphoric acid is increased up to a factor of 1000.

As already mentioned, formation of chemical compounds can occur as a result of high dose implantation. The formation of Si_3N_4 and SiC, through implantation of nitrogen and carbon at room temperature as well as at 600 to 850 °C, has been studied by Edelmann *et al.* [211]. The compounds were formed after annealing at 850 °C. Examination of the layers was done with electron microscopy as well as infrared spectroscopy. Further studies on Si_3N_4 formation have been made by Freeman *et al.* [258]. The production of SiC has been investigated by Rothemund and Fritzsche [597] as well as Borders *et al.* [95] by measurement of transmission in the infrared region. Experiments on SiO_2 formation by oxygen irradiation of silicon have been made by Watanabe and Tooi [759], Freeman *et al.* [258], and Pavlov and Shitova [560]. Good insulating properties were found, but capacitance-voltage measurement results were unsatisfactory, and indicated that unannealed radiation damage probably plays an important role. The layers were unsuitable for semiconductor applications.

It is also possible through ion implantation to increase the adhesion of metal films on glass substrates [148], to polish glass surfaces using oblique ion irradiation [63], [563] (by ion sputtering), to sensitize insulators for selective electroless plating or selective evaporation of metal layers [147], [711], or to produce samples for Mössbauer experiments [751].

By implanting oxygen into aluminium or titanium, layers with specific resistivities between 10^{-5} and 10^{12} Ωcm can be produced by dose variation between zero and about 10^{17} cm^{-2} [564]. The electrical characteristics of the implanted samples have good reproducibility and are not annealing dependent. On the other hand, it is also possible to fabricate cermet thin-film resistors with values between 10 and 10^9 Ω [565] by irradiation of silicon-oxide layers (composition: SiO_x; $x \approx 1.5$) with aluminium or titanium. In this special case, the metal atoms are brought into the oxide layer by a secondary implantation (recoil implantation, see Section 3.7.2 and Figure 3.38).

Resonators for acoustic surface waves can be fabricated by helium implantation into quartz, with a dose of 1.5×10^{16} cm^{-2} and an energy of 100 keV [320]. The alteration in the optical impedance caused by the implantation changes the reflection of the surface waves.

Besides the non-semiconductor uses of ion implantation mentioned in this chapter, many further possibilities have been investigated experimentally; publications on this research can be found in the specialized literature of the individual disciplines. Therefore, the survey just presented should by no means be seen as a complete one. The examples mentioned here have only been intended to provide a general view, beyond the strict limits of semiconductor technology, into the many sided possibilities that ion implantation offers.

9
Appendix

In this appendix, a collection of semiconductor and insulator material parameters of importance to ion implantation is presented, including data on diffusion coefficients, solubilities, and impurity atom ranges, as well as vapour-pressure tables. Also provided are tables of the complementary error function as well as the abundance of isotopes of all elements.

9.1 DATA ON SEMICONDUCTORS AND INSULATORS

In Table 9.1, all data are given for 300 K if not otherwise specified. The mobility values given are for intrinsic or very low-doped material. In the case of higher doping concentrations, the mobility is usually lower (compare, for example, the relationship for silicon given in equation (5.34)). For insulators, the data fluctuate considerably because depending on the fabrication process (thermal oxidation, pyrolysis, growth of single crystals), different structures result, as well as slightly different stoichiometries in some cases.

9.2 DIFFUSION COEFFICIENTS

In Figures 9.1 to 9.4, diffusion coefficients are shown for a number of elements in silicon, germanium and SiO_2. Diffusion coefficients for several elements in III–V semiconductors are given in Table 9.2. In the cases of silicon and germanium, the values are relatively reliable; they are to be understood as intrinsic diffusion coefficients, while at higher doping concentrations an increase in the diffusion coefficient can occur (see Section 3.9.1). However, the values for SiO_2 as well as the III–V semiconductors are highly dependent on the experimental conditions, and can only provide rough approximations. For example, with Zn in GaAs at a constant temperature, the diffusion coefficient varies according to the concentration and the diffusion atmosphere by five orders of magnitude, probably due to differences in the relative influences of lattice diffusion and interstitial diffusion.

Table 9.1 Data on important semiconductors and insulators

Semi-conductor	Periodical group	Molecular weight	Symmetry‡	Density (g/cm²)	Heat conductivity (W cm⁻¹ K⁻¹)	Band gap (eV)	Melting point (°C)	Atomic density (10^{22} cm^{-3})
Si	IV	28.09	d	2.33	1.4	1.11	1 416±4	5
Ge		72.6	d	5.3	0.61	0.664	937	4.42
SiC		40.1	k, z, h	3.2	4.9	2.5; 2.994	2 830	4.81
Se	VI	78.96	h	4.8	0.02	1.74	220	3.66
Te		127.61	h	6.3	0.03	0.32	455	2.97
AlAs	III–V	101.9	k, z	3.6	0.08	2.16	1 750	2.12
AlN		41	h, w	3.18–3.26	0.3	5.9	2 400†	4.78
AlP		57.95	k, z	2.4	0.9	2.45	2 000†	2.49
BN		24.82	k, z; h, w	2.25	0.8	8.0	2 700; 3 000	5.46
GaAs		144.64	k, z	5.3	0.54	1.43	1 238	2.18
GaN		83.73	h, w	6.1	—	3.5	600	4.38
GaP		100.7	k, z	4.13	1.1	2.26	1 467	2.47
GaSb		191.48	k, z	5.6	0.35	0.7	712	1.76
InAs		189.74	k, z	5.7	0.26	0.36	943	1.81
InN		128.83	h, w	6.88	—	2.0	1 200†	3.21
InP		145.49	k, z	4.79	0.7	1.34	1 070	1.98
InSb		236.58	k, z	5.8	0.18	0.18	525	1.47
PbS	IV–VI	239.28	k, s	7.6	0.02	0.41	1 114	1.91
PbSe		286.17	—	8.3	0.04	0.29	1 065	1.74
PbTe		334.82	k, s	8.2	0.08	0.32	917	1.47
HgTe	II–VI	328.22	z	8.12	0.026	—0.15	670	1.49
ZnTe		192.99	z	5.6	0.11	2.25	1 300	1.74
CdSe		191.37	z	5.8	0.063	1.74	1 240	1.82
CdS		144.48	z	4.8	0.2	2.42	1 475	2
ZnS		97.45	z	4.1	0.26	3.66	1 830	2.53
ZnSe		144.34	z	5.3	0.3	2.67	1 520	2.21
CdTe		240.02	z	5.8	0.07	1.5	1 100	1.45
SiO_2		60.08	a	1.8–2.4* 2.24–2.27 (thermal oxide) 2.65 (single crystal)	0.014	~8	~1 700	2.3
Al_2O_3		101.96	a	3.8–4.0	0.25	8.7	2 050	2.25
TiO_2		79.9	a	2.40–2.49*			1 850	1.85
Si_3N_4		140.28	a	2.75–3.0* 3.18 (single crystal)			1900†	1.22

* Values obtained depend on the fabrication method (pyrolytic, evaporation, sputtering).
† Dissociated.
‡ a = amorphous, h = hexagonal, k = cubic, d = diamond, s = rock salt, w = wurtzite, z = zinc blende.

Table 9.1—*contd*

Dielectric constant	Dielectric strength (V/cm)	Intrinsic Mobility: Electrons (cm^2/Vs)	Intrinsic Mobility: Holes (cm^2/Vs)	Refractive index	Thermal expansion coefficient ($10^{-6}/K$)
11.9	30×10^6	1 880	600	3.45	2.44–2.5
16	8×10^6	4 000	2 000	5.6	5.5–5.8
9.72; 10.32	—	300	50	2.5; 2.6	2.9
8.5	—	17	—	3.7	—
5.0	—	1 000	—	2.7	—
12.0	—	1 200	200	3.3; 3.1	—
9.14	10^7	—	14	1.99	4.03–6.09
11.6	—	80	—	3.4	—
7.1; 3.8	2×10^6	—	—	2.2	3.5; $a_0 = -2.9$, $c_0 = 40.5$
11.1	3.5×10^6	8 500	400	4.03	5.9–6.0
		1.6×10^5	4 000 (77 K)		
13.18	—	400	—	2.0	$a_0 = 5.59$; $c_0 = 3.17$
9.036	—	150	120	3.45	5.81
		2 100 (77 K)	1 000 (77 K)		
14.0	—	5 000	1 000	3.8	6.7
11.7	—	2.3×10^4	200	4.56	5.19
—	—	50	—	—	—
12.35	—	4 600	650	3.45	—
17.72	300	10^5	1 700	4.22	5.04
		10^6 (77 K)	7000 (77 K)		
175	—	610	700	3.7	—
250	—	10	950	—	—
400	—	1 730	840	3.8	18–21
20	—	2.2×10^4	500	3.7	4
9	—	340	110	3.6	—
10.6	—	800	10	2.9	—
5.3–10	—	300	50	2.5	—
5.13	—	165	5	2.4	—
8.4	—	608	28	2.9	—
10.9	—	1 000	80	2.8	5.5
1.7–4.4*	6×10^6	—	—	1.32–1.50* 1.460–1.466 (thermal oxide)	0.5
7.5–9.6*	3×10^6	—	—	1.74	—
to 82*	—	—	—	1.64	—
6–9*	6×10^6	—	—	1.9–2.1	3.85–4*

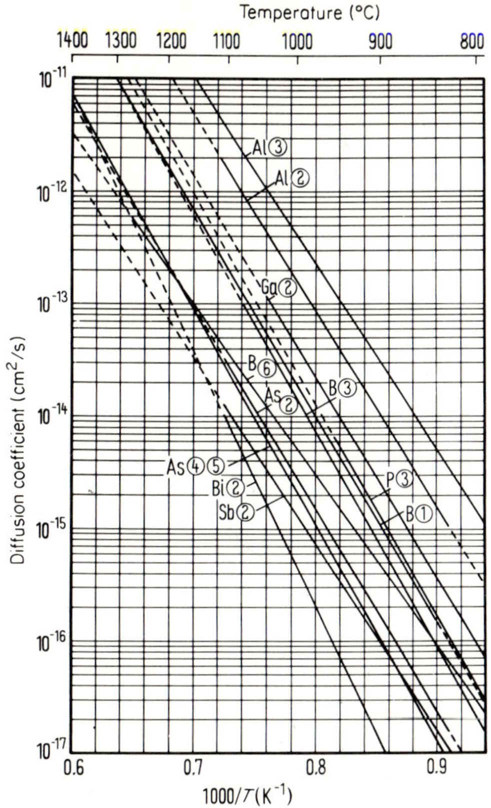

Figure 9.1 Diffusion coefficients of slow-diffusing elements in silicon, according to Fair (1) [232], Fuller and Dietzenberger (2) [265], Neuberger (3) [537a], Masters and Fairfield (4) [478a], Ohkawa *et al.* (5) [545a] and Hirayama and Shohno (6) [335a]

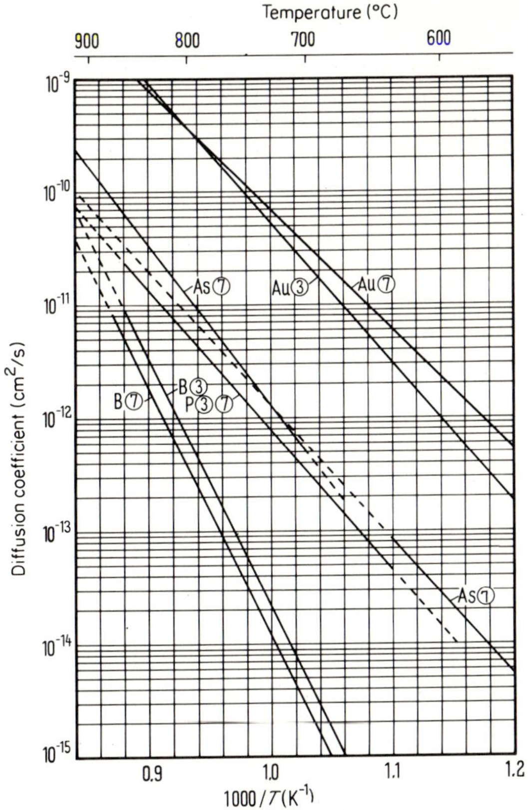

Figure 9.2 Diffusion coefficients of various elements in germanium, according to compilations by Gürs and Gürs [307b] (7) as well as Neuberger [537a] (3)

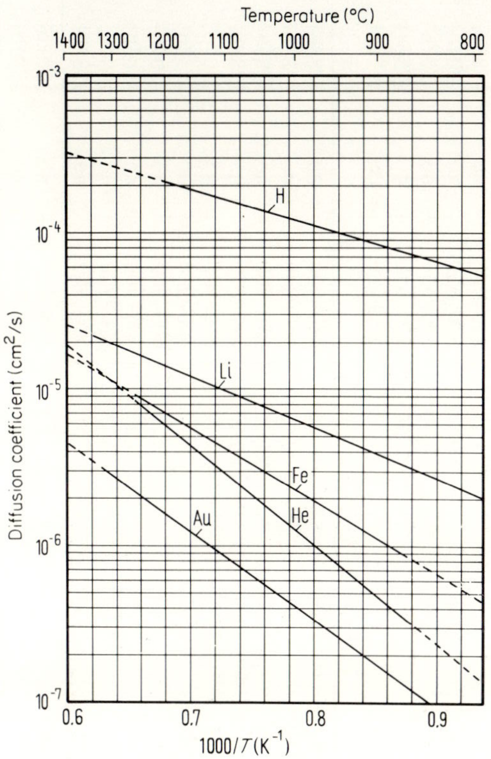

Figure 9.3 Diffusion coefficients of rapidly-diffusing elements in silicon, according to a compilation by Gürs and Gürs [307b]

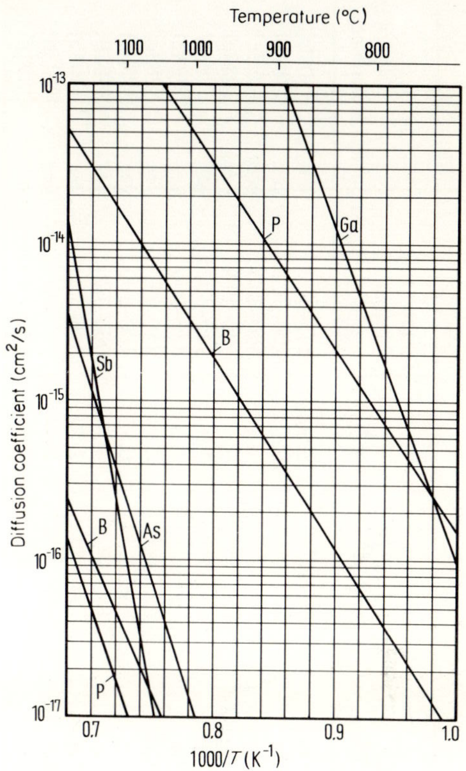

Figure 9.4 Diffusion coefficients of several elements in SiO_2, according to a compilation by Ghezzo [285]. For boron and phosphorus, the minimum as well as the maximum published values are given

Table 9.2 Diffusion coefficients of several elements in III–V semiconductors; $D = D_0 \exp(-E_a/kT)$

Semi-conductor	Element	D_0 (cm^2/s)	E_a (eV)	Temperature range (°C)	Anomaly	References
GaAs	Ga	1×10^7	5.60	1125–1225	no	[307b]
	As	4×10^{21}	10.2	1200–1225	yes	[307b]
	Ag (slow	2.5×10^{-3}	1.5	500–1160	yes	[788c]
	Ag (fast	4×10^{-4}	0.8	500–1160	yes	[788c]
	Au	1×10^{-3}	1.0	740–1025	yes	[788c]
	Be	7.3×10^{-6}	1.2	880–990	yes	[788c]
	Cd	5×10^{-2}	2.6	850–1150	yes	[788c]
	Cu	3×10^{-2}	0.53	100–500	yes	[788c]
	Se	3×10^3	4.16	1000–1200	yes	[307b]
	Sn	6×10^{-4}	2.50	1096–1201	no	[307b]
	Zn	15	2.49		yes	[307b]
	S	1.85×10^{-2}	2.6	900–1200	no	[791a]
GaP	Zn	1.0	2.1	700–1300	yes	[788c]
	S	3.2×10^3	4.7	1100–1300	no	[791a]
GaSb	Ga	3.2×10^3	3.14	658–700	no	[788c]
	Sb	3.4×10^4	3.43	658–700	no	[788c]
	Sb	8.7×10^{-3}	1.13	220–650	no	[788b]
	In	1.2×10^{-7}	0.53	320–650	no	[307b]
	Sn	2.4×10^{-5}	0.80	320–650	no	[307b]
	Te	3.8×10^{-4}	1.2	320–650	no	[788c]
	Zn	$D = (1.8-8)10^{-11}$ cm^2/s		560–580	yes	[166a]
InAs	Ag	7.3×10^{-4}	0.26	450–900	no	[788b]
	Au	5.8×10^{-3}	0.65	600–840	no	[788b]
	Cd	7.4×10^{-4}	1.15	650–900	no	[788b]
	Cu	3.6×10^{-3}	0.52	240–510	no	[788b]
	Ge	3.74×10^{-6}	1.17	600–900	no	[307b]
	Mg	1.98×10^{-6}	1.17	600–900	no	[788b]
	Se	12.55	2.20	600–900	no	[307b]
	Sn	1.5×10^{-6}	1.17	600–900	no	[788b]
	Te	3.34×10^{-5}	1.28	600–900	no	[307b]
	Zn	4.2×10^{-3}	0.96	600–900	no	[788b]
InP	In	1×10^5	3.85	894–920	no	[307b]
	P	7×10^{10}	5.65	904–980	no	[307b]
InSb	In	5×10^{-2}	1.81			[788c]
	Cd	1.26	1.75	400–500	no	[308a]
	Sn	5.5×10^{-8}	0.75	390–512	no	[712a]
	Zn	1.6×10^6	2.3	350–500	no	[307b]
	Sb	1.4×10^{-6}	0.75	300–500	no	[307b]
	Te	1.7×10^{-7}	0.57	300–500	no	[307b]
	Au	7.0×10^{-4}	0.32	140–510	no	[788b]
	Cd	1.0×10^{-5}	1.1	250–500	no	[788b]
AlSb	Cu	3.5×10^{-3}	0.36	150–550		[788c]

9.3 SOLUBILITIES OF ELEMENTS IN SILICON AND GERMANIUM

In Figures 9.5 and 9.6, chemical solubilities are shown for various elements in silicon and germanium, according to References [307b] and [731]. This chemical solubility is not to be considered as identical to an electrical activation limit, since complex formation [235], [243] is not taken into account in this case. The values can therefore serve only as approximations for determining the upper electrical activation limit.

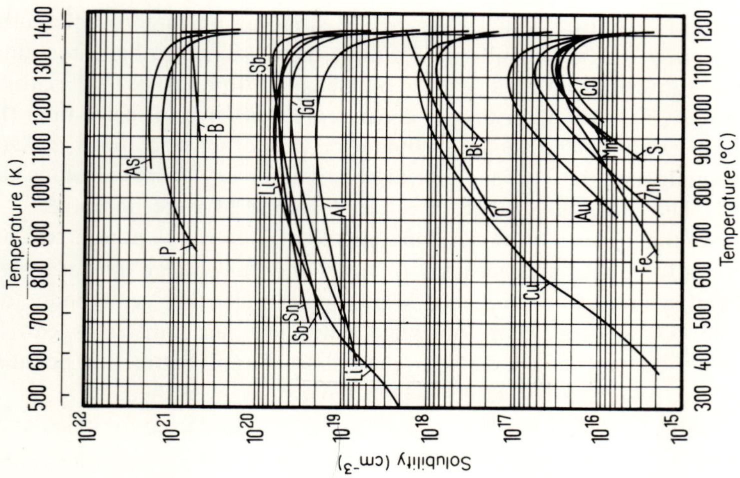

Figure 9.6 Chemical solubility of various elements in silicon [307b], [731]

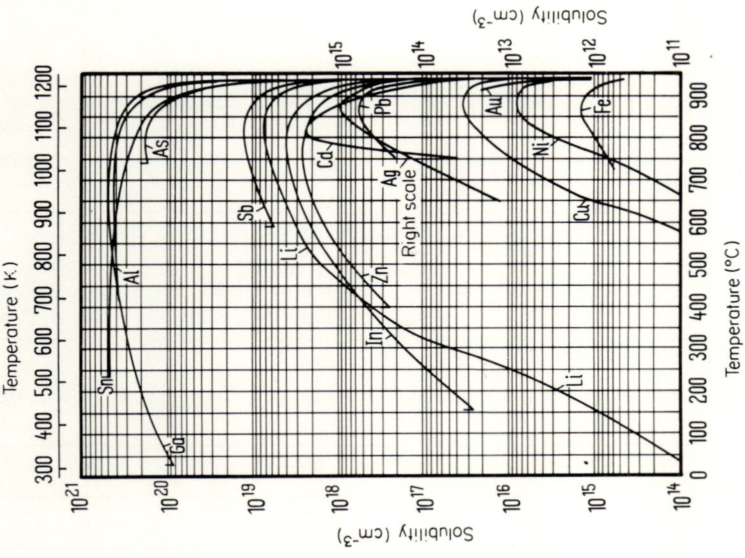

Figure 9.5 Chemical solubility of various elements in germanium [307a], [731]

9.4 RANGE TABLES

In Tables 9.5 to 9.11, values are given for the range (R), the projected range (R_p) and the corresponding range straggling (ΔR and ΔR_p), as well as the electronic stopping and nuclear stopping, according to a recent theory by Biersack [79], [79a] for the most important ion and substrate combinations. In addition, range-parameter curves for several elements in SiO_2, Si_3N_4 and Al are shown in Figures 3.25 to 3.27.

Briefly outlined, the theory by Biersack [79], [79a] is as follows: the average directional cosine is calculated for projection of the particle paths in the direction of incidence, with the aid of a diffusion model. With the LSS theory, in contrast, transport equations are used which describe the stopping by integro-differential equations. This new method results in a simple expression for this directional cosine, which in turn results in an algorithm by means of which $R(E + \delta E)$ can be calculated from $R(E)$. This algorithm is considerably easier and faster than the LSS momentum method, and has the advantage that other and more realistic screening potentials than the Thomas–Fermi potential for ion–atom collisions can be used easily. The tables given here are based on the Molière potential.

Table 9.3 lists the ion–substrate combinations for which range parameters are given. The tables for GaAs and germanium apply without change for ZnSe and, with multiplicative factors for R_p and ΔR_p, for several other semiconductors listed in Table 9.4 as well. The ion-range data in other substrates can be approximated on the basis of the density relationship, or still better, can be taken from the detailed collections of tables which have been published.

Before each table (Tables 9.5 to 9.11), the numerical values used for calculation purposes are listed. C_K is a multiplicative correction factor for the constants k and k' of the electronic stopping according to the LSS theory (*cf.* equations (2.16) and (2.17)).

Table 9.3 Ion-substrate combinations for which range parameters are given in Tables 9.5 to 9.11

Substrate	Ions		
Silicon	As, Ar, Au, B, F, Ne, P, Sb	Si_3N_4	As, B
Germanium, GaAs	Al, As, B, Be, Cd, H, Mg, O, P, S, Se, Te, Zn	Al_2O_3	As, B
		Photoresist AZ 111	As, B, P, Sb
SiO_2	As, B, P, Zn	Photoresist KTFR	As, B, P, Sb

Table 9.4 Correction factors for use of Ge and GaAs range values for ZnSe, CdS, AlSb and InP

ZnSe	CdS	AlSb	InP
1	0.92	0.78	0.9

Table 9.5 Range parameters of As, Ar, Au, B, F, Ne, P and Sb in silicon ($Z_2 = 14$; $M_2 = 28.09$; $d_2 = 2.33$ gcm^{-3})

E (keV)	S_e (keV/μm)	S_n (keV/μm)	R (μm)	ΔR (μm)	R_p (μm)	ΔR_p (μm)
Antimony (Sb)	$Z_1 = 51$; $M_1 = 120.904$; $C_K = 1$					
10	87.48	1365	0.0102	0.0032	0.0092	0.0030
20	123.7	1628	0.0163	0.0048	0.0147	0.0045
30	151.5	1757	0.0218	0.0063	0.0195	0.0059
40	175.0	1833	0.0269	0.0076	0.0241	0.0072
50	195.6	1882	0.0318	0.0089	0.0284	0.0084
60	214.3	1914	0.0365	0.0102	0.0327	0.0097
70	231.4	1936	0.0412	0.0114	0.0369	0.0108
80	247.4	1951	0.0458	0.0127	0.0410	0.0120
90	262.4	1961	0.0503	0.0139	0.0451	0.0131
100	276.6	1967	0.0548	0.0150	0.0491	0.0143
110	290.1	1970	0.0592	0.0162	0.0532	0.0154
120	303.0	1971	0.0636	0.0174	0.0572	0.0165
130	315.4	1970	0.0680	0.0185	0.0611	0.0176
140	327.3	1968	0.0724	0.0197	0.0651	0.0187
150	338.8	1964	0.0767	0.0208	0.0691	0.0198
160	349.9	1960	0.0810	0.0219	0.0730	0.0209
170	360.7	1955	0.0854	0.0230	0.0769	0.0219
180	371.1	1949	0.0897	0.0242	0.0809	0.0230
190	381.3	1943	0.0940	0.0253	0.0848	0.0241
200	391.2	1936	0.0983	0.0264	0.0887	0.0251
220	410.3	1922	0.1069	0.0286	0.0966	0.0272
240	428.5	1907	0.1154	0.0307	0.1044	0.0293
260	446.0	1891	0.1240	0.0329	0.1123	0.0314
280	462.9	1875	0.1326	0.0350	0.1202	0.0335
300	479.1	1858	0.1411	0.0371	0.1280	0.0355
320	494.8	1842	0.1497	0.0393	0.1359	0.0375
340	510.1	1825	0.1582	0.0414	0.1438	0.0396
360	524.9	1809	0.1668	0.0434	0.1517	0.0416
380	539.2	1792	0.1754	0.0455	0.1596	0.0436
400	553.2	1776	0.1840	0.0476	0.1675	0.0456
420	566.9	1760	0.1926	0.0496	0.1755	0.0476
440	580.2	1744	0.2012	0.0517	0.1835	0.0496
460	593.3	1729	0.2098	0.0537	0.1914	0.0515
480	606.0	1713	0.2184	0.0557	0.1994	0.0535
500	618.5	1698	0.2270	0.0577	0.2074	0.0554

Table 9.5—*contd* Range parameters of Ar in silicon

E (keV)	S_e (keV/μm)	S_n (keV/μm)	R (μm)	ΔR (μm)	R_p (μm)	ΔR_p (μm)
Argon (Ar) $Z_1 = 18; M_1 = 39.948; C_K = 1$						
10	90.47	611.0	0.0180	0.0063	0.0118	0.0060
20	129.7	616.3	0.0318	0.0107	0.0215	0.0102
30	160.2	601.2	0.0450	0.0149	0.0312	0.0143
40	186.0	581.9	0.0581	0.0188	0.0410	0.0181
50	208.9	562.2	0.0711	0.0227	0.0509	0.0219
60	229.7	543.1	0.0840	0.0264	0.0609	0.0256
70	248.8	525.1	0.0970	0.0299	0.0711	0.0291
80	266.7	508.2	0.1099	0.0334	0.0813	0.0326
90	283.6	492.4	0.1228	0.0368	0.0916	0.0360
100	299.6	477.7	0.1356	0.0400	0.1020	0.0393
110	314.8	464.0	0.1485	0.0432	0.1124	0.0426
120	329.3	451.1	0.1613	0.0463	0.1229	0.0457
130	343.3	439.0	0.1741	0.0492	0.1335	0.0488
140	356.8	427.7	0.1869	0.0521	0.1441	0.0518
150	369.9	417.0	0.1996	0.0549	0.1547	0.0547
160	382.5	407.0	0.2123	0.0577	0.1654	0.0576
170	394.7	397.5	0.2249	0.0603	0.1760	0.0604
180	406.7	388.5	0.2375	0.0629	0.1867	0.0631
190	418.2	380.1	0.2501	0.0654	0.1974	0.0658
200	429.6	372.0	0.2626	0.0678	0.2081	0.0684
220	451.4	357.0	0.2874	0.0725	0.2294	0.0734
240	472.3	343.4	0.3121	0.0769	0.2507	0.0782
260	492.3	331.0	0.3365	0.0811	0.2720	0.0828
280	511.7	319.6	0.3607	0.0851	0.2931	0.0871
300	530.4	309.1	0.3846	0.0888	0.3142	0.0913
320	548.5	299.4	0.4083	0.0924	0.3352	0.0953
340	566.0	290.4	0.4318	0.0958	0.3561	0.0991
360	583.1	282.0	0.4550	0.0991	0.3768	0.1028
380	599.7	274.2	0.4780	0.1022	0.3974	0.1063
400	616.0	266.8	0.5008	0.1051	0.4179	0.1096
420	631.8	259.9	0.5233	0.1080	0.4382	0.1129
440	647.3	253.5	0.5456	0.1107	0.4584	0.1160
460	662.4	247.3	0.5677	0.1133	0.4784	0.1190
480	677.2	241.6	0.5896	0.1157	0.4983	0.1219
500	691.7	236.1	0.6113	0.1181	0.5181	0.1246

Table 9.5—*contd* Range parameters of As in silicon

E (keV)	S_e (keV/μm)	S_n (keV/μm)	R (μm)	ΔR (μm)	R_p (μm)	ΔR_p (μm)
Arsenic (As)	$Z_1 = 33; M_1 = 74.922; C_K = 1$					
10	89.57	1078	0.0121	0.0040	0.0099	0.0037
20	126.7	1193	0.0201	0.0064	0.0164	0.0060
30	155.1	1235	0.0274	0.0087	0.0225	0.0081
40	179.1	1250	0.0345	0.0108	0.0284	0.0101
50	200.3	1253	0.0415	0.0129	0.0343	0.0121
60	219.4	1249	0.0483	0.0149	0.0400	0.0140
70	237.0	1242	0.0551	0.0169	0.0458	0.0159
80	253.3	1233	0.0618	0.0188	0.0516	0.0177
90	268.7	1222	0.0685	0.0208	0.0573	0.0196
100	283.2	1210	0.0752	0.0227	0.0631	0.0214
110	297.1	1197	0.0819	0.0246	0.0689	0.0232
120	310.3	1184	0.0886	0.0265	0.0747	0.0250
130	322.9	1172	0.0953	0.0284	0.0805	0.0268
140	335.1	1159	0.1020	0.0302	0.0863	0.0286
150	346.9	1146	0.1087	0.0321	0.0921	0.0304
160	358.3	1133	0.1154	0.0339	0.0980	0.0321
170	369.3	1121	0.1221	0.0357	0.1039	0.0339
180	380.0	1108	0.1288	0.0375	0.1097	0.0356
190	390.4	1096	0.1356	0.0393	0.1157	0.0373
200	400.6	1084	0.1423	0.0411	0.1216	0.0391
220	420.1	1061	0.1558	0.0446	0.1335	0.0425
240	438.8	1039	0.1693	0.0481	0.1454	0.0458
260	456.7	1018	0.1828	0.0515	0.1575	0.0491
280	473.9	997.7	0.1964	0.0549	0.1696	0.0524
300	490.6	978.3	0.2100	0.0582	0.1817	0.0557
320	506.7	959.7	0.2237	0.0615	0.1939	0.0589
340	522.3	941.9	0.2373	0.0648	0.2061	0.0620
360	537.4	924.9	0.2510	0.0679	0.2184	0.0652
380	552.1	908.6	0.2647	0.0711	0.2307	0.0683
400	566.5	892.9	0.2784	0.0742	0.2431	0.0713
420	580.5	877.9	0.2921	0.0773	0.2555	0.0743
440	594.1	863.5	0.3058	0.0803	0.2679	0.0773
460	607.5	849.6	0.3195	0.0832	0.2804	0.0802
480	620.5	836.2	0.3332	0.0862	0.2928	0.0831
500	633.3	823.3	0.3470	0.0891	0.3053	0.0860

Table 9.5—*contd* Range parameters of B in silicon

E (keV)	S_e (keV/μm)	S_n (keV/μm)	R (μm)	ΔR (μm)	R_p (μm)	ΔR_p (μm)
Boron (B)	$Z_1 = 5; M_1 = 11.009; C_K = 1.5$					
10	102.2	83.13	0.0637	0.0149	0.0310	0.0193
20	144.0	67.01	0.1141	0.0226	0.0634	0.0318
30	175.8	56.75	0.1592	0.0279	0.0956	0.0414
40	202.3	49.59	0.2005	0.0318	0.1270	0.0492
50	225.4	44.27	0.2388	0.0348	0.1574	0.0556
60	246.1	40.13	0.2748	0.0372	0.1868	0.0610
70	264.9	36.80	0.3088	0.0392	0.2153	0.0657
80	282.2	34.06	0.3412	0.0409	0.2429	0.0699
90	298.4	31.75	0.3721	0.0424	0.2696	0.0735
100	313.5	29.78	0.4018	0.0436	0.2956	0.0768
110	327.7	28.07	0.4305	0.0448	0.3209	0.0798
120	341.1	26.57	0.4581	0.0458	0.3456	0.0825
130	353.9	25.24	0.4849	0.0467	0.3697	0.0849
140	366.0	24.06	0.5109	0.0475	0.3933	0.0872
150	377.7	23.00	0.5362	0.0482	0.4163	0.0893
160	388.8	22.04	0.5608	0.0489	0.4389	0.0913
170	399.4	21.17	0.5849	0.0496	0.4610	0.0931
180	409.7	20.37	0.6084	0.0502	0.4828	0.0948
190	419.6	19.64	0.6314	0.0507	0.5041	0.0964
200	429.1	18.96	0.6539	0.0512	0.5251	0.0979
220	447.2	17.76	0.6977	0.0522	0.5661	0.1007
240	464.1	16.72	0.7400	0.0530	0.6059	0.1032
260	480.0	15.80	0.7810	0.0537	0.6446	0.1055
280	495.0	15.00	0.8207	0.0544	0.6824	0.1076
300	509.2	14.28	0.8594	0.0550	0.7193	0.1095
320	522.7	13.63	0.8972	0.0556	0.7553	0.1113
340	535.4	13.05	0.9341	0.0561	0.7906	0.1130
360	547.6	12.52	0.9701	0.0566	0.8253	0.1145
380	559.2	12.03	1.0055	0.0571	0.8593	0.1160
400	570.3	11.59	1.0402	0.0575	0.8928	0.1173
420	580.9	11.18	1.0743	0.0580	0.9257	0.1186
440	591.0	10.80	1.1078	0.0583	0.9581	0.1198
460	600.7	10.45	1.1407	0.0587	0.9901	0.1210
480	610.0	10.12	1.1732	0.0591	1.0216	0.1221
500	618.9	9.818	1.2052	0.0594	1.0527	0.1231

Table 9.5—*contd* Range parameters of F in silicon

E (keV)	S_e (keV/μm)	S_n (keV/μm)	R (μm)	ΔR (μm)	R_p (μm)	ΔR_p (μm)
Fluorine (F) $Z_1 = 9$; $M_1 = 18.998$; $C_K = 1$						
10	39.59	229.5	0.0400	0.0139	0.0191	0.0138
20	53.64	205.6	0.0778	0.0261	0.0400	0.0263
30	64.05	185.4	0.1172	0.0381	0.0635	0.0389
40	72.61	169.1	0.1579	0.0498	0.0890	0.0515
50	80.01	155.9	0.1998	0.0612	0.1161	0.0640
60	86.59	144.9	0.2426	0.0723	0.1448	0.0763
70	92.56	135.6	0.2862	0.0830	0.1746	0.0884
80	98.05	127.6	0.3303	0.0933	0.2054	0.1003
90	103.2	120.6	0.3748	0.1033	0.2372	0.1119
100	107.9	114.5	0.4196	0.1128	0.2697	0.1233
110	112.4	109.1	0.4647	0.1220	0.3028	0.1343
120	116.7	104.3	0.5099	0.1309	0.3364	0.1451
130	120.7	99.90	0.5552	0.1394	0.3706	0.1556
140	124.6	95.95	0.6005	0.1476	0.4051	0.1658
150	128.3	92.34	0.6458	0.1555	0.4399	0.1758
160	131.8	89.04	0.6911	0.1631	0.4751	0.1855
170	135.2	86.01	0.7364	0.1704	0.5105	0.1949
180	138.5	83.21	0.7815	0.1775	0.5461	0.2041
190	141.7	80.61	0.8266	0.1843	0.5818	0.2131
200	144.8	78.19	0.8715	0.1908	0.6177	0.2218
220	150.6	73.83	0.9609	0.2033	0.6899	0.2387
240	156.2	70.00	1.0497	0.2151	0.7623	0.2547
260	161.4	66.60	1.1377	0.2260	0.8349	0.2699
280	166.4	63.56	1.2251	0.2364	0.9075	0.2845
300	171.1	60.82	1.3117	0.2461	0.9801	0.2984
320	175.7	58.34	1.3975	0.2553	1.0526	0.3117
340	180.0	56.09	1.4826	0.2640	1.1249	0.3244
360	184.2	54.02	1.5669	0.2723	1.1970	0.3367
380	188.2	52.12	1.6505	0.2802	1.2689	0.3484
400	192.1	50.37	1.7334	0.2877	1.3406	0.3597
420	195.8	48.74	1.8155	0.2949	1.4119	0.3706
440	199.5	47.23	1.8969	0.3017	1.4830	0.3810
460	203.0	45.83	1.9777	0.3083	1.5538	0.3911
480	206.3	44.51	2.0577	0.3146	1.6243	0.4009
500	209.6	43.28	2.1371	0.3206	1.6945	0.4103

Table 9.5—*contd* Range parameters of Au in silicon

E (keV)	S_e (keV/μm)	S_n (keV/μm)	R (μm)	ΔR (μm)	R_p (μm)	ΔR_p (μm)
Gold (Au)	$Z_1 = 79; M_1 = 196.967; C_K = 1$					
10	82.91	1476	0.0090	0.0027	0.0086	0.0027
20	117.3	1906	0.0145	0.0040	0.0138	0.0039
30	143.6	2155	0.0191	0.0050	0.0181	0.0049
40	165.8	2323	0.0233	0.0059	0.0220	0.0058
50	185.4	2445	0.0272	0.0068	0.0256	0.0066
60	203.1	2538	0.0309	0.0077	0.0291	0.0074
70	219.4	2612	0.0345	0.0085	0.0325	0.0082
80	234.5	2671	0.0380	0.0092	0.0357	0.0089
90	248.7	2720	0.0414	0.0100	0.0389	0.0097
100	262.2	2760	0.0447	0.0108	0.0420	0.0104
110	275.0	2795	0.0480	0.0115	0.0451	0.0111
120	287.2	2823	0.0512	0.0122	0.0482	0.0118
130	298.9	2848	0.0544	0.0130	0.0512	0.0125
140	310.2	2869	0.0576	0.0137	0.0541	0.0132
150	321.1	2887	0.0607	0.0144	0.0571	0.0139
160	331.6	2903	0.0638	0.0151	0.0600	0.0145
170	341.9	2917	0.0669	0.0158	0.0629	0.0152
180	351.8	2928	0.0700	0.0165	0.0658	0.0159
190	361.4	2938	0.0730	0.0171	0.0686	0.0165
200	370.8	2947	0.0760	0.0178	0.0715	0.0172
220	388.9	2960	0.0820	0.0192	0.0771	0.0185
240	406.2	2970	0.0880	0.0205	0.0827	0.0198
260	422.8	2976	0.0939	0.0218	0.0883	0.0210
280	438.7	2980	0.0998	0.0231	0.0939	0.0223
300	454.1	2981	0.1056	0.0244	0.0994	0.0236
320	469.0	2981	0.1114	0.0257	0.1049	0.0248
340	483.4	2980	0.1172	0.0270	0.1103	0.0261
360	497.5	2977	0.1229	0.0282	0.1158	0.0273
380	511.1	2973	0.1287	0.0295	0.1212	0.0285
400	524.4	2968	0.1344	0.0308	0.1267	0.0297
420	537.3	2963	0.1402	0.0320	0.1321	0.0310
440	550.0	2957	0.1459	0.0333	0.1375	0.0322
460	562.3	2950	0.1516	0.0345	0.1429	0.0334
480	574.4	2943	0.1572	0.0357	0.1483	0.0346
500	586.3	2935	0.1629	0.0370	0.1537	0.0358

Table 9.5—*contd* Range parameters of Ne in silicon

E (keV)	S_e (keV/μm)	S_n (keV/μm)	R (μm)	ΔR (μm)	R_p (μm)	ΔR_p (μm)
Neon (Ne)	$Z_1 = 10; M_1 = 19.992; C_K = 1$					
10	8.532	262.1	0.0392	0.0146	0.0179	0.0139
20	14.14	238.6	0.0774	0.0285	0.0377	0.0272
30	18.99	217.4	0.1184	0.0430	0.0602	0.0414
40	23.41	199.8	0.1619	0.0582	0.0851	0.0562
50	27.52	185.2	0.2079	0.0737	0.1122	0.0715
60	31.41	173.0	0.2558	0.0895	0.1413	0.0873
70	35.12	162.5	0.3056	0.1053	0.1722	0.1033
80	38.69	153.4	0.3570	0.1211	0.2047	0.1195
90	42.12	145.4	0.4097	0.1369	0.2387	0.1358
100	45.45	138.4	0.4636	0.1524	0.2740	0.1521
110	48.68	132.1	0.5184	0.1677	0.3105	0.1684
120	51.83	126.4	0.5741	0.1828	0.3481	0.1846
130	54.90	121.4	0.6306	0.1976	0.3866	0.2007
140	57.90	116.8	0.6876	0.2120	0.4260	0.2166
150	60.83	112.5	0.7450	0.2261	0.4662	0.2323
160	63.71	108.6	0.8029	0.2398	0.5070	0.2478
170	66.54	105.1	0.8610	0.2532	0.5485	0.2630
180	69.31	101.8	0.9194	0.2662	0.5905	0.2781
190	72.04	98.67	0.9779	0.2788	0.6330	0.2928
200	74.72	95.80	1.0365	0.2911	0.6759	0.3073
220	79.96	90.61	1.1538	0.3146	0.7627	0.3355
240	85.05	86.03	1.2710	0.3367	0.8507	0.3626
260	90.01	81.96	1.3876	0.3575	0.9394	0.3886
280	94.85	78.31	1.5035	0.3770	1.0286	0.4135
300	99.57	75.02	1.6185	0.3953	1.1181	0.4374
320	104.2	72.03	1.7326	0.4126	1.2076	0.4602
340	108.7	69.30	1.8455	0.4288	1.2972	0.4821
360	113.2	66.80	1.9572	0.4441	1.3866	0.5031
380	117.5	64.50	2.0677	0.4584	1.4757	0.5232
400	121.8	62.37	2.1770	0.4720	1.5644	0.5424
420	126.0	60.40	2.2850	0.4848	1.6527	0.5609
440	130.1	58.57	2.3916	0.4969	1.7405	0.5786
460	134.2	56.85	2.4970	0.5083	1.8277	0.5956
480	138.1	55.25	2.6010	0.5191	1.9144	0.6119
500	142.1	53.75	2.7038	0.5294	2.0005	0.6276

Table 9.5—*contd* Range parameters of P in silicon

E (keV)	S_e (keV/μm)	S_n (keV/μm)	R (μm)	ΔR (μm)	R_p (μm)	ΔR_p (μm)
Phosphorus (P)	$Z_1 = 15; M_1 = 30.974; C_K = 1$					
10	83.02	475.0	0.0217	0.0076	0.0128	0.0074
20	119.9	463.8	0.0392	0.0132	0.0243	0.0129
30	148.6	442.8	0.0562	0.0184	0.0359	0.0181
40	173.1	421.4	0.0730	0.0234	0.0479	0.0232
50	194.8	401.6	0.0898	0.0281	0.0601	0.0280
60	214.6	383.5	0.1066	0.0326	0.0726	0.0327
70	232.8	367.1	0.1233	0.0369	0.0852	0.0372
80	249.9	352.2	0.1399	0.0410	0.0979	0.0415
90	266.0	338.7	0.1565	0.0449	0.1108	0.0458
100	281.3	326.3	0.1730	0.0487	0.1238	0.0498
110	295.9	315.0	0.1894	0.0523	0.1368	0.0538
120	309.8	304.5	0.2057	0.0557	0.1499	0.0575
130	323.3	294.9	0.2219	0.0590	0.1631	0.0612
140	336.2	285.9	0.2381	0.0622	0.1762	0.0648
150	348.7	277.6	0.2541	0.0652	0.1894	0.0682
160	360.9	269.8	0.2700	0.0681	0.2026	0.0715
170	372.7	262.6	0.2858	0.0709	0.2157	0.0747
180	384.1	255.7	0.3015	0.0736	0.2289	0.0778
190	395.3	249.3	0.3171	0.0762	0.2420	0.0808
200	406.2	243.3	0.3325	0.0787	0.2550	0.0837
220	427.2	232.2	0.3631	0.0834	0.2811	0.0893
240	447.4	222.3	0.3932	0.0878	0.3070	0.0945
260	466.8	213.3	0.4228	0.0918	0.3326	0.0994
280	485.5	205.1	0.4520	0.0956	0.3581	0.1041
300	503.5	197.7	0.4807	0.0992	0.3833	0.1084
320	521.1	190.8	0.5090	0.1025	0.4083	0.1126
340	538.1	184.5	0.5369	0.1057	0.4330	0.1165
360	554.6	178.6	0.5644	0.1086	0.4575	0.1202
380	570.8	173.2	0.5915	0.1114	0.4817	0.1238
400	586.5	168.2	0.6182	0.1140	0.5057	0.1272
420	601.8	163.4	0.6445	0.1165	0.5294	0.1304
440	616.9	159.0	0.6704	0.1189	0.5529	0.1334
460	631.6	154.9	0.6960	0.1211	0.5762	0.1364
480	646.0	150.9	0.7213	0.1233	0.5992	0.1392
500	660.1	147.3	0.7462	0.1253	0.6220	0.1419

Table 9.6 Range parameters of Al, As, B, Be, Cd, H, Mg, O, P, S, Se, Te and Zn in GaAs and germanium ($Z_2 = 32$; $M_2 = 72.59$; $d_2 = 5.35$ gcm^{-3})

E (keV)	S_e (keV/μm)	S_n (keV/μm)	R (μm)	ΔR (μm)	R_p (μm)	ΔR_p (μm)
Aluminium (Al)	$Z_1 = 13$; $M_1 = 26.982$; $C_K = 1$					
10	107.1	374.0	0.0272	0.0081	0.0082	0.0086
20	151.5	378.2	0.0468	0.0133	0.0159	0.0147
30	185.6	369.7	0.0652	0.0178	0.0239	0.0204
40	214.3	358.3	0.0830	0.0220	0.0321	0.0258
50	239.6	346.6	0.1002	0.0259	0.0405	0.0309
60	262.4	335.1	0.1171	0.0296	0.0491	0.0358
70	283.5	324.3	0.1337	0.0330	0.0579	0.0406
80	303.0	314.1	0.1500	0.0363	0.0669	0.0452
90	321.4	304.6	0.1661	0.0393	0.0759	0.0497
100	338.8	295.7	0.1820	0.0422	0.0850	0.0540
110	355.3	287.3	0.1976	0.0450	0.0943	0.0582
120	371.1	279.5	0.2131	0.0476	0.1036	0.0623
130	386.3	272.1	0.2284	0.0502	0.1130	0.0662
140	400.8	265.2	0.2435	0.0526	0.1224	0.0701
150	414.9	258.7	0.2584	0.0549	0.1318	0.0738
160	428.5	252.6	0.2732	0.0571	0.1413	0.0774
170	441.7	246.8	0.2878	0.0592	0.1509	0.0809
180	454.5	241.3	0.3022	0.0612	0.1604	0.0844
190	466.9	236.1	0.3165	0.0631	0.1699	0.0877
200	479.1	231.1	0.3307	0.0650	0.1795	0.0910
220	502.4	222.0	0.3586	0.0685	0.1986	0.0972
240	524.8	213.6	0.3859	0.0718	0.2177	0.1032
260	546.2	206.0	0.4127	0.0749	0.2368	0.1089
280	566.8	198.9	0.4391	0.0777	0.2558	0.1143
300	586.7	192.5	0.4650	0.0804	0.2748	0.1195
320	605.9	186.5	0.4904	0.0829	0.2936	0.1245
340	624.5	180.9	0.5155	0.0853	0.3124	0.1293
360	642.6	175.8	0.5401	0.0875	0.3310	0.1339
380	660.2	170.9	0.5644	0.0896	0.3496	0.1383
400	677.3	166.4	0.5882	0.0916	0.3680	0.1425
420	694.0	162.1	0.6118	0.0935	0.3863	0.1465
440	710.3	158.1	0.6350	0.0953	0.4045	0.1505
460	726.2	154.3	0.6578	0.0970	0.4226	0.1543
480	741.8	150.7	0.6804	0.0987	0.4405	0.1579
500	757.1	147.4	0.7027	0.1002	0.4584	0.1614

Table 9.6—*contd* Range parameters of As in GaAs and germanium

E (keV)	S_e (keV/μm)	S_n (keV/μm)	R (μm)	ΔR (μm)	R_p (μm)	ΔR_p (μm)
Arsenic (As)	$Z_1 = 33$; $M_1 = 74.922$; $C_K = 1$					
10	127.9	1254	0.0105	0.0040	0.0054	0.0038
20	180.8	1430	0.0171	0.0062	0.0090	0.0060
30	221.5	1504	0.0231	0.0082	0.0124	0.0080
40	255.8	1542	0.0288	0.0100	0.0158	0.0098
50	285.9	1561	0.0343	0.0118	0.0191	0.0116
60	313.2	1569	0.0396	0.0136	0.0224	0.0133
70	338.3	1572	0.0449	0.0153	0.0257	0.0150
80	361.7	1570	0.0501	0.0169	0.0290	0.0167
90	383.6	1565	0.0553	0.0186	0.0323	0.0183
100	404.4	1558	0.0604	0.0202	0.0356	0.0199
110	424.1	1550	0.0655	0.0218	0.0389	0.0215
120	443.0	1541	0.0705	0.0234	0.0422	0.0231
130	461.1	1531	0.0755	0.0249	0.0455	0.0247
140	478.5	1520	0.0806	0.0264	0.0489	0.0262
150	495.3	1510	0.0856	0.0280	0.0522	0.0278
160	511.5	1499	0.0905	0.0295	0.0556	0.0293
170	527.3	1487	0.0955	0.0309	0.0590	0.0308
180	542.5	1476	0.1005	0.0324	0.0624	0.0323
190	557.4	1465	0.1054	0.0339	0.0658	0.0338
200	571.9	1454	0.1104	0.0353	0.0692	0.0353
220	599.8	1431	0.1202	0.0382	0.0761	0.0382
240	626.5	1409	0.1300	0.0410	0.0830	0.0411
260	652.0	1387	0.1399	0.0437	0.0900	0.0439
280	676.7	1366	0.1497	0.0464	0.0970	0.0468
300	700.4	1346	0.1594	0.0491	0.1040	0.0496
320	723.4	1326	0.1692	0.0517	0.1111	0.0523
340	745.6	1306	0.1790	0.0543	0.1182	0.0550
360	767.3	1288	0.1887	0.0568	0.1254	0.0577
380	788.3	1269	0.1984	0.0594	0.1326	0.0604
400	808.8	1252	0.2081	0.0618	0.1398	0.0630
420	828.7	1235	0.2178	0.0643	0.1471	0.0656
440	848.2	1218	0.2275	0.0667	0.1544	0.0682
460	867.3	1202	0.2372	0.0690	0.1617	0.0708
480	885.9	1186	0.2469	0.0714	0.1690	0.0733
500	904.2	1171	0.2565	0.0736	0.1764	0.0758

Table 9.6—*contd* Range parameters of Be in GaAs and germanium

E (keV)	S_e (keV/μm)	S_n (keV/μm)	R (μm)	ΔR (μm)	R_p (μm)	ΔR_p (μm)
Beryllium (Be)	$Z_1 = 4; M_1 = 9.012; C_K = 1$					
10	64.66	50.11	0.1107	0.0177	0.0277	0.0280
20	91.45	43.30	0.1906	0.0263	0.0584	0.0481
30	112.0	38.19	0.2608	0.0323	0.0907	0.0650
40	129.3	34.30	0.3245	0.0368	0.1236	0.0798
50	144.6	31.25	0.3834	0.0404	0.1566	0.0928
60	158.3	28.77	0.4385	0.0433	0.1895	0.1045
70	171.0	26.72	0.4905	0.0457	0.2221	0.1150
80	182.8	24.98	0.5398	0.0478	0.2543	0.1246
90	193.8	23.50	0.5869	0.0496	0.2861	0.1333
100	204.3	22.20	0.6320	0.0511	0.3174	0.1413
110	214.2	21.06	0.6753	0.0525	0.3483	0.1488
120	223.7	20.06	0.7170	0.0537	0.3787	0.1557
130	232.8	19.15	0.7574	0.0548	0.4086	0.1621
140	241.5	18.34	0.7964	0.0558	0.4381	0.1681
150	249.9	17.61	0.8344	0.0567	0.4672	0.1738
160	258.1	16.94	0.8712	0.0575	0.4958	0.1791
170	266.0	16.32	0.9071	0.0583	0.5240	0.1841
180	273.6	15.76	0.9421	0.0590	0.5519	0.1888
190	281.0	15.24	0.9762	0.0596	0.5793	0.1933
200	288.2	14.75	1.0096	0.0603	0.6064	0.1976
220	302.1	13.89	1.0742	0.0614	0.6594	0.2055
240	315.4	13.13	1.1363	0.0623	0.7111	0.2127
260	328.0	12.46	1.1961	0.0632	0.7616	0.2193
280	340.1	11.87	1.2539	0.0640	0.8109	0.2254
300	351.8	11.33	1.3098	0.0647	0.8591	0.2311
320	363.0	10.85	1.3641	0.0653	0.9063	0.2363
340	373.9	10.41	1.4168	0.0659	0.9525	0.2412
360	384.4	10.01	1.4682	0.0664	0.9978	0.2458
380	394.5	9.646	1.5183	0.0669	1.0423	0.2501
400	404.3	9.309	1.5672	0.0674	1.0860	0.2541
420	413.9	8.998	1.6150	0.0678	1.1289	0.2579
440	423.2	8.709	1.6618	0.0682	1.1711	0.2615
460	432.2	8.441	1.7077	0.0686	1.2127	0.2649
480	441.0	8.190	1.7526	0.0689	1.2536	0.2682
500	449.5	7.956	1.7967	0.0693	1.2939	0.2713

Table 9.6—*contd* Range parameters of B in GaAs and germanium

E (keV)	S_e (keV/μm)	S_n (keV/μm)	R (μm)	ΔR (μm)	R_p (μm)	ΔR_p (μm)
Boron (B) $Z_1 = 5; M_1 = 11.009; C_K = 1$						
10	72.45	75.79	0.0847	0.0158	0.0215	0.0223
20	102.1	67.63	0.1474	0.0243	0.0450	0.0387
30	124.6	60.84	0.2037	0.0306	0.0699	0.0530
40	143.4	55.42	0.2557	0.0356	0.0956	0.0657
50	159.8	51.01	0.3045	0.0397	0.1217	0.0772
60	174.4	47.37	0.3507	0.0432	0.1480	0.0877
70	187.8	44.29	0.3948	0.0462	0.1743	0.0973
80	200.0	41.65	0.4370	0.0487	0.2005	0.1062
90	211.4	39.36	0.4776	0.0510	0.2266	0.1144
100	222.1	37.35	0.5168	0.0531	0.2526	0.1221
110	232.2	35.56	0.5547	0.0549	0.2783	0.1293
120	241.7	33.97	0.5915	0.0566	0.3039	0.1361
130	250.7	32.54	0.6273	0.0581	0.3292	0.1425
140	259.3	31.24	0.6622	0.0595	0.3544	0.1485
150	267.5	30.06	0.6962	0.0608	0.3793	0.1543
160	275.4	28.97	0.7294	0.0620	0.4040	0.1597
170	282.9	27.98	0.7619	0.0631	0.4285	0.1649
180	290.2	27.06	0.7938	0.0641	0.4527	0.1698
190	297.2	26.21	0.8250	0.0651	0.4768	0.1746
200	303.9	25.42	0.8556	0.0660	0.5006	0.1791
220	316.6	24.00	0.9153	0.0677	0.5477	0.1876
240	328.6	22.74	0.9731	0.0692	0.5940	0.1955
260	339.8	21.63	1.0292	0.0706	0.6397	0.2028
280	350.4	20.64	1.0838	0.0718	0.6846	0.2097
300	360.4	19.75	1.1371	0.0730	0.7289	0.2161
320	369.9	18.94	1.1891	0.0741	0.7726	0.2221
340	378.9	18.20	1.2400	0.0751	0.8158	0.2279
360	387.4	17.53	1.2899	0.0760	0.8584	0.2333
380	395.6	16.91	1.3388	0.0769	0.9005	0.2384
400	403.4	16.34	1.3869	0.0777	0.9421	0.2433
420	410.8	15.81	1.4341	0.0785	0.9833	0.2480
440	417.9	15.32	1.4806	0.0792	1.0241	0.2524
460	424.8	14.86	1.5265	0.0799	1.0644	0.2567
480	431.3	14.43	1.5716	0.0806	1.1044	0.2608
500	437.6	14.03	1.6162	0.0812	1.1441	0.2648

Table 9.6—*contd* Range parameters of Cd in GaAs and germanium

E (keV)	S_e (keV/μm)	S_n (keV/μm)	R (μm)	ΔR (μm)	R_p (μm)	ΔR_p (μm)
Cadmium (Cd)	$Z_1 = 48; M_1 = 113.904; C_K = 1$					
10	131.3	1709	0.0080	0.0032	0.0054	0.0029
20	185.7	2062	0.0129	0.0048	0.0086	0.0044
30	227.5	2240	0.0171	0.0061	0.0114	0.0058
40	262.7	2348	0.0210	0.0074	0.0141	0.0070
50	293.7	2419	0.0248	0.0087	0.0167	0.0082
60	321.7	2468	0.0284	0.0099	0.0192	0.0093
70	347.5	2503	0.0320	0.0110	0.0217	0.0104
80	371.5	2528	0.0354	0.0121	0.0242	0.0115
90	394.0	2545	0.0389	0.0133	0.0267	0.0126
100	415.3	2558	0.0422	0.0143	0.0291	0.0136
110	435.6	2566	0.0456	0.0154	0.0315	0.0147
120	455.0	2571	0.0489	0.0165	0.0340	0.0157
130	473.6	2573	0.0522	0.0176	0.0364	0.0167
140	491.4	2574	0.0555	0.0186	0.0388	0.0177
150	508.7	2572	0.0587	0.0196	0.0412	0.0187
160	525.4	2570	0.0620	0.0207	0.0436	0.0197
170	541.5	2566	0.0652	0.0217	0.0460	0.0206
180	557.2	2561	0.0684	0.0227	0.0484	0.0216
190	572.5	2556	0.0716	0.0237	0.0508	0.0226
200	587.4	2549	0.0748	0.0247	0.0532	0.0235
220	616.1	2535	0.0812	0.0267	0.0580	0.0254
240	643.5	2520	0.0875	0.0286	0.0628	0.0273
260	669.7	2503	0.0938	0.0306	0.0676	0.0292
280	695.0	2485	0.1001	0.0325	0.0724	0.0310
300	719.4	2467	0.1064	0.0344	0.0772	0.0329
320	743.0	2449	0.1127	0.0363	0.0821	0.0347
340	765.9	2430	0.1189	0.0382	0.0869	0.0365
360	788.1	2411	0.1252	0.0400	0.0918	0.0383
380	809.7	2392	0.1314	0.0418	0.0967	0.0401
400	830.7	2374	0.1377	0.0437	0.1016	0.0419
420	851.2	2355	0.1439	0.0455	0.1065	0.0437
440	871.2	2336	0.1501	0.0473	0.1114	0.0454
460	890.8	2318	0.1564	0.0491	0.1163	0.0472
480	910.0	2300	0.1626	0.0508	0.1212	0.0489
500	928.7	2282	0.1688	0.0526	0.1262	0.0507

Table 9.6—*contd* Range parameters of Mg in GaAs and germanium

E (keV)	S_e (keV/μm)	S_n (keV/μm)	R (μm)	ΔR (μm)	R_p (μm)	ΔR_p (μm)
Magnesium (Mg)	$Z_1 = 12; M_1 = 23.985; C_K = 1$					
10	106.4	321.7	0.0303	0.0086	0.0088	0.0094
20	150.5	321.9	0.0524	0.0141	0.0172	0.0162
30	184.3	312.3	0.0730	0.0189	0.0260	0.0223
40	212.9	301.1	0.0928	0.0232	0.0351	0.0282
50	238.0	289.9	0.1120	0.0272	0.0444	0.0338
60	260.7	279.3	0.1307	0.0309	0.0539	0.0391
70	281.6	269.3	0.1490	0.0344	0.0636	0.0442
80	301.0	260.1	0.1670	0.0376	0.0734	0.0492
90	319.3	251.6	0.1847	0.0406	0.0834	0.0539
100	336.5	243.6	0.2021	0.0434	0.0934	0.0585
110	353.0	236.2	0.2192	0.0461	0.1036	0.0629
120	368.7	229.4	0.2360	0.0486	0.1137	0.0672
130	383.7	222.9	0.2526	0.0511	0.1240	0.0714
140	398.2	216.9	0.2690	0.0533	0.1343	0.0754
150	412.1	211.3	0.2851	0.0555	0.1446	0.0792
160	425.7	206.0	0.3011	0.0576	0.1549	0.0830
170	438.8	201.0	0.3168	0.0595	0.1653	0.0867
180	451.5	196.3	0.3323	0.0614	0.1756	0.0902
190	463.8	191.8	0.3477	0.0632	0.1859	0.0936
200	475.9	187.6	0.3628	0.0649	0.1963	0.0970
220	499.1	179.8	0.3926	0.0682	0.2169	0.1034
240	521.3	172.7	0.4218	0.0712	0.2375	0.1095
260	542.5	166.3	0.4503	0.0739	0.2579	0.1152
280	563.0	160.4	0.4782	0.0765	0.2783	0.1207
300	582.7	155.0	0.5056	0.0789	0.2985	0.1259
320	601.8	150.0	0.5324	0.0811	0.3186	0.1309
340	620.3	145.4	0.5588	0.0832	0.3386	0.1356
360	638.3	141.0	0.5847	0.0852	0.3584	0.1402
380	655.7	137.0	0.6101	0.0870	0.3781	0.1446
400	672.7	133.3	0.6352	0.0888	0.3976	0.1487
420	689.3	129.8	0.6598	0.0904	0.4170	0.1528
440	705.5	126.4	0.6840	0.0920	0.4362	0.1566
460	721.3	123.3	0.7079	0.0935	0.4552	0.1603
480	736.8	120.4	0.7314	0.0949	0.4741	0.1639
500	751.9	117.6	0.7545	0.0962	0.4929	0.1674

Table 9.6—*contd* Range parameters of P in GaAs and germanium

E (keV)	S_e (keV/μm)	S_n (keV/μm)	R (μm)	ΔR (μm)	R_p (μm)	ΔR_p (μm)
Phosphorus (P)	$Z_1 = 15; M_1 = 30.974; C_K = 1$					
10	112.1	460.7	0.0232	0.0072	0.0073	0.0076
20	158.6	474.0	0.0397	0.0118	0.0140	0.0128
30	194.2	468.6	0.0551	0.0160	0.0208	0.0176
40	224.3	458.2	0.0700	0.0198	0.0278	0.0222
50	250.8	446.4	0.0845	0.0234	0.0349	0.0267
60	274.7	434.4	0.0987	0.0267	0.0422	0.0309
70	296.7	422.7	0.1127	0.0300	0.0496	0.0351
80	317.2	411.4	0.1265	0.0330	0.0571	0.0391
90	336.4	400.6	0.1402	0.0360	0.0648	0.0431
100	354.6	390.4	0.1537	0.0388	0.0725	0.0469
110	371.9	380.7	0.1670	0.0415	0.0803	0.0506
120	388.5	371.5	0.1802	0.0441	0.0882	0.0542
130	404.3	362.8	0.1933	0.0466	0.0961	0.0578
140	419.6	354.6	0.2063	0.0490	0.1041	0.0612
150	434.3	346.7	0.2192	0.0513	0.1122	0.0646
160	448.6	339.3	0.2319	0.0535	0.1203	0.0679
170	462.4	332.3	0.2445	0.0556	0.1284	0.0711
180	475.8	325.5	0.2571	0.0577	0.1366	0.0743
190	488.8	319.1	0.2695	0.0597	0.1447	0.0774
200	501.5	313.0	0.2818	0.0617	0.1529	0.0804
220	526.0	301.6	0.3062	0.0654	0.1694	0.0862
240	549.3	291.1	0.3302	0.0688	0.1859	0.0918
260	571.8	281.5	0.3538	0.0721	0.2024	0.0972
280	593.3	272.6	0.3771	0.0752	0.2189	0.1023
300	614.1	264.3	0.4000	0.0781	0.2354	0.1072
320	634.3	256.6	0.4226	0.0808	0.2518	0.1120
340	653.8	249.4	0.4449	0.0834	0.2683	0.1166
360	672.7	242.7	0.4669	0.0859	0.2846	0.1210
380	691.1	236.4	0.4886	0.0882	0.3009	0.1252
400	709.1	230.5	0.5100	0.0905	0.3172	0.1294
420	726.6	225.0	0.5312	0.0926	0.3334	0.1333
440	743.6	219.7	0.5521	0.0947	0.3495	0.1372
460	760.3	214.7	0.5727	0.0966	0.3655	0.1409
480	776.7	210.0	0.5931	0.0985	0.3815	0.1445
500	792.7	205.5	0.6132	0.1003	0.3973	0.1480

Table 9.6—*contd* Range parameters of O in GaAs and germanium

E (keV)	S_e (keV/μm)	S_n (keV/μm)	R (μm)	ΔR (μm)	R_p (μm)	ΔR_p (μm)
Oxygen (O) $Z_1 = 8; M_1 = 15.999; C_K = 1$						
10	91.82	164.6	0.0493	0.0116	0.0131	0.0141
20	129.6	156.4	0.0860	0.0186	0.0265	0.0244
30	158.5	146.7	0.1198	0.0244	0.0407	0.0338
40	182.8	137.7	0.1518	0.0293	0.0555	0.0424
50	204.0	129.8	0.1823	0.0336	0.0707	0.0505
60	223.2	122.8	0.2117	0.0374	0.0861	0.0580
70	240.7	116.6	0.2402	0.0409	0.1017	0.0651
80	256.9	111.2	0.2677	0.0439	0.1175	0.0718
90	272.1	106.3	0.2945	0.0467	0.1333	0.0781
100	286.3	101.9	0.3206	0.0493	0.1492	0.0841
110	299.3	97.86	0.3461	0.0517	0.1651	0.0898
120	312.7	94.22	0.3709	0.0538	0.1810	0.0952
130	325.0	90.89	0.3952	0.0559	0.1968	0.1004
140	336.7	87.83	0.4190	0.0577	0.2126	0.1054
150	348.0	85.01	0.4424	0.0595	0.2284	0.1102
160	358.9	82.39	0.4652	0.0611	0.2441	0.1147
170	369.3	79.96	0.4877	0.0627	0.2598	0.1191
180	379.5	77.69	0.5098	0.0642	0.2753	0.1233
190	389.3	75.57	0.5315	0.0655	0.2908	0.1274
200	398.8	73.58	0.5528	0.0668	0.3063	0.1313
220	417.0	69.95	0.5945	0.0692	0.3369	0.1387
240	434.2	66.72	0.6350	0.0714	0.3671	0.1456
260	450.6	63.83	0.6744	0.0734	0.3971	0.1522
280	466.2	61.21	0.7128	0.0752	0.4267	0.1583
300	481.1	58.83	0.7502	0.0769	0.4560	0.1641
320	495.4	56.66	0.7869	0.0785	0.4849	0.1695
340	509.2	54.66	0.8227	0.0799	0.5135	0.1747
360	522.4	52.83	0.8578	0.0813	0.5419	0.1797
380	535.1	51.13	0.8923	0.0825	0.5699	0.1844
400	547.4	49.55	0.9261	0.0837	0.5977	0.1889
420	559.3	48.08	0.9593	0.0849	0.6251	0.1932
440	570.7	46.71	0.9919	0.0859	0.6523	0.1973
460	581.9	45.43	1.0241	0.0869	0.6792	0.2013
480	592.7	44.22	1.0557	0.0879	0.7059	0.2050
500	603.1	43.09	1.0869	0.0888	0.7324	0.2087

Table 9.6—*contd* Range parameters of S in GaAs and germanium

E (keV)	S_e (keV/μm)	S_n (keV/μm)	R (μm)	ΔR (μm)	R_p (μm)	ΔR_p (μm)
Sulphur (S)	$Z_1 = 16; M_1 = 31.972; C_K = 1$					
10	116.0	495.3	0.0220	0.0069	0.0070	0.0072
20	164.0	513.3	0.0374	0.0113	0.0132	0.0121
30	200.8	509.7	0.0518	0.0152	0.0196	0.0166
40	231.9	500.2	0.0656	0.0188	0.0261	0.0209
50	259.3	488.8	0.0791	0.0222	0.0328	0.0251
60	284.0	476.8	0.0924	0.0255	0.0395	0.0291
70	306.8	464.9	0.1054	0.0285	0.0464	0.0330
80	328.0	453.4	0.1183	0.0315	0.0534	0.0368
90	347.9	442.3	0.1310	0.0343	0.0605	0.0405
100	366.7	431.7	0.1436	0.0370	0.0677	0.0441
110	384.6	421.6	0.1561	0.0396	0.0750	0.0477
120	401.7	411.9	0.1684	0.0421	0.0823	0.0511
130	418.1	402.8	0.1807	0.0446	0.0897	0.0545
140	433.8	394.1	0.1928	0.0469	0.0971	0.0578
150	449.1	385.8	0.2048	0.0491	0.1046	0.0610
160	463.8	377.9	0.2168	0.0513	0.1121	0.0641
170	478.1	370.4	0.2286	0.0534	0.1196	0.0672
180	491.9	363.2	0.2403	0.0555	0.1272	0.0702
190	505.4	356.4	0.2520	0.0574	0.1348	0.0731
200	518.5	349.8	0.2635	0.0593	0.1425	0.0760
220	543.8	337.5	0.2864	0.0630	0.1578	0.0816
240	568.0	326.2	0.3089	0.0664	0.1732	0.0869
260	591.2	315.8	0.3311	0.0697	0.1886	0.0921
280	613.5	306.1	0.3530	0.0727	0.2040	0.0970
300	635.0	297.1	0.3746	0.0756	0.2194	0.1018
320	655.8	288.8	0.3960	0.0784	0.2348	0.1064
340	676.0	280.9	0.4170	0.0810	0.2502	0.1108
360	695.6	273.6	0.4378	0.0835	0.2655	0.1151
380	714.6	266.7	0.4583	0.0859	0.2808	0.1193
400	733.2	260.2	0.4785	0.0881	0.2961	0.1233
420	751.3	254.1	0.4985	0.0903	0.3113	0.1271
440	768.9	248.3	0.5183	0.0924	0.3264	0.1309
460	786.2	242.8	0.5379	0.0943	0.3415	0.1345
480	803.1	237.6	0.5572	0.0962	0.3565	0.1380
500	819.6	232.6	0.5763	0.0981	0.3714	0.1414

Table 9.6—*contd* Range parameters of Se in GaAs and germanium

E (keV)	S_e (keV/μm)	S_n (keV/μm)	R (μm)	ΔR (μm)	R_p (μm)	ΔR_p (μm)
Selenium (Se)	$Z_1 = 34; M_1 = 79.917; C_K = 1$					
10	126.3	1305	0.0102	0.0039	0.0054	0.0037
20	178.6	1497	0.0166	0.0060	0.0090	0.0058
30	218.7	1580	0.0223	0.0079	0.0124	0.0077
40	252.6	1622	0.0277	0.0097	0.0157	0.0095
50	282.4	1645	0.0330	0.0114	0.0190	0.0112
60	309.3	1657	0.0381	0.0131	0.0222	0.0128
70	334.1	1662	0.0432	0.0148	0.0254	0.0144
80	357.2	1662	0.0482	0.0164	0.0286	0.0160
90	378.8	1658	0.0531	0.0179	0.0318	0.0176
100	399.3	1653	0.0580	0.0195	0.0350	0.0191
110	418.8	1646	0.0628	0.0210	0.0383	0.0206
120	437.4	1637	0.0677	0.0226	0.0415	0.0221
130	455.3	1628	0.0725	0.0241	0.0447	0.0236
140	472.5	1618	0.0773	0.0255	0.0480	0.0251
150	489.1	1608	0.0820	0.0270	0.0512	0.0266
160	505.1	1597	0.0868	0.0285	0.0545	0.0280
170	520.6	1586	0.0916	0.0299	0.0578	0.0295
180	535.7	1575	0.0963	0.0313	0.0611	0.0309
190	550.4	1564	0.1010	0.0327	0.0644	0.0324
200	564.7	1553	0.1058	0.0341	0.0677	0.0338
220	592.3	1531	0.1152	0.0369	0.0744	0.0366
240	618.6	1509	0.1246	0.0397	0.0811	0.0394
260	634.9	1487	0.1340	0.0423	0.0878	0.0421
280	668.2	1466	0.1434	0.0450	0.0946	0.0448
300	691.6	1445	0.1527	0.0476	0.1014	0.0475
320	714.3	1425	0.1621	0.0502	0.1083	0.0502
340	736.3	1405	0.1714	0.0527	0.1152	0.0528
360	757.6	1386	0.1808	0.0552	0.1221	0.0554
380	778.4	1367	0.1901	0.0577	0.1291	0.0580
400	798.6	1349	0.1994	0.0601	0.1360	0.0605
420	818.3	1331	0.2087	0.0625	0.1431	0.0631
440	837.6	1314	0.2180	0.0649	0.1501	0.0655
460	856.4	1297	0.2273	0.0672	0.1572	0.0680
480	874.8	1281	0.2366	0.0695	0.1643	0.0705
500	892.9	1265	0.2459	0.0718	0.1714	0.0729

Table 9.6—*contd* Range parameters of Te in GaAs and germanium

E (keV)	S_e (keV/μm)	S_n (keV/μm)	R (μm)	ΔR (μm)	R_p (μm)	ΔR_p (μm)
Tellurium (Te)	$Z_1 = 52; M_1 = 129.907; C_K = 1$					
10	129.0	1809	0.0076	0.0030	0.0055	0.0027
20	182.4	2216	0.0122	0.0045	0.0086	0.0041
30	223.4	2428	0.0161	0.0058	0.0114	0.0054
40	257.9	2560	0.0198	0.0070	0.0140	0.0065
50	288.4	2650	0.0232	0.0081	0.0165	0.0076
60	315.9	2714	0.0266	0.0092	0.0190	0.0086
70	341.2	2761	0.0298	0.0102	0.0214	0.0096
80	364.8	2796	0.0330	0.0112	0.0237	0.0106
90	386.9	2822	0.0362	0.0123	0.0261	0.0115
100	407.8	2842	0.0393	0.0132	0.0284	0.0125
110	427.7	2857	0.0423	0.0142	0.0307	0.0134
120	446.8	2868	0.0454	0.0152	0.0330	0.0143
130	465.0	2876	0.0484	0.0162	0.0353	0.0152
140	482.5	2881	0.0513	0.0171	0.0375	0.0161
150	499.5	2884	0.0543	0.0180	0.0398	0.0170
160	515.9	2885	0.0573	0.0190	0.0420	0.0179
170	531.7	2885	0.0602	0.0199	0.0443	0.0188
180	547.2	2884	0.0631	0.0208	0.0465	0.0197
190	562.2	2881	0.0660	0.0217	0.0488	0.0205
200	576.8	2878	0.0689	0.0226	0.0510	0.0214
220	604.9	2869	0.0747	0.0244	0.0555	0.0231
240	631.8	2857	0.0804	0.0262	0.0600	0.0248
260	657.6	2844	0.0862	0.0280	0.0645	0.0265
280	682.4	2830	0.0919	0.0297	0.0689	0.0282
300	706.4	2815	0.0975	0.0315	0.0734	0.0298
320	729.5	2798	0.1032	0.0332	0.0779	0.0315
340	752.0	2782	0.1089	0.0349	0.0824	0.0331
360	773.8	2765	0.1145	0.0366	0.0869	0.0347
380	795.0	2747	0.1202	0.0383	0.0914	0.0364
400	815.7	2730	0.1258	0.0400	0.0959	0.0380
420	835.8	2712	0.1315	0.0416	0.1004	0.0396
440	855.5	2694	0.1371	0.0433	0.1050	0.0412
460	874.7	2677	0.1427	0.0449	0.1095	0.0428
480	893.5	2659	0.1484	0.0465	0.1141	0.0443
500	911.9	2642	0.1540	0.0482	0.1186	0.0459

Table 9.6—*contd* Range parameters of H in GaAs and germanium

E (keV)	S_e (keV/μm)	S_n (keV/μm)	R (μm)	ΔR (μm)	R_p (μm)	ΔR_p (μm)
Hydrogen (H)	$Z_1 = 1; M_1 = 1.008; C_K = 1$					
10	49.68	1.077	0.3418	0.0052	0.1135	0.0716
20	69.14	0.7676	0.5072	0.0059	0.2180	0.1013
30	82.96	0.6096	0.6373	0.0062	0.3137	0.1206
40	93.54	0.5113	0.7497	0.0064	0.4028	0.1349
50	101.9	0.4433	0.8515	0.0065	0.4872	0.1463
60	108.5	0.3931	0.9461	0.0067	0.5681	0.1557
70	113.7	0.3542	1.0357	0.0068	0.6465	0.1638
80	117.9	0.3232	1.1218	0.0069	0.7229	0.1709
90	121.1	0.2977	1.2052	0.0070	0.7980	0.1773
100	123.5	0.2763	1.2868	0.0070	0.8721	0.1832
110	125.3	0.2581	1.3670	0.0071	0.9456	0.1886
120	126.5	0.2425	1.4463	0.0072	1.0188	0.1937
130	127.2	0.2288	1.5249	0.0072	1.0918	0.1985
140	127.6	0.2167	1.6033	0.0073	1.1649	0.2031
150	127.6	0.2060	1.6815	0.0074	1.2381	0.2074
160	127.4	0.1964	1.7598	0.0075	1.3118	0.2117
170	126.9	0.1877	1.8383	0.0075	1.3858	0.2158
180	126.2	0.1798	1.9172	0.0076	1.4605	0.2198
190	125.3	0.1727	1.9966	0.0077	1.5358	0.2237
200	124.4	0.1661	2.0766	0.0077	1.6118	0.2275
220	122.1	0.1546	2.2387	0.0079	1.7664	0.2351
240	119.5	0.1446	2.4040	0.0080	1.9246	0.2425
260	116.7	0.1360	2.5731	0.0082	2.0868	0.2498
280	113.9	0.1285	2.7464	0.0084	2.2534	0.2570
300	111.0	0.1218	2.9241	0.0085	2.4247	0.2643
320	108.0	0.1158	3.1066	0.0087	2.6008	0.2716
340	105.2	0.1105	3.2940	0.0089	2.7820	0.2790
360	102.4	0.1056	3.4866	0.0091	2.9684	0.2865
380	99.61	0.1012	3.6845	0.0094	3.1602	0.2940
400	96.94	0.09721	3.8878	0.0096	3.3574	0.3017
420	94.36	0.09353	4.0967	0.0099	3.5603	0.3096
440	91.87	0.09015	4.3113	0.0101	3.7688	0.3176
460	89.47	0.08702	4.5317	0.0104	3.9832	0.3258
480	87.16	0.08412	4.7579	0.0107	4.2034	0.3341
500	84.94	0.08143	4.9902	0.0110	4.4295	0.3427

Table 9.6—*contd* Range parameters of Zn in GaAs and germanium

E (keV)	S_e (keV/μm)	S_n (keV/μm)	R (μm)	ΔR (μm)	R_p (μm)	ΔR_p (μm)
Zinc (Zn)	$Z_1 = 30$; $M_1 = 63.925$; $C_K = 1$					
10	129.9	1113	0.0116	0.0043	0.0053	0.0042
20	183.7	1250	0.0190	0.0067	0.0091	0.0067
30	225.0	1303	0.0257	0.0089	0.0127	0.0089
40	259.8	1327	0.0321	0.0110	0.0163	0.0110
50	290.4	1336	0.0384	0.0130	0.0199	0.0130
60	318.1	1338	0.0444	0.0150	0.0234	0.0150
70	343.6	1335	0.0504	0.0168	0.0270	0.0169
80	367.4	1329	0.0564	0.0187	0.0306	0.0188
90	389.6	1321	0.0622	0.0205	0.0342	0.0206
100	410.7	1311	0.0681	0.0223	0.0378	0.0225
110	430.8	1301	0.0739	0.0240	0.0414	0.0243
120	449.9	1290	0.0796	0.0258	0.0451	0.0261
130	468.3	1279	0.0853	0.0275	0.0487	0.0279
140	486.0	1267	0.0911	0.0291	0.0524	0.0296
150	503.0	1256	0.0968	0.0308	0.0561	0.0314
160	519.5	1244	0.1024	0.0324	0.0598	0.0331
170	535.5	1233	0.1081	0.0340	0.0635	0.0348
180	551.0	1221	0.1137	0.0356	0.0673	0.0365
190	566.1	1210	0.1194	0.0372	0.0711	0.0382
200	580.8	1198	0.1250	0.0388	0.0748	0.0398
220	609.2	1176	0.1362	0.0418	0.0825	0.0431
240	636.3	1155	0.1474	0.0448	0.0901	0.0464
260	662.3	1134	0.1586	0.0478	0.0979	0.0495
280	687.3	1114	0.1697	0.0506	0.1056	0.0527
300	711.4	1095	0.1808	0.0535	0.1135	0.0558
320	734.7	1076	0.1918	0.0562	0.1214	0.0588
340	757.3	1058	0.2029	0.0589	0.1293	0.0618
360	779.3	1041	0.2139	0.0616	0.1372	0.0648
380	800.6	1024	0.2248	0.0642	0.1452	0.0677
400	821.4	1008	0.2358	0.0668	0.1533	0.0706
420	841.7	992.8	0.2467	0.0693	0.1613	0.0734
440	861.5	977.8	0.2576	0.0717	0.1694	0.0762
460	880.9	963.4	0.2684	0.0741	0.1775	0.0790
480	899.8	949.5	0.2793	0.0765	0.1857	0.0817
500	918.4	936.0	0.2901	0.0788	0.1938	0.0844

Table 9.7 Range parameters of As, B, P and Zn in SiO_2 ($d_2 = 2.65$ gcm^{-3})

E (keV)	E_e (keV/μm)	S_n (keV/μm)	R (μm)	ΔR (μm)	R_p (μm)	ΔR_p (μm)
Arsenic (As)	$Z_1 = 33; M_1 = 74.922; C_K = 1$					
10	112.9	1391	0.0093	0.0028	0.0080	0.0027
20	159.7	1526	0.0155	0.0046	0.0132	0.0043
30	195.6	1570	0.0213	0.0062	0.0182	0.0059
40	225.8	1583	0.0269	0.0077	0.0231	0.0074
50	252.5	1581	0.0324	0.0092	0.0279	0.0088
60	276.6	1572	0.0378	0.0107	0.0326	0.0103
70	298.8	1559	0.0432	0.0122	0.0374	0.0117
80	319.4	1544	0.0486	0.0136	0.0421	0.0131
90	338.8	1527	0.0539	0.0151	0.0469	0.0145
100	357.1	1509	0.0593	0.0165	0.0517	0.0158
110	374.5	1491	0.0647	0.0179	0.0564	0.0172
120	391.2	1473	0.0700	0.0193	0.0612	0.0185
130	407.2	1455	0.0754	0.0206	0.0660	0.0199
140	422.5	1437	0.0808	0.0220	0.0709	0.0212
150	437.4	1419	0.0862	0.0234	0.0757	0.0226
160	451.7	1401	0.0915	0.0247	0.0806	0.0239
170	465.6	1384	0.0969	0.0261	0.0854	0.0252
180	479.1	1367	0.1024	0.0274	0.0903	0.0265
190	492.2	1351	0.1078	0.0287	0.0952	0.0278
200	505.0	1335	0.1132	0.0301	0.1001	0.0291
220	529.7	1304	0.1241	0.0327	0.1100	0.0316
240	553.2	1274	0.1350	0.0352	0.1200	0.0341
260	575.8	1246	0.1460	0.0378	0.1300	0.0366
280	597.5	1220	0.1570	0.0403	0.1400	0.0391
300	618.5	1194	0.1680	0.0427	0.1501	0.0415
320	638.8	1170	0.1790	0.0452	0.1603	0.0439
340	658.5	1147	0.1901	0.0476	0.1704	0.0463
360	677.5	1125	0.2012	0.0499	0.1806	0.0486
380	696.1	1104	0.2123	0.0522	0.1909	0.0509
400	714.2	1084	0.2234	0.0545	0.2012	0.0532
420	731.8	1064	0.2345	0.0568	0.2115	0.0555
440	749.1	1046	0.2457	0.0590	0.2218	0.0577
460	765.9	1028	0.2568	0.0612	0.2321	0.0599
480	782.4	1011	0.2680	0.0633	0.2425	0.0620
500	798.5	994.8	0.2791	0.0655	0.2529	0.0641

Table 9.7—*contd* Range parameters of B in SiO_2

E (keV)	S_e (keV/μm)	S_n (keV/μm)	R (μm)	ΔR (μm)	R_p (μm)	ΔR_p (μm)
Boron (B)	$Z_1 = 5; M_1 = 11.009; C_K = 1$					
10	89.94	115.5	0.0516	0.0143	0.0283	0.0155
20	126.7	89.82	0.0990	0.0235	0.0604	0.0272
30	154.7	74.62	0.1439	0.0303	0.0935	0.0366
40	178.0	64.41	0.1863	0.0355	0.1265	0.0443
50	198.3	57.00	0.2265	0.0397	0.1589	0.0509
60	216.5	51.33	0.2647	0.0431	0.1905	0.0564
70	233.1	46.83	0.3012	0.0459	0.2214	0.0613
80	248.3	43.15	0.3362	0.0483	0.2515	0.0655
90	262.5	40.08	0.3699	0.0504	0.2808	0.0693
100	275.8	37.48	0.4024	0.0522	0.3094	0.0726
110	288.2	35.23	0.4338	0.0539	0.3374	0.0757
120	300.1	33.27	0.4642	0.0553	0.3647	0.0785
130	311.3	31.55	0.4938	0.0566	0.3914	0.0810
140	321.9	30.02	0.5226	0.0578	0.4176	0.0833
150	332.1	28.64	0.5507	0.0589	0.4432	0.0855
160	341.9	27.41	0.5781	0.0599	0.4684	0.0875
170	351.2	26.28	0.6048	0.0608	0.4931	0.0893
180	360.2	25.26	0.6311	0.0617	0.5174	0.0911
190	368.9	24.33	0.6567	0.0625	0.5413	0.0927
200	377.2	23.46	0.6819	0.0632	0.5648	0.0943
220	393.1	21.93	0.7310	0.0646	0.6108	0.0971
240	407.9	20.61	0.7784	0.0658	0.6555	0.0996
260	421.8	19.45	0.8244	0.0669	0.6990	0.1019
280	435.0	18.43	0.8691	0.0679	0.7415	0.1040
300	447.4	17.53	0.9126	0.0688	0.7830	0.1059
320	459.2	16.71	0.9551	0.0697	0.8237	0.1077
340	470.3	15.98	0.9967	0.0705	0.8635	0.1093
360	481.0	15.32	1.0374	0.0712	0.9026	0.1109
380	491.1	14.71	1.0773	0.0719	0.9411	0.1123
400	500.8	14.16	1.1165	0.0725	0.9789	0.1137
420	510.0	13.65	1.1550	0.0731	1.0161	0.1149
440	518.8	13.18	1.1929	0.0737	1.0528	0.1161
460	527.3	12.74	1.2302	0.0742	1.0890	0.1172
480	535.4	12.33	1.2670	0.0747	1.1247	0.1183
500	543.2	11.96	1.3033	0.0752	1.1599	0.1193

Table 9.7—*contd* Range parameters of P in SiO_2

E (keV)	S_e (keV/μm)	S_n (keV/μm)	R (μm)	ΔR (μm)	R_p (μm)	ΔR_p (μm)
Phosphorus (P)	$Z_1 = 15; M_1 = 30.974; C_K = 1$					
10	114.8	647.5	0.0155	0.0053	0.0107	0.0050
20	162.3	620.6	0.0284	0.0093	0.0202	0.0089
30	198.8	585.1	0.0411	0.0131	0.0299	0.0127
40	229.6	551.8	0.0539	0.0168	0.0399	0.0163
50	256.7	521.9	0.0667	0.0202	0.0501	0.0198
60	281.2	495.4	0.0796	0.0236	0.0604	0.0231
70	303.7	471.7	0.0925	0.0268	0.0709	0.0264
80	324.7	450.6	0.1054	0.0299	0.0816	0.0296
90	344.4	431.6	0.1183	0.0329	0.0923	0.0326
100	363.0	414.4	0.1311	0.0357	0.1031	0.0356
110	380.7	398.8	0.1440	0.0384	0.1140	0.0384
120	397.6	384.5	0.1568	0.0410	0.1249	0.0412
130	413.9	371.4	0.1695	0.0435	0.1358	0.0439
140	429.5	359.4	0.1822	0.0459	0.1468	0.0464
150	444.6	348.2	0.1949	0.0483	0.1577	0.0489
160	459.2	337.8	0.2075	0.0505	0.1687	0.0513
170	473.3	328.1	0.2200	0.0526	0.1796	0.0537
180	487.0	319.1	0.2324	0.0547	0.1905	0.0559
190	500.4	310.7	0.2448	0.0567	0.2014	0.0581
200	513.4	302.7	0.2571	0.0586	0.2123	0.0602
220	538.4	288.2	0.2814	0.0622	0.2339	0.0642
240	562.4	275.2	0.3055	0.0656	0.2554	0.0680
260	585.3	263.6	0.3292	0.0687	0.2768	0.0715
280	607.4	253.0	0.3526	0.0716	0.2979	0.0749
300	628.7	243.4	0.3757	0.0744	0.3189	0.0781
320	649.4	234.6	0.3985	0.0770	0.3396	0.0811
340	669.3	226.5	0.4209	0.0794	0.3602	0.0839
360	688.7	219.1	0.4431	0.0817	0.3806	0.0866
380	707.6	212.2	0.4650	0.0839	0.4007	0.0891
400	726.0	205.8	0.4866	0.0860	0.4207	0.0916
420	743.9	199.8	0.5079	0.0879	0.4405	0.0939
440	761.4	194.2	0.5290	0.0898	0.4600	0.0961
460	778.5	189.0	0.5498	0.0916	0.4794	0.0982
480	795.3	184.0	0.5703	0.0932	0.4986	0.1002
500	811.7	179.4	0.5906	0.0948	0.5175	0.1022

Table 9.7—*contd* Range parameters of Zn in SiO_2

E (keV)	S_e (keV/μm)	S_n (keV/μm)	R (μm)	ΔR (μm)	R_p (μm)	ΔR_p (μm)
Zinc (Zn)	$Z_1 = 30; M_1 = 63.925; C_K = 1$					
10	116.7	1296	0.0097	0.0030	0.0079	0.0028
20	165.0	1396	0.0163	0.0050	0.0135	0.0047
30	202.1	1420	0.0226	0.0068	0.0188	0.0064
40	233.4	1419	0.0287	0.0085	0.0240	0.0081
50	260.9	1408	0.0347	0.0102	0.0292	0.0097
60	285.8	1392	0.0407	0.0119	0.0343	0.0113
70	308.7	1373	0.0467	0.0135	0.0395	0.0129
80	330.0	1353	0.0526	0.0152	0.0447	0.0145
90	350.1	1332	0.0586	0.0168	0.0499	0.0160
100	369.0	1311	0.0645	0.0184	0.0551	0.0176
110	387.0	1291	0.0705	0.0200	0.0603	0.0191
120	404.2	1271	0.0764	0.0215	0.0656	0.0206
130	420.7	1251	0.0824	0.0231	0.0709	0.0221
140	436.6	1232	0.0884	0.0246	0.0762	0.0236
150	451.9	1213	0.0944	0.0261	0.0815	0.0251
160	466.8	1195	0.1004	0.0276	0.0869	0.0266
170	481.1	1177	0.1064	0.0291	0.0922	0.0280
180	495.1	1160	0.1125	0.0306	0.0976	0.0295
190	508.6	1144	0.1185	0.0321	0.1030	0.0309
200	521.8	1128	0.1246	0.0335	0.1085	0.0323
220	547.3	1097	0.1367	0.0364	0.1194	0.0351
240	571.7	1068	0.1489	0.0392	0.1304	0.0379
260	595.0	1041	0.1611	0.0420	0.1414	0.0406
280	617.5	1016	0.1733	0.0447	0.1525	0.0433
300	639.1	991.4	0.1856	0.0474	0.1637	0.0459
320	660.1	968.7	0.1979	0.0500	0.1749	0.0485
340	680.4	947.1	0.2101	0.0525	0.1861	0.0511
360	700.1	926.6	0.2224	0.0550	0.1974	0.0536
380	719.3	907.2	0.2347	0.0575	0.2086	0.0560
400	738.0	888.7	0.2470	0.0599	0.2199	0.0584
420	756.2	871.1	0.2593	0.0623	0.2312	0.0608
440	774.0	854.4	0.2716	0.0646	0.2426	0.0631
460	791.4	838.3	0.2839	0.0668	0.2539	0.0654
480	808.4	823.0	0.2962	0.0691	0.2653	0.0676
500	825.1	808.4	0.3084	0.0713	0.2766	0.0698

Table 9.8 Range parameters of As and B in Si_3N_4 ($d_2 = 3.45$ gcm^{-3})

E (keV)	S_e (keV/μm)	S_n (keV/μm)	R (μm)	ΔR (μm)	R_p (μm)	ΔR_p (μm)
Arsenic (As)	$Z_1 = 33$; $M_1 = 74.922$; $C_K = 1$					
10	146.1	1790	0.0072	0.0022	0.0061	0.0020
20	206.6	1966	0.0120	0.0035	0.0103	0.0033
30	253.0	2023	0.0165	0.0048	0.0142	0.0045
40	292.1	2040	0.0209	0.0060	0.0179	0.0057
50	326.6	2038	0.0251	0.0072	0.0217	0.0068
60	357.8	2027	0.0293	0.0083	0.0254	0.0079
70	386.5	2011	0.0335	0.0095	0.0290	0.0090
80	413.1	1991	0.0377	0.0106	0.0327	0.0101
90	438.2	1969	0.0418	0.0117	0.0364	0.0111
100	461.9	1947	0.0460	0.0128	0.0401	0.0122
110	484.4	1924	0.0501	0.0139	0.0438	0.0132
120	506.0	1900	0.0543	0.0150	0.0475	0.0143
130	526.6	1877	0.0584	0.0161	0.0513	0.0153
140	546.5	1854	0.0626	0.0171	0.0550	0.0163
150	565.7	1831	0.0668	0.0182	0.0587	0.0173
160	584.3	1809	0.0710	0.0192	0.0625	0.0184
170	602.2	1787	0.0751	0.0203	0.0663	0.0194
180	619.7	1765	0.0793	0.0213	0.0700	0.0204
190	636.7	1744	0.0835	0.0224	0.0738	0.0214
200	653.2	1723	0.0877	0.0234	0.0777	0.0224
220	685.1	1684	0.0962	0.0254	0.0853	0.0243
240	715.6	1646	0.1046	0.0274	0.0930	0.0263
260	744.8	1610	0.1131	0.0294	0.1007	0.0282
280	772.9	1575	0.1216	0.0313	0.1085	0.0301
300	800.0	1543	0.1301	0.0332	0.1163	0.0319
320	826.3	1512	0.1387	0.0351	0.1241	0.0338
340	851.7	1482	0.1472	0.0370	0.1320	0.0356
360	876.4	1454	0.1558	0.0388	0.1399	0.0374
380	900.4	1426	0.1644	0.0406	0.1478	0.0392
400	923.8	1401	0.1730	0.0424	0.1558	0.0409
420	946.6	1376	0.1816	0.0442	0.1637	0.0427
440	968.9	1352	0.1902	0.0459	0.1717	0.0444
460	990.7	1329	0.1989	0.0476	0.1797	0.0461
480	1012	1307	0.2075	0.0493	0.1877	0.0477
500	1033	1286	0.2161	0.0509	0.1957	0.0494

Table 9.8—*contd* Range parameters of B in Si_3N_4

E (keV)	S_e (keV/μm)	S_n (keV/μm)	R (μm)	ΔR (μm)	R_p (μm)	ΔR_p (μm)
Boron (B)	$Z_1 = 5; M_1 = 11.009; C_K = 1$					
10	115.8	148.1	0.0402	0.0111	0.0219	0.0120
20	163.2	115.3	0.0771	0.0182	0.0467	0.0211
30	199.1	95.85	0.1120	0.0235	0.0723	0.0284
40	229.2	82.79	0.1449	0.0275	0.0978	0.0344
50	255.3	73.30	0.1762	0.0307	0.1228	0.0394
60	278.7	66.04	0.2059	0.0333	0.1473	0.0438
70	300.0	60.27	0.2342	0.0355	0.1712	0.0475
80	319.7	55.55	0.2614	0.0374	0.1945	0.0508
90	337.9	51.61	0.2876	0.0390	0.2172	0.0537
100	355.0	48.27	0.3128	0.0404	0.2393	0.0564
110	371.1	45.38	0.3372	0.0416	0.2609	0.0587
120	386.3	42.87	0.3609	0.0428	0.2821	0.0609
130	400.7	40.65	0.3838	0.0438	0.3028	0.0628
140	414.4	38.68	0.4062	0.0447	0.3231	0.0647
150	427.5	36.92	0.4280	0.0455	0.3429	0.0663
160	440.1	35.33	0.4493	0.0463	0.3624	0.0679
170	452.2	33.88	0.4701	0.0470	0.3816	0.0693
180	463.7	32.57	0.4904	0.0477	0.4004	0.0707
190	474.9	31.37	0.5104	0.0483	0.4189	0.0719
200	485.6	30.26	0.5299	0.0489	0.4372	0.0731
220	506.0	28.29	0.5680	0.0499	0.4728	0.0753
240	525.1	26.58	0.6049	0.0509	0.5074	0.0773
260	543.0	25.10	0.6406	0.0517	0.5412	0.0790
280	560.0	23.78	0.6753	0.0525	0.5741	0.0807
300	575.9	22.62	0.7091	0.0532	0.6063	0.0821
320	591.1	21.57	0.7422	0.0539	0.6379	0.0835
340	605.5	20.62	0.7744	0.0545	0.6688	0.0848
360	619.2	19.77	0.8061	0.0550	0.6991	0.0860
380	632.2	18.99	0.8371	0.0556	0.7290	0.0871
400	644.6	18.27	0.8675	0.0561	0.7583	0.0881
420	656.5	17.62	0.8974	0.0565	0.7872	0.0891
440	667.9	17.01	0.9268	0.0570	0.8156	0.0900
460	678.8	16.45	0.9558	0.0574	0.8437	0.0909
480	689.3	15.92	0.9844	0.0578	0.8714	0.0917
500	699.3	15.44	1.1026	0.0582	0.8988	0.0925

Table 9.9 Range parameters of As and B in Al_2O_3 ($d_2 = 3.965$ gcm^{-3})

E (keV)	S_e (keV/μm)	S_n (keV/μm)	R (μm)	ΔR (μm)	R_p (μm)	ΔR_p (μm)
Arsenic (As)	$Z_1 = 33$; $M_1 = 74.922$; $C_K = 1$					
10	166.1	2051	0.0063	0.0019	0.0053	0.0019
20	235.0	2249	0.0105	0.0031	0.0090	0.0029
30	287.8	2313	0.0144	0.0042	0.0124	0.0040
40	332.3	2331	0.0182	0.0053	0.0157	0.0050
50	371.5	2328	0.0220	0.0063	0.0190	0.0060
60	407.0	2314	0.0256	0.0073	0.0223	0.0070
70	439.6	2295	0.0293	0.0083	0.0255	0.0079
80	469.9	2271	0.0330	0.0093	0.0288	0.0089
90	498.4	2246	0.0366	0.0103	0.0320	0.0098
100	525.4	2220	0.0402	0.0112	0.0353	0.0108
110	551.0	2193	0.0439	0.0122	0.0385	0.0117
120	575.5	2165	0.0475	0.0131	0.0418	0.0126
130	599.0	2138	0.0512	0.0141	0.0451	0.0135
140	621.7	2112	0.0548	0.0150	0.0484	0.0144
150	643.5	2085	0.0585	0.0159	0.0517	0.0153
160	664.6	2059	0.0622	0.0169	0.0550	0.0162
170	685.0	2034	0.0659	0.0178	0.0583	0.0171
180	704.9	2009	0.0695	0.0187	0.0616	0.0180
190	724.2	1985	0.0732	0.0196	0.0650	0.0189
200	743.0	1961	0.0769	0.0205	0.0683	0.0198
220	779.3	1915	0.0843	0.0223	0.0751	0.0215
240	813.9	1871	0.0918	0.0240	0.0819	0.0232
260	847.2	1830	0.0992	0.0258	0.0887	0.0249
280	879.2	1790	0.1067	0.0275	0.0956	0.0266
300	910.0	1753	0.1142	0.0291	0.1024	0.0283
320	939.9	1717	0.1217	0.0308	0.1094	0.0299
340	968.8	1683	0.1293	0.0324	0.1163	0.0315
360	996.9	1650	0.1368	0.0340	0.1233	0.0331
380	1024	1619	0.1444	0.0356	0.1303	0.0347
400	1051	1590	0.1519	0.0372	0.1373	0.0362
420	1077	1561	0.1595	0.0387	0.1443	0.0378
440	1102	1534	0.1671	0.0402	0.1513	0.0393
460	1127	1508	0.1747	0.0417	0.1584	0.0407
480	1151	1483	0.1823	0.0432	0.1655	0.0422
500	1175	1458	0.1899	0.0447	0.1725	0.0437

Table 9.9—*contd* Range parameters of B in Al_2O_3

E (keV)	S_e (keV/μm)	S_n (keV/μm)	R (μm)	ΔR (μm)	R_p (μm)	ΔR_p (μm)
Boron (B)	$Z_1 = 5; M_1 = 11.009; C_K = 1$					
10	132.4	168.2	0.0353	0.0097	0.0193	0.0106
20	186.5	130.6	0.0677	0.0160	0.0412	0.0186
30	227.7	108.4	0.0983	0.0206	0.0638	0.0250
40	262.0	93.56	0.1273	0.0241	0.0863	0.0303
50	291.9	82.76	0.1547	0.0269	0.1084	0.0347
60	318.6	74.50	0.1807	0.0292	0.1300	0.0385
70	343.0	67.95	0.2056	0.0311	0.1510	0.0417
80	365.4	62.60	0.2294	0.0328	0.1715	0.0446
90	386.3	58.14	0.2523	0.0342	0.1915	0.0472
100	405.8	54.35	0.2744	0.0354	0.2110	0.0494
110	424.2	51.08	0.2958	0.0365	0.2300	0.0515
120	441.6	48.24	0.3166	0.0375	0.2486	0.0534
130	458.1	45.73	0.3367	0.0383	0.2668	0.0551
140	473.8	43.51	0.3563	0.0391	0.2847	0.0567
150	488.8	41.51	0.3754	0.0399	0.3021	0.0581
160	503.1	39.71	0.3940	0.0406	0.3193	0.0595
170	516.9	38.09	0.4122	0.0412	0.3361	0.0607
180	530.1	36.60	0.4300	0.0417	0.3526	0.0619
190	542.9	35.24	0.4475	0.0423	0.3689	0.0630
200	555.1	33.99	0.4646	0.0428	0.3849	0.0640
220	578.5	31.77	0.4980	0.0437	0.4162	0.0659
240	600.3	29.85	0.5302	0.0445	0.4466	0.0677
260	620.8	28.17	0.5615	0.0453	0.4762	0.0692
280	640.1	26.69	0.5919	0.0459	0.5051	0.0706
300	658.4	25.38	0.6215	0.0465	0.5333	0.0719
320	675.7	24.20	0.6504	0.0471	0.5610	0.0731
340	692.2	23.14	0.6787	0.0476	0.5881	0.0742
260	707.8	22.17	0.7064	0.0481	0.6147	0.0753
380	722.7	21.30	0.7335	0.0486	0.6409	0.0762
400	736.9	20.49	0.7601	0.0490	0.6666	0.0771
420	750.5	19.75	0.7863	0.0494	0.6919	0.0780
440	763.5	19.07	0.8121	0.0498	0.7168	0.0788
460	776.0	18.44	0.8375	0.0501	0.7414	0.0795
480	787.9	17.85	0.8624	0.0505	0.7657	0.0803
500	799.4	17.30	0.8871	0.0508	0.7897	0.0810

Table 9.10 Range parameters of Sb, As, B and P in AZ 1350 photoresist ($d_2 = 1.33$ gcm^{-3})

E (keV)	S_e (keV/μm)	S_n (keV/μm)	R (μm)	ΔR (μm)	R_p (μm)	ΔR_p (μm)
Antimony (Sb)	$Z_1 = 51; M_1 = 120.904; C_K = 1$					
10	66.36	1061	0.0147	0.0029	0.0144	0.0029
20	93.85	1252	0.0227	0.0044	0.0223	0.0043
30	114.9	1344	0.0298	0.0057	0.0293	0.0056
40	132.7	1396	0.0365	0.0070	0.0358	0.0069
50	148.4	1429	0.0429	0.0082	0.0421	0.0081
60	162.6	1450	0.0492	0.0094	0.0483	0.0092
70	175.6	1463	0.0553	0.0105	0.0543	0.0103
80	187.7	1472	0.0614	0.0116	0.0602	0.0115
90	199.1	1477	0.0674	0.0128	0.0661	0.0126
100	209.9	1479	0.0733	0.0139	0.0720	0.0137
110	220.1	1479	0.0792	0.0150	0.0778	0.0147
120	229.9	1478	0.0851	0.0160	0.0835	0.0158
130	239.3	1475	0.0909	0.0171	0.0893	0.0169
140	248.3	1471	0.0968	0.0182	0.0950	0.0179
150	257.0	1467	0.1026	0.0192	0.1007	0.0189
160	265.4	1462	0.1084	0.0203	0.1064	0.0200
170	273.6	1457	0.1141	0.0213	0.1121	0.0210
180	281.5	1451	0.1199	0.0224	0.1177	0.0220
190	289.3	1445	0.1257	0.0234	0.1234	0.0231
200	296.8	1439	0.1314	0.0244	0.1291	0.0241
220	311.3	1426	0.1430	0.0265	0.1404	0.0261
240	325.1	1412	0.1545	0.0285	0.1517	0.0281
260	338.4	1398	0.1660	0.0305	0.1630	0.0301
280	351.1	1385	0.1775	0.0325	0.1743	0.0321
300	363.5	1371	0.1890	0.0345	0.1857	0.0341
320	375.4	1357	0.2006	0.0365	0.1970	0.0360
340	386.9	1343	0.2121	0.0385	0.2084	0.0380
360	398.1	1329	0.2237	0.0404	0.2198	0.0399
380	409.0	1316	0.2353	0.0423	0.2312	0.0418
400	419.7	1302	0.2469	0.0443	0.2426	0.0437
420	430.0	1289	0.2585	0.0462	0.2541	0.0456
440	440.1	1277	0.2702	0.0481	0.2655	0.0475
460	450.0	1264	0.2818	0.0500	0.2770	0.0494
480	459.7	1252	0.2935	0.0519	0.2885	0.0512
500	469.2	1240	0.3052	0.0538	0.3000	0.0531
550	492.0	1210	0.3345	0.0584	0.3289	0.0577
600	513.9	1183	0.3639	0.0630	0.3579	0.0622
650	534.9	1156	0.3934	0.0675	0.3870	0.0667
700	555.0	1131	0.4230	0.0719	0.4162	0.0711
750	574.5	1108	0.4527	0.0763	0.4455	0.0754
800	593.3	1085	0.4825	0.0806	0.4749	0.0797
850	611.5	1063	0.5123	0.0848	0.5043	0.0839
900	629.2	1043	0.5422	0.0890	0.5338	0.0880
950	646.4	1023	0.5721	0.0931	0.5633	0.0921
1000	663.2	1004	0.6021	0.0971	0.5929	0.0962

Table 9.10—*contd* Range parameters of As in AZ 1350 photoresist

E (keV)	S_e (keV/μm)	S_n (keV/μm)	R (μm)	ΔR (μm)	R_p (μm)	ΔR_p (μm)
Arsenic (As)	$Z_1 = 33; M_1 = 74.922; C_K = 1$					
10	71.76	902.3	0.0147	0.0035	0.0142	0.0034
20	101.5	983.9	0.0243	0.0056	0.0235	0.0055
30	124.3	1009	0.0333	0.0077	0.0322	0.0075
40	143.5	1014	0.0420	0.0096	0.0407	0.0094
50	160.5	1011	0.0506	0.0115	0.0490	0.0112
60	175.8	1003	0.0591	0.0133	0.0573	0.0130
70	189.9	993.3	0.0676	0.0152	0.0655	0.0148
80	203.0	981.9	0.0760	0.0170	0.0737	0.0166
90	215.3	969.8	0.0844	0.0188	0.0819	0.0184
100	226.9	957.3	0.0929	0.0206	0.0901	0.0201
110	238.0	944.7	0.1013	0.0223	0.0983	0.0218
120	248.6	932.2	0.1098	0.0241	0.1065	0.0236
130	258.7	919.7	0.1183	0.0258	0.1148	0.0253
140	268.5	907.4	0.1268	0.0275	0.1230	0.0269
150	277.9	895.4	0.1353	0.0292	0.1313	0.0286
160	287.0	883.6	0.1438	0.0309	0.1396	0.0303
170	295.8	872.1	0.1524	0.0326	0.1480	0.0319
180	304.4	860.9	0.1609	0.0343	0.1563	0.0336
190	312.8	849.9	0.1695	0.0359	0.1647	0.0352
200	320.9	839.3	0.1781	0.0376	0.1731	0.0368
220	336.5	818.8	0.1954	0.0408	0.1900	0.0401
240	351.5	799.4	0.2128	0.0440	0.2069	0.0432
260	365.8	781.0	0.2302	0.0472	0.2239	0.0463
280	379.6	763.5	0.2476	0.0503	0.2410	0.0494
300	392.9	746.9	0.2652	0.0534	0.2581	0.0525
320	405.8	731.2	0.2827	0.0564	0.2753	0.0554
340	418.3	716.2	0.3003	0.0594	0.2925	0.0584
360	430.4	701.9	0.3180	0.0623	0.3097	0.0613
380	442.2	688.3	0.3357	0.0652	0.3271	0.0641
400	453.6	675.3	0.3534	0.0680	0.3444	0.0670
420	464.8	662.8	0.3711	0.0708	0.3618	0.0697
440	475.7	650.9	0.3888	0.0736	0.3791	0.0725
460	486.4	639.5	0.4066	0.0763	0.3965	0.0751
480	496.8	628.6	0.4244	0.0790	0.4140	0.0778
500	507.1	618.1	0.4421	0.0816	0.4314	0.0804
550	531.8	593.7	0.4866	0.0880	0.4750	0.0867
600	555.3	571.4	0.5310	0.0941	0.5186	0.0928
650	577.9	551.0	0.5753	0.1000	0.5621	0.0987
700	599.7	532.4	0.6195	0.1057	0.6056	0.1043
750	620.6	515.1	0.6636	0.1111	0.6490	0.1098
800	640.9	499.2	0.7076	0.1164	0.6922	0.1150
850	660.5	484.3	0.7514	0.1214	0.7353	0.1200
900	679.5	470.5	0.7949	0.1263	0.7782	0.1249
950	698.0	457.6	0.8383	0.1310	0.8209	0.1296
1000	716.0	445.5	0.8815	0.1355	0.8634	0.1341

Table 9.10—*contd* Range parametersof B in AZ 1350 photoresist

E (keV)	S_e (keV/μm)	S_n (keV/μm)	R (μm)	ΔR (μm)	R_p (μm)	ΔR_p (μm)
Boron (B)	$Z_1 = 5; M_1 = 11.009; C_K = 1$					
10	70.92	91.33	0.0618	0.0165	0.0490	0.0164
20	100.3	67.50	0.1227	0.0276	0.1014	0.0283
30	122.8	54.60	0.1807	0.0356	0.1533	0.0371
40	141.7	46.33	0.2354	0.0416	0.2033	0.0440
50	158.3	40.50	0.2872	0.0462	0.2512	0.0495
60	173.3	36.13	0.3362	0.0500	0.2971	0.0539
70	187.0	32.72	0.3828	0.0530	0.3411	0.0577
80	199.7	29.97	0.4273	0.0556	0.3833	0.0608
90	211.6	27.69	0.4699	0.0578	0.4241	0.0636
100	222.8	25.78	0.5109	0.0597	0.4633	0.0659
110	233.4	24.14	0.5504	0.0614	0.5014	0.0681
120	243.4	22.73	0.5886	0.0629	0.5382	0.0699
130	253.0	21.48	0.6256	0.0642	0.5740	0.0716
140	262.1	20.39	0.6615	0.0654	0.6088	0.0731
150	270.9	19.41	0.6965	0.0665	0.6427	0.0745
160	279.3	18.53	0.7305	0.0675	0.6758	0.0758
170	287.4	17.73	0.7636	0.0684	0.7081	0.0770
180	295.1	17.01	0.7960	0.0692	0.7398	0.0780
190	302.6	16.35	0.8277	0.0700	0.7707	0.0790
200	309.9	15.75	0.8588	0.0707	0.8011	0.0800
220	323.7	14.68	0.9190	0.0720	0.8601	0.0817
240	336.5	13.76	0.9771	0.0732	0.9171	0.0832
260	348.7	12.96	1.0333	0.0742	0.9724	0.0845
280	360.1	12.25	1.0878	0.0752	1.0260	0.0857
300	370.9	11.63	1.1407	0.0760	1.0782	0.0868
320	381.1	11.07	1.1924	0.0768	1.1292	0.0878
340	390.8	10.57	1.2428	0.0775	1.1789	0.0887
360	400.0	10.12	1.2921	0.0782	1.2276	0.0896
380	408.8	9.704	1.3403	0.0788	1.2754	0.0904
400	417.1	9.327	1.3877	0.0794	1.3222	0.0911
420	425.1	8.981	1.4341	0.0800	1.3682	0.0918
440	432.8	8.661	1.4798	0.0805	1.4134	0.0925
460	440.1	8.366	1.5248	0.0810	1.4580	0.0931
480	447.2	8.093	1.5690	0.0815	1.5019	0.0937
500	453.9	7.838	1.6127	0.0819	1.5451	0.0942
550	469.7	7.271	1.7192	0.0829	1.6508	0.0955
600	484.2	6.788	1.8225	0.0839	1.7534	0.0967
650	497.6	6.370	1.9229	0.0847	1.8532	0.0978
700	510.0	6.004	2.0210	0.0855	1.9507	0.0987
750	521.6	5.682	2.1168	0.0863	2.0460	0.0996
800	532.5	5.395	2.2107	0.0869	2.1394	0.1004
850	542.7	5.138	2.3028	0.0876	2.2310	0.1012
900	552.4	4.906	2.3933	0.0882	2.3211	0.1019
950	561.5	4.696	2.4823	0.0887	2.4097	0.1026
1000	570.2	4.505	2.5699	0.0893	2.4970	0.1032

Table 9.10—*contd* Range parameters of P in AZ 1350 photoresist

E (keV)	S_e (keV/μm)	S_n (keV/μm)	R (μm)	ΔR (μm)	R_p (μm)	ΔR_p (μm)
Phosphorus (P)	$Z_1 = 15; M_1 = 30.974; C_K = 1$					
10	78.87	467.1	0.0210	0.0062	0.0192	0.0059
20	111.5	439.1	0.0392	0.0111	0.0359	0.0106
30	136.6	408.7	0.0574	0.0158	0.0528	0.0151
40	157.7	381.6	0.0759	0.0204	0.0700	0.0195
50	176.3	358.1	0.0945	0.0247	0.0874	0.0237
60	193.2	337.7	0.1133	0.0289	0.1051	0.0278
70	208.6	319.8	0.1322	0.0329	0.1229	0.0317
80	223.0	304.1	0.1511	0.0367	0.1408	0.0355
90	236.5	290.0	0.1701	0.0404	0.1588	0.0391
100	249.3	277.5	0.1891	0.0439	0.1768	0.0426
110	261.5	266.1	0.2081	0.0473	0.1949	0.0459
120	273.1	255.9	0.2270	0.0505	0.2129	0.0491
130	284.2	246.5	0.2459	0.0536	0.2310	0.0522
140	294.9	237.9	0.2647	0.0565	0.2490	0.0551
150	305.2	230.0	0.2834	0.0593	0.2669	0.0580
160	315.2	222.7	0.3020	0.0621	0.2848	0.0607
170	324.9	215.9	0.3206	0.0647	0.3026	0.0633
180	334.3	209.6	0.3390	0.0672	0.3204	0.0658
190	343.4	203.7	0.3573	0.0696	0.3380	0.0683
200	352.3	198.2	0.3756	0.0719	0.3556	0.0706
220	369.4	188.2	0.4117	0.0763	0.3904	0.0750
240	385.7	179.3	0.4473	0.0804	0.4249	0.0792
260	401.3	171.3	0.4825	0.0841	0.4590	0.0830
280	416.4	164.1	0.5172	0.0877	0.4926	0.0866
300	430.9	157.6	0.5514	0.0910	0.5258	0.0900
320	444.8	151.7	0.5851	0.0941	0.5587	0.0932
340	458.4	146.2	0.6184	0.0970	0.5911	0.0961
360	471.5	141.2	0.6513	0.0998	0.6231	0.0990
380	484.2	136.6	0.6837	0.1023	0.6548	0.1016
400	496.6	132.3	0.7157	0.1048	0.6860	0.1042
420	508.7	128.3	0.7473	0.1071	0.7169	0.1065
440	520.4	124.6	0.7785	0.1093	0.7474	0.1088
460	531.9	121.1	0.8093	0.1114	0.7776	0.1110
480	543.1	117.8	0.8398	0.1134	0.8074	0.1130
500	554.1	114.8	0.8699	0.1153	0.8368	0.1150
550	580.4	107.8	0.9436	0.1197	0.9091	0.1195
600	605.4	101.8	1.0152	0.1236	0.9795	0.1236
650	629.3	96.50	1.0850	0.1272	1.0481	0.1273
700	652.1	91.78	1.1530	0.1304	1.1151	0.1307
750	673.9	87.56	1.2195	0.1334	1.1806	0.1338
800	694.9	83.75	1.2844	0.1361	1.2446	0.1366
850	715.1	80.30	1.3479	0.1386	1.3073	0.1392
900	734.5	77.16	1.4102	0.1409	1.3687	0.1417
950	753.3	74.28	1.4712	0.1431	1.4290	0.1439
1000	771.3	71.64	1.5310	0.1451	1.4882	0.1460

Table 9.11 Range parameters of Sb, As, B and P in PMMA photoresist ($d_2 = 1.19$ gcm^{-3})

E (keV)	S_e (keV/μm)	S_n (keV/μm)	R (μm)	ΔR (μm)	R_p (μm)	ΔR_p (μm)
Antimony (Sb)	$Z_1 = 51; M_1 = 120.904; C_K = 1$					
10	60.57	961.1	0.0162	0.0032	0.0161	0.0032
20	85.65	1137	0.0251	0.0048	0.0248	0.0048
30	104.9	1221	0.0329	0.0063	0.0325	0.0062
40	121.1	1269	0.0402	0.0077	0.0397	0.0076
50	135.4	1300	0.0473	0.0090	0.0467	0.0089
60	148.4	1320	0.0542	0.0103	0.0535	0.0101
70	160.2	1333	0.0610	0.0115	0.0601	0.0114
80	171.3	1341	0.0676	0.0127	0.0667	0.0126
90	181.7	1346	0.0742	0.0140	0.0731	0.0138
100	191.5	1348	0.0807	0.0152	0.0796	0.0150
110	200.9	1349	0.0872	0.0163	0.0859	0.0161
120	209.8	1348	0.0936	0.0175	0.0923	0.0173
130	218.4	1346	0.1000	0.0187	0.0986	0.0185
140	226.6	1343	0.1064	0.0199	0.1049	0.0196
150	234.6	1339	0.1128	0.0210	0.1111	0.0207
160	242.3	1335	0.1191	0.0221	0.1174	0.0219
170	249.7	1331	0.1254	0.0233	0.1236	0.0230
180	256.9	1326	0.1318	0.0244	0.1299	0.0241
190	264.0	1321	0.1381	0.0255	0.1361	0.0252
200	270.8	1315	0.1444	0.0267	0.1423	0.0263
220	284.1	1304	0.1570	0.0289	0.1547	0.0285
240	296.7	1292	0.1696	0.0311	0.1672	0.0307
260	308.8	1279	0.1822	0.0333	0.1796	0.0329
280	320.5	1267	0.1948	0.0354	0.1920	0.0350
300	331.7	1255	0.2074	0.0376	0.2044	0.0372
320	342.6	1242	0.2200	0.0397	0.2169	0.0393
340	353.1	1230	0.2326	0.0419	0.2293	0.0414
360	363.3	1218	0.2453	0.0440	0.2418	0.0435
380	373.3	1206	0.2579	0.0461	0.2543	0.0456
400	383.0	1194	0.2706	0.0482	0.2668	0.0477
420	392.4	1182	0.2833	0.0503	0.2793	0.0497
440	401.7	1171	0.2960	0.0523	0.2919	0.0518
460	410.7	1159	0.3087	0.0544	0.3045	0.0538
480	419.5	1148	0.3215	0.0564	0.3171	0.0558
500	428.2	1137	0.3342	0.0584	0.3297	0.0578
550	449.1	1111	0.3662	0.0635	0.3613	0.0628
600	469.0	1086	0.3983	0.0684	0.3930	0.0677
650	488.1	1062	0.4305	0.0733	0.4248	0.0726
700	506.5	1040	0.4628	0.0781	0.4567	0.0773
750	524.3	1018	0.4952	0.0828	0.4887	0.0820
800	541.4	997.6	0.5277	0.0875	0.5208	0.0866
850	558.1	978.0	0.5602	0.0921	0.5529	0.0912
900	574.2	959.3	0.5927	0.0966	0.5852	0.0957
950	589.9	941.4	0.6254	0.1010	0.6174	0.1001
1000	605.2	924.3	0.6580	0.1054	0.6497	0.1045

Table 9.11—*contd* Range parameters of As in PMMA photoresist

E (keV)	S_e (keV/μm)	S_n (keV/μm)	R (μm)	ΔR (μm)	R_p (μm)	ΔR_p (μm)
Arsenic (As)	$Z_1 = 33; M_1 = 74.922; C_K = 1$					
10	65.56	820.2	0.0162	0.0038	0.0158	0.0037
20	92.72	896.3	0.0268	0.0062	0.0261	0.0060
30	113.6	919.8	0.0366	0.0083	0.0357	0.0082
40	131.1	925.6	0.0462	0.0104	0.0450	0.0102
50	146.6	923.4	0.0556	0.0125	0.0542	0.0122
60	160.6	917.1	0.0649	0.0145	0.0633	0.0142
70	173.5	908.5	0.0741	0.0165	0.0723	0.0162
80	185.4	898.5	0.0834	0.0184	0.0813	0.0181
90	196.7	887.8	0.0926	0.0204	0.0903	0.0200
100	207.3	876.8	0.1018	0.0223	0.0993	0.0219
110	217.4	865.6	0.1111	0.0242	0.1084	0.0238
120	227.1	854.4	0.1203	0.0261	0.1174	0.0256
130	236.4	843.3	0.1295	0.0280	0.1264	0.0275
140	245.3	832.3	0.1388	0.0298	0.1355	0.0293
150	253.9	821.5	0.1481	0.0317	0.1446	0.0311
160	262.2	810.9	0.1574	0.0335	0.1537	0.0329
170	270.3	800.6	0.1667	0.0353	0.1628	0.0347
180	278.1	790.5	0.1761	0.0371	0.1720	0.0365
190	285.7	780.6	0.1855	0.0389	0.1811	0.0382
200	293.2	771.0	0.1948	0.0407	0.1903	0.0400
220	307.5	752.5	0.2137	0.0442	0.2088	0.0434
240	321.1	735.0	0.2326	0.0476	0.2273	0.0469
260	334.2	718.3	0.2516	0.0510	0.2459	0.0502
280	346.8	702.5	0.2706	0.0544	0.2645	0.0536
300	359.0	687.5	0.2897	0.0577	0.2832	0.0568
320	370.7	673.2	0.3088	0.0610	0.3020	0.0601
340	382.1	659.6	0.3280	0.0642	0.3208	0.0632
360	393.2	646.6	0.3472	0.0673	0.3397	0.0664
380	404.0	634.2	0.3664	0.0705	0.3586	0.0694
400	414.4	622.4	0.3857	0.0735	0.3775	0.0725
420	424.6	611.1	0.4050	0.0765	0.3965	0.0755
440	434.6	600.3	0.4243	0.0795	0.4155	0.0784
460	444.4	589.9	0.4437	0.0824	0.4345	0.0813
480	453.9	579.9	0.4630	0.0853	0.4535	0.0842
500	463.2	570.4	0.4824	0.0881	0.4725	0.0870
550	485.8	548.0	0.5307	0.0950	0.5201	0.0938
600	507.3	527.7	0.5791	0.1016	0.5677	0.1004
650	527.9	509.1	0.6273	0.1080	0.6152	0.1067
700	547.8	492.0	0.6755	0.1141	0.6626	0.1128
750	566.9	476.2	0.7235	0.1200	0.7099	0.1187
800	585.4	461.6	0.7714	0.1257	0.7571	0.1244
850	603.3	448.0	0.8190	0.1312	0.8041	0.1298
900	620.7	435.3	0.8665	0.1364	0.8509	0.1351
950	637.5	423.4	0.9137	0.1415	0.8976	0.1401
1000	654.0	412.3	0.9607	0.1464	0.9440	0.1450

Table 9.11—*contd* Range parameters of B in PMMA photoresist

E (keV)	S_e (keV/μm)	S_n (keV/μm)	R (μm)	ΔR (μm)	R_p (μm)	ΔR_p (μm)
Boron (B)	$Z_1 = 5; M_1 = 11.009; C_K = 1$					
10	65.67	85.32	0.0664	0.0173	0.0539	0.0170
20	92.83	63.10	0.1318	0.0290	0.1112	0.0294
30	113.6	51.06	0.1942	0.0374	0.1677	0.0386
40	131.1	43.33	0.2532	0.0438	0.2220	0.0457
50	146.5	37.89	0.3090	0.0487	0.2741	0.0514
60	160.3	33.81	0.3618	0.0526	0.3238	0.0560
70	173.0	30.62	0.4121	0.0559	0.3716	0.0599
80	184.7	28.04	0.4601	0.0586	0.4174	0.0631
90	195.7	25.92	0.5062	0.0610	0.4615	0.0659
100	206.0	24.13	0.5504	0.0630	0.5041	0.0684
110	215.7	22.60	0.5931	0.0648	0.5454	0.0706
120	225.0	21.27	0.6344	0.0664	0.5853	0.0725
130	233.7	20.11	0.6744	0.0678	0.6241	0.0742
140	242.1	19.08	0.7132	0.0691	0.6619	0.0758
150	250.1	18.17	0.7510	0.0702	0.6987	0.0772
160	257.8	17.35	0.7878	0.0713	0.7346	0.0786
170	265.2	16.60	0.8237	0.0723	0.7696	0.0798
180	272.3	15.93	0.8588	0.0732	0.8040	0.0809
190	279.1	15.31	0.8931	0.0740	0.8376	0.0819
200	285.7	14.75	0.9267	0.0748	0.8705	0.0829
220	298.2	13.74	0.9921	0.0762	0.9347	0.0846
240	309.8	12.88	1.0551	0.0775	0.9967	0.0862
260	320.7	12.13	1.1161	0.0786	1.0567	0.0876
280	330.9	11.47	1.1753	0.0796	1.1151	0.0888
300	340.5	10.89	1.2330	0.0805	1.1720	0.0900
320	349.6	10.37	1.2892	0.0814	1.2276	0.0910
340	358.2	9.900	1.3441	0.0822	1.2819	0.0920
360	366.3	9.475	1.3979	0.0829	1.3350	0.0929
380	374.0	9.089	1.4506	0.0836	1.3872	0.0937
400	381.3	8.735	1.5023	0.0843	1.4384	0.0945
420	388.3	8.411	1.5532	0.0849	1.4888	0.0952
440	394.9	8.112	1.6032	0.0855	1.5384	0.0959
460	401.3	7.836	1.6524	0.0860	1.5872	0.0966
480	407.4	7.580	1.7010	0.0865	1.6354	0.0972
500	413.2	7.341	1.7489	0.0870	1.6829	0.0978
550	426.8	6.811	1.8659	0.0882	1.7991	0.0992
600	439.1	6.358	1.9797	0.0892	1.9121	0.1004
650	450.5	5.967	2.0905	0.0902	2.0224	0.1016
700	461.0	5.625	2.1988	0.0911	2.1301	0.1026
750	470.8	5.323	2.3049	0.0919	2.2356	0.1035
800	480.0	5.054	2.4089	0.0927	2.3392	0.1045
850	488.6	4.813	2.5111	0.0934	2.4410	0.1053
900	496.7	4.597	2.6117	0.0941	2.5411	0.1061
950	504.4	4.400	2.7107	0.0948	2.6397	0.1068
1000	511.8	4.221	2.8082	0.0954	2.7369	0.1075

Table 9.11—*contd* Range parameters of P in PMMA photoresist

E (keV)	S_e (keV/μm)	S_n (keV/μm)	R (μm)	ΔR (μm)	R_p (μm)	ΔR_p (μm)
Phosphorus (P)	$Z_1 = 15$; $M_1 = 30.974$; $C_K = 1$					
10	72.31	428.8	0.0230	0.0067	0.0213	0.0063
20	102.3	404.0	0.0427	0.0119	0.0397	0.0114
30	125.2	376.6	0.0626	0.0169	0.0583	0.0162
40	144.6	352.0	0.0826	0.0218	0.0771	0.0209
50	161.7	330.7	0.1028	0.0264	0.0962	0.0254
60	177.1	312.1	0.1232	0.0309	0.1155	0.0298
70	191.3	295.7	0.1437	0.0351	0.1350	0.0340
80	204.5	281.3	0.1643	0.0392	0.1546	0.0380
90	216.9	268.5	0.1849	0.0431	0.1742	0.0418
100	228.6	257.0	0.2055	0.0469	0.1939	0.0455
110	239.7	246.6	0.2260	0.0505	0.2136	0.0491
120	250.3	237.1	0.2466	0.0539	0.2333	0.0525
130	260.5	228.5	0.2671	0.0572	0.2530	0.0558
140	270.3	220.6	0.2875	0.0604	0.2727	0.0590
150	279.8	213.3	0.3078	0.0634	0.2922	0.0620
160	289.0	206.6	0.3280	0.0663	0.3117	0.0649
170	297.8	200.3	0.3482	0.0691	0.3312	0.0677
180	306.4	194.5	0.3682	0.0718	0.3505	0.0704
190	314.8	189.1	0.3881	0.0744	0.3698	0.0730
200	322.9	184.0	0.4079	0.0769	0.3889	0.0755
220	338.6	174.8	0.4471	0.0816	0.4269	0.0802
240	353.5	166.5	0.4858	0.0860	0.4645	0.0846
260	367.8	159.2	0.5240	0.0900	0.5017	0.0888
280	381.6	152.5	0.5617	0.0938	0.5384	0.0926
300	394.8	146.5	0.5989	0.0974	0.5746	0.0962
320	407.6	141.0	0.6356	0.1007	0.6104	0.0996
340	420.0	136.0	0.6718	0.1039	0.6458	0.1028
360	432.0	131.3	0.7075	0.1068	0.6807	0.1059
380	443.7	127.0	0.7428	0.1096	0.7152	0.1087
400	455.0	123.1	0.7776	0.1123	0.7493	0.1114
420	466.0	119.4	0.8120	0.1148	0.7830	0.1140
440	476.8	115.9	0.8460	0.1172	0.8163	0.1165
460	487.2	112.7	0.8795	0.1194	0.8492	0.1188
480	497.5	109.7	0.9126	0.1216	0.8817	0.1210
500	507.5	106.8	0.9454	0.1236	0.9138	0.1231
550	531.5	100.4	1.0256	0.1284	0.9927	0.1280
600	554.3	94.79	1.1037	0.1326	1.0696	0.1324
650	576.0	89.86	1.1798	0.1365	1.1445	0.1363
700	596.7	85.48	1.2539	0.1400	1.2176	0.1400
750	616.6	81.56	1.3264	0.1432	1.2891	0.1433
800	635.6	78.03	1.3972	0.1461	1.3590	0.1464
850	653.9	74.82	1.4665	0.1489	1.4275	0.1492
900	671.5	71.90	1.5345	0.1514	1.4947	0.1518
950	688.4	69.23	1.6011	0.1538	1.5606	0.1543
1000	704.8	66.77	1.6665	0.1560	1.6253	0.1566

9.5 ABUNDANCE OF ISOTOPES

In Figure 9.7, isotope abundances for elements are shown. Isotopes with abundances below 1 per cent are only indicated by a short line beneath the data on their relative abundance; the other isotopes are represented by lines of corresponding lengths. In an ion beam, the individual isotopes occur in these approximate proportions, since their ionization hardly depends on the mass.

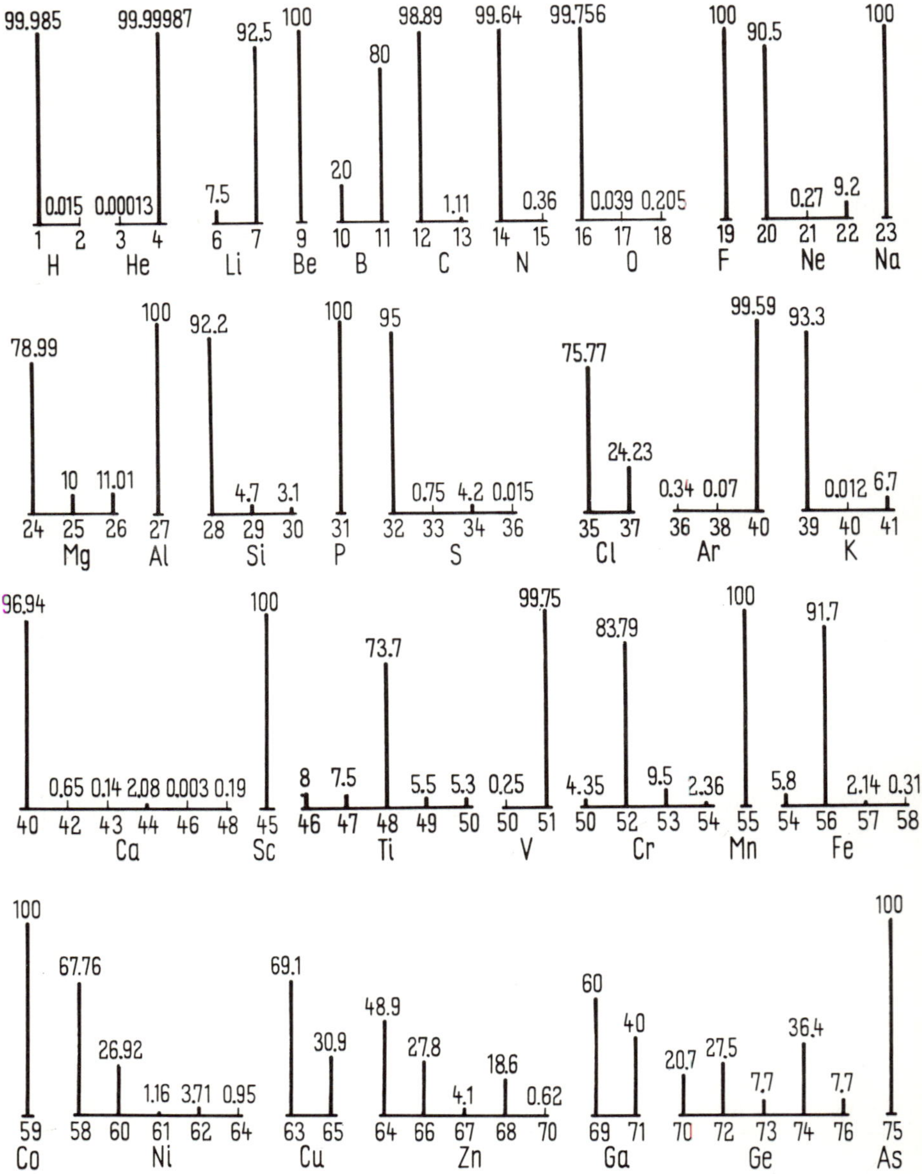

Figure 9.7 Abundance of isotopes

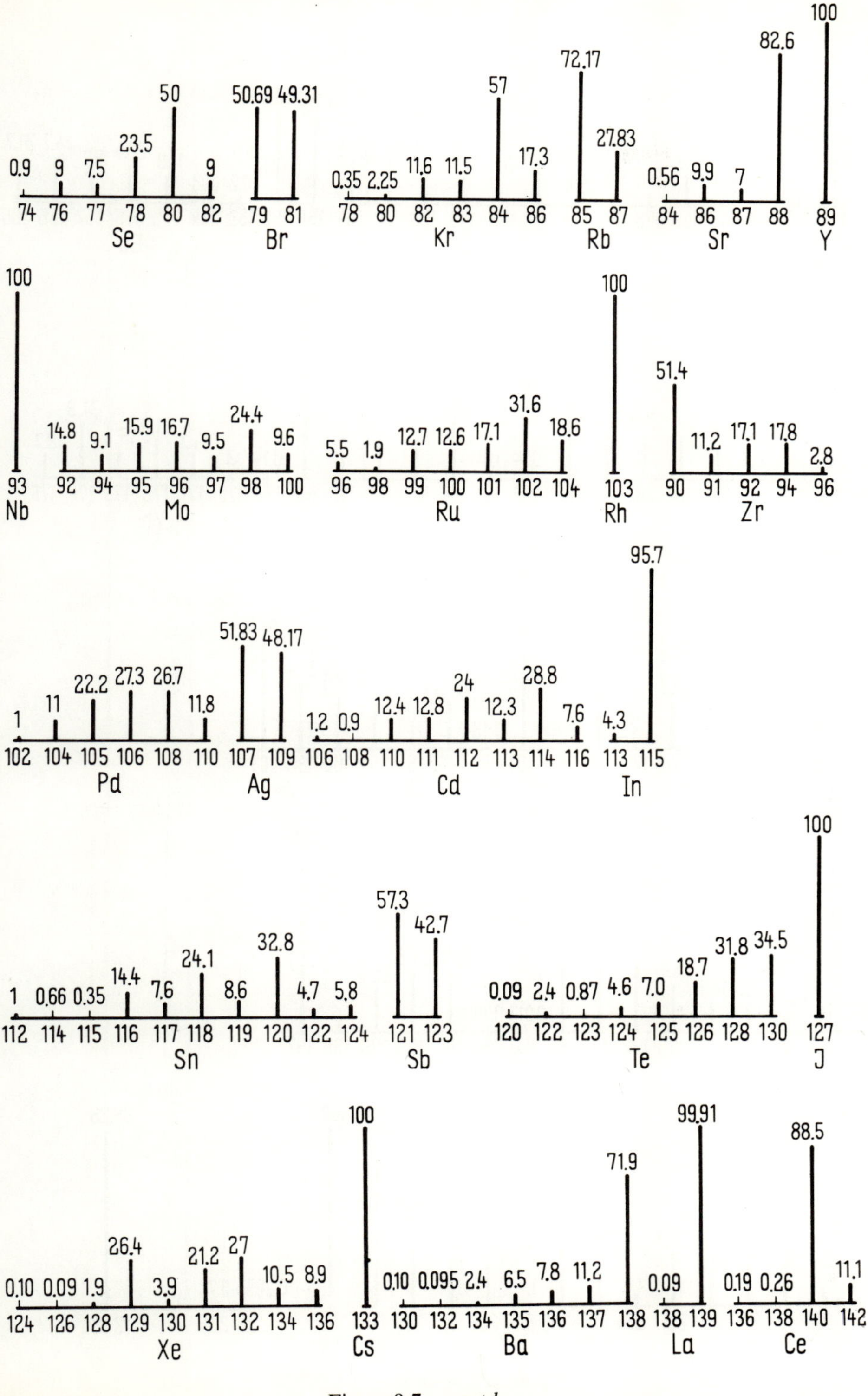

Figure 9.7—*contd*

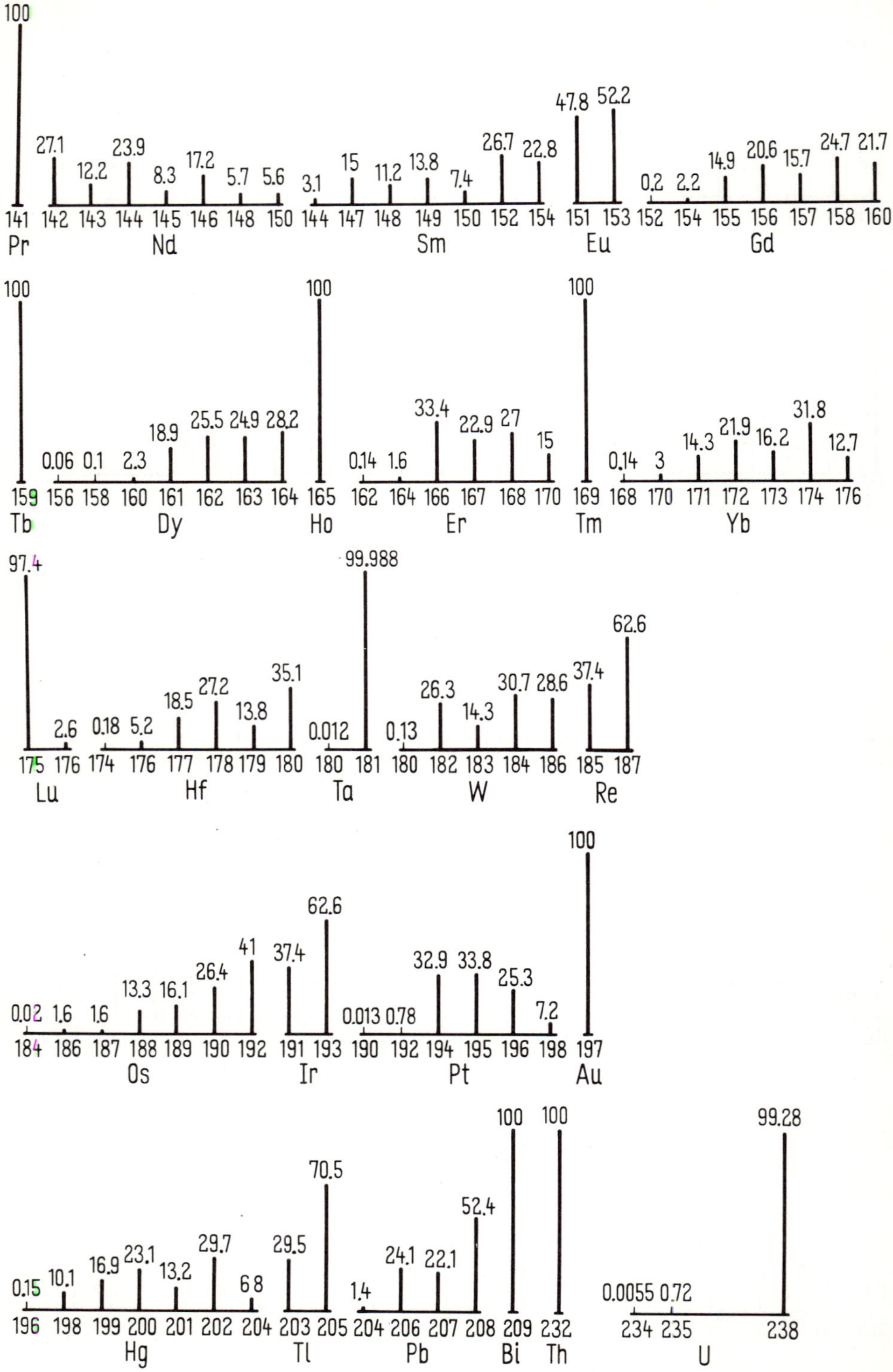

Figure 9.7—*contd*

Thus, a simple means is available for identification of the individual elements in an ion beam. If molecular ions occur, all combinations between the individual isotopes are contained in the beam. The data given, of course, do not apply if enriched isotopes or activated elements are implanted. Often, the masses are also shown proportionally to the magnetic field, which corresponds to a square-root dependence of the mass and distorts the distributions in the measured spectrum accordingly.

9.6 VAPOUR PRESSURE

In the operation of ion sources, elements should preferably be used through which a high beam current can be obtained (*cf.* Section 4.1.5). A list of recommended substances has already been given in Table 4.1. To determine the suitability of various elements for direct ionization, vapour-pressure curves are given in Figures 9.8 to 9.10 depending on temperature, according to Honig [345]. For direct ionization, elements with vapour pressures between 10^{-1} and 10 Pa are suitable, according to the probability of their ionization (which is generally unknown).

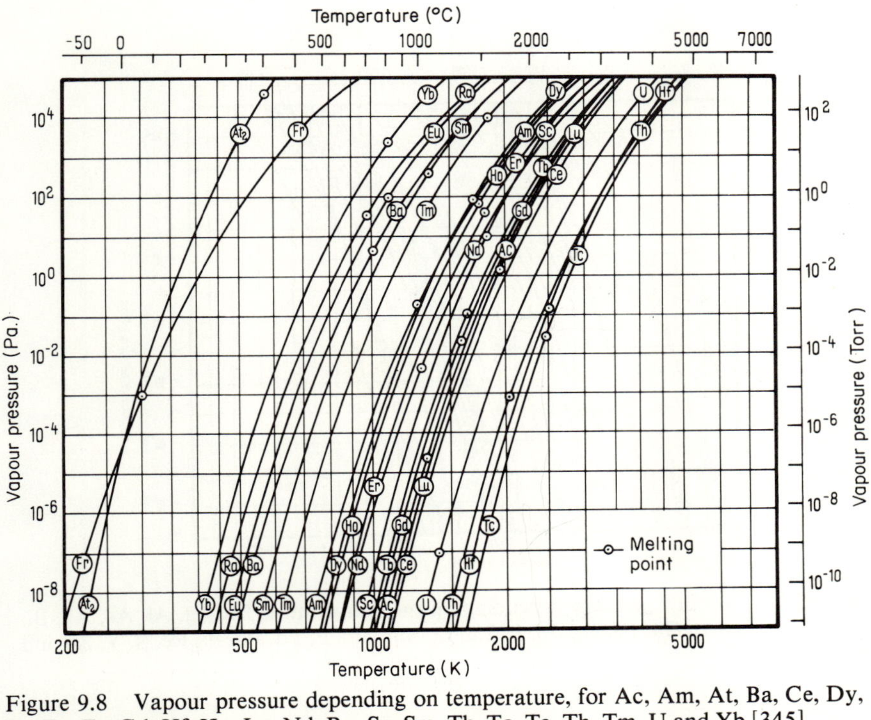

Figure 9.8 Vapour pressure depending on temperature, for Ac, Am, At, Ba, Ce, Dy, Er, Eu, Fr, Gd, Hf, Ho, Lu, Nd, Ra, Sc, Sm, Tb, Tc, Te, Th, Tm, U and Yb [345]

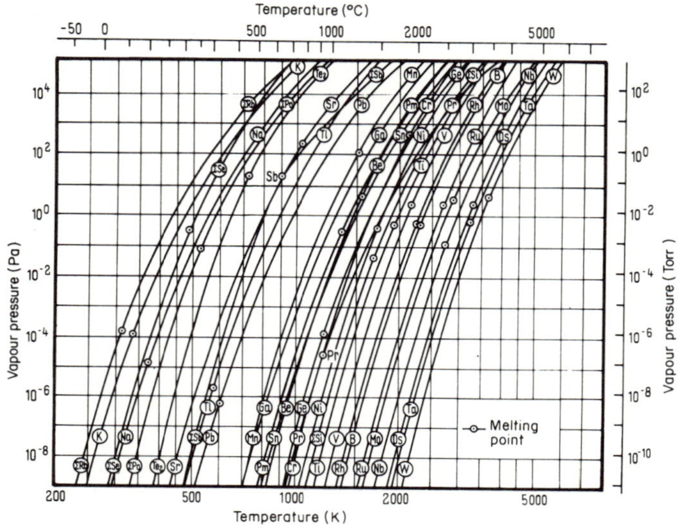

Figure 9.9 Vapour pressure depending on temperature, for B, Be, Cr, Ga, Ge, K, Mn, Mo, Na, Nb, Ni, Os, Pb, Pm, Po, Pr, Rb, Rh, Ru, Sb, Si, Sn, Sr, Ta, Te, Ti, Tl, V and W [345]

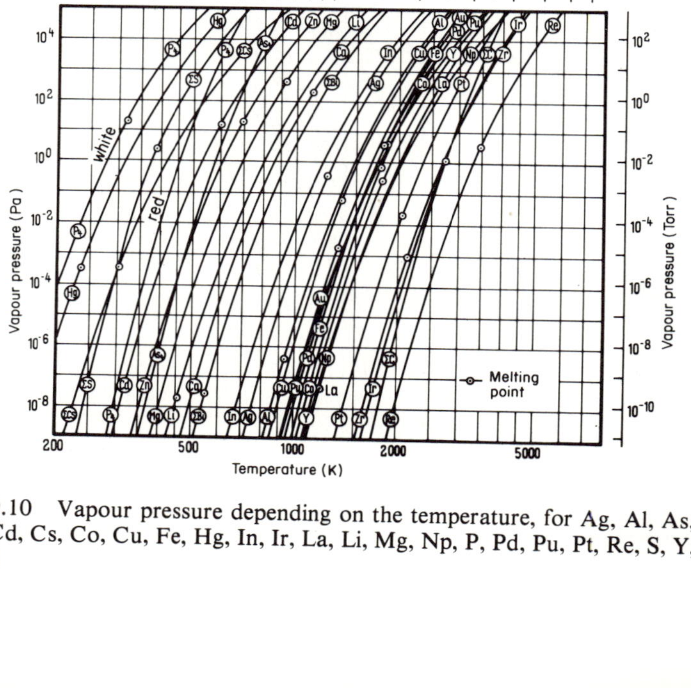

Figure 9.10 Vapour pressure depending on the temperature, for Ag, Al, As, Au, Bi, C, Ca, Cd, Cs, Co, Cu, Fe, Hg, In, Ir, La, Li, Mg, Np, P, Pd, Pu, Pt, Re, S, Y, Zn and Zr [345]

9.7 ERROR FUNCTION

The error function (erf x), as well as the complementary error function (erfc x), occur frequently in the solution of diffusion problems. They are defined as follows:

$$\operatorname{erf} x = \frac{2}{\sqrt{\pi}} \int_0^x e^{-t^2}\, dt$$

$$\operatorname{erfc} x = \frac{2}{\sqrt{\pi}} \int_x^\infty e^{-t^2}\, dt = 1 - \operatorname{erf} x \tag{9.1}$$

Several authors use either the above notation without the factor $2/\sqrt{\pi}$ or $\Phi(x)$ instead of erf x. Values of the complementary error function between $x = 0$ and $x = 4.5$ are given in Table 9.12. For a detailed treatment, reference can be made e.g., to the work by Abramowitz |32|. Several error-function properties are listed in Table 9.13.

Table 9.12 Values of the complementary error function for $x = 0$ to $x = 4.5$

x	erfc x	x	erfc x	x	erfc x
0.00	+1.00000	0.27	+0.70258	0.54	+0.44506
0.01	+0.98871	0.28	+0.69211	0.55	+0.43667
0.02	+0.97743	0.29	+0.68171	0.56	+0.42838
0.03	+0.96615	0.30	+0.67137	0.57	+0.42018
0.04	+0.95488	0.31	+0.66109	0.58	+0.41207
0.05	+0.94362	0.32	+0.65087	0.59	+0.40406
0.06	+0.93237	0.33	+0.64077	0.60	+0.39614
0.07	+0.92114	0.34	+0.63063	0.61	+0.38831
0.08	+0.90992	0.35	+0.62061	0.62	+0.38058
0.09	+0.89871	0.36	+0.61067	0.63	+0.37295
0.10	+0.88753	0.37	+0.60079	0.64	+0.36541
0.11	+0.87637	0.38	+0.59099	0.65	+0.35797
0.12	+0.86524	0.39	+0.58126	0.66	+0.35062
0.13	+0.85413	0.40	+0.57160	0.67	+0.34337
0.14	+0.84305	0.41	+0.56203	0.68	+0.33621
0.15	+0.83200	0.42	+0.55253	0.69	+0.32915
0.16	+0.82098	0.43	+0.54311	0.70	+0.32219
0.17	+0.81000	0.44	+0.53377	0.71	+0.31532
0.18	+0.79906	0.45	+0.52451	0.72	+0.30856
0.19	+0.78816	0.46	+0.51534	0.73	+0.30189
0.20	+0.77729	0.47	+0.50625	0.74	+0.29532
0.21	+0.76647	0.48	+0.49725	0.75	+0.28884
0.22	+0.75570	0.49	+0.48833	0.76	+0.28246
0.23	+0.74497	0.50	+0.47950	0.77	+0.27617
0.24	+0.73429	0.51	+0.47075	0.78	+0.26998
0.25	+0.72367	0.52	+0.46210	0.79	+0.26389
0.26	+0.71310	0.53	+0.45353	0.80	+0.25789

Table 9.12—*contd* Values of the complementary error function for $x = 0$ to $x = 4.5$

x	erfc x	x	erfc x	x	erfc x
0.81	+0.25199	1.31	+0.063936	1.81	+0.010475
0.82	+0.24618	1.32	+0.061934	1.82	+0.010056
0.83	+0.24047	1.33	+0.059984	1.83	+0.009653
0.84	+0.23485	1.34	+0.058086	1.84	+0.009264
0.85	+0.22935	1.35	+0.056237	1.85	+0.008888
0.86	+0.22389	1.36	+0.054438	1.86	+0.008527
0.87	+0.21856	1.37	+0.052687	1.87	+0.008179
0.88	+0.21331	1.38	+0.050983	1.88	+0.007843
0.89	+0.20815	1.39	+0.049326	1.89	+0.007520
0.90	+0.20309	1.40	+0.047714	1.90	+0.007209
0.91	+0.19811	1.41	+0.046147	1.91	+0.006910
0.92	+0.19323	1.42	+0.044623	1.92	+0.006621
0.93	+0.18843	1.43	+0.043142	1.93	+0.006344
0.94	+0.18372	1.44	+0.041703	1.94	+0.006077
0.95	+0.17910	1.45	+0.040304	1.95	+0.005820
0.96	+0.17457	1.46	+0.038946	1.96	+0.005573
0.97	+0.17012	1.47	+0.037627	1.97	+0.005336
0.98	+0.16576	1.48	+0.036345	1.98	+0.005107
0.99	+0.16149	1.49	+0.035102	1.99	+0.004888
1.00	+0.15729	1.50	+0.033894	2.00	+0.004677
1.01	+0.15318	1.51	+0.032723	2.01	+0.004475
1.02	+0.14916	1.52	+0.031586	2.02	+0.004280
1.03	+0.14521	1.53	+0.030483	2.03	+0.004093
1.04	+0.14135	1.54	+0.029414	2.04	+0.003914
1.05	+0.13756	1.55	+0.028377	2.05	+0.003741
1.06	+0.13385	1.56	+0.027371	2.06	+0.003576
1.07	+0.13022	1.57	+0.026397	2.07	+0.003417
1.08	+0.12667	1.58	+0.025452	2.08	+0.003265
1.09	+0.12319	1.59	+0.024537	2.09	+0.003119
1.10	+0.11979	1.60	+0.023651	2.10	+0.002979
1.11	+0.11646	1.61	+0.022793	2.11	+0.002845
1.12	+0.11321	1.62	+0.021961	2.12	+0.002716
1.13	+0.11002	1.63	+0.021157	2.13	+0.002592
1.14	+0.10691	1.64	+0.020378	2.14	+0.002474
1.15	+0.10337	1.65	+0.019624	2.15	+0.002361
1.16	+0.10090	1.66	+0.018895	2.16	+0.002252
1.17	+0.09799	1.67	+0.018189	2.17	+0.002148
1.18	+0.09516	1.68	+0.017507	2.18	+0.002049
1.19	+0.09239	1.69	+0.016847	2.19	+0.001954
1.20	+0.08968	1.70	+0.016209	2.20	+0.001862
1.21	+0.08704	1.71	+0.015592	2.21	+0.001775
1.22	+0.08446	1.72	+0.014997	2.22	+0.001692
1.23	+0.08194	1.73	+0.014421	2.23	+0.001612
1.24	+0.079494	1.74	+0.013865	2.24	+0.001535
1.25	+0.077099	1.75	+0.013328	2.25	+0.001462
1.26	+0.074764	1.76	+0.012809	2.26	+0.001392
1.27	+0.072486	1.77	+0.012309	2.27	+0.001326
1.28	+0.070265	1.78	+0.011825	2.28	+0.001262
1.29	+0.068101	1.79	+0.011359	2.29	+0.001201
1.30	+0.065992	1.80	+0.010909	2.30	+0.001143

Table 9.12—*contd* Values of the complementary error function for $x = 0$ to $x = 4.5$

x	erfc x	x	erfc x	x	erfc x
2.31	+0.001087	2.81	+0.0000706	3.31	+0.000002854
2.32	+0.001034	2.82	+0.0000666	3.32	+0.000002663
2.33	+0.000983	2.83	+0.0000627	3.33	+0.000002485
2.34	+0.000935	2.84	+0.0000591	3.34	+0.000002318
2.35	+0.000889	2.85	+0.0000556	3.35	+0.000002162
2.36	+0.000845	2.86	+0.0000524	3.36	+0.000002016
2.37	+0.000803	2.87	+0.0000493	3.37	+0.000001880
2.38	+0.000763	2.88	+0.0000464	3.38	+0.000001752
2.39	+0.000724	2.89	+0.0000436	3.39	+0.000001633
2.40	+0.000688	2.90	+0.0000410	3.40	+0.000001521
2.41	+0.000653	2.91	+0.0000386	3.41	+0.000001417
2.42	+0.000620	2.92	+0.0000363	3.42	+0.000001320
2.43	+0.000589	2.93	+0.0000341	3.43	+0.000001229
2.44	+0.000559	2.94	+0.0000321	3.44	+0.000001145
2.45	+0.000530	2.95	+0.0000302	3.45	+0.000001066
2.46	+0.000503	2.96	+0.0000282	3.46	+0.000000992
2.47	+0.000477	2.97	+0.0000266	3.47	+0.000000923
2.48	+0.0004527	2.98	+0.0000250	3.48	+0.000000859
2.49	+0.0004292	2.99	+0.0000235	3.49	+0.000000799
2.50	+0.0004069	3.00	+0.0000220	3.50	+0.000000743
2.51	+0.0003857	3.01	+0.0000207	3.51	+0.000000690
2.52	+0.0003654	3.02	+0.0000194	3.52	+0.000000642
2.53	+0.0003462	3.03	+0.0000182	3.53	+0.000000597
2.54	+0.0003280	3.04	+0.0000171	3.54	+0.000000554
2.55	+0.0003106	3.05	+0.0000160	3.55	+0.000000515
2.56	+0.0002941	3.06	+0.0000150	3.56	+0.000000478
2.57	+0.0002784	3.07	+0.0000141	3.57	+0.000000444
2.58	+0.0002636	3.08	+0.0000132	3.58	+0.000000412
2.59	+0.0002494	3.09	+0.0000124	3.59	+0.000000383
2.60	+0.0002360	3.10	+0.000011648	3.60	+0.000000355
2.61	+0.0002232	3.11	+0.000010915	3.61	+0.000000330
2.62	+0.0002111	3.12	+0.000010225	3.62	+0.000000306
2.63	+0.0001997	3.13	+0.000009577	3.63	+0.000000284
2.64	+0.0001888	3.14	+0.000008969	3.64	+0.000000263
2.65	+0.0001784	3.15	+0.000008398	3.65	+0.000000244
2.66	+0.0001686	3.16	+0.000007861	3.66	+0.000000226
2.67	+0.0001593	3.17	+0.000007358	3.67	+0.000000210
2.68	+0.0001505	3.18	+0.000006885	3.68	+0.000000194
2.69	+0.0001422	3.19	+0.000006441	3.69	+0.000000180
2.70	+0.0001343	3.20	+0.000006025	3.70	+0.000000167
2.71	+0.0001268	3.21	+0.000005635	3.71	+0.000000154
2.72	+0.0001197	3.22	+0.000005269	3.72	+0.00000014337
2.73	+0.0001130	3.23	+0.000004926	3.73	+0.00000013274
2.74	+0.0001066	3.24	+0.000004604	3.74	+0.00000012288
2.75	+0.0001006	3.25	+0.000004302	3.75	+0.00000011373
2.76	+0.0000949	3.26	+0.000004020	3.76	+0.00000010524
2.77	+0.0000895	3.27	+0.000003755	3.77	+0.00000009736
2.78	+0.0000844	3.28	+0.000003507	3.78	+0.00000009005
2.79	+0.0000795	3.29	+0.000003275	3.79	+0.00000008328
2.80	+0.0000750	3.30	+0.000003057	3.80	+0.00000007701

Table 9.12—*contd* Values of the complementary error function for $x = 0$ to $x = 4.5$

x	erfc x	x	erfc x	x	erfc x
3.81	+0.00000007118	4.05	+0.00000001019	4.28	+0.00000000142
3.82	+0.00000006579	4.06	+0.00000000937	4.29	+0.00000000130
3.83	+0.00000006080	4.07	+0.00000000862	4.30	+0.00000000119
3.84	+0.00000005617	4.08	+0.00000000793	4.31	+0.00000000109
3.85	+0.00000005188	4.09	+0.00000000729	4.32	+0.00000000100
3.86	+0.00000004792	4.10	+0.00000000670	4.33	+0.00000000092
3.87	+0.00000004425	4.11	+0.00000000616	4.34	+0.00000000084
3.88	+0.00000004085	4.12	+0.00000000566	4.35	+0.00000000077
3.89	+0.00000003770	4.13	+0.00000000520	4.36	+0.00000000070
3.90	+0.00000003479	4.14	+0.00000000477	4.37	+0.00000000064
3.91	+0.00000003210	4.15	+0.00000000438	4.38	+0.00000000059
3.92	+0.00000002961	4.16	+0.00000000403	4.39	+0.00000000053
3.93	+0.00000002731	4.17	+0.00000000370	4.40	+0.00000000049
3.94	+0.00000002518	4.18	+0.00000000339	4.41	+0.00000000045
3.95	+0.00000002322	4.19	+0.00000000311	4.42	+0.00000000041
3.96	+0.00000002140	4.20	+0.00000000286	4.43	+0.00000000037
3.97	+0.00000001972	4.21	+0.00000000262	4.44	+0.00000000034
3.98	+0.00000001817	4.22	+0.00000000240	4.45	+0.00000000031
3.99	+0.00000001674	4.23	+0.00000000220	4.46	+0.00000000028
4.00	+0.00000001542	4.24	+0.00000000202	4.47	+0.00000000026
4.01	+0.00000001420	4.25	+0.00000000185	4.48	+0.00000000024
4.02	+0.00000001307	4.26	+0.00000000170	4.49	+0.00000000021
4.03	+0.00000001203	4.27	+0.00000000155	4.50	+0.00000000020
4.04	+0.00000001107				

Table 9.13 Several properties of the error function

$$\text{erf}(-x) = -\text{erf}\,x$$

$$\text{erf}\,0 = 0; \quad \text{erf}\,\infty = 1$$

$$\text{erfc}\,0 = 1; \quad \text{erfc}\,\infty = 0$$

$$\text{erf}\,x \approx \frac{2}{\sqrt{\pi}}x \quad \text{for } x \leqslant 1$$

$$\text{erfc}\,x \approx \frac{2}{\sqrt{\pi}}\frac{\exp(-x^2)}{x} \quad \text{for } x \gg 1$$

$$\text{erfc}\,x = 1 - \text{erfc}\,x = \frac{2}{\sqrt{\pi}}\int_x^{\infty} e^{-t^2}\,dt$$

$$\frac{d(\text{erf}\,x)}{dx} = \frac{2}{\sqrt{\pi}}\exp(-x)^2$$

$$\int_0^{\infty} \text{erfc}\,t\,dt = x\,\text{erfc}\,x + \frac{1}{\sqrt{\pi}}|1 - \exp(x^{-2})|$$

$$\int_0^{\infty} \text{erf}\,t\,dt = \frac{1}{\sqrt{\pi}}$$

$$\text{erf}\,x = \frac{2}{\sqrt{\pi}}\int_x^{x} e^{-t^2}\,dt$$

$$\frac{2}{\sqrt{\pi}}\int_a^{b} e^{-t^2}\,dt = \text{erfc}\,a - \text{erfc}\,b$$

References

Books and Review Articles

|1| Brice, D. K.: *Ion Implantation Range and Energy Deposition Distributions*, Vol. 1, *High Incident Ion Energies*, New York 1975

|1a| Carter, G., Grant, W. A.: *Ion Implantation of Semiconductors*, London 1976

|2| Dearnaley, G., Freeman, J. H., Nelson, R. S., Stephen, J.: *Ion Implantation*, Amsterdam 1973

|3| Degen, P. L.: *Phys. Stat. Sol.* **16** (1973) 9

|4| Dill, H. G., Finnila, R. M., Leupp, A. M., Toombs, T. N.: *Solid State Technol.* **15** (1972) 27

|5| Gibbons, J. F., Johnson, W. S., Mylroie, S. W.: *Projected Range Statistics*, Stroudsburg, USA 1975

|6| Gibbons, J. F.: *Proc. IEEE* **56** (1968) 295

|7| Gibbons, J. F.: *Proc. IEEE* **60** (1972) 1062

|8| Lee, D. H., Mayer, J. W.: *Proc. IEEE* **62** (1974) 1241

|9| Mayer, J. W., Eriksson, L., Davies, J. A.: *Ion Implantation in Semiconductors*, New York 1970

|10| Ruge, I., Müller, H., Ryssel, H.: *Festkörperprobleme XII*, Braunschweig 1972, 23

|11| Schultz, M.: *Appl. Phys.* **4** (1974) 91

|12| Stroud, P. T.: *Thin Solid Films* **11** (1972) 1

|12a| Townsend, P. D., Kelly, J. C., Hartley, N. E. W.: *Ion Implantation, Sputtering and their Applications*, London, New York, San Francisco 1976

|13| Wilson, R. G., Brewer, G. R.: *Ion Beams*, New York 1973

|13a| Hirvonen, J. K.: *Ion Implantation*, Academic Press, New York 1981

Bibliographies

|14| Agajanian, A. H.: *Ion Implantation—An Annotated Bibliography. Radiation Effects* **23** (1974) 73

|14a| Agajanian, A. H.: *Ion Implantation in Microelectronics—A Comprehensive Bibliography*, Computer Science Information Guides Vol. 1, IFI–Plenum, New York–Washington–London 1981.

|15| Mazzio, J.: *Ion Implantation: A Selective Bibliography*, Sandia Laboratories, Albuquerque, N. Mex. NTIS Report No. SC-B-710148 (1971)

[16] Morgan, R., Greenhalgh, K. R.: *Ion Implantation: A Bibliography*, AERE Harwell, England, AERE-Bib-176 (1972)
[17] Plunkett, J. C., Stone, J. L.: *A Selected Bibliography on Ion Implantation in Solid State Technology*, Solid State Technol. Dec. 1975
[18] Seager, D. K.: *Ion Implantation: A Bibliography*, Sandia Laboratories, Albuquerque, N. Mex. NTIS Report Nr. SC-B-71048, Suppl. I (1973)

Conference Reports

[19] *Proc. Int. Conf. Appl. Ion Beams Semiconductor Technology* (Ed. Glotin, P.) Grenoble 1967
[20] *European Conference on Ion Implantation. Reading* (Sept. 1970) Stevenage, England 1970
[21] *Ion Implantation* (Eds Eisen, F. H., Chadderton, C. S.), Gordon and Breach, London 1971
[22] *Ion Implantation in Semiconductors* (Eds Ruge, I., Graul, J.), Springer-Verlag, Berlin–Heidelberg–New York 1971
[23] *Ion Implantation in Semiconductors and other Materials* (Ed. Crowder, B. L.) New York 1973
[24] *Ion Implantation in Semiconductors* (Ed. Namba, S.) New York 1975
[25] *Ion Implantation in Semiconductors and other Materials* (Ed. Chernow, F., Borders, J. A., Brice, D. W.) New York 1977
[25a] *Proc. Int. Conf. on Ion Beam Modification of Materials*, Parts 1–3 (Eds Gyulai, J., Lohner, T., Pásztor, E.) Budapest, Hungary 1978
[25b] *Ion Beam Modification of Materials*, Parts 1–2 (Eds Benenson, R. E., Kaufmann, E. N., Miller, G. L., Scholz, W. W.), North-Holland, Amsterdam 1981
[26] *Proc. US-Japan Seminar Ion Implantation in Semiconductors* (Ed. Namba, S.) Kyoto (Aug. 1971) Jap. Soc. for the Promotion of Science (1972)
[27] *Proc. Int. Conference on Ion Implantation in Semiconductors* (Eds Rzewuski, H., Fiderkiewicz, A., Werner, Z., Tom, M., Zak, C., Lada, A.) Inst. of Nuclear Research Swierk, Polen (1974)
[28] *Proc. Int. Conference on Ion Implantation in Semiconductors* (Ed. Gyulai, J.) Hungarian Acad. of Sciences, Budapest (1976)
[28a] *Low-energy Ion Beams*, Inst. of Physics Conf. Ser. **38**, Bristol and London 1977
[28b] Conf. on Ion Implantation Equipment, Trento, Italy 1978
[28c] *Ion Implantation Equipment and Techniques* (Eds McKenna, Ch. M., Scanlon, P. J., Winnard, J. R.) North-Holland, Amsterdam 1981
[28d] *Ion Implantation Techniques* (Eds Ryssel, H., Glawischnig, H.), Springer Series in Electrophysics **10** (1982).
[28e] *Ion Implantation: Equipment and Techniques* (Eds Ryssel, H., Glawischnig, H.), Springer Series in Electrophysics **11** (1983).
[29] *Application of Ion Beams to Metals* (Eds Picraux, S. T., EerNisse, E. P., Vook, F. L.) New York 1974
[30] *Application of Ion Beams to Materials 1975* (Eds Carter, G., Colligon, J. S., Grant, W. A.) Inst. of Phys. Conf. Ser. **28** (1976)
[30a] *Ion Implantation into Metals* (Eds Ashworth, V., Grant, W. A., Proctor, R. P. M.), *Proc. of the 3rd Int. Conf. on Modification of Surface Properties of Metals by Ion Implantation*, Manchester 1981, Pergamon Press, Oxford 1982

Technical Articles

|31| Aboaf, J. A., Kerr, D. R., Bassons, E.: *J. Electrochem. Soc.* **120** (1973) 1103

|32| Abramowitz, M., Stegun, I. A. (Eds): *Handbook of Mathematical Functions*, New York 1970

|33| Addamiano, A., Anderson, G. W., Comas, J., Hughes, H. L., Lucke, W.: *J. Electrochem. Soc.* **119** (1972) 1355

|34| Akasaka, Y., Horie, K., Nomura, K., Kawazu, S.: *Suppl. J. Jap. Soc. Appl. Phys.* **43** (1974) 493

|34a| *Defects and Radiation Effects in Semiconductors 1978* (Ed. Albany, J. H.) Inst. Physics Conf. Ser. **46**, Inst. of Physics, London 1979

|35| Allen, W. G., Atkinson, C.: *Solid-State Electronics* **16** (1973) 1283

|36| Alton, G. D., Love, L. O.: *Can. J. Phys.* **46** (1968) 695

|37| Alväger, T., Hansen, N. J.: *Rev. Sci. Inst.* **33** (1962) 367

|38| Amsel, A., Nadai, J. P., D'Artemare, E., David, D., Girard, E., Moulin, J.: *Nucl. Inst. and Methods* **92** (1971) 481

|39| Andersen, H. H., Bay, H. L.: *J. Appl. Phys.* **46** (1975) 1919

|40| Andersen, H. St., Sigmund, P.: *Nucl. Instr. and Meth.* **38** (1965) 238

|40a| Anderson, C. L., Celler, G. K., Rozgonyi, G. A. (Eds): *Laser and Electron Beam Processing of Electronic Materials*, 156th Meeting of Electrochem. Soc., Los Angeles 1979

|41| Anderson, W. W., Swanson, R. M.: *J. Appl. Phys.* **42** (1971) 5125

|41a| Appleton, B. R., Feldmann, L. C., Brown, W. L.: *Solid State Research with Accelerators* (Brookhaven), Brookhaven Report No. 50083 (1968) 45

|41b| Antill, J. E., Bennet, N. J., Dearnaley, G., Fern, F. H., Goode, P. D., Turner, J. F.: In: [23] 415

|41c| Appleton, B. R., White, C. W., Larson, B. C., Wilson, S. R., Narayan, J., 5th Conf. Use of Small Accelerators for Res. and Industry, 1978

|41d| Appleton, B. R., Larson, B. C., White, C. W., Narayan, J., Wilson, S. R., Pronko, P. P.: In: *Laser Solid Interactions and Laser Processing 1978* (Eds Ferri, S. D., Leamy, H. J., Poate, J. M.) New York 1979

|41e| Antoniadis, D. A., Dutton, R. W., *IEEE* **ED-26** (1979) 490

|41f| Antoniadis, D. A., Hansen, S. E., Dutton, R. W., Stanford University, *Technical Report No. 5019–2* (1978)

|42| Archer, J. A.: *Solid-State Electronics* **17** (1974) 387

|43| v. Ardenne, M.: *Tabellen für Elektronen, Ionenphysik und Übermikroskopie*, Berlin 1956

|43a| Arsenault, R. J. (Ed.): *Proc. of the 1973 Int. Conf. on Defects Clusters in B.C.C. Metals and Their Alloys*, National Bureau of Standards, Gaithersburg, Md. 1973

[43b] Ashworth, V., Carber, G., Grant, W. A., Jones, P. D., Proctor, R. P. M., Sayegh, N. N., Street, A. D.: In: [23] 443

[43c] Ashworth, V., Baxter, D., Grant, W. A., Proctor, R. P. M., Wellington, T. C.: In: [24] 367

[44] Atalla, M. M., Tannenbaum, E.: *Bell Syst. Techn. J.* **39** (1960) 933

[45] Aubuchon, K. G.: *Int. Conf. on Prop. and Use of MIS Structures*, Grenoble (1969)

[45a] Auston, D. H., Surko, C. M., Venkatesan, T. N. C., Slusher, R. E., Golovchenko, J. A.: *Appl. Phys. Lett.* **33** (1978) 437

[45b] Auston, D. H., Golovchenko, J. A., Simons, A. L., Slusher, R. E., Smith, P. R., Surco, C. M., Venkatesan, T. N. C.: In: *Laser Solid Interactions and Laser Processing 1978* (Eds Ferri, S. D., Leamy, H. J., Poate, J. M.) New York 1979, 11

[46] Axmann, A.: *Appl. Phys. Lett.* **23** (1973) 645
[47] Axmann, A., Schulz, M., Fritsche, C. R.: *Appl. Phys.* **12** (1977) 173
[48] Baccarini, G., Ostoga, P.: *Solid-State Electronics* **18** (1975) 579
[50] Bader, R., Kalbitzer, S.: *Radiation Effects* **6** (1970) 211
[51] Bader, R., Kalbitzer, S.: *Appl. Phys. Lett.* **16** (1970) 13
[51a] Baeri, P., Campisano, S. U., Foti, G., Rimini, E., *Phys. Rev. Lett.* **41** (1978) 1246
[51b] Baeri, R., Campisano, S. U., Foti, G., Rimini, E., *J. Appl. Phys.* **50** (1979) 789
[52] Bäuerlein, R.: In: *Radiation Damage in Solids* (Ed. Billington, D. S.) New York 1962, 358
[53] Balarin, M., Rattke, R., Zetschke, A.: *Phys. Stat. Sol.* **22** (1967) 123
[54] Baldo, E., Cappellani, F., Restelli, G.: *Rad. Effects* **19** (1973) 271
[55] Baranova, E. C., Gusev, J. M., Martynenko, Yu. V., Starinin, C. V., Hailbullin, E. B.: In: [23] 59
[56] Barbe, D. F.: *Proc. IEEE* **63** (1975) 38
[57] Barnoski, M. K., Hunsperger, R. G., Lee, A.: *Appl. Phys. Lett.* **24** (1974) 627
[58] Barnoski, M. K., Loper, D. D.: *Solid-State Electronics* **16** (1973) 433 and 441
[59] Baron, R., Shifrin, G. A., Marsh, O. J., Mayer, J. W.: *J. Appl. Phys.* **40** (1969) 3702
[60] Baruch, P., Monnier, J., Blanchard, B., Castaing, C.: *Appl. Phys. Lett.* **26** (1975) 77
[61] Baruch, P., Constantin, C., Phister, J. C., Saintesprit, R.: *Discussions Faraday Soc.* **31** (1962) 76
[62] Bauer, W., Thomas, G. J.: *J. Nucl. Mater.* **53** (1974) 127
[63] Bayly, A. R., Townsend, P. D.: *Optics and Laser Technol.* **2** (1970) 117
[64] Bayly, A. R., Townsend, P. D.: In: [20] 120
[65] Bean, K. E., Gleim, P. S., Yeakley, R. L.: *J. Electrochem. Soc.* **114** (1967) 733
[66] Beanland, D. G.: In: [25] 31
[67] Bell, E. C., Glaccum, A. E., Hemment, P. L. F. H., Sealy, D. J.: *Rad. Effects* **22** (1974) 253
[67a] Bell, R. O., Toulemonde, M., Siffert, P.: *Appl. Phys.* **19** (1979) 313
[68] Belyi, I. M., Gumanski, G. A., Karas', V. I., Lomako, V. M., Tashlykov, I. S., Tishkov, V. S.: *Sov. Phys. Semicond.* **9** (1976) 1326
[69] Bennett, J. R. J.: *Proc. Int. Conf. on Ion Sources*, I.N.S.T-N.-Saclay (1969) 571
[69a] Benninghoven, A.: *Surface Science* **53** (1975) 596
[70] Berchtold, K., Krumpholz, B., Suri, J.: *Appl. Phys. Lett.* **26** (1975) 585
[71] Bernas, R., Kaluszyner, L., Dryaux, J.: *J. de Physique et le Radium* **15** (1954) 273
[72] Bernheim, M.: *4th Int. Conf. on Electron and Ion Beam Science and Technology*, Los Angeles (1970)
[73] Bertolotti, M., Sette, D., Stagni, L., Vitali, G.: *Appl. Phys. Lett.* **18** (1971) 257
[74] Bethe, H.: *Ann. Phys.* **5** (1930) 325
[75] Bethe, H.: *Z. Phys.* **76** (1932) 293
[76] Bethe, H., Livingstone, M. S.: *Rev. Mod. Phys.* **9** (1937) 265
[77] Betz, H.: Not published
[78] Bicknell, R. W.: *Phil. Mag.* **26** (1972) 273
[79] Biersack, J. P., *Nucl. Inst. Meth.* **182/183** (1981) 199
[79a] Biersack, J. P., Ziegler, J. F.: In: *Ion Implantation Techniques* (Eds Ryssel, H., Glawischnig, H.) Springer Series in Electrophysics, **10**, 122 (1982)
[80] Björkquist, K., Domeij, B., Eriksson, L., Fladda, G., Fontell, A., Mayer, J. W.: *Appl. Phys. Lett.* **13** (1968) 379
[81] Blamires, N. G.: In: [20] 52
[82] Blamires, N. G., Osborne, D. N., Owen, R. B., Stephen, J.: In: [19] 669
[83] Blanc, D., Degeith, A.: *J. de Physique et le Radium* **22** (1961) 230

[84] Bloch, F.: *Z. Phys.* **81** (1933) 363
[84a] Bloembergen, N.: In: *Laser Solid Interactions and Laser Processing, 1978* (Eds Ferri, S. D., Leamy, H. J., Poate, J. M.) New York 1979
[85] Blood, P., Dearnaley, G., Wilkins, M. A.: *Rad. Effects* **21** (1974) 245
[86] Blood, P., Dearnaley, G., Wilkins, M. A.: *J. Appl. Phys.* **45** (1974) 5123
[87] Blood, P., Dearnaley, G., Wilkins, M. A.: In: [23] 75
[88] Blum, J. M., McGroddy, J. C., McMullin, P. G., Shih, K. K., Smith, A. W., Ziegler, J. F.: *IBM Publication RC 5195* (1975)
[89] Bogardus, F. H., Poponiak, M. R.: *Appl. Phys. Lett.* **23** (1973) 553
[90] Bogenschütz, A. F.: *Ätzpraxis für Halbleiter*, München 1967
[91] Bohr, N.: *Kgl. Danske Videnskab. Selskab., Mat-Fys. Medd.* **18** (1948) No. 8
[92] Bohr, N.: *Phil. Mag.* **25** (1913) 10
[93] Bohr, N.: *Phil. Mag.* **30** (1915) 581
[94] Bohr, N., Lindhard, J.: *Kgl. Danske Videnskap. Selskab., Mat. Fys. Medd.* **28** (1954) No. 7
[94a] Bomke, H. A., Berkowitz, H. L., Harmatz, M., Kronenberg, S., Lux, R.: *Appl. Phys. Lett.* **33** (1978) 955
[95] Borders, J. A., Picraux, S. I., Beezhold, W.: *Appl. Phys. Lett.* **18** (1971) 509
[96] Bourgoin, J. C., Corbett, J. W.: In: [365] 149
[97] Bower, R. W., Dill, H. G.: Int. Electron Devices Meeting, Washington (1966)
[98] Bower, R. W., Dill, H. G., Aubuchon, K. G., Thomson, S. A.: *IEEE Trans. Electron Devices* **ED-15** (1968) 757
[99] Boyle, W. S., Smith, G. E.: *Bell Syst. Tech. J.* **49** (1970) 587
[100] Brewer, G. R.: *IEEE Spectrum* **8** (1971) 23
[101] Brice, D. K.: *Appl. Phys. Lett.* **16** (1970) 103
[102] Brice, D. K.: In: [23] 171
[103] Brice, D. K.: *Rad. Effects* **11** (1971) 227
[104] Brice, D. K.: In: [21] 101
[105] Brook, P., Whitehead, C. S.: *Electronics Lett.* **4** (1968) 335
[106] Brower, K. L., Beezhold, W.: In: [22] 7
[107] Brower, K. L., Beezhold, W.: *J. Appl. Phys.* **43** (1972) 3499
[108] Brownson, J.: *J. Electrochem. Soc.* **111** (1964) 919
[109] Brown, D. M., Gray, P. V., Heumann, F. K., Phillipp, H. R., Taft, E. A.: *J. Electrochem. Soc.* **115** (1968) 311
[110] Buck, T. M., Poate, J. M., Pickar, K. A., Hsieh, C. M.: *Appl. Phys. Lett.* **21** (1972) 485
[111] Buck, T. M., Poate, J. M., Pickar, K. A., Hsieh, C. M.: *Surface Science* **35** (1973) 362
[112] Buckel, W., Dietrich, M., Heim, G., Kessler, J.: *Z. Phys.* **245** (1971) 283
[113] Buehler, M. W., Pearson, G. L.: *Solid-State Electronics* **9** (1966) 395
[114] Bulthuis, K.: *Phys. Lett.* **27A** (1968) 493
[115] Busen, K. M., Linzey, P.: *Trans. Mat. Soc. AIME* **236** (1966) 306
[115a] Butcher, D. N.: M.Sc. Thesis Brighton Polytechnic (1974)
[116] v. Buttlar, H.: *Einführung in die Grundlagen der Kernphysik*, Frankfurt 1964
[117] Campbell, A. B., Shewchun, J., Thompson, D. A., Davies, J. A., Mitchell, J. B.: In: [24] 291
[118] Cappelani, F., Restelli, G., Spinoni, I.: *J. Phys. C.: Solid State Phys.* **7** (1974) 650
[119] Carter, G., Baruah, J. N., Grant, W. A.: *Rad. Effects* **16** (1972) 107
[120] Carter, G., Colligon, J. S., Leck, J. H.: *Proc. Phys. Soc.* **79** (1962) 299
[122] Carter, G., Whitton, J. L.: *Rad. Effects* **15** (1972) 143
[123] Cass, T. C., Reddi, V. G. K.: *Appl. Phys. Lett.* **23** (1973) 268
[124] Caughey, D. M., Thomas, R. E.: *Proc. IEEE* **55** (1967) 2192

[124a] Celler, G. K., Poate, J. M., Kimerling, L. S.: *Appl. Phys. Lett.* **32** (1978) 464
[124b] Celler, G. K.: In: Rimini, E. (Ed.) *Laser Effects on Ion Implanted Semiconductors*, 13, Istituto di Struttura della Materia, Università di Catania, 1978
[124c] Celler, G. K., Borutta, R., Brown, W. L., Poate, J. M., Rozgonyi, G. A., Sheng, T. T.: In: *Laser Solid Interactions and Laser Processing, 1978* (Eds Ferri, S. D., Leamy, H. J., Poate, J. M.) New York 1979, 381
[125] Cembali, F., Galloni, R., Mousty, F., Rosa, R., Zagnani, F.: *Rad. Effects* **21** (1974) 255
[126] Cembali, F., Galloni, R., Zignani, Z.: *Rad. Effects* **26** (1975) 61
[127] Cairns, J. A., Nelson, R. S., Holloway, D. F.: Report AERE-R 6408 (1970)
[128] Cairns, J. A.: *Nucl. Instr. and Methods* **92** (1971) 507
[129] Cairns, J. A., Holloway, D. F., Nelson, R. S.: Report AERE-R 6490 (1970)
[130] Chang, L. L., Pearson, G. L.: *J. Phys. Chem. Solids* **25** (1964) 23
[131] Chatterjee, P. K., Streetman, B. G., Keune, D. L., Herzog, A. H.: Intern. Electron Devices Meeting, Washington (1975)
[132] Chatterjee, P. K., Vaidyanathan, K. V., McLevige, W. V., Streetman, B. G.: *Appl. Phys. Lett.* **27** (1975) 567
[133] Chen, W. H., Chen, W. I.: *J. Electrochem. Soc.* **114** (1967) 1297
[134] Cheng, L. J., Covelli, J. C., Corbett, J. C., Watkins, G. D.: *Phys. Rev.* **152** (1966) 761
[135] Chernow, F., Eldridge, G., Ruse, G., Wahlin, L.: *Appl. Phys. Lett.* **12** (1968) 339
[136] Cheshire, I. M., Dearnaley, G., Poate, J. M.: *Phys. Lett.* **27A** (1968) 304
[137] Chiu, T. L., Ghosh, H. N.: *IBM. J. Res. Develop.* **15** (1971) 472
[138] Cho, A. Y., Reinhart, F. K.: *J. Appl. Phys.* **45** (1974) 1812
[139] Chu, W. K., Mayer, J. W., Nicolet, M. A.: *Backscattering Spectrometry*, New York–San Francisco–London 1977
[140] Chu, W. K., Kelm, jr., R. G.: *J. Electrochem. Soc.* **122** (1975) 995
[141] Chu, W. K., Müller, H., Mayer, J. W., Sigmon, T. W.: *Appl. Phys. Lett.* **25** (1974) 297
[142] Chu, T. L.: *J. Vac. Sci. Technol.* **6** (1969) 25
[143] Chu, T. L., Szedon, J. R., Lee, Ch.: *J. Electrochem. Soc.* **114** (1968) 318
[144] Clark, A. H., Manchester, K. F.: *Trans. Met. Soc. AIME* **242** (1968) 1173
[145] Colby, J. W., Katz, L. E.: *J. Electrochem. Soc.* **123** (1976) 409
[146] Colligon, J. S.: *Vacuum* **11** (1961) 272
[147] Collins, L. E., O'Connell, P. A., Perkins, J. G., Pontet, F. R., Stroud, P. T.: *Nucl. Instr. Methods* **92** (1971) 455
[148] Collins, L. E., Perkins, J. G., Stroud, D. T.: *Thin Solid Films* **4** (1969) 41
[149] Comer, J. J., Roosild, S. A.: *Rad. Effects* **25** (1975) 275
[150] Copeland, J. A.: *IEEE Trans. Electron Devices* **ED-16** (1969) 445
[151] Copeland, J. A.: *IEEE Trans. Electron Devices* **ED-17** (1970) 404
[152] Cooper, J. A., Ward, E. R., Schwartz, J. R.: *Solid-State Electronics* **15** (1972) 1219
[153] Corbett, J. W.: In: [21] 1
[154] Corbett, J. W.: *Electron Radiation Damage in Semiconductors and Metals*, New York 1966
[154a] Corbett, J. W., Ianiello, L. C. (Eds): *Proc. of Int. Conf. on Radiation Induced Voids in Metals*, Albany, N.Y. (1971) CONF-710601 (1972)
[155] Crawford, B.: *Electronics* **45** (April 24 1972) 85
[156] Croset, M., Petreanu, E., Samuel, D.: *J. Electrochem. Soc.* **118** (1971) 717
[157] Crowder, B. L.: *J. Electrochem. Soc.* **118** (1971) 943
[158] Crowder, B. L.: *J. Electrochem. Soc.* **117** (1970) 671

[159] Crowder, B. L., Fairfield, J. M.: *J. Electrochem. Soc.* **112** (1970) 363
[159a] Crowder, B. L., Tan, S. I.: *IBM Technical Disclosure Bulletin* **14** (1971) 198
[160] Crowder, B. L., Morehead, jr., F. F.: *Appl. Phys. Lett.* **14** (1969) 313
[161] Crowder, B. L.: In: [26] 63
[162] Crowder, B. L., Title, R. S., Brodsky, M. H., Pettit, G. D.: *Appl. Phys. Lett.* **16** (1970) 205
[162a] Crowder, B. L., Ziegler, J. F., Cole, G. W.: In: [23] 257
[163] Csepregi, L., Mayer, J. W., Sigmon, T. W.: *Phys. Lett.* **54** (1975) 157
[164] Csepregi, L., Chu, W. K., Müller, H., Mayer, J. W.: *Rad. Effects* **28** (1976) 227
[165] Csepregi, L., Mayer, J. W., Sigmon, T. W.: *Appl. Phys. Lett.* **29** (1976) 92
[166] Cussins, W. D.: *Proc. Phys. Soc.* **368** (1955) 213
[166a] Da Cunha, S. F., Bougnot, J.: *Phys. Stat. Sol.* (**a**), **22** (1974) 205
[167] Daly, D. F., Pickar, K. A.: *Appl. Phys. Lett.* **15** (1969) 267
[167a] Das, S. K., Kaminsky, M., Fenske, G. A.: In: [30], 293
[167b] Das, S. K., Kaminsky, M.: In: [29] 543
[167c] Danilbiko, Y. K., Libidyaeva, T. P., Manenkov, A. A., Sidorin, A. V.: *Zh. Eksp. Teor. Fiz. SSSR* **74** (1978) 765
[168] D'Asaro, L. A.: *J. Luminescence* **7** (1973) 310
[169] Davies, D. E.: *Solid-State Electronics* **13** (1970) 229
[170] Davies, D. E.: *Appl. Phys. Lett.* **14** (1969) 227
[171] Davies, D. E., Kennedy, J. K., Lowe, L. F.: *Electr. Lett.* **11** (1975) 462
[172] Davies, D. E., Roosild, S., Lowe, L.: *Solid-State Electronics* **18** (1975) 733
[173] Davies, J. A., Ball, G. C., Brown, F., Domeij, B.: *Can. J. Phys.* **42** (1964) 1070
[174] Davies, J. A., Denhartog, J., Eriksson, L., Mayer, J. W.: *Can. J. Phys.* **45** (1967) 4053
[175] Davies, J. A., Eriksson, L., Mayer, J. W.: *Appl. Phys. Lett.* **12** (1968) 255
[176] Davies, J. A., Foti, G., Howe, L. M., Mitchell, J. B., Winterbon, K. B.: *Phys. Rev. Lett.* **34** (1975) 1441
[177] Davies, J. A., Friesen, J., McIntyre, J. D.: *Can. J. Chem.* **38** (1960) 1526
[178] Davies, J. A., Jespersgård, P.: *Can. J. Phys.* **44** (1966) 1631
[179] Davies, R. E., Johnson, W. E., Lark-Horowitz, K., Siegel, S.: *Phys. Rev.* **74** (1948) 1255
[180] Davies, E. D., Kennedy, J. K., Ludington, C. E.: *J. Electrochem. Soc.* **122** (1975) 1374
[181] Deal, B. E.: *J. Electrochem. Soc.* **121** (1974) 198 C
[182] Deal, B. E., Grove, A. S.: *J. Appl. Phys.* **36** (1965) 3770
[182a] Dearnaley, G.: In: *New Uses of Ion Accelerators* (Ed. Ziegler, J. F.) New York 1975, Ch. 5
[182b] Dearnaley, G., Goode, P. D., Miller, W. S., Turner, J. F.: In: [23] 405
[182c] Dearnaley, G.: In: [29] 63
[183] Dearnaley, G., Freeman, J. H., Gard, G. A., Wilkins, M. A.: *Can. J. Phys.* **46** (1968) 587
[184] Dearnaley, G., Gard, G. A., Temple, W., Wilkins, M. A.: *Appl. Phys. Lett.* **27** (1975) 17
[185] Dearnaley, G.: *IEEE Trans. Nucl. Sci.* **11** (1964) 249
[186] Dearnaley, G., Wilkins, M. A., Goode, P. D., Freeman, J. H., Gard, G. A.: Report AERE-R 6197 (1969)
[186a] Dearnaley, G., Goode, P. D.: In: *Ion Implantation Equipment and Techniques* (Eds McKenna, Ch. M., Scanlon, P. J., Winnard, J. R.) North-Holland, Amsterdam 1981
[186b] Dearnaley, G.: In: *Ion Implantation: Equipment and Techniques* (Eds Ryssel, H., Glawischnig, H.) Springer Series in *Electrophysics* **11** (1983)

[186c] Degen, P. L.: *Phys. Stat. Sol.* **16** (1973) 9
[187] Dennis, J. R., Woodward, G. K., Hale, E. B.: In: [365] 467
[188] Dewald, J. F.: *J. Electrochem. Soc.* **104** (1956) 244
[189] Dienes, G. J., Vineyard, G. H.: *Radiation Effects in Solids*, New York 1957
[189a] Dill, H. G., Finnila, R. M., Leupp, A. M., Toombs, T. N.: *Solid State Technol.* (Dec. 1972) 27
[190] Dill, H. G., Toombs, T. N., Bauer, L. O.: In: [22] 315
[191] Dill, H. G., Bower, R. W., Toombs, T. N.: In: [21] 349
[192] Dolan, R., Roosild, S., Buchanan, B.: *Proc. 2nd Conf. Microelectronics Meetings of the I.N.E.A.*, München 1966, 207
[193] Donnelly, J. P., Harman, T. C., Foyt, A. G.: *Appl. Phys. Lett.* **18** (1971) 259
[194] Donnelly, J. P., Foyt, A. G., Hinkley, E. D., Lindley, W. T., Dimmock, J. O.: *Appl. Phys. Lett.* **12** (1968) 303
[196] Donnelly, J. P., Foyt, A. G., Lindley, W. T., Iseler, G. W.: *Solid-State Electronics* **13** (1970) 755
[197] Donnelly, J. P., Harman, T. C., Foyt, A. G., Lindley, W. T.: *Appl. Phys. Lett.* **20** (1972) 279
[198] Donnelly, J. P., Harman, T. C., Foyt, A. G., Lindley, W. T.: *J. Nonmetals* **1** (1973) 123
[199] Donnelly, J. P., Harman, T. C., Foyt, A. G.: *Proc. IRIS Detector Special Group Meeting*, 17–18 March 1971, San Diego, Calif.
[200] Donnelly, J. P., Harman, T. C.: *Solid-State Electronics* **18** (1975) 288
[201] Donnelly, J. P., Calawa, A. R., Harman, T. C., Foyt, A. G., Lindley, W. T.: *Solid-State Electronics* **15** (1972) 403
[202] Donnelly, J. P., Harman, T. C., Foyt, A. G., Lindley, W. T.: *Solid-State Electronics* **16** (1973) 529
[202a] Donnelly, J. P., Holloway, H.: *Appl. Phys. Lett.* **23** (1973) 682
[202b] Donnelly, J. P: *Nucl. Inst. Meth.* **182/183** (1981) 553
[203] Douglas, E. C., Dingwall, A. G. F.: *IEEE Trans. Electron Devices* **ED-22** (1975) 849
[204] Douglas, E. C., Dingwall, A. G. F.: *IEEE Trans. Electron Devices* **ED-21** (1974) 324
[204a] Douglas, E. C., D'Aiello, R. V.: *IEEE Trans. Electron Devices* **ED-27** (1980) 792
[205] Drum, C. M.: Electrochem. Soc. Meeting (1975)
[206] Drum, C. M., Miller, P.: Int. Electron Devices Meeting, Washington (1971)
[207] Duffek, E. F., Benjamin, E. A., Mylroie, C.: *Electrochem. Technol.* **3** (1965) 75
[208] Dunlap, H. L., Hunsperger, R. G., Marsh, O. J: Report NAS 12-124, N70-17314 (1969)
[209] Dyment, J. C., D'Asaro, L. A., North, J. C., Miller, B. I., Ripper, J. E.: *Proc. IEEE* **60** (1972) 726
[210] Dyment, J. C., North, J. C., D'Asaro, L. A.: *J. Appl. Phys.* **44** (1973) 207
[211] Edelmann, F. L., Kuznetsov, O. N., Lezheiko, L. V., Lubopytova, E. V.: *Rad. Effects* **29** (1976) 13
[212] EerNisse, E. P.: *Appl. Phys. Lett.* **18** (1971) 581
[213] EerNisse, E. P.: *J. Appl. Phys.* **42** (1971) 480
[214] EerNisse, E. P., Norris, C. B.: In: [24] 437
[215] EerNisse, E. P., Norris, C. B.: *J. Appl. Phys.* **45** (1974) 5196
[215a] Eggermont, G. E. J., Tamminga, Y., Hofker, W. K.: In: *Laser Solid Interactions and Laser Processing, 1978* (Eds Ferri, S. D., Leamy, H. J., Poate, J. M.) New York 1979, 321
[216] Ehrstein, J. R. (Ed.): *Spreading Resistance Symposium. NBS Special Publication 400-10*, US Department of Commerce, Washington (1974)
[217] Eisen, F. H., Higgins, J. A., Zucca, R. R.: Report AFRCL-TR-74-0192 (1974)

[218] Eisen, F. H., Higgins, J. A., Zucca, R.: Report AFRCL-TR-74-0712 (1974)
[219] Eisen, F. H.: In: [24] 3
[220] Eisen, F. H., Welch, B.: In: [21] 459
[221] Eisen, F. H., Welch, B., Westmoreland, J. E., Mayer, J. W.: In: *Proc. Int. Conf. Atomic Collision Phenomena in Solids* (Eds Palmer, D. W., Thomson, M. W., Townsend, P. D.) New York 1970, 111
[222] Eisen, F. H.: *Can. J. Phys.* **46** (1968) 561
[223] Eldridge, G., Govind, P. K., Nieman, D. A., Chernow, F.: In: [20] 143
[223a] El-Hoshy, A. H., Gibbons, J. F.: *Phys. Rev.* **173** (1968) 454
[224] Elkin, E. L., Watkins, G. D.: *Phys. Rev.* **174** (1968) 881
[225] Eloy, J. F.: Proc. Int. Conf. on Ion Sources I.N.S.I.N.-Saclay (1969) 619
[226] Enge, H. A.: In: *Focusing of Charged Particles* (Ed. Septier, A.) New York 1967, vol. 2, Ch. 4.2
[227] Enge, H. A.: *Rev. Sci. Instr.* **35** (1964) 278
[228] Eriksson, L., Davies, J. A., Jespersgård, P.: *Phys. Rev.* **161** (1967) 219
[229] Eriksson, L., Davies, J. A., Johannsson, N. G. E., Mayer, J. W.: *J. Appl. Phys.* **40** (1969) 842
[230] Ermanis, F. Schwarz, B.: *J. Electrochem. Soc.* **121** (1974) 1665
[231] Fahrner, W., Goetzberger, A.: *Appl. Phys. Lett.* **21** (1972) 329
[232] Fair, R. B.: *J. Electrochem. Soc.* **122** (1975) 800
[233] Fair, R. B.: *Solid-State Electronics* **17** (1974) 17
[233a] Fair, R. B.: In: *Semiconductor Silicon* (Ed. Huff, H. R.), 968 *Electrochem. Soc.* (1977)
[234] Fair, R. B., Tsai, J. C. C.: *J. Electrochem. Soc.* **122** (1975) 1689
[235] Fair, R. B., Weber, G. R.: *J. Appl. Phys.* **44** (1973) 273
[236] Fairfield, J. M., Crowder, B. L.: *Trans-Met. Soc. AIME* **245** (1969) 469
[237] Fang, F. F., Rupprecht, H. S.: *IEEE J. Solid State Circuits* **SC-10** (1975) 205
[238] Favennec, P. N., Diguet, D.: *Appl. Phys. Lett.* **23** (1973) 546
[239] Favennec, P. N., Pelvus, G. P., Binet, M., Bandet, P.: In: [23] 621
[240] Fermi, E.: *Z. Phys.* **48** (1928) 73
[240a] Ferri, S. D., Leamy, H. J., Poate, J. M. (Eds): *Laser Solid Interactions and Laser Processing, 1978*, Am. Inst. Phys., New York 1979
[241] Fiorito, G., Gasparrini, G., Svelto, F.: *Appl. Phys. Lett.* **23** (1973) 448
[242] Firsov, O. B.: *Sov. Phys. JETP* **36** (1959) 1076
[243] Fistul', V. J.: *Heavily Doped Semiconductors*, New York 1969
[244] Fladda, G., Bjorkquist, K., Eriksson, E., Sigurd, D.: *Appl. Phys. Lett.* **16** (1970) 313
[245] Folkmann, F.: *J. Phys. E.: Sci. Instr.* **8** (1975) 429
[246] Forbes, L.: *IEEE J. Solid State Circuits* **SC-8** (1973) 226
[246a] Foti, G., Rimini, E., Tseng, W. S., Mayer, J. W.: *Appl. Phys.* **15** (1978) 365
[246b] Foti, G., Rimini, E., Campisano, S. U.: *Phys. Stat. Sol.* **(a) 47** (1978) 213
[247] Foyt, A. G.: *Appl. Phys. Lett.* **16** (1970) 335
[248] Foyt, A. G., Donnelly, J. P., Lindley, W. T.: *Appl. Phys. Lett.* **14** (1969) 372
[249] Foyt, A. G., Herman, T. C., Donnelly, J. P.: *Appl. Phys. Lett.* **18** (1971) 321
[250] Foyt, A. G., Lindley, W. T., Donnelly, J. P.: *Appl. Phys. Lett.* **16** (1970) 335
[251] Foyt, A. G., Lindley, W. T., Wolfe, C. M., Donnelly, J. P.: *Solid-State Electronics* **12** (1969) 209
[252] Frank, H.: *Phys. Stat. Sol.* **18** (1966) 401
[253] Frank, F. C., Read, W. T.: *Phys. Rev.* **79** (1950) 723
[254] Freeman, J. H.: In: [30] 340
[255] Freeman, J. H., Caldecourt, L. R., Done, K. C. W., Francis, R. J.: Report AERE-R 6496 (1970)
[257] Freeman, J. H.: *Nucl. Instr. and Methods* **22** (1963) 306

[258] Freeman, J. H., Gard, G. A., Mazey, D. J., Stephen, J. H., Whiting, F. B.: In: [20] 74
[259] Freeman, J. H., Sidenius, G.: *Proc. Second Int. Conf. Ion Sources*, 11–15 Sept. Wien (1972)
[260] Fritsche, C., Goetzberger, A., Axmann, A., Rothemund, W., Sixt, G.: *Rad. Effects* **7** (1971) 87
[261] Fritsche, C. R., Rothemund, W.: *Appl. Phys.* **7** (1975) 39
[262] Fritsche, C. R., Rothemund, W.: *J. Electrochem. Soc.* **119** (1972) 1243
[263] Fritsche, C. R., Rothemund, W.: *J. Electrochem. Soc.* **120** (1973) 1603
[264] Fujimoto, F., Komaki, K., Nakayama, H., Ishii, M.: *Rad. Effects* **13** (1972) 43
[265] Fuller, C. S., Ditzenberger, J. A.: *J. Appl. Phys.* **27** (1956) 544
[266] Furukawa, S., Ishiwara, H.: *J. Appl. Phys.* **43** (1972) 1268
[267] Furukawa, S., Ishiwara, H.: *Jap. J. Appl. Phys.* **11** (1972) 1062
[268] Furukawa, S., Matsumura, H., Ishiwara, H.: *Jap. J. Appl. Phys.* **11** (1972) 134
[268a] Furukawa, S., Ishiwara, H.: *Proc. 2nd Conf. on Solid State Devices*, Tokyo (1970), Suppl. *J. Jap. Soc. Appl. Phys.* **40** (1971) 3
[269] Furukawa, S., Matsumura, H., Ishiwara, H.: In: [26] 73
[270] Gabovich, M. D.: *Zh. Tekh. Fiz.* **28** (1958) 872
[271] Gabovich, M. D., Budernaya, L. D., Poritskii, V. Y., Protsenko, I. M.: All Soviet Meeting on Ion Beam Physics, Kiev (1974)
[272] Galaktionova, I. A., Gusev, V. M., Naumenko, V. G., Titov, V. V.: *Sov. Phys. Sem.* **2** (1968) 656
[273] Gamo, K., Iwaki, H., Masuda, K., Namba, S., Ishiwara, S., Kimura, I.: In: [22] 459
[274] Gamo, K., Iwaki, M., Masuda, K., Namba, S.: *Jap. J. Appl. Phys.* **10** (1971) 523
[275] Gamo, K., Iwaki, M., Masuda, K., Namba, S., Ishiwara, S., Kimura, I., Mitchel, I. V., Ilic, G., Whitton, J. L., Davies, J. A.: *Jap. J. Appl. Phys.* **12** (1973) 735
[276] Gamo, K., Masuda, K., Namba, S., Ishiwara, S., Kimura, I.: *Appl. Phys. Lett.* **17** (1970) 391
[277] Gamo, K., Takai, M., Lin, M. S., Masuda, K., Namba, S.: In: [24] 35
[277a] Gamo, K., Ukegawa, T., Namba, S.: *Jap. J. Appl. Phys.* **19** (1980) L 379
[278] Gardner, E. E., Hallenback, F. J., Schumann, P. A.: *Solid-State Electronics* **6** (1963) 311
[279] Garmire, E., Stoll, H., Yariv, A., Hunsperger, R. W.: *Appl. Phys. Lett.* **21** (1972) 87
[279a] Gat, A., Gibbons, J. F.: *Appl. Phys. Lett.* **32** (1978) 142
[279b] Gat, A., Gibbons, J. F., Magee, T. J., Peng, J., Deline, V. R., Williams, P., Evans, C. A., Jr.: *Appl. Phys. Lett.* **32** (1978) 276
[280] Gavulov, A. A., Kachurin, G. A., Pridachin, N. B., Smirnov, L. S.: *Sov. Phys. Sem.* **8** (1975) 1455
[281] Geerk, J., Langguth, K. G.: Report GfK Karlsruhe (1974/75)
[282] George, J., Chruma, J.: *Solid State Technology* **16** (1973) 43
[283] Gettings, M., Stephens, K. G.: *Rad. Effects* **22** (1974) 53
[284] Gettings, M., Meyer, O., Linke, G.: *Rad. Effects* **21** (1974) 51
[285] Ghezzo, M., Brown, D. M.: *J. Electrochem. Soc.* **120** (1973) 146
[286] Gibbons, P. E.: In: [22] 410
[287] Gibbons, J. F.: In: [26] 79
[288] Gibbons, J. F., Mylroie, S.: *Appl. Phys. Lett.* **22** (1973) 568
[289] Gibbons, J. F.: In: [19] 561
[289a] Gibbons, J. F., Gat, A., Gerzberg, L., Lietoila, A., Regolini, J. L., Sigmon, T. W., Pease, R. F. W., Magee, T. J., Peng, J., Hong, J., Deline, V., Katz, W., Williams, P., Evans, C. A., Jr.: In: *Laser Solid Interactions and Laser processing, 1978* (Eds Ferri, S. D., Leamy, H. J., Poate, J. M.) New York 1979, 365

|290| Glotin, P., Bernard, J., Monfret, A.: *Rad. Effects* **7** (1971) 65

|291| Glotin, P.: *Can. J. Phys.* **46** (1968) 705

|292| Glotin, P., Grapa, J., Monfret, A.: In: [19] 619

|293| Goel, J. E., Standley, R. D., Gibson, W. M., Rodgers, J. W.: *Appl. Phys. Lett.* **21** (1972) 72

|294| Goetzberger, A., Bartelink, D. J., McVittie, J. P., Gibbons, J. F.: *Appl. Phys. Lett.* **29** (1976) 259

|295| Goetzberger, A., Shockley, W.: *J. Appl. Phys.* **31** (1960) 1821

|296| Goldstein, H.: *Classical Mechanics*, Reading, Mass. 1956

|296a| Golovchenko, J. A., Venkatesan, T. N. C.: *Appl. Phys. Lett.* **32** (1978) 147

|297| Gonda, S., Makita, Y.: *Appl. Phys. Lett.* **27** (1975) 392

|298| Gonda, S., Makita, Y., Maekawa, S.: *IEEE Trans. Electron Devices* **ED-22** (1975) 712

|298a| Goode, P. D.: In: |30| 154

|299| Good, D. K., Dearnaley, G.: *J. Vac. Sci. Technol.* **12** (1975) 463

|300| Goode, P. D.: Report AERE-R 6401 (1970)

|301| Goode, P. D., Wilkins, M. A., Dearnaley, G.: In: [21] 187

|301a| Goodwin, C. A.: In: *Laser Solid Interactions and Laser Processing, 1978* (Eds Ferri, S. D., Leamy, H. J., Poate, J. M.) New York 1979, 411

|302| Goto, K., Yanagisawa, S., Wada, O., Takanashi, H.: *Jap. J. Appl. Phys.* **13** (1974) 1127

|302a| Grant, W. A.: In: |30| 127

|302b| Götzlich, J., Haberger, K., Ryssel, H., Kranz, H., Traumüller, E.: *Rad. Effects* **47** (1980) 203

|303| Graul, J., Kaiser, H., Wilhelm, W. J., Ryssel, H.: *IEEE J. Solid State Circuits* **SC-10** (1975) 201

|304| Gray, T. J., Lear, R., Dexter, R. J., Schwettmann, F. N., Wiemer, K. C.: *Thin Solid Films* **19** (1973) 103

|305| Green, M. A., Gunn, M. W.: *Solid-State Electronics* **14** (1971) 1167

|305a| Greenwald, A. G., Kirkpatrick, A. R., Little, R. G., Mircucci, J. A.: *J. Appl. Phys.* **50** (1979) 783

|306| Grove, A. S., Leistiko, O., Sah, C. T.: *J. Appl. Phys.* **35** (1964) 1695

|307| Grove, A. S.: *Physics and Technology of Semiconductor Devices*, New York 1967

|307a| Grove, A. S., Leistiko, jr., O., Sah, C. T.: *J. Phys. Chem. Sol.* **25** (1964) 985

|307b| Gürs, V., Gürs, K.: In: Landolt/Börnstein: IV 2 c, *Leichtmetalle, Sonderwerkstoffe, Halbleiter, Korrosion* (Eds Borchers, V. H., Schmidt, E.) Berlin–Heidelberg–New York 1965

|308| Gunn, J. B.: *Solid State Comm.* **1** (1963) 88

|308a| Gusev, I. A., Murin, A. N., Soregin, P. P.: *Sov. Phys.-Solid State* **6** (1964) 1491

|309| Gusev, V. M., Titov, V. V., Guseva, M. I., Kurimiyi, V. I.: *Soviet Phys.-Solid State* **7** (1966) 1673

|310| Guthrie, A., Wakerling, R. K. (Eds): *Electromagnetic Separation of Isotopes in Commercial Quantities*, N.N.E.S. 1–4, T.I.D. 5217 (1949)

|311| Gyulai, J., Csepregi, L., Nagy, T., Mayer, J. W., Müller, H., Le Vide **174** (1974) 416

|312| Gyulai, J., Mayer, J. W., Mitchell, I. V., Rodriguez, V.: *Rad. Effects* **17** (1970) 332

|313| Haas, G. A., Gray, H. F.: *J. Appl. Phys.* **46** (1975) 3885

|314| Hall, R. N.: *Phys. Rev.* **87** (1952) 387

|314a| Hanson, G. R., Siegel, B. M.: Electrochem. Soc. Meeting 1980, St Louis

|315| Harris, J. I., Eisen, F. H., Welch, B., Haskell, J. D., Pashley, R. D., Mayer, J. W.: *Appl. Phys. Lett.* **21** (1972) 601

|316| Harris, J. S.: In: [22] 157

[317] Harth, W.: *Halbleitertechnologie*, Stuttgart 1972, Teubner Studienskripten Bd. 54
[318] Hartley, N. E. W.: In: [30] 210
[319] Hartley, N. E. W., Dearnaley, G., Turner, J. F.: In: [23] 423
[320] Hartmann, P.: *Appl. Phys. Lett.* **28** (1976) 73
[321] Hasegawa, H., Forward, K., Hartnagel, H.: *Thin Solid Films* **32** (1975) 65
[322] Hasegawa, S., Ichida, K., Shimizu, T.: *Jap. J. Appl. Phys.* **12** (1973) 1181
[323] Hasegawa, S., Karimotó, H., Shimizu, T.: *Jap. J. Appl. Phys.* **12** (1973) 1190
[324] Hasegawa, S., Forward, K. E., Hartnagel, H.: *Electronics Lett.* **11** (1975) 53
[325] Haskell, J. D., Grant, W. A., Stephans, G. A., Whitton, J. L.: In: [22] 193
[325a] Hauffe, K.: *Oxidation of Metals*, New York 1965
[326] Heim, G., Stritzker, B.: *Appl. Phys.* **7** (1975) 239
[327] Hemment, P. L. F., Sealey, B. J., Stephens, K. G.: In: [24] 27
[327a] Herriot, D. R., Coller, R. J., Alles, D. S., Stafford, J. W.: *IEEE Trans. Electron Devices* **ED-22** (1975) 385
[328] Herzer, H., Kalbitzer, S.: In: [22] 307
[329] Herzog, R.: *Z. f. Physik* **89** (1954) 447
[330] Hesse, K., Strack, H.: *Solid-State Electronics* **15** (1972) 767
[331] Hickmott, T. W.: *J. Appl. Phys.* **43** (1972) 2339
[332] Hickmott, T. W.: *Appl. Phys. Lett.* **22** (1973) 267
[333] Higgins, J. A., Welch, M., Eisen, F. H., Robinson, G. D.: *Electr. Lett.* **12** (1976) 17
[334] Hilibrand, J., Gold, R. D.: *RCA Rev.* **21** (1960) 245
[334a] Hill, C.: In: *Laser Solid Interactions and Laser Processing, 1978* (Eds Ferri, S. D., Leamy, H. J., Poate, J. M.) New York 1979, 419
[335] Hirata, M., Saito, H.: *J. Phys. Soc. Jap.* **27** (1969) 405
[335a] Hirayama, M., Shohno, K.: *J. Electrochem. Soc.* **122** (1975) 1671
[336] Höfflinger, B., Gabler, L.: In: [24] 717
[337] Hofker, W. K., Oosthoek, D. P., Koeman, N. J., de Grefte, H. A. M.: *Rad. Effects* **24** (1975) 223
[338] Hofker, W. K.: *Philips Research Reports, Suppl. No. 8* (1975)
[339] Hofker, W. K., Werner, H. W., Oosthoek, D. P., Koeman, N. J.: *Appl. Phys.* **4** (1974) 125
[340] Hofker, W. K., Werner, H. W., Oosthoek, D. P., de Grefte, H. A. M.: In: [23] 133
[340a] Hofker, W. K., Eggermont, G. E. J., Tamminga, K., Oosthoek, D. P.: In: *Laser Solid Interactions and Laser Processing, 1978* (Eds Ferri, S. D., Leamy, H. J., Poate, J. M.) New York 1979, 425
[341] Holm, R.: *Electric Contacts Handbook*, Berlin–Heidelberg–New York 1967
[342] Holmén, G., Burén, A., Högberg, P.: *Rad. Effects* **24** (1975) 51
[343] Holmén, G., Peterström, S., Burén, A.: *Rad. Effects* **24** (1975) 45
[344] Holmén, G.: *Rad. Effects* **24** (1975) 7
[345] Honig, R. E., Kramer, D. A.: *RCA Rev.* **23** (1962) 4
[347] Hou, S. L., Beck, K., Marley, J. A.: *Appl. Phys. Lett.* **14** (1969) 151
[348] Hou, S. L., Marley, J. A.: *Appl. Phys. Lett.* **16** (1970) 467
[349] Howes, J. H., Knill, G.: In: [20] 97
[350] Hsieh, C. M., Matthews, J., Seidel, H. D., Pickar, K. A., Drum, C. M.: *Appl. Phys. Lett.* **22** (1973) 238
[351] Hu, S. M.: *J. Appl. Phys.* **39** (1968) 3844
[352] Hu, S. M.: *Solid-State Electronics* **15** (1972) 809
[352a] Hu, S. M., Schmidt, S.: *J. Appl. Phys.* **39** (1968) 4272
[353] Huang, C., van der Ziel, A.: *Solid-State Electronics* **18** (1975) 509
[355] Hunsperger, R. G., Dunlap, H. L., Marsh, O. J.: *Development of Ion Implantation Techniques for Microelectronics*, NAS 12-124 and N69-24439 (1968)

[356] Hunsperger, R. G., Hirsch, N.: *Electronics Lett.* **9** (1973) 577
[357] Hunsperger, R. G., Hirsch, N.: *Solid-State Electronics* **18** (1975) 349
[358] Hunsperger, R. G., Marsh, O. J.: *Appl. Phys. Lett.* **14** (1971) 327
[359] Hunsperger, R. G., Marsh, O. J.: *J. Electrochem. Soc.* **116** (1969) 488
[360] Hunsperger, R. G., Marsh, O. J., Mead, C. A.: *Appl. Phys. Lett.* **13** (1968) 295
[361] Hunsperger, R. G., Marsh, O. J.: *Rad. Effects* **6** (1970) 263
[363] Hunsperger, R. G., Marsh, O. J.: *Metallurgical Transactions* **1** (1970) 603
[364] Hunsperger, R. G., Wilson, R. G., Jamba, D. M.: *J. Appl. Phys.* **45** (1972) 1318
[365] Huntley, F. A. (Ed.): *Lattice Defects in Semiconductors 1974*, Inst. Phys. Conf. Ser. **23**, Inst. of Physics, London (1975)
[366] Hurrle, A.: Dissertation, Univ. Freiburg (1975)
[367] Hurwitz, C. E., Donnelly, J. P.: *Solid-State Electronics* **18** (1975) 753
[368] Inada, T., Ohnuki, Y.: *Appl. Phys. Lett.* **25** (1974) 228
[369] Ing, S. W., Morrison, R. E., Alt, R. E., Aldrich, R. W.: *J. Electrochem. Soc.* **110** (1963) 533
[370] Irvin, J. C.: *Bell System Tech. J.* **41** (1962) 387
[371] Ishino, S., Nakazawa, F., Hasiguti, R. R.: *J. Phys. Chem. Sol.* **24** (1963) 1033
[372] Ishitani, T., Shimizu, R.: *Phys. Lett.* **46A** (1974) 487
[373] Ishitani, T., Shimizu, E., Murata, K.: *Phys. Stat. Sol.* (b) **50** (1972) 681
[374] Ishitani, T., Shimizu, R., Murata, K.: *Jap. J. Appl. Phys.* **11** (1972) 125
[375] Ishihara, H., Furukawa, S., Yamada, J., Kawamura, M.: In: [24] 423
[376] Itoh, T., Inada, T., Kanekawa, K.: *Appl. Phys. Lett.* **12** (1968) 244
[377] Itoh, T., Kushiro, Y.: *J. Appl. Phys.* **42** (1971) 5120
[378] Itoh, T., Oana, Y.: *Appl. Phys. Lett.* **24** (1974) 320
[379] Itoh, T., Shinada, K.: *Jap. J. Appl. Phys.* **14** (1975) 1627
[380] Itoh, T., Ohdomari, I.: In: [26] 43
[381] Itoh, T., Oana, Y.: *J. Appl. Phys.* **44** (1973) 4982
[382] Iwaki, M., Gamo, K., Masuda, K., Namba, S., Ishihara, S., Kimura, I.: In: [23] 111
[382a] Iwaki, M., Gamo, K., Masuda, K., Namba, S., Ishihara, S., Kimura, I.: In: [24] 163
[383] Jain, R. K., Van Overstraeten, R. J.: *J. Electrochem. Soc.* **122** (1975) 552
[384] Johansson, N. G. E., Mayer, J. W., Marsh, O. J.: *Solid-State Electronics* **13** (1970) 317
[385] Johansson, N. G. E., Mayer, J. W.: *Solid-State Electronics* **13** (1970) 123
[385a] Johnson, N. M., Gold, R. B., Lietoila, A., Gibbons, J. F.: In: *Laser Solid Interactions and Laser Processing, 1978* (Eds Ferri, S. D., Leamy, H. J., Poate, J. M.) New York 1979, 550
[386] Johnson, W. A., North, J. C., Wolfe, R.: *J. Appl. Phys.* **44** (1973) 4753
[387] Johnson, W. S., Gibbons, J. F.: *Projected Range Statistics in Semiconductors*, Stanford, Cal. 1969
[388] Johnson, W. S.: Thesis SU-SEL-69-014, Stanford University (1969)
[388a] Johnston, W. G., Rosolowski, J. H.: In: [30] 228
[389] Jones, K. C., Stevens, P. R. C.: *Electr. Lett.* **5** (1969) 499
[390] Jonscher, A. K.: *Principles of Semiconductor Devices Operation*, London 1960, 154
[391] Jorgensen, P. J.: *J. Chem. Phys.* **37** (1962) 874
[392] Kachurin, G. A., Pridachin, N. B., Smirnov, L. S.: In: [365] 461
[393] Kachurin, G. A., Zelevinskaya, V. M., Smirnov, L. S.: *Sov. Phys. Sem.* **2** (1969) 1527
[393a] Kachurin, G. A., Nidaev, E. V., Khodychikh, A. V., Kovaleva, L. A.: *Sov. Phys. Semicond.* **10** (1976) 1128

[394] Kalbitzer, S., Bader, R., Herzer, H., Bethge, K.: *Z. Physik* **203** (1967) 117
[395] Kalbitzer, S., Bader, R., Melzer, W., Stumpfi, W.: *Nucl. Inst. Methods* **54** (1967) 323
[396] Kaminsky, M.: *Adv. Mass. Spectrom.* **3** (1964) 69
[397] Kanaya, K., Koga, K., Toki, K.: *J. Phys. E: Sci. Instr.* **5** (1972) 541
[398] Kass, S.: *Corrosion of Zirconium Alloys*, ASTM, Publication No. 368 (1964), 3
[399] Kato, Y., Katayama, Y., Kobayashi, K. L. I., Komatsubara, K. F.: *J. Appl. Phys.* **46** (1975) 4614
[400] Kato, T. K., Nishi, Y.: *Jap. J. Appl. Phys.* **3** (1964) 377
[401] Kellett, C. M., King, W. J., Leith, F. A.: US Air Force Report AF 19 (628)-4970 AD 635267 (1966)
[402] Kellner, W., Kniepkamp, H., Ristow, D., Boroffka, H.: Int. Electron Devices Meeting, Technical Digest, Washington (1975) 238
[403] Kelly, R., Sanders, J. B.: *Nucl. Instr. and Methods* **132** (1976) 335
[404] Kelson, G., Stellrecht, H. H., Perloff, D. S.: *IEEE J. Solid State Circuits* **SC-8** (1973) 336
[405] Kennedy, D. P., Murley, P. C., Kleinfelder, W.: *IBM J. Res. Dev.* **12** (1968) 399
[406] Kennedy, D. P., O'Brien, R. R.: *IBM J. Res. Dev.* **13** (1969) 212
[407] Kern, W.: Nat. Aerospace and Electronics Conf. (1975), 93
[408] Kerr, J. A., Large, L. N.: In: [19] 601
[409] Kimerling, L. C., Poate, J. M.: In: [365] 126
[410] Kinchin, G. H., Pease, R. S.: *Rep. Progr. Phys.* **18** (1955) 1
[411] King, W. J., Burrel, J. T., Harrison, S., Martin, F., Kellett, C. M.: *Nuclear Instr. and Methods* **38** (1965) 178
[411a] King, M., Rose, P. H.: *Nucl. Inst. Meth.* **189** (1981) 169
[412] Kleinfelder, W. J.: Techn Report K701-1- Stanford Electr. Labs., Calif. (1967)
[413] Koch, J.: *Electromagnetic Isotope Separators and Applications of Electromagnetically Enriched Isotopes*, Amsterdam 1958
[414] Komoshida, M., Kudoh, O.: *Appl. Phys. Lett.* **24** (1974) 501
[415] Komarov, V. L., Tsepakin, S. G., Chemayakin, G. V.: *Proc. Int. Conf. on Ion Sources*, I.N.S.S.N.-Saclay (1969), 383
[415a] Kosonocky, W. F., Carnes, J. E.: *RCA Rev.* **36** (1975) 566
[416] Kostka, A., Kalbitzer, S.: *Rad. Effects* **19** (1973) 77
[417] Kostka, A., Kalbitzer, S.: In: [24] 689
[418] Kräutle, H., Kalbitzer, S.: In: [22] 499
[419] Krimmel, E. F., Pfleiderer, H.: *Rad. Effects* **19** (1973) 83
[419a] Kulcinski, G. L.: In: [29] 613
[419b] Kubena, R. L., Anderson, C. L., Seliger, R. L., Jullens, R. A., Stevens, E. H.: *J. Vac. Sci. Technol.* **19** (1981) 916
[420] Kulkarni, M. V., Harson, J. C., James, G. A.: *IEEE Trans. Electron Devices* **ED-19** (1972) 1098
[420a] Kutukova, O. G., Strel'tsov, L. N.: *Sov. Phys. Semicond.* **10** (1976) 265
[421] Kuwano, J.: *Jap. J. Appl. Phys.* **8** (1969) 876
[422] Ladany, I., Kressel, H.: *RCA Rev.* **33** (1972) 517
[423] Laegsgaard, E., Martin, F. W., Gibson, W. M.: *Nucl. Instr. and Methods* **60** (1968) 24
[424] Langmuir, S., Kingdom, K. N.: *Phys. Rev.* **21** (1923) 380
[424a] Lau, S. S., von Almen, M., Golecki, I., Nicolet, M. A., Kennedy, E. F., Tseng, W. F.: *Appl. Phys. Lett.* **35** (1979) 327
[425] Lecroisnier, D. P., Pelous, G. P.: *IEEE Trans. Electron Devices* **ED-21** (1974) 113
[425a] Lecroisnier, D. P., Pelous, G. P., Henoc, P.: In: [365] 487
[426] Lee, D. H.: *Proc. IEEE* **61** (1973) 666

[427] Lee, D. H., Ying, R. S., Yamba, D. M.: *Proc. IEEE* **62** (1974) 1025
[427a] Lee, D. H., Mayer, J. W.: *Proc. IEEE* **62** (1974) 1241
[428] Lee, D. H., Ying, R. S: *Proc. IEEE* **62** (1974) 1295
[429] Lee, G. A.: In: [216] 75
[430] Leith, F. A., King, W. J., McNally, P., Davies, E., Kellett, C. M.: Report AFCRL-67-0123 AD 651313 (1967)
[431] Lehovec, K., Slobodskoy, A.: *Solid-State Electronics* **3** (1961) 45
[432] Lepselter, M. P., MacRae, A. U., MacDonald, R. W.: *Proc. IEEE* **57** (1969) 812
[433] Lerach, L.: *Phys. Stat. Sol.* **30** (1975) 625
[434] Liebl, H.: *J. Appl. Phys.* **38** (1967) 5277
[435] Liebl, H.: *J. Phys. E: Sci. Instr.* **8** (1975) 797
[436] Liebmann, G.: *Proc. Phys. Soc. Lond.* **1362** (1949) 753
[437] Lindley, W. T., Phelan, jr., R. J., Wolfe, C. M., Foyt, A. G.: *Appl. Phys. Lett.* **14** (1969) 197
[438] Lindhard, J., Winter, A.: *Kgl. Danske Videnskab. Selskab., Mat.-Fys. Medd.* **34** (1964) No. 4
[439] Lindhard, J., Nielsen, V., Scharff, M.: *Kgl. Danske Videnskab. Selskab. Mat.-Fys. Medd.* **36** (1968) No. 10
[440] Lindhard, J.: *Kgl. Danske Videnskab. Selskab., Mat.-Fys. Medd.* **34** (1965) No. 14
[441] Lindhard, J., Scharff, H.: *Phys. Rev.* **124** (1961) 128
[442] Lindhard, J., Scharff, M., Schiøtt, H. E.: *Kgl. Danske Videnskab. Selskab., Mat.-Fys. Medd.* **33** (1963) No. 14
[443] Lindhard, J., Scharff, M.: *Kgl. Danske Videnskab. Selskab., Mat.-Fys. Medd.* **27** (1953) No. 15
[444] Littlejohn, M. A., Hauser, J. R., Montheith, L. K.: *Rad. Effects* **10** (1971) 185
[445] Logan, M. A.: *Bell Syst. Tech. J.* **40** (1961) 885
[446] MacDougal, J. D., Manchester, K. E., Roughan, P. E.: *IEEE* **57** (1969) 1538
[447] Maciver, B. A.: *Electronics Lett.* **11** (1975) 484
[448] MacNally, P. J., King, W. J.: 16th Nat. Infrared Information Symposium, Ft. Monmouth (1968)
[449] MacNally, P. J.: *Rad. Effects* **6** (1970) 149
[450] MacPherson, M. R.: *Appl. Phys. Lett.* **18** (1971) 502
[451] MacRae, A. U.: *Rad. Effects* **7** (1971) 59
[452] McCaldin, J. O.: *Nucl. Instr. and Meth.* **38** (1965) 153
[453] McCargo, M., Davies, J. A., Brown, F.: *Can. J. Phys.* **41** (1963) 1231
[454] McGill, T. C., Kurtin, S. L., Shifrin, G. A.: *J. Appl. Phys.* **41** (1970) 246
[455] Madden, P. K., Davidson, S. M.: *Rad. Effects* **14** (1972) 271
[456] Mader, S., Michel, A.: *J. Vac. Sci. Technol.* **13** (1976) 391
[456a] Mader, S., Michel, A.: *Phys. Stat. Sol.* (a) **33** (1976) 793
[457] Maekawa, S., Oshida, T.: *J. Phys. Soc. Jap.* **19** (1964) 253
[458] Maekawa, S.: *J. Phys. Soc. Jap.* **17** (1962) 1592
[459] Magee, T. J., Lehmann, M.: In: [30] 112
[460] Makita, Y., Gonda, S., Ijuin, M., Tsurushima, T., Tanoue, H., Maekawa, S.: *Appl. Phys. Lett.* **28** (1976) 103
[461] Manara, A., Ostidich, A., Pedroli, G., Restelli, G.: *Thin Solid Films* **8** (1971) 359
[462] Manchester, K. E.: *Electronics* **40** (1967) 116
[463] Manchester, K. E., Silbey, C. B., Alton, G.: *Nucl. Inst. and Methods* **38** (1965) 169
[463a] Maracas, G. N., Harris, G. L., Lee, C. A., McFarlane, R. A.: *Appl. Phys. Lett.* **33** (1978) 453
[464] Marcatili, E. A.: *Bell Syst. Tech. J.* **48** (1969) 2103
[466] Marine, J., Motte, C.: *Appl. Phys. Lett.* **23** (1973) 450

[467] Marine, J.: In: [20] 153
[468] Marsh, O. J., Baron, R., Shifrin, G. A., Mayer, J. W.: *J. Appl. Phys.* **13** (1968) 199
[469] Marsh, O. J., Dunlap, H. L.: *Rad. Effects* **6** (1970) 301
[470] Marsh, O. J., Mayer, J. W., Shifrin, G. A.: In: [19] 513
[471] Martin, F.: Ion Physics, Report 607 TR-255 (1966)
[472] Martin, F. W., King, W. J., Harrison, S.: *IEEE Trans. Nucl. Sci.* **NS-11** (1964) 280
[473] Martin, F. W., King, W. J., Harrison, S.: *IEEE Trans. Nucl. Sci.* **NS-13** (1966) 22
[474] Maruska, H. P., Stevenson, D. A., Pankove, J. I.: *Appl. Phys. Lett.* **22** (1975) 303
[475] Masic, R., Warnecke, R. J., Sautter, J. M.: *Proc. Int. Conf. on Ion Sources*, I.N.S.T.N.-Saclay, Frankreich (1969), 387
[476] Masuda, K., Gamo, K., Imada, A., Namba, J.: In: [22] 455
[477] Masuhara, T., Itoh, J.: *IEEE Trans. Electron Devices* **ED-21** (1974) 799
[478] Masters, B. J., Fairfield, J. M., Crowder, B. L.: In: [21] 81
[478a] Masters, B. J., Fairfield, J. M.: *J. Appl. Phys.* **40** (1969) 2390
[479] Matsumura, H., Furukawa, S.: *Jap. J. Appl. Phys.* **14** (1975) 1783
[480] Matzke, H. J., Königer, M.: *Phys. Stat. Sol.* (a), **1** (1970) 469
[481] Maul, J. L.: Dissertation, TU München (1974)
[482] Mayer, J. W., Csepregi, L., Gyulai, J., Nagy, T., Mezey, G., Revesz, P., Kotai, E.: *Thin Solid Films* **32** (1976) 303
[483] Mayer, J. W., Davies, J. A., Eriksson, E.: *Appl. Phys. Lett.* **11** (1967) 365
[484] Mayer, J. W., Eriksson, L., Picraux, S. T., Davies, J. A.: *Can. J. Phys.* **46** (1968) 663
[485] Mayer, J. W., Marsh, O. J.: *Appl. Solid State Science* (Ed. Wolfe, R.) New York 1969, 239
[486] Mayer, J. W., Marsh, O. J., Mankarious, R., Bower, R.: *J. Appl. Phys.* **38** (1967) 1975
[487] Mayer, J. W., Marsh, O. J., Shifrin, G. A., Baron, R.: *Can. J. Phys.* **45** (1967) 4073
[487a] Mazey, D. J., Nelson, R. S., Barnes, R. S.: *Phil. Mag.* **17** (1968) 1145
[488] Mazur, R. G., Dickey, D. H.: *J. Electrochem. Soc.* **113** (1966) 255
[489] Meek, R. L., Gibson, W. M., Sellschop, J. P. F.: *Appl. Phys. Lett.* **18** (1971) 535
[490] Merz, J. L., Sadowski, F. A., Rodgers, J. W.: *Solid State Comm.* **9** (1971) 1037
[491] Mets, E. J.: *J. Electrochem. Soc.* **112** (1965) 420
[492] Meyer, L.: *Phys. Stat. Sol.* (b) **44** (1971) 253
[493] Meyer, N. I., Guldbrandsen, T.: *IEEE* **51** (1963) 1631
[494] Meyer, O.: *IEEE Trans. Nucl. Sci.* **NS-15** (1968) 232
[495] Meyer, O., Linker, G.: *KFK-Nachr.* **6** (1974) 1
[496] Meyer, O.: In: [30] 168
[496a] Mezey, G., Szökefalvi-Nagy, Z., Badinka, Cs.: *Thin Solid Films* **19** (1973) 173
[496b] Meyer, O.: In: *Ion Implantation*, Academic Press, New York 1981, 415
[497] Michel, A. E., Fang, F. F., Pan, E. S.: *J. Appl. Phys.* **45** (1974) 2991
[498] Miller, G. L.: *IEEE Trans. Electron Devices* **ED-19** (1972) 1103
[499] Miller, S. E.: *Bell Syst. Tech. J.* **48** (1969) 2059
[500] Minear, R. L., Nelson, D. G., Gibbons, J. F.: *J. Appl. Phys.* **43** (1972) 3468
[501] Misawa, T. S., Moline, R. A., Tretola, A. R.: *Solid State Electronics* **15** (1972) 189
[502] Miyazaki, T., Tamura, M.: In: [24] 41
[503] Mizutani, T., Kurumada, K.: *Electr. Lett.* **11** (1975) 638
[504] Moline, R. A., Buckley, R. R., Haszko, S. E., MacRae, A. U.: *IEEE Trans. Electron Devices* **ED-20** (1973) 840
[505] Moline, R. A., Cullis, A. G.: *Appl. Phys. Lett.* **26** (1975) 551

[506] Moline, R. A.: *J. Appl. Phys.* **42** (1971) 3553
[507] Moline, R. A.: *J. Appl. Phys.* **42** (1971) 2471
[508] Moline, R. A., Liebermann, R., Simpson, J., MacRae, A. U.: *J. Electrochem. Soc.* **121** (1974) 1362
[509] Moline, R. A., Reutlinger, G. W.: In: [22] 58
[510] Moline, R. A., Reutlinger, G. W., North, J. C.: In: *Atomic Collisions in Solids*, Vol. 1 (Eds Datz, J., Appleton, B. R., Moak, C. D.) New York 1975
[511] Monfret, A., Bernard, J.: In: [22] 389
[512] Monteith, L. K., Littlejohn, M. A., Hauser, J. R., Hendricks, H. D.: *Rad. Effects* **16** (1972) 133
[513] Morehead, F. F., Crowder, B. L., Title, R. J.: *J. Appl. Phys.* **43** (1972) 1112
[514] Morehead, F. F., Crowder, B. L.: In: [21] 25
[515] Morris, B. L., Langer, P. H., White, jr., J. C.: In: [216] 63
[515a] Müller, K., Henkelmann, R., Boroffka, H.: *Nucl. Instr. and Meth.* **128** (1975) 417
[516] Müller, H., Chu, W. K., Gyulai, J., Mayer, J. W., Sigmon, T. W., Cass, T. R.: *Appl. Phys. Lett.* **26** (1975) 292
[517] Müller, H.: Dissertation TU München (1973)
[518] Müller, H., Eisen, F. H., Mayer, J. W.: *J. Electrochem. Soc.* **122** (1975) 651
[519] Müller, H., Gyulai, J., Chu, W. K., Mayer, J. W., Sigmon, T. W.: *J. Electrochem. Soc.* **122** (1975) 1234
[520] Müller, H., Gyulai, J., Mayer, J. W., Eisen, F. H., Welch, B.: In: [24] 19
[521] Müller, H., Kranz, H., Ryssel, H., Schmid, K.: *Appl. Phys.* **4** (1974) 115
[522] Müller, H., Ryssel, H., Schmid, K.: *J. Appl. Phys.* **43** (1972) 2006
[523] Müller, H., Ryssel, H., Ruge, I.: In: [22] 85
[523a] Muhl, S., Collins, R. A., Dearnaley, G.: In: [30] 147
[524] Munro, P. C., Thompson jr., H. W.: *J. Electrochem. Soc.* **122** (1975) 127
[525] Murarka, S. P.: *Phys. Rev. B.* **12** (1975) 2502
[526] Murashima, S., Kanamori, H., Ishibashi, F.: *Jap. J. Appl. Phys.* **9** (1970) 58
[527] Mylroie, S., Gibbons, J. F.: In: [23] 243
[528] Nakamura, M., Kato, T., Oi, N.: *Jap. J. Appl. Phys.* **7** (1968) 512
[528a] Namba, S., Masuda, K.: *Adv. Electronics and Electron Phys.* **37** (1975) 263
[529] Namba, S., Masuda, K., Gamo, K., Iwaki, M.: In: [26] 1
[529a] Narayan, J., Young, R. T., White, C. W.: *J. Appl. Phys.* **49** (1978) 3912
[530] *Nat. Bureau of Standards (NBS): Special Publication 400-17*
[531] *Nat. Bureau of Standards (NBS): Special Publication 400-25*
[532] Nelson, D. G., Gibbons, J. F., Johnson, W. S.: *Appl. Phys. Lett.* **15** (1969) 246
[533] Nelson, R. S.: In: [20] 212
[534] Nelson, R. S., Cairns, J. A., Blamires, N.: *Rad. Effects* **6** (1970) 131
[535] Nelson, R. S.: *Radiation Damage and Defects in Semiconductors* (Ed. Whitehouse, J. E.), *Inst. Phys. Conf. Ser.* **16** (1973)
[536] Nelson, R. S., Thomson, M. W.: *Phil. Mag.* **8** (1963) 1677
[536a] Nelson, R. S., Mazey, D. J.: In: *Radiation Damage in Reactor Materials*, Vol. II, Wien IAEA (1969), 157
[537] Neuberger, M.: III–V Semiconducting Compounds, In: *Handbook of Electronic Materials*, Vol. 2, New York 1971
[537a] Neuberger, M.: Group IV Semiconducting Materials, In: *Handbook of Electronic Materials*, Vol. 5, New York 1971
[538] Nicholas, K. H.: *ACTA Electronica* **19** (1976) 95
[539] Nicholas, K. H., Ford, R. A.: *IEEE Internat. Electron Devices Meeting, Technical Digest*, Washington (1973) 51
[540] Nishimatsu, S., Natsuaki, N., Warabisako, T., Tokujama, T.: *Suppl. J. Jap. Soc. Appl. Phys.* **40** (1971) 29
[541] Nomura, K., Hirose, Y., Akasaka, Y., Horie, K., Kawazu, S.: In: [24] 681

[542] North, J. C., Gibson, W. M.: *Appl. Phys. Lett.* **16** (1970) 126
[543] Novak, R. L.: *Bull. Am. Phys. Soc.* **8** (1965) 235
[544] Oetzmann, H., Feuerstein, A., Grahmann, H., Kalbitzer, S.: *Ion Beam Surface Layer Analysis* (Eds Meyer, O., Linker, G., Köppeler, F.) New York 1976, 245
[545] Ohdomari, I., Itoh, T.: *Jap. J. Appl. Phys.* **11** (1972) 1709
[545a] Ohkawa, S., Nakajima, J., Furukawa, Y.: *Jap. J. Appl. Phys.* **14** (1975) 458
[546] Ohl, R.: *Bell Syst. Techn. J.* **31** (1952) 104
[547] Ohmura, Y., Mimura, S., Kanazawa, M., Abe, T., Konaka, M.: *Rad. Effects* **15** (1972) 167
[548] Ohmura, Y., Zohta, Y., Kanazawa, M.: *Solid State Comm.* **11** (1972) 263
[549] Ohmura, Y., Zohta, Y., Kanazawa, M.: *Phys. Stat. Sol.* **15** (1973) 93
[550] Ohmura, Y., Koike, K., Kobayashi, H.: In: [24] 183
[551] Ono, Y., Saito, K., Shiraki, Y.: *Jap. J. Appl. Phys.* **14** (1975) 1489
[552] Oosthoek, D. P., den Boer, J. A., Hofker, W. K.: In: [20] 88
[552a] Orloff, J., Swanson, L. W., 14th Symp. on Electron, Ion and Photon Beam Technol. 1977, USA
[553] Panish, M. P.: *IEEE Trans. Microwave Theory and Techniques* **MTT-23** (1975) 20
[554] Pabst, W.: *Nucl. Instr. and Methods* **120** (1974) 593
[555] Pabst, W.: *Nucl. Instr. and Methods* **124** (1975) 143
[556] Pashley, R. D.: *Rad. Effects* **11** (1971) 1
[557] Pavlov, A. V., Pavlov, P. V., Zorin, E. I., Tetel'baum, D. I.: *Proc. All Soviet Meeting on Ion Beam Physics*, Kiev (1974)
[558] Pavlov, P. V., Tetel'baum, D. I., Zorin, E. I., Alekseev, V. I.: *Sov. Phys.-Solid State* **8** (1967) 2141
[559] Pavlov, P. V., Vasil'yev, K., Zorin, E. I., Tetel'baum, I., Tulovchikov, V. S., Chigirinskaya, T. Yu.: *Izvestiya vus S.S.S.R., Radioelektronika* **13** (1966) 493
[560] Pavlov, P. V., Shitova, E. V.: *Sov. Phys.-Doklady* **12** (1967) 11
[561] Payne, R. S., Scavuzzo, R. J.: IEEE Int. Electron Devices Meeting, Washington (1971)
[562] Payne, R. S., Scavuzzo, R. J., Olson, K. H., Nacci, J. M., Moline, R. A.: *IEEE Trans. Electron Devices* **ED-21** (1974) 273
[563] Pearson, A. D., Hartell, W. B.: *Mat. Res. Bull.* **7** (1972) 567
[564] Perkins, J. G.: *Thin Solid Films* **9** (1972) 257
[565] Perkins, J. G.: *J. Non-Crystalline Solids* **3** (1972) 349
[566] Perloff, D. S.: *J. Electrochem. Soc.* **120** (1973) 1135
[567] Pertritz, R. L.: *Phys. Rev.* **110** (1958) 1254
[568] Pickar, K. A., Dalton, J. V., Seidel, H. D., Matheys, J. R.: *Appl. Phys. Lett.* **19** (1971) 43
[569] Picraux, S. T.: *Rad. Effects* **17** (1973) 261
[570] Picraux, S. T., Vook, F. L.: In: [22] 1
[571] Picraux, S. T., Vook, F. L.: *Rad. Effects* **11** (1971) 179
[572] Picraux, S. T., Westmoreland, J. E., Mayer, J. W., Hart, R. R., Marsh, O. J.: *Appl. Phys. Lett.* **14** (1969) 7
[572a] Picraux, S. T., EerNisse, E. P., Vook, F. L.: *Application of Ion Beams to Metals, Proc. Int. Conf. Albuquerque, N. Mex.* (1973) New York 1974
[572b] Picraux, P. S. T.: In: [30] 183
[573] Pliskin, W. A., Quall, R. P.: *J. Electrochem. Soc.* **111** (1964) 872
[574] Ponpon, J. P., Grob, J. J., Stuck, R., Burger, P., Siffert, P.: In: [29] 420
[575] Powell, R. J., Ligenza, J. R., Schneider, M. S.: *IEEE Trans. Electron Devices* **ED-21** (1974) 636
[576] Prince, J. L., Schwettmann, F. N.: *J. Electrochem. Soc.* **121** (1974) 705

[576a] *Proc. of ASTM Conf. on Irradiation Effects on Structural Alloys for Nuclear Reactor Applications*, Niagara Falls, N.Y. (1970) ASTM-STP-484 (1971)
[576b] *Proc. of ASTM Conf. on Effects of Radiation on Substructure and Mechanical Properties of Metals and Alloys, Los Angeles, Calif.* (1972) ASTM-STP-529
[577] Pruniaux, B. R., North, J. C., Miller, G. L.: In: [22] 212
[578] Pruniaux, B. R., North, J. C., Payer, A. V.: *IEEE Trans. Electron Devices* **ED-19** (1972) 672
[579] Prussin, S., Fern, A. M.: *J. Electrochem. Soc.* **122** (1975) 830
[579a] Prussin, S.: In: [24] 449
[580] Prussin, S.: *J. Appl. Phys.* **45** (1974) 1635
[581] Przyborski, W., Roed, J., Lippert, J.: *Rad. Effects* **1** (1969) 33
[582] Putley, E. H.: *The Hall Effect and Related Phenomena*, London 1960
[582a] Ramin, M., Ryssel, H., Kranz, H.: *Appl. Phys.* **22** (1980) 393
[583] Rand, M. J., Roberts, J. F.: *J. Electrochem. Soc.* **120** (1973) 446
[584] Reddi, V. G. K., Sansbury, J. D.: *Appl. Phys. Lett.* **20** (1972) 30
[585] Reddi, V. G. K., Yu, A. Y. C.: *Solid State Technology* **15** (1972) 35
[586] Regolini, J. L., Sigmon, T. W., Gibbons, J. F.: *Appl. Phys. Lett.* **35** (1979) 114
[586a] Révész, P., Farkas, G., Mezey, G., Gyulai, J.: *Appl. Phys. Lett.* **33** (1978) 431
[587] Revez, A. G., Eband, R. J.: *J. Phys. Chem. Solids* **30** (1969) 551
[588] Revez, A. G.: *Phys. Stat. Sol.* **19** (1967) 193
[589] Reyer, O., Scherber, W.: *J. Phys. Chem. Sol.* **32** (1971) 1909
[590] Rickards, J., Dearnaley, G.: In: [29] 101
[591] Rideout, V. L., Gänsslen, F. H., LeBlanc, A.: *IBM J. Res. Develop.* **19** (1975) 50
[591a] Rimini, E. (Ed.): *Laser Effects in Ion Implanted Semiconductors*, Istituto di Struttura della Materia, Università di Catania, 1978
[592] Ripper, J. E., Dyment, J. C., D'Asaro, L. A., Paoli, T. L.: *Appl. Phys. Lett.* **18** (1971) 155
[593] Robertson, G. I.: *J. Electrochem. Soc.* **122** (1975) 796
[594] Robinson, M. T., Oen, O. S.: *Le Bombardment Ionique* (Conf. Proc., Ed. Trillat, J. J.) Paris 1962
[595] Roosild, S., Dolan, R., Brickmanan, B.: *J. Electrochem. Soc.* **115** (1968) 307
[596] Rosendahl, K.: *Rad. Effects* **7** (1971) 95
[596a] Roth, J.: In: [30] 280
[597] Rothemund, W., Fritsche, C. R.: *J. Electrochem. Soc.* **121** (1974) 586
[598] Roughan, P. E., MacDougall, J. D., Clark, A. H., Manchester, K. E., Anderson, F. W.: Electrochem. Soc., Boston Meeting (1968)
[599] Rourke, F. M., Sheffield, J. C., White, F. A.: *Rev. Sci. Instrum.* **32** (1961) 455
[600] Ruegg, H. W.: *IEEE Trans. Electron Devices* **ED-14** (1967) 239
[601] Runge, H., Krimmel, E. F.: *Solid-State Electronics* **18** (1975) 149
[602] Runge, H.: In: [24] 703
[602a] Runge, H.: *Phys. Stat. Sol. (a)* **39** (1977) 595
[603] Runyan, W. R.: *Silicon Semiconductor Technology*, New York 1965
[604] Rutherford, E.: *Phil. Mag.* **21** (1911) 669
[605] Ryding, G., Wittkower, A. B.: *IEEE Trans. Manufacturing Technology* **MFT-4** (1975) 21
[606] Ryssel, H.: Dissertation, TU München (1973)
[607] Ryssel, H., Kranz, H.: *Appl. Phys.* **7** (1975) 11
[608] Ryssel, H., Kranz, H., Eichinger, P.: In: [30] 1
[609] Ryssel, H., Müller, H., Schmidt, K., Ruge, I.: In: [23] 215
[610] Ryssel, H., Kranz, H., Schmid, K., Ruge, I.: In: [24] 169
[610a] Ryssel, H., Lang, G., Biersack, J. P., Müller, K., Krüger, W.: *IEEE Trans. Electron Devices* **ED-27** (1980) 58

[611] Ryssel, H., Schmid, K., Müller, H.: *J. Phys. E: Sci. Inst.* **6** (1973) 492
[611a] Ryssel, H., Kranz, H., Biersack, J., Müller, K., Henkelmann, R.: In: [25] 727
[612] Ryssel, H.: Not published
[612a] Ryssel, H., Haberger, K.: *Microcircuit Engineering 81* (Ed. Oosenbrug, A.), Lausanne (1981) 299
[613] Sah, C. F., Reddi, V. G. K.: *IEEE Trans. Electron Devices* **ED-11** (1964) 345
[614] Sah, C. T., Noyce, R. N., Shockley, W.: *Proc. IRE* **45** (1957) 1228
[615] Sanders, J. B.: *Can. J. Phys.* **46** (1968) 445
[616] Sandhu, J. S., Reuter, J. L.: *IBM J. Dev.* **15** (1971) 464
[617] Sansbury, J. D., Gibbons, J. F.: *Appl. Phys. Lett.* **14** (1969) 311
[618] Sansbury, J. D., Gibbons, J. F.: *Rad. Effects* **6** (1970) 269
[619] Sato, H.: *Jap. J. Appl. Phys.* **12** (1973) 242
[620] Sattler, A. R., Vook, F. L.: *Rad. Effects in Semiconductors* (Ed. Vook, F. L.) New York 1968, 243
[621] Satya, A. V. S., Palanki, H. R.: In: [24] 405
[622] Schiøtt, H. E.: *Kgl. Danske Videnskab. Selskab., Mat. Fys. Medd.* **35** (1966) No. 9
[623] Schiøtt, H. E.: *Rad. Effects* **6** (1970) 107
[623a] Schmid, K.: Personal communication
[624] Schmid, K., Fischer, G., Müller, H., Ryssel, H.: *Rad. Effects* **23** (1974) 145
[625] Schmid, K., Kranz, H., Ryssel, H., Müller, W., Dathe, J.: *Phys. Stat. Sol.* (a) **23** (1974) 523
[626] Schmidt, P. F., Owen, A. E.: *J. Electrochem. Soc.* **111** (1964) 682
[627] Schnable, G. L., Kern, W., Comizzeli, R. B.: *J. Electrochem. Soc.* **122** (1975) 1093
[628] Schneider, J.: Master thesis, TU München (1974)
[629] Schneider, I., Marrone, M., Kabler, M. N.: *Appl. Opt.* **9** (1970) 1163
[630] Schottky, W.: *Naturwissenschaften* **26** (1938) 843
[631] Schottky, W.: *Z. Phys.* **113** (1939) 367
[632] Schroeder, J. B., Dieselmann, H. D.: *Proc. IEEE* **55** (1967) 125
[633] Schroen, W.: In: [216] 235
[633a] Schulz, J. C., Collins, R. J.: *Appl. Phys. Lett.* **34** (1979) 84
[634] Schultz, M.: *Appl. Phys. Lett.* **23** (1973) 31
[635] Schulz, M., Goetzberger, A., Franz, I., Langheinrich, W.: *Appl. Phys.* **3** (1974) 275
[636] Schumann, jr., P. A., Gardner, E. E.: *Solid-State Electronics* **12** (1969) 371
[637] Schumann, jr., P. A.: *J. Electrochem. Soc.* **115** (1968) 1197
[638] Schwartz, B., Ermanis, T., Brastad, M. H.: *J. Electrochem. Soc.* **123** (1976) 1089
[639] Schwettmann, F. N.: *Appl. Phys. Lett.* **22** (1973) 570
[640] Schwettmann, F. N.: *J. Appl. Phys.* **45** (1974) 1919
[641] Schwuttke, G. H., Brack, K., Gorey, E. F., Kahan, A., Lowe, L. F.: *Rad. Effects* **6** (1970) 103
[642] Schwuttke, G. H., Brack, K., Gorey, E. F., Kahan, A., Lowe, L. F., Euler, F.: *Phys. Stat. Sol.* **14** (1972) 107
[643] Schwuttke, G. H., Brack, K.: In: *Silicon Carbide 1973, Proc. 3rd Int. Conf. SiC.* (Eds Marshall, R. C., Faust, jr., J. W., Ryan, C. E.) Columbia, S.C. 1974, 626
[644] Schwuttke, G. H., Brack, K.: *Trans. Met. Soc. A.I.M.E.* **245** (1969) 475
[645] Sealy, B. J., Hemment, P. L. F.: *Thin Solid Films* **22** (1974) 539
[646] Sealy, B. J., D'Cruz, A. D. E.: *Electronics Letters* **11** (1975) 323
[647] Sealy, B. J., Surridge, R. K.: *Thin Solid Films* **26** (1975) L19
[647a] Sealy, B. J., Kular, S. S., Stephens, K. G., Croft, R., Palmer, A.: *Electr. Lett.* **14** (1978) 720

[648] Seeger, A., Chik, K. P.: *Phys. Stat. Sol.* **29** (1968) 455
[649] Seidel, T. E.: In: [22] 47
[650] Seidel, T. E., Gibson, W. C.: *IEEE Trans. Electron Devices* **ED-20** (1973) 744
[651] Seidel, T. E., Iglesias, D. E., Niehaus, W. C.: *IEEE Trans. Electron Devices* **ED-21** (1974) 523
[652] Seidel, T. E., MacRae, A. U.: In: [21] 149
[653] Seidel, T. F., Meek, R. L., Cullis, A. G.: In: [365] 494
[654] Seidel, T. F., Meek, R. L., Cullis, A. G.: *J. Appl. Phys.* **46** (1975) 600
[655] Seliger, R. L., Fleming, W. P.: *J. Appl. Phys.* **45** (1974) 1416
[656] Seliger, R. L.: *J. Appl. Phys.* **43** (1972) 2352
[656a] Seliger, R. L., Kubena, R. L., Olney, R. D., Ward, J. W., Wang, V.: *J. Vac. Sci. Technol.* **16** (1979) 1610
[657] Senechal, R. R., Basinsky, J.: *J. Appl. Phys.* **39** (1968) 3723, 4581
[658] Séquin, C. H.: *Bell Syst. Techn. J.* **51** (1972) 1923
[659] Seshan, K., Washburn, J.: *Rad. Effects* **26** (1975) 31
[660] Severin, P. J.: *Philips Res. Repts* **26** (1971) 279
[661] Severin, P. J., Bulle, H.: *J. Electrochem. Soc.* **122** (1975) 134
[662] Shannon, J. M., Ford, R. A., Gard, G. A.: *Rad. Effects* **6** (1970) 217
[663] Shannon, J. M., Stephen, J., Freeman, J. H.: *Electronics* **42** (1969) 96
[664] Shaw, D.: *Phys. Stat. Sol.* (b) **72** (1975) 11
[665] Shaw, D. (Ed.): *Atomic Diffusion in Semiconductors*, London 1973
[666] Shifrin, G. A., Jamba, D. M., Jones, W. R., Marsh, O. J., Wauk, M. T., Wilson, R. G.: Tech. Report 2, NR 251-001, AD 702778 (1969)
[667] Shifrin, G. A., Zanio, K. R., Jamba, D. M., Jones, W. R., Marsh, O. J., Wilson, R. G.: Report N00014-69-C-0171, AD 693154 (1969)
[668] Shiraki, Y., Shimada, T., Ikezu, T., Komatsubara, K. F.: Suppl. *J. Jap. Soc. Appl. Phys.* **42** (1973) 269
[669] Shiraki, Y., Shimada, T., Ono, Y., Komatsubara, K. F.: *Jap. J. Appl. Phys.* **14** (1975) 1495
[670] Shimada, T., Shiraki, Y., Kato, Y., Komatsubara, K. F.: In: [365] 446
[671] Shimizu, S., Iwamatsu, S., Ono, M.: *Appl. Phys. Lett.* **22** (1973) 286
[672] Shimizu, T., Hasegawa, S., Karimoto, H.: In: [24] 525
[673] Shockley, W.: U.S. Patent No. 2787, 564 (1957)
[674] Shockley, W.: *Bell Syst. Tech. J.* **28** (1949) 435
[675] Shockley, W., Read, W. T.: *Phys. Rev.* **87** (1952) 835
[675a] Shtyrkov, E. I., Khaibullin, I. B., Zaripov, M. M., Galyatudinov, M. F., Bayazitov, R. M.: *Sov. Phys. Semicond.* **9** (1975) 1309
[676] Sidenius, G.: *Nucl. Instr. and Methods* **38** (1965) 19
[677] Sigmon, T. W., Chu, W. K., Müller, H., Mayer, J. W.: *Appl. Phys.* **5** (1975) 347
[678] Sigmon, T. W., Chu, W. K., Müller, H., Mayer, J. W.: In: [24] 633
[679] Sigmon, T. W.: *Int. Electron Devices Meeting, Techn. Digest, Washington* (1973), 387
[680] Sigmon, T. W.: *Proc. IEEE* **63** (1975) 1619
[681] Sigmund, P., Mathies, M. T., Phillips, D. L.: *Rad. Effects* **11** (1971) 39
[682] Sigmund, P.: *Appl. Phys. Lett.* **14** (1969) 114
[683] Sigmund, P.: *Phys. Rev.* **184** (1969) 383
[684] Sigmund, P.: *Rev. Roum. Phys.* **17** (1972) 823, 969, 1079
[685] Sigmund, P., Sanders, J. B.: In: [19] 215
[686] Sigournay, N.: *IEE Colloqu. MOS Integr. Circuits* (1971)
[687] Sigurd, D., Fladda, G., Eriksson, L., Björkquist, K.: *Rad. Effects* **3** (1970) 145
[688] Singer, B. M., Kostelec, J.: *IEEE Trans. Electron Devices* **ED-21** (1974) 84
[689] Sirtl, E., Adler, A.: *Z. Metallkunde* **52** (1961) 529

[690] Skolnik, L. M., Spitzer, W. G., Kahan, A., Hunsperger, M. G.: *J. Appl. Phys.* **42** (1971) 5223

[691] Smith, B. J., Stephen, J., Hammersley, P. J.: *Rad. Effects* **26** (1975) 17

[692] Smits, F. H.: *Bell Syst. Tech. J.* **37** (1958) 711

[693] Somekh, S., Garmire, E., Yariv, A., Garvin, H. L., Hunsperger, R. G.: *Appl. Phys. Lett.* **22** (1973) 46

[694] Sosin, A., Bauer, W.: *Studies in Radiation Effects*, Vol. 3, New York 1969

[695] Spenke, E.: *Elektronische Halbleiter*, Berlin–Heidelberg–New York 1965

[696] Spitzer, W. G. In: *Festkörperprobleme XI* (Ed. Madelung, O.) Braunschweig 1971, 1

[697] Spitzer, S. M., Schwartz, B., Weigle, G. D.: *J. Electrochem. Soc.* **122** (1975) 391

[698] Spitzer, S. M., North, J. C.: *J. Appl. Phys.* **44** (1973) 214

[699] Spiwak, R. R.: *IEEE Trans. Inst. and Measurements* **IM-18** (1969) 197

[700] Standley, R. D., Gibson, W. M., Rodgers, J. W.: *Appl. Optics* **11** (1972) 1313

[701] Starodubtsev, S. V., Romanov, A. M.: *The Passage of Charged Particles Through Matter*, Israel Programme for Scientific Translation, Jerusalem 1965

[702] Steffen, K. G.: *High Energy Beam Optics*, New York 1965

[703] Stein, H. J., Vook, F. L., Borders, J. A.: *Appl. Phys. Lett.* **16** (1970) 106

[704] Stephen, J., Grimshaw, G. A.: *Rad. Effects* **7** (1971) 73

[705] Stephen, J., Smith, B. J., Ninder, G. W.: In: [24] 665

[706] Stoll, H., Yariv, A., Hunsperger, R. G., Tangonam, G. L.: *Appl. Phys. Lett.* **23** (1973) 664

[707] Stolte, C. A.: *IEEE Int. Electron Devices Meeting, Techn. Digest, Washington* (1975) 585

[708] Strack, H.: *J. Appl. Phys.* **34** (1963) 2405

[709] Streetman, B. G., Anderson, R. E., Wolford, D. J.: *J. Appl. Phys.* **45** (1974) 574

[710] Stritzker, B.: In: [30] 160

[711] Stroud, P. T.: *Thin Solid Films* **9** (1972) 273

[712] Swanson, R. S., Meindl, J. D.: *IEEE J. Solid-State Circuits* **SC-7** (1972) 146

[712a] Sze, S. M., Wei, L. Y.: *Phys. Rev.* **124** (1961) 84

[713] Sze, S. M.: *Physics of Semiconductor Devices*, New York 1969

[713a] Takagi, T., Yamada, I., Kimura, H.: In: [24] 335

[714] Takai, M., Gamo, K., Masuda, K., Namba, S.: *Jap. J. Appl. Phys.* **12** (1973) 1926

[715] Takai, M., Gamo, K., Masuda, K., Namba, S.: *Jap. J. Appl. Phys.* **14** (1975) 1935

[715a] Takai, M., Ryssel, H.: *Appl. Phys. Lett.* **35** (1979) 696

[715b] Takai, M., Ryssel, H.: In: *Laser and Electron Beam Processing of Electronic Materials* (Eds Anderson, C. L., Celler, G. K., Rozgonyi, G. A.), 156th Meeting of Electrochem. Soc., Los Angeles 1979, 478

[716] Takusagawa, M., Funayama, T., Nishizawa, J., Demizu, K., Nakano, T.: *J. Appl. Phys.* **38** (1967) 4084

[716a] Taminga, Y., Eggermont, G. E. J., Hofker, W. K., Hoonhout, D., Garrett, R., Saris, F. W.: *Phys. Lett.* **69A** (1979) 436

[717] Tamura, M.: *Appl. Phys. Lett.* **23** (1973) 651

[718] Tamura, M., Ikeda, T., Yoshihiro, N.: *Suppl. J. Jap. Soc. Appl. Phys.* **40** (1971) 9

[719] Tamura, M., Yoshihiro, N., Ikeda, T.: *Appl. Phys. Lett.* **27** (1975) 427

[720] Tanoue, H., Tsurushima, T.: In: [24] 285

[721] Tarui, Y., Komiya, Y., Teshima, H., Takahashi, R.: In: [26] 99

[722] Templeton, L. F.: *Solid State Technology* **18** (1975) 46

[723] Tenney, A. S., Ghezzo, M.: *J. Electrochem. Soc.* **120** (1973) 1091

[724] Thomas, G. O., Kahng, D., Manz, R. C.: *J. Electrochem. Soc.* **109** (1962) 1055

[725] Thomas, L. H.: *Proc. Cambr. Phil. Soc.* **23** (1927) 524

[726] Thompson, M. W.: *Phil. Mag.* **18** (1968) 377

[726a] Thompson, W. M.: In: [20] 109

[727] Thurmond, C. D.: In: *Properties of Elemental and Compound Semiconductors* (Ed. Gatos, H. C.) New York 1960

[728] Tinsley, A. W., Jones, K. C.: In: [20] 187

[729] Tinsley, A. W., Grant, W. A., Carter, G., Nobles, M. J.: In: [22] 199

[730] Tinsley, A. W., Stephens, G. A., Nobles, M. J., Grant, W. A.: *Rad. Effects* **23** (1974) 165

[730a] Towler, C., Collins, R. A., Dearnaley, G.: *J. Vac. Sci. Technol.* **12** (1975) 520

[730b] Trillat, J. J., Haymann: In: Trillat, J. J.: *Le bombardement ionique. Theories et applications*, Paris 1961

[731] Trumbore, F. A.: *Bell Syst. Techn. J.* **39** (1960) 205

[732] Tsai, J. C. C., Marabito, J. M.: *Surface Sci.* **44** (1974) 247

[733] Tsai, J. C. C., Marabito, J. M., Lewis, R. K.: In: [23] 87

[734] Tsuchimoto, T., Tokuyama, T.: In: [21] 237

[735] Tsuchimoto, T., Tokuyama, T.: *Rad. Effects* **6** (1970) 121

[736] Tsukada, T., Nakashima, H., Umeda, J., Nakamura, S., Chinone, N., Ito, R., Nakada, O.: *Appl. Phys. Lett.* **20** (1972) 344

[737] Tsurushima, T., Tanoue, H., Gibbons, J. F.: Electrochem. Soc. Meeting, Los Angeles (1970)

[738] Tsurushima, T., Tanoue, H.: In: [24] 429

[739] Tsurushima, T., Tanoue, H.: *J. Phys. Soc. Jap.* **31** (1971) 1965

[740] Tokuyama, T., Ikeda, T., Tsuchimoto, T.: *4th Int. Conf. Microelectronics*, München (1970)

[741] Tung-ho, C., Hsien-tsan, T.: In: [30] 96

[742] Upadhyayula, L. I., Naraya, S. Y., Douglas, E. C.: *Electr. Lett.* **11** (1975) 201

[742a] Urli, N. B., Corbett, J. W. (Eds): *Radiation Effects in Semiconductors 1976*, Inst. Phys. Conf. Ser. **31**, Inst. of Physics, London (1977)

[743] Valdes, L. B.: *Proc. IRG* **42** (1954) 420

[744] Van der Pauw, L. J.: *Philips Tech. Rdsch.* **20** (1958/59) 230

[745] Van der Pauw, L. J.: *Philips Res. Repts* **13** (1958) 1

[745a] Verbeck, H., Eckstein, W.: In: [29] 597

[745b] Verheijke, M. L.: *Philips Tech. Rev.* **34** (1974) 330

[745c] Varnell, G. L.: *Proc. Kodak Interface 1972* (1973) 56

[746] Vermilyea, D. A.: *J. Electrochem. Soc.* **112** (1965) 1232

[748] Vodicka, V. W., Zuleeg, R.: *IEEE Int. Electron Devices Meeting, Technical Digest, Washington* (1975) 625

[748a] Vogg, H.: *Halbleiter-Gammaspektren zur Neutronen-Aktivierungsanalyse*, München 1971

[749] Vook, F. L.: *Radiation Damage and Defects in Semiconductors* (Ed. Whithouse, J. E.), *Inst. Phys. Conf. Ser.* **16** (1973) 60

[750] Vook, F. L., Stein, J. H.: *Rad. Effects* **2** (1969) 23

[751] de Waand, H., Feldman, L. C.: In: [24] 317

[752] Wada, Y., Ashikawa, M.: *Jap. J. Appl. Phys.* **15** (1976) 389

[753] Wagner, S.: *J. Electrochem. Soc.* **119** (1972) 1570

[754] Walker, R. S., Thomson, D. A.: *Nucl. Instr. and Methods* **135** (1976) 489

[755] Wallace, R., Litton, G. M., Steward, P. G.: *2nd Symp. on Accelerator Dosimetry and Experience*, Stanford (1969)

[755a] Wang, J. C., Wood, R. F., Pronko, P. P.: *Appl. Phys. Lett.* **33** (1978) 455

[756] Wang, K. L., Gray, P. V.: *IEEE Trans. Electron Devices* **ED-22** (1975) 354

[757] Wang, P. P., Spencer, O. S.: *IBM J. Res. Dev.* **19** (1975) 530

[758] Wash, A. G., Sarma, N., Bhattacharya, P. K.: *Phys. Stat. Sol.* (a) **32** (1975) 63

[759] Watanabe, M., Tooi, A.: *Jap. J. Appl. Phys.* **5** (1966) 737

[760] Watkins, G. D.: *J. Phys. Soc. Jap. Suppl. II* **18** (1963) 22

[761] Waycoat, Application note

[761a] Weber, E. V., Yourke, H. S.: *Electronics* Nov. 10 (1977) 96
[762] Weber, M.: Master thesis, TU München (1973)
[762a] Webber, R. F., Thorn, R. S., Large, L. N.: *Int. J. Electronics* **26** (1969) 163
[763] Wei, D. T. Y., Lee, W. W., Bloom, L. R.: *Appl. Phys. Lett.* **25** &1975) 329
[764] Welch, B. M., Eisen, F. H., Higgins, J. A.: *J. Appl. Phys.* **45** (1974) 3685
[765] Werner, H. W.: *ACTA Electronica* **19** (1976) 53
[766] Westmoreland, J. E., Mayer, J. W., Eisen, F. H., Welch, B.: *Appl. Phys. Lett.* **15** (1969) 308
[767] Westmoreland, J. E., Marsh, O. J., Hunsperger, R. G.: *Rad. Effects* **5** (1970) 245
[768] Whalin, W.: In: *Handbuch der Physik*, Bd. 34. Berlin–Göttingen–Heidelberg 1958
[768a] White, C. W., Christie, W. H., Appleton, B. R., Wilson, S. R., Pronko, P. P.: *Appl. Phys. Lett.* **33** (1978) 662
[768b] White, C. W., Christie, W. H., Pronko, P. P., Appleton, B. R., Wilson, S. R., Young, R. T., Wang, J. C., Wood, R. F., Narayan, J., Magee, C. W.: In: *Proc. Int. Conf. on Ion Beam Modification of Materials*, Parts 1–3 (Eds Gyulai, J., Lohner, T., Pásztor, E.) Budapest, Hungary 1978, 839
[769] Whitton, J. L., Bellavance, G. R.: *Rad. Effects* **9** (1971) 127
[770] Whitton, J. L., Carter, G., Baruah, J. N., Grant, W. A.: *Rad. Effects* **16** (1972) 101
[771] Whitton, J. L., Carter, G., Freeman, J. M., Gard, G. A.: *J. Mat. Science* **4** (1969) 208
[772] Whitton, J. L., Davies, J. A.: *J. Electrochem. Soc.* **111** (1964) 1347
[773] Whitton, J. L.: *J. Appl. Phys.* **36** (1965) 3917
[774] Wiemer, K. C., Dexter, R. J., Morgan, I. H.: Int. Electron Devices Meeting, Washington (1972)
[775] Wien, W.: *Verhandlungen Physik. Gesellsch.* **16** (1897) 165
[776] Williams, E. J.: *Rev. Mod. Phys.* **17** (1945) 217
[776a] Williams, J. S., Brown, W. L., Leamy, H. J., Poate, J. M., Rodgers, J. W., Rousseau, D., Rozgonyi, G. A., Shelnutt, J. A., Sheng, T. T.: *Appl. Phys. Lett.* **33** (1978) 542
[777] Wilmsen, C. W., Vasbinder, G. C., Chau, Y. K.: *J. Vac. Sci. Technol.* **12** (1975) 56
[778] Wilson, R. G., Jamba, D. M.: *J. Appl. Phys.* **38** (1967) 1967
[779] Winterbon, K. B.: *Rad. Effects* **13** (1972) 215
[780] Wittkower, A. B., Rose, P. H., Ryding, G.: *Solid State Technology* **18** (1975) 41
[781] Wittmaack, K., Schulz, F., Hietel, B.: In: [24] 193
[782] Wittmaack, K., Maul, J., Schulz, F.: *Int. J. Mass. Spectrom. and Ion Phys.* **11** (1973) 23
[783] Wohlleben, K., Beck, W.: *Z. Naturforschung* **21a** (1966) 1057
[784] Wolfe, R., North, J. C., Barns, R. L., Robinson, M., Levenstein, H. J.: *Appl. Phys. Lett.* **19** (1971) 298
[785] Wolfe, R., North, J. C.: *Bell Syst. Tech. J.* **51** (1972) 1436
[785a] Wolfstirn, K. B.: *J. Phys. Chem. Sol.* **31** (1970) 601
[786] Wollnick, H.: *Nucl. Instr. and Meth.* **53** (1967) 197
[787] Woodcock, J. M., Shannon, J. M., Clark, D. J.: *Solid-State Electronics* **18** (1975) 267
[787a] Wu, C. P., Magee, C. W.: *Appl. Phys. Lett.* **34** (1979) 737
[788] Wu, C. P., Douglas, E. C., Mueller, C. W.: *IEEE Trans. Electron Devices* **ED-22** (1975) 319
[788a] Yamaguchi, M., Hirayama, T.: *Jap. J. Appl. Phys.* **15** (1976) 365
[788b] Yarbrough, D. W.: *Solid State Technol.* **11** (1968) 23
[788c] Yarbrough, D. W.: Research and Development Technical Report, ECOM-2692 (1968)

[789] Yariv, A.: *Laser Focus Magazine*, Dez. (1972) 40
[790] Yen, E. T., Masters, B. J., Kastl, R.: In: [24] 501
[791] Yoshihiro, N., Ikeda, T., Tamura, M., Tokuyama, T., Tsuchimoto, T.: *Proc. 3rd Conf. on Solid State Devices, Tokyo* (1971), *Suppl. Oyo Buturi* **41** (1972) 225
[791a] Young, D. B. Y., Pearson, G. L.: *J. Phys. Chem. Sol.* **21** (1970) 517
[791b] Young, R. T., Narayan, J., White, C. W., Wood, R. F., Cleland, J. W., Westbrook, R. D., Mooney, P. M.: In: *Proc. Int. Conf. on Ion Beam Modification of Materials*, Parts 1–3 (Eds Gyulai, J., Lohner, T., Pasztor, E.) Budapest, Hungary 1978, 831
[792] Yuba, Y., Gamo, K., Masuda, K., Namba, S.: *Jap. J. Appl. Phys.* **13** (1974) 641
[793] Zelevinskaya, V. M., Kachurin, G. A., Pridachin, N. B.: *Sov. Phys. Sem.* **8** (1974) 252
[793a] Ziegler, J. F. (Ed.): *New Uses of Ion Accelerators*, New York 1975
[794] Ziegler, J. F., Cole, G. W., Baglin, J. E. E.: *Appl. Phys. Lett.* **21** (1972) 177
[795] Ziegler, J. F., Crowder, B. L., Kleinfelder, W. J.: *IBM J. Res. Dev.* **15** (1971) 452
[796] Zölch, R.: Dissertation, TU München (1976)
[797] Zorin, E. I., Pavlov, P. V., Tetel'baum, D. I.: *Sov. Phys. Sem.* **2** (1968) 111

Index